The Story of Science

Newton at the Center

In the eighteenth century, experiments with electricity led to some shocking discoveries. What's a little pain in the interest of science? Get plugged in to the power of electricity in chapter 31.

The Story of Science

Newton at the Center

Joy Hakim

Smithsonian Books
Washington and New York

Published by Smithsonian Books

Produced by American Historical Publications

Byron Hollinshead	President
Sabine Russ	Managing Editor and Picture Editor
Lorraine Jean Hopping	Editor
Kate Davis	Copy Editor
Marleen Adlerblum	Designer and Illustrator

Library of Congress Cataloging-in-Publication Data

The story of science: Newton at the center/Joy Hakim.
p.cm.
Includes index.
ISBN 1-58834-161-5
1. Science—History. I. Hakim, Joy.

Q125.S776 2005
509—dc22

2004058465

British Library Cataloguing-in-Publication Data available

Manufactured in the United States of America
10 09 08 07 06 05 1 2 3 4 5

Grateful acknowledgment is made to publishers who have granted permission to quote from their copyrighted works (see page 442).

This book is for Casey Gray Hakim and Eli Thomas Hakim, with love.

Contents

KEY FOR FEATURE SECTIONS

Science ∞ *Math* *Technology*  *Geography* *Philosophy/Religion* *Art* *Music*

The Scientific Quest...and This Book

Read this book and you'll know more science than Isaac Newton did. And since Newton was one of the smartest humans ever, that's saying something. Like all good scientists, Newton understood that he was involved in a quest that can never be finished. He knew that his work would get trimmed and topped.

Artists and literary figures don't think like that. No one wants to change or add to Shakespeare. But science keeps building, and when new blocks are put in place, that often means throwing out the old or adapting it before heading on—to new heights. So there's nothing dull or static about science's story. It's a tale of exploration, one that will stretch your mind to its limits.

Scientists want to know how the universe works. From the time of the ancient Greeks to now, that's been the scientific quest. In the sixteenth century that investigation was heating up. There were questions that needed answers.

Francis Bacon died as the result of a scientific experiment (or so the story goes). He was stuffing a duck with snow—to see if cold is a preservative. He caught a chill, and that was that.

It seemed obvious then that heat was a substance, an invisible something related to the Greek element fire. But Francis Bacon (a contemporary of Shakespeare and Queen Elizabeth I) had another idea. He wrote, "heat... is Motion and nothing else." Heat is only motion? When your daily warmth comes from a fireplace, that concept seems like a stretch.

The dispute was a big one. The *heat-is-motion* idea

attracted Robert Hooke (a man Isaac Newton detested), who described heat as "being nothing else but a very brisk and vehement agitation of the parts of a body." Heat is not a *thing*? That idea seemed bizarre. The other folks, the *heat-is-something* crowd, called their tangible heat stuff "caloric" and thought it was an element.

This wasn't the only disagreement. Newton said that light is made up of "corpuscles"; by that, he meant tiny particles. But a Dutchman, Christiaan Huygens, said light isn't made of particles; it's wave action.

How important were these disagreements? Very. Before they could finally be settled, it was determined that all matter—from your toenails to the stars in the heavens—*is* made of particles, that those particles are in constant motion, and that most of them carry an electric charge. Those *particles in motion* produce heat, exert magnetic force, describe atoms, and, in this technological age, carry messages. In other words, particles in motion underlie just about everything. Harnessed as electricity, they move modern civilization. But the wave folks, like Huygens, weren't wrong either. Science is more complex than anyone had imagined.

We might never have learned any of this without the Scientific Revolution that was gestating in Francis Bacon's day. (He was born in 1561.) Besides his mental grasp of heat, Bacon figured out that it takes time for light to travel, *time that can be measured*. (Almost everyone else thought light traveled instantaneously.) But aside from those astonishing insights, Bacon wasn't much of a scientist; he was a great propagandist who inspired others with his vision of what science could accomplish. The science he foresaw was *not* based on common sense or theology (religious belief) or Aristotle. It started with observation and experimentation

Are Saturn's rings gas clouds, liquid flows, solid bands . . . or something else? James Clerk Maxwell used paper and pencil to prove that they're swarms of tiny particles. Count on page 360 to tell you how.

and advanced to proofs, something we now call the "scientific method." Bacon thought of scientific research as a collective enterprise aimed at benefiting all humankind.

The Scientific Revolution evolved from Bacon, from the French philosopher René Descartes, and from a few other original thinkers. Why did it happen in the West? It was, in part, a freedom thing. European nations, in varying degrees, managed to provide enough freedom to allow for creative ferment and question asking.

Those who enlightened us, who gave us a scientific base—something we now call "classical physics"—were a colorful bunch, ranging from charismatic Galileo Galilei to lonely Isaac Newton to a slightly unsavory American spy who happened to be way ahead of his time when it came to understanding heat. What they had in common was a passion to know how the world works. They, along with lots of others, established a foundation for modern science. That's the story this book attempts to tell.

By the way, this cast of characters did *not* call themselves "scientists." That word was invented by William Whewell, an English classics scholar. In 1840, he wrote in *The Philosophy of Inductive Sciences*, "We need very much a name to describe a cultivator of science in general. I should incline to call him a Scientist." Before Whewell, cultivators of the physical sciences were known as "natural philosophers." By the twentieth century, *scientist* had taken root.

Dealing with all this—the stories and the scientific information—in a book as beautifully illustrated and conceived as this one, has been the work of an unusually talented team. Byron Hollinshead led the project. Sabine Russ researched and chose all the pictures, asked cogent questions, and dealt with the big and small decisions involved in putting a book together. Lorraine Jean Hopping, the editor, has been much more than just that. She's added to the writing (credited as LJH on some sections) and brought real expertise in scientific education to the mix. Marleen Adlerblum created the design. Turn the pages and you'll see how good she is. Kate Davis did knowledgeable, amazing copyediting. Together they have been enthused and dedicated to a degree that few writers are lucky enough to enjoy. I couldn't be more grateful.

At the same time, Doug MacIver, Maria Garriott, and Cora Teter, of the Talent Development Middle School project at Johns Hopkins University, have created a coordinated "minds on" component to help make this book effective in classrooms (as they did with my history books). The crew at the Smithsonian—Larry Small, Don Fehr, Carolyn Gleason, Julie

McCarroll, and Stephanie Norby—sponsored and encouraged.

Then there have been readers—individuals with backgrounds and expertise in science who read and commented on the manuscript, mostly because they wanted to help with a project intended to further science literacy. Mordechai Feingold, a history professor at Caltech, provided both reassurance and insightful comments, particularly on his special friends—Galileo and Newton. John Hubisz, professor of physics at North Carolina State University (and former president of the American Association of Physics Teachers), brought expertise in science and education. Science educator Juliana Texley, lead reviewer for *NSTA Recommends*, offered detailed, knowledgeable suggestions. Hans Christian von Baeyer helped get me started. After I read one of his beautifully written books, I telephoned him and he invited me to his classroom at the College of William & Mary (Thomas Jefferson's alma mater). He was soon teaching me a lot of physics. Tom Lough, associate professor of science education at Kentucky's Murray State University (and winner of an NSTA Distinguished Teaching Award) read the final manuscript carefully and provided constructive criticism. Edwin Taylor, Senior Research Scientist Emeritus at MIT and the author of several outstanding physics texts, read about the project and sent an E-mail with suggestions on writing science. I got out my lasso, and he was soon roped in. Taylor offered insightful comments and checked pictures and captions as we were finishing this volume. The lasso came out again when I got an E-mail from Robert Fleck, an astrophysicist and professor of physical sciences at Embry-Riddle Aeronautical University with a background in science history and an interest in education. He read with an eye for detail and zeal for the subject. At the National Science

What's the fate of that pretty white bird inside the glass? It's clearly written on the faces of two of these onlookers. The story behind *An Experiment on a Bird in the Air Pump* is on page 219.

Teachers Association (NSTA), David Beacom encouraged and Gerry Wheeler (author of a physics text I use often) answered questions when I asked.

This list is long, but I was learning as I wrote, so I needed all the help I could get, and the generosity of the scientists I approached was heartening. (When you're writing for young readers, the very best people want to help you.) Rocky Kolb, at the Fermi Labs, read an early version and commented; so did Neil de Grasse Tyson, the Frederick P. Rose Director of the Hayden Planetarium. (Both are terrific writers; check them out.) Richard Schwartz, a mathematician at the University of Maryland, shared his knowledge of calculus and of Fermat's Last Theorem. Frederick Seitz, a distinguished scientist and former president of Rockefeller University, lent his support when it was most needed. Barbara Hass, a research librarian, helped me become more Web-savvy. "If you absolutely positively have to know, ask a librarian," quotes the American Library Association. I agree. With all that expertise, this book should be errorless, but I've learned that rarely happens. I'd like to blame publishing gremlins, but errors are my responsibility. If you find any, let me know, and we'll correct them in the next printing.

In addition to all those good people, I'd like to add special thanks to all the teachers/educators who have inspired and encouraged me. As to administrators, attend a monthly administrators' meeting in Los Angeles's District 6—as I was fortunate enough to do—and you'll not only talk kids and books and ideas, you'll find yourself inspired and energized. Great teachers make a lifetime impact on their students; they are national treasures.

As always, my husband, Sam, was wonderfully supportive. My brother, Roger (and his terrific wife, Patti), kept me posted on science articles and books. Ellen, our whiz of a daughter, helped me be organized and in touch with the twenty-first century. Jeff, a math professor, patiently and cheerfully answered his mom's questions about numbers and then added suggestions on calculus, Boolean math, and more. Danny, a writer, provided a sounding board. Todd, Haya, Liz, Natalie, Sammy, Casey, Eli, and you, dear reader, have helped inspire the books.

—Joy Hakim

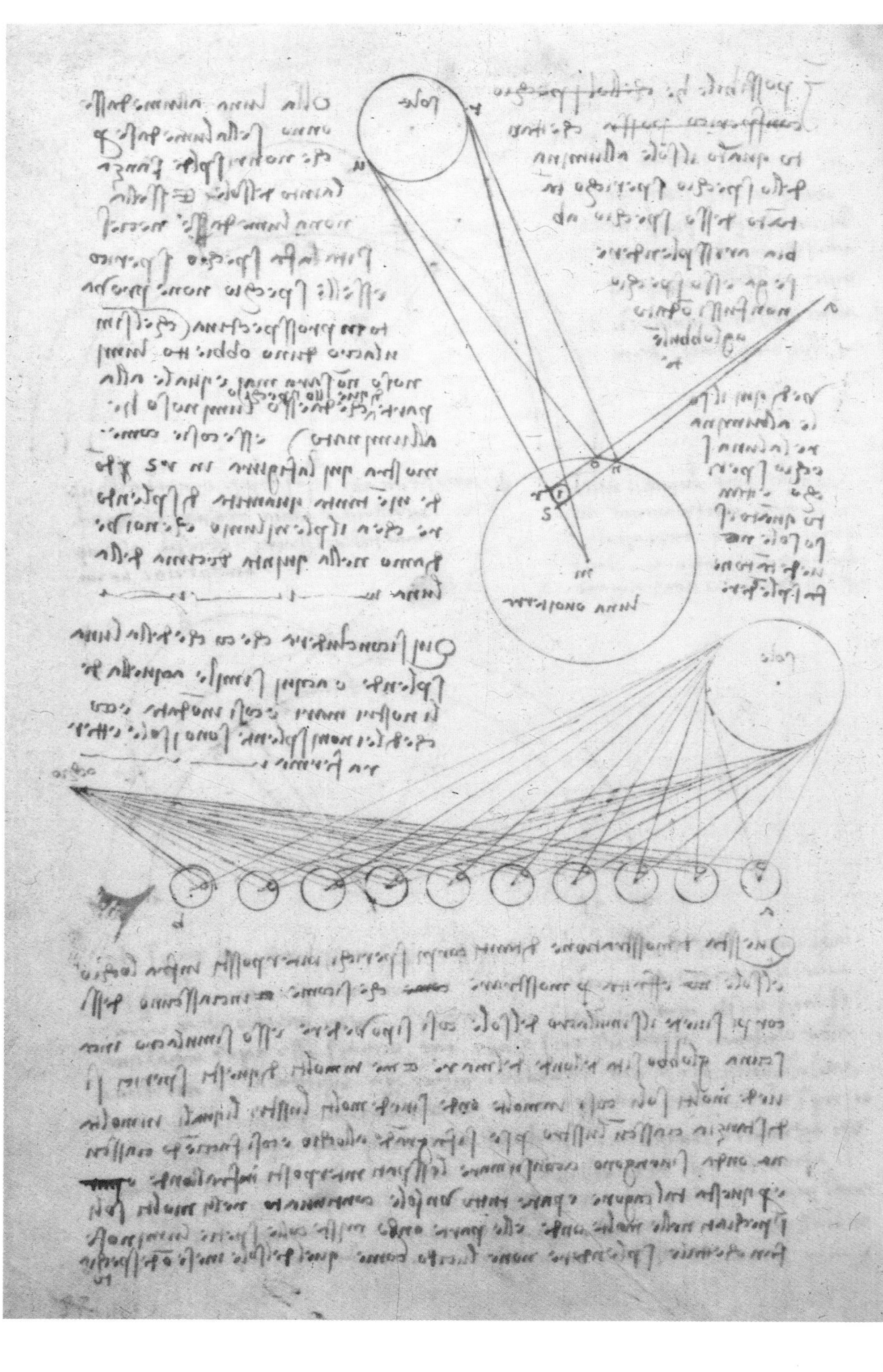

1 Off-Center? It Can't Be!

It must be considered that nothing is more difficult to carry out, nor more doubtful of success, nor more dangerous to handle, than to initiate a new order of things.

—Niccolò Machiavelli (1469–1527), Italian statesman, *The Prince*

Two voices speak for the future, the voice of science and the voice of religion. Science and religion are two great human enterprises that endure through the centuries and link us with our descendants.

—Freeman Dyson (1923–), British-born American physicist, *Imagined Worlds*

For centuries, Europe has been filled with vast forests—like those of the Hansel and Gretel story—inhabited by wolves, bears, deer, and foxes. But now it is the fifteenth century and woodcutters have chopped down so many trees that the forests are almost gone and logs for building or cooking are in short supply. The wild game that once roamed the forests and provided peasants with a high-protein diet has just about disappeared. Malnutrition is a growing problem.

It isn't the only problem. The English and French have been fighting each other off and on since 1337. For nearly three centuries, the English have held parts of France. In 1429, Joan of Arc (who says she was sent by God) dons armor, leads a French army, and routs the hated English conquerors in a big battle at the city of Orléans. But war being what it is, she goes too far for some French generals, is turned over to the English, and is burned at the stake. A final battle, fought at Castillon (on the Dordogne River) in 1453, ends the English adventure in France. It is also the end of what is known as the Hundred Years' War.

If any one person symbolized the new order that burst forth in the fifteenth century, it was the polymath Leonardo da Vinci, born in 1452. (Polymath? It means someone who is learned on many subjects.) Leonardo left some 5,000 notebook pages filled with thoughts and sketches—on light, astronomy, anatomy, engineering, and more. Perhaps because he was a lefty, or because he wanted to hide his ideas, he wrote his notes from right to left, which is backward to most of us. To read the facing page, you'd not only need to know Italian, you would have to use a mirror. See if you can spot the Italian words for Sun (*sole*) and Moon (*luna*).

Good-bye Forests

Technological marvels sometimes come in surprising forms. In the medieval world, the moldboard plow was cutting edge (literally). It could take hard soil and make it useful for farming. But the plow was heavy and hard to turn and needed to be pulled by an ox. The old small square fields (which demanded lots of turns) no longer made sense. When horseshoes were invented and horses began pulling those plows, it didn't make much difference. Long strip fields were the way to go. And that meant cutting down forests.

Some people were happy to see the forests disappear. Think of those old fairy tales, like "Little Red Riding Hood"; Europe's woods were scary. In a classic book, *The Forest and the Sea*, Marston Bates describes those thick forests as a "silent world by day.... The various songbirds so common in our modern woods were then confined to the forest margins.... The birds of the forest were owls, ravens, eagles,

Progress? Above, you can see it in the form of cleared land and farm activity. That painting is one of the gorgeous illuminations from an 831-page manuscript called *Grimani Breviary* (ca. 1500; a breviary is a book of prayers, hymns, and psalms). A hundred years later, Flemish artist Gillis van Coninxloo painted the forest as a place of mystery and beauty (left). Big question: Will the world's forests continue to be threatened by population growth?

pigeons. . . . There was plenty of noise in the forest at night: hooting owls, howling wolves, screaming pumas. They were mostly unearthly, frightening noises."

The opening of Europe's forests to farming and settlement in the Middle Ages is sometimes compared to the opening of the American West many centuries later. The pioneer farmer was seen as the heroic tamer of a stubborn wilderness. Providing more farmland meant more food for more people. But there was another side to the picture: deforestation—the permanent destruction of forests and open woodlands. It created a shortage of firewood for cooking and heating and of wood for building. It altered climate patterns too.

By the fifteenth century deforestation was a very old story. Read *The Epic of Gilgamesh*, written about 4,000 years ago, for an ancient account. Humans with axes and fire have been cutting down forests and changing their content (planting different trees for special purposes) for thousands of years. Today, tropical rain forests are being destroyed at a rate that has some experts predicting there will be almost none left by 2100. Does it matter?

Trees (and other plants) take carbon dioxide (CO_2) from the atmosphere and release oxygen (O_2). When forests are cleared, it not only changes the air we breathe (oxygen is essential to life), it changes global weather and destroys plant and animal habitats.

The East Asia and Pacific region (some of which is shown in the lower right corner of the satellite image at right) has lost 95 percent of its primary forests, according to a 2004 World Bank report. Note eastern Thailand (below), where the brown color tells of massive deforestation next to an uncut patch of dark green. Without trees to hold the soil, silt (muddy runoff) has made some waterways unusable.

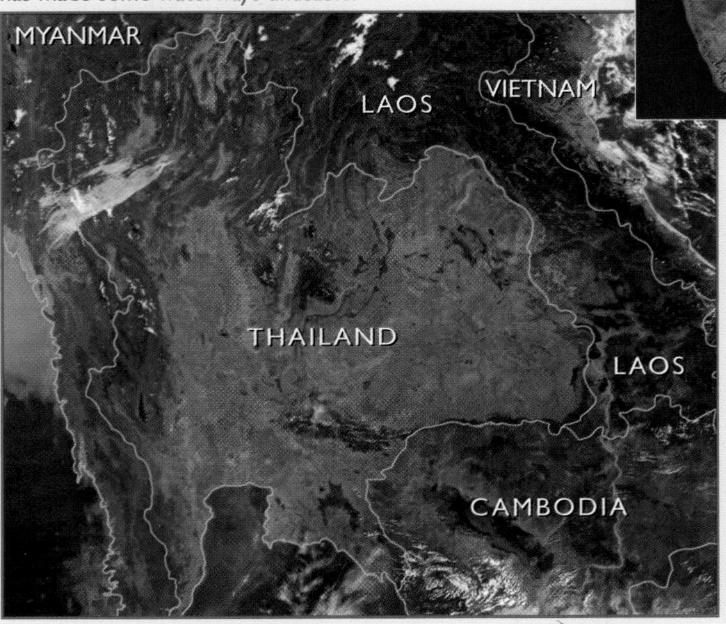

Deforestation is a global crisis. Biologist Wangari Maathai of Kenya, Africa, initiated the planting of 30 million trees. Maathai, who won the 2004 Nobel Peace Prize, said, "Deforestation causes rivers to dry up and rainfall patterns to shift, which, in turn, result in much lower crop yields and less land for grazing. . . . Unless we properly manage resources like forests, water, land, minerals and oil, we will not win the fight against poverty. And there will not be peace."

The Hundred Years' War actually lasted 116 years (1337–1453). That doesn't mean there was fighting for all those years. But there were plenty of battles and lots of nastiness. Once in a while, knights in armor would put on tournaments and jousts to pretty things up a bit, but for most people it was brutal and horrendous and an awful waste of resources and money—to say nothing of people. Shakespeare's plays *Henry V* and *Henry VI* describe the war from an English point of view. You can rely on Shakespeare to tell a good story. For exact history, look elsewhere. To the right is an anonymous French artist's version. That's Joan of Arc announcing the liberation of Orléans to Charles VII. The sixteenth-century map (far right) is of Constantinople (now Istanbul, Turkey). Below is a fresco of the Siege of Constantinople, painted on an exterior wall of a Romanian church.

That very year, 1453, the Ottoman Turks batter down the thick walls that surround the city of Constantinople (on Europe's Asian border), conquer that capital of the Eastern Christian world, and send its Greek-speaking citizens fleeing west to Italy and beyond. It is the end of

WHAT'S IN A NAME?

Byzantium was a Greek city on the European side of the Bosporus Strait. That strait separates Europe from Asia Minor, which means the city was in a key location. Emperor Constantine I thought so. He moved his capital there from Rome and named the city Constantinople, after himself (in 330). Thanks to Constantine, Christianity was to have two official capitals: Rome and Constantinople. Religious practices became slightly different in each, though some people didn't think the differences were slight. One became the Eastern Empire, sometimes called the Byzantine Empire; the other remained, for a while, the Holy Roman Empire.

In modern times (1930) Constantinople was named Istanbul. Today that ancient city is Turkey's biggest and busiest. But the capital of Turkey is Ankara.

an era. Constantinople has been Europe's largest, wealthiest, and most splendid city. During centuries when most of Europe forgot the astonishing heritage of ancient Greece and Rome, scholars in far-off Constantinople kept it alive.

Now some of those scholars head west, bringing books, artworks, and knowledge of the past with them. Their timing is right. Europeans have been building universities; they are eager to learn. Greek and Roman art and ideas come from a time when there was much intellectual freedom. They burst

The Villa d'Este water garden, built about 1550 near Rome, is a marvel of pneumatic (air-powered) and hydraulic (water-powered) devices inspired by the ancient inventor Hero of Alexandria. British author James Burke writes in his book *Connections*: "The slope of a hill was used to supply fountains and dozens of grottoes where water-powered figures moved and played and spouted. Within fifty years of the Roman example there were water gardens all over Europe, and princes and bishops would delight in switching secret valves to deluge their guests with soot or spray them with water."

like cannonballs in the narrow confines of the medieval world. It is no surprise that scholars and ordinary people too are awed by the splendid sculptures, poems, plays, and scientific writings from the ancient world.

That rediscovered learning is about to gain an easy way to spread. In the same year that Constantinople falls, a German goldsmith named Johannes Gutenberg is perfecting a new printing process. Books that have been hand-copied one at a time can be printed. It is the beginning of mass production, and presses will soon be creaking and groaning with books and broadsheet newspapers. Ideas will now have vehicles for direct travel. Someday, historians will see this weighty year, 1453, as the end of the Middle Ages and the beginning of the Renaissance and modern times.

Of course, it actually takes several hundred years to make that leap. But you have to begin somewhere, and 1453 is a good place to start.

Before PERIPATETIC (per-uh-puh-TET-ik) was an adjective for "worldly wanderers," it was a noun. The Peripatetics were followers of the Greek philosopher Aristotle, who taught and debated while strolling around the walkways of the Lyceum at Athens.

Imagine yourself living in Holland in these energetic, changing times. Holland is an affluent crossroads and a center of new music. Dutch composers are writing complex, startling works in four parts that sing with harmony and contrast. Those composers, who are in demand in courts and cathedrals, are sometimes called peripatetic, which means they roam from place to place (as ancient Greek musicians and storytellers did). Several of them have been to Florence in Italy and have come back telling of a cathedral with a magnificent, octagonal ribbed dome that is the world's largest; it is bigger than Rome's vast Pantheon, and can be seen from 24 kilometers (15 miles) away.

In 1453, China was the world's leading culture. The Chinese had invented gunpowder, printing (Gutenberg built on their work), the compass, and the idea of a meritocracy—hiring government workers based on merit, or ability. All these inventions made their way to Europe, where, says historian William McNeill, they "were the principal secret to the rise of the West.... [They were] exploited in more radical and far-reaching ways than the Chinese themselves had done." What may look like a spoon (above) is actually an ancient Chinese compass called a *sinan*. The handle of this magnetized-iron ladle swivels around and points to the south, instead of the north.

The Florentine dome, with high windows letting in shafts of sunlight, is the work of Filippo Brunelleschi (broo-nuh-LES-kee). That revered architect (he died in 1446) figured out a way to build a giant dome without any scaffolding to support its center. Sandstone beams—some weighing 770 kilograms (1,700 pounds)—and huge marble slabs had to be lifted nearly 100 meters (about 300 feet) off the ground to create and face the structure. Brunelleschi had designed what he called an "unheard-of machine" to do the lifting—a giant hoist turned by oxen. Visitors who flock to Florence come to see the ox-hoist as well as the dome. Brunelleschi meant to have his architecture recapture the grandeur of ancient Rome. He did that, and more.

Not far from Holland, in the busy city of Ghent, you can see a controversial 24-panel altarpiece painted by the Flemish

To the left is Brunelleschi's famous *Duomo* ("Dome") in Florence. Brunelleschi and his mathematician-mapmaker friend, Paolo Toscanelli, poked a hole in the dome, letting in a shaft of light. The light projected an image of the Sun on the cathedral floor. The image moved according to the Sun's position in the sky, but because the hole was so high, sunlight only reached the floor for a few weeks around the summer solstice. In smaller cathedrals, moving-Sun images could be watched for a full year.

Van Eyck's masterpiece at the Cathedral of St. Bavo in Ghent (now in Belgium) features folding panels of quiet, sculpture-like portraits. Only on feast days was the altarpiece opened (as above) and its dazzling brilliance displayed. The lower central panel is based on part of the vision of St. John the Evangelist. St. John the Baptist (shown on the top row in a green robe) called Christ the "lamb of God." In the figures of Adam and Eve (top row, outer images), the artist presents two fallible humans who paid attention to the serpent, not to God.

artist Jan van Eyck (van IKE). Artists make their own paints, grinding minerals and plants between two stones and then mixing the powder with a liquid—usually egg. Van Eyck has substituted oil for egg, working with something new—oil paints—which give depth and glow to colors. Because oil dries slowly, van Eyck is able to rework and add to his paintings. The altarpiece is shocking because it seems so real: Adam and Eve are usually painted as if they were gorgeous models; van Eyck has made them very human. Some viewers are offended. Others are awed.

Here in the Low Countries, where the sea laps the land, mapmaking is an art. World maps are patched with regions labeled "unknown." It makes many who are curious ache to explore.

Those living in the midst of these transforming times can't step back and see this era as it will seem to later generations. Like most of us, they accept the practices of their time. Schools, religion, politics, and science are all controlled by the same authority—the church. The idea of separation of church and state—an American idea—would seem strange, even unthinkable, to people of the times. But the ferment of the age is leading some to ask questions.

In 1453 most Europeans can't read or write. (Printed books will soon provide an incentive to do so.) But since you've imagined yourself into these times, put yourself among the privileged and you'll go to school. There, you're taught grammar and to read the Bible in Latin. You learn the "new" arithmetic, which means Arabic numbers (1, 2, 3, etc.)

as well as Roman ones (I, II, III, etc.). You don't use the symbols + or – or = because they are not yet invented. You study two other subjects that are considered essential for a sound education—music and astronomy. Here are the "facts" of astronomy that you are taught:

Ah, Dante! You need to read his great poem, *The Divine Comedy*, to understand why he is so revered. This painting (left) of Dante and his influential work was done in the fifteenth century by Domenico di Michelino. Note the heavenly spheres at the top (*Paradiso*), the mountain known as Purgatory (*Purgatorio*), and scenes from hell (*Inferno*) on the left.

- Earth is a sphere standing still at the center of the heavens. Its habitable surface forms a circle, with Jerusalem as its navel.
- The stars and planets are made of a perfect substance not found on Earth. It's called *aether* (EE-thuhr). It's the fifth essence (or quintessence), after earth, air, fire, and water.
- The Moon and Sun and stars are held in place by invisible crystal spheres that revolve around the Earth, one inside another.
- Heaven has its own sphere, above the stars.
- Hell, Satan's home (called the Underworld), is at Earth's bottom.

German artist Matthias Grünewald (ca. 1475–1528) and Dutch artist Hieronymus Bosch (ca. 1450–1516) painted bizarre fantasies when most artists were striving for realism. Perhaps they were influenced by the witchcraft and plagues of the Middle Ages. They were certainly motivated by religious beliefs of their times. A panel from Grünewald's famous Isenheim Altar (near right) shows a saint being tempted and frightened by creatures from hell. In Bosch's painting (far right), blessed souls are guided toward a tube of light leading into heaven.

TIME TRAVEL

Picture a wide ribbon, like a banner, that cuts through time. On one side of the ribbon the landscape is medieval—the towns are walled and so is the thinking; on the other side the view is open and modern. The ribbon has some holes in it—ideas sift back and forth—but fewer than you might think. The modern world is very different from that which came before. When it comes to science, the people on each side of the ribbon have a different basic idea.

On one side of the ribbon, the medieval thinkers in Europe are obsessed with the idea of perfection. That idea comes from the Greek philosopher Plato and from Christian beliefs. The medieval thinkers are taught that everything is perfect in God's created world (except on Earth, where there are clearly imperfections). Since they are looking for perfect planets and perfect elements and perfect scientific laws—that's what they find. They see no need to ask questions. They think all the scientific knowledge that men and women should have is in the Bible.

On the other side of the ribbon, the modern thinkers ask questions, observe, and experiment. You can call their system of thinking both *empirical* (based on experiments and observation) and *rational* (based on thinking). They believe the Bible deals with matters of faith and morality. In contrast, science is a search to understand how the universe works.

Separating the two quests was wrenching to people and institutions, but it helped bring modernism.

Other things made a difference, too. In the fourteenth century, the plague wiped out 25 million people in Europe in 5 years. (Europe's population in 1347 is estimated at 75 million.) So many died so quickly that the dead couldn't even get buried. The gravediggers died, too. A horror like that made many people reexamine their deepest thoughts.

Much of the questioning came from the new universities, appearing first in the eleventh century. By the fifteenth century, you could find a university in almost every major city.

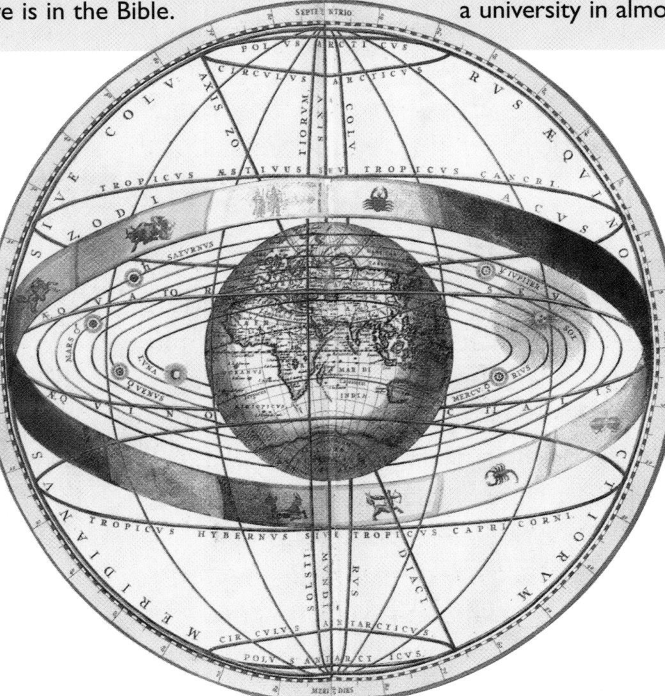

Here's the cosmos, gorgeously pictured in 1661 by Andreas Cellarius of Amsterdam. The round, oversized Earth is in the center, circled by a band of zodiac constellations (shown in symbolic form). The other planets and the Sun orbit in between. The scheme is all based on Ptolemy's teachings, and it seemed to work well enough—as long as sailors, who might have relied on it for navigational purposes, stayed close to home.

As I said, you are lucky. You have a teacher who teaches you of the ancient Greeks, especially of Aristotle, who lived in Athens in the fourth century B.C.E., and of Claudius Ptolemy (TOL-uh-mee), who lived in Alexandria, Egypt, in the second century C.E. You study the stars and consider the heavens. But when it comes to studying nature or observing the world around you (as Aristotle did), no one, including your teacher, thinks that worthwhile. Nature is thought to be God's responsibility; yours is to consider your soul. So there isn't much discussion about science.

Almost everyone in the fifteenth and sixteenth centuries agrees on these things:

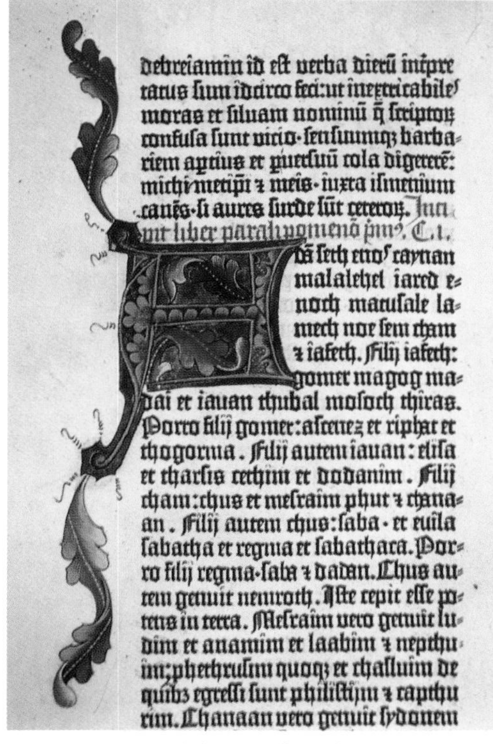
debreiamim id est verba dierũ interpre
tatus sum idcirco feci: ut inextricabiles
moras et silvam nominũ q̄ scriptorũ
confusa sunt vicio · sensuumqȝ barba-
riem apertius et p versuũ cola digererē:
michi ĩmetipi ⁊ meis · iuxta ismenium
canēs · si aures surde sũt ceterorũ. Inci
pit liber paralipomenõ prim⁹. C. i.
Adã seth enos caynan
malalehel iared e-
noch matusale la-
mech noe sem cham
⁊ iafeth. Filii iafeth:
gomer magog ma-
dai et iavan thubal mosoch thiras.
Porro filii gomer: ascenez et riphat et
thogorma. Filii autem iavan: elisa
et tharsis cethim et dodanim. Filii
cham: chus et mesraim phut ⁊ chanaan. Filii autem chus: saba · et evila
sabatha et regma et sabathaca. Porro filii regma · saba ⁊ dadan. Chus autem genuit nemroth. Iste cepit esse po-
tens in terra. Mesraim vero genuit lu-
dim et anamim et laabim ⁊ nepthu-
im: phethrusim quoqȝ et chasluim de
quibȝ egressi sunt philistiim ⁊ capthu
rim. Chanaan vero genuit sydonem

Johannes Gutenberg was an inventor, a goldsmith, and a meticulous craftsman who wanted to mechanically reproduce the gorgeous colors used in hand-colored and woodblock printing—and did so. In 1454, he used movable type to produce 180 copies of a Bible that many people believe has never been surpassed in beauty. He and other craftsmen perfected the new method of printing, working out problems with inks and paper as well as type. Their process was far more sophisticated than anything in the world and was used well into the twentieth century. (As to Gutenberg's personal life, it is sad.) Above is a decorated letter *A* from Gutenberg's Bible.

- Earth is the center of a created world that includes stars, planets, comets, and the Sun.
- Earth doesn't move.
- The heavenly bodies, except for the five planets, travel around Earth in perfect, circular paths called orbits.
- The planets—whose orbits each sometimes move in a short, backward loop—govern the variable events in the world.

It is easy to confirm this. Look at the sky on a starry night. The stars appear to be moving; Earth seems to be standing still. To think anything else is to reject common sense and the senses.

But in Poland, a church official does just that. He comes to believe that our senses can't always be trusted. Sometimes they fool us. He dares to think the unthinkable: *Earth is a moving planet that turns on its axis as it circles the Sun.*

The man's name is Nicolaus Copernicus (koh-PER-nuh-kuhs), and he is born in 1473, which is 20 years after Johannes Gutenberg's invention of movable type, the fall of Constantinople, and the end of the Hundred Years' War.

GEOGRAPHY CLASS—WAKE UP!

As to geography, this is what you learn in school in the fifteenth century:

All of Earth's lands—which to Europeans means Asia, Europe, and Africa—are clustered inside a great ring of forbidding water called the Ocean River. A world map looks like a wheel. Its edge is a river shaped like an O; inside is a T (the Mediterranean Sea is the stem of the T). East is on top; north is to the left; south to the right. The first of the O-T maps was drawn by the Greeks in the fifth century B.C.E. Delphi, the Greek holy city, was right in the center of those maps. In 1400, it is Jerusalem that sits at Earth's navel.

Elsewhere on the globe, maps and stories are similar. If you live in the Valley of Mexico, you *know* that the magnificent Aztec city of Tenochtitlán (modern Mexico City), with its temples and gorgeous gardens, is the very center of the universe. You feel the same way if

This famous *mappamundi* ("map of the world") was originally made around 1234 for the monastery at Ebstorf in Germany. Christ is holding the world. His head (at the top) is in Asia near the Garden of Eden. His left hand is in southern Africa where, as Europeans believed, monstrous creatures lived. Christ's right hand is in the home of the powerful Amazons. Jerusalem is at the center. The Ebstorf map was destroyed during World War II; this is a 1339 facsimile. The mapmaker (thought to be Gervase of Tilbury, England) notes in one corner, "It can be seen that [this work] is of no small utility to its readers, giving directions for travellers, and the things on the way that most pleasantly delight the eye."

you live in powerful Hanyong (modern Seoul), the capital of the Korean peninsula. There, the Yi dynasty has introduced a new 24-letter phonetic alphabet that makes it easy for almost everyone to learn to read. When Koreans draw maps, they put themselves right in the middle of a flat Earth.

A scholar from Florence, Italy, finds a manuscript of Ptolemy's *Geographia* in Constantinople. In 1410, he translates it. Ptolemy has details on Africa and Asia that fill in big blanks in the O-T map. That information had been forgotten during long centuries when general knowledge seemed pointless. Ptolemy's Ocean River can serve as a highway, not a barrier. His maps use a grid and lines of longitude and latitude to plot out the land and sea. With grids and geometry and knowledge of the stars, an explorer can know where he is even without familiar landmarks.

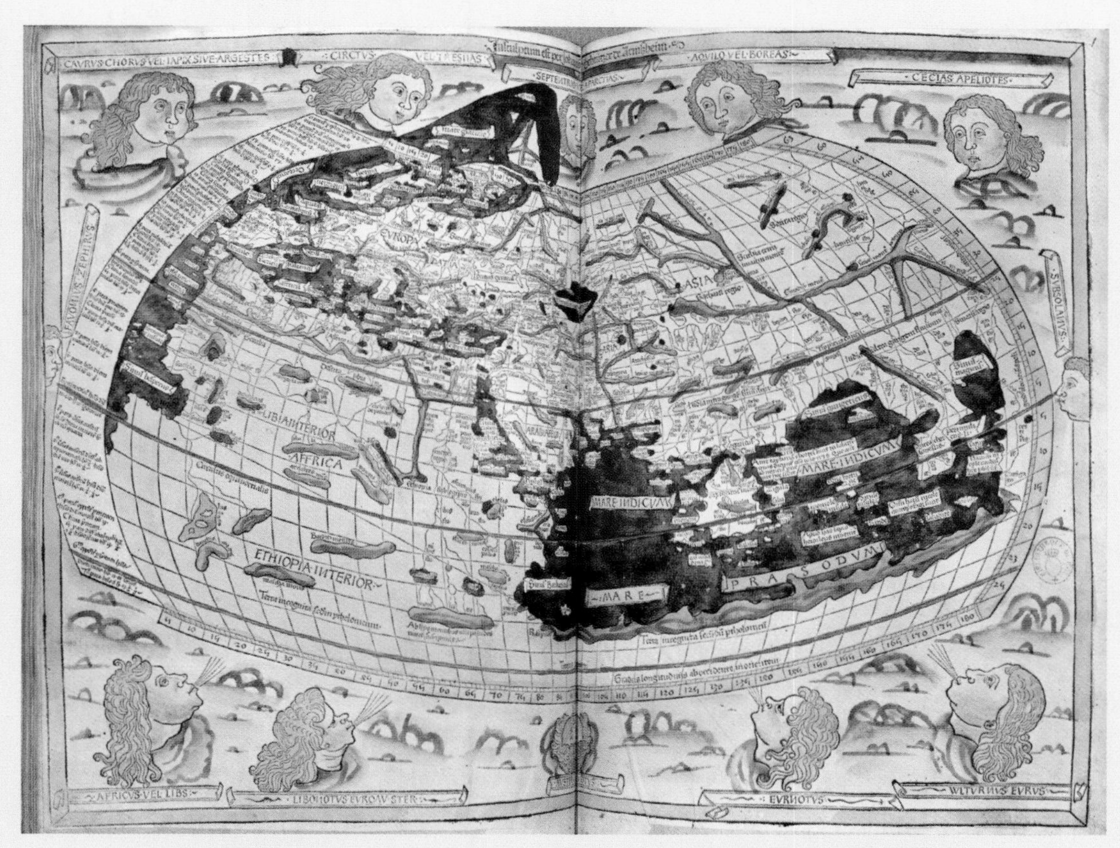

In the fifteenth century, when printing became available, Ptolemy's *Geographia* (written in the second century C.E.) became very popular. But none of Ptolemy's maps had survived in the West. European mapmakers, like Francesco Berlinghieri (his map is above), adapted versions found in the Byzantine world. Note the curved lines of longitude and latitude. As to the heads surrounding the map, they represent the winds.

A New Age: Bringing New Ways of Seeing

The sun does not move.... The earth is not in the center of the circle of the sun, nor in the center of the universe.
—Leonardo da Vinci (1452–1519), Italian artist and inventor, *The Notebooks of Leonardo da Vinci*

The universe is not only queerer than we suppose, but queerer than we *can* suppose.
—J. B. S. Haldane (1892–1964), British-Indian biologist, *Possible Worlds and Other Essays*

Copernicus (1473–1543) is born at a propitious time for someone who likes to read. In earlier years, handwritten books were often found chained to tables in monasteries lucky enough to have them. By 1480, there are 30 presses in Germany alone, and ordinary people are buying and reading books. At about the same time, Chinese technology is making inexpensive paper available. Printers can barely keep up with the demand.

Propitious (pruh-PISH-uhs) means favorable.

Two of the first books to be printed in Europe are by Johannes Müller, a German astronomer and mathematician who calls himself Regiomontanus. He has updated the Greek astronomer Ptolemy's masterwork, *Almagest*, adding a table that shows new data on the movements of the planets. His other book is about trigonometry (the mathematics of measuring triangles).

Leonardo da Vinci designed a simple printing press that was especially good for printing woodcuts and engravings. This model was built later, based on his drawings.

Young Copernicus reads hungrily at his home in

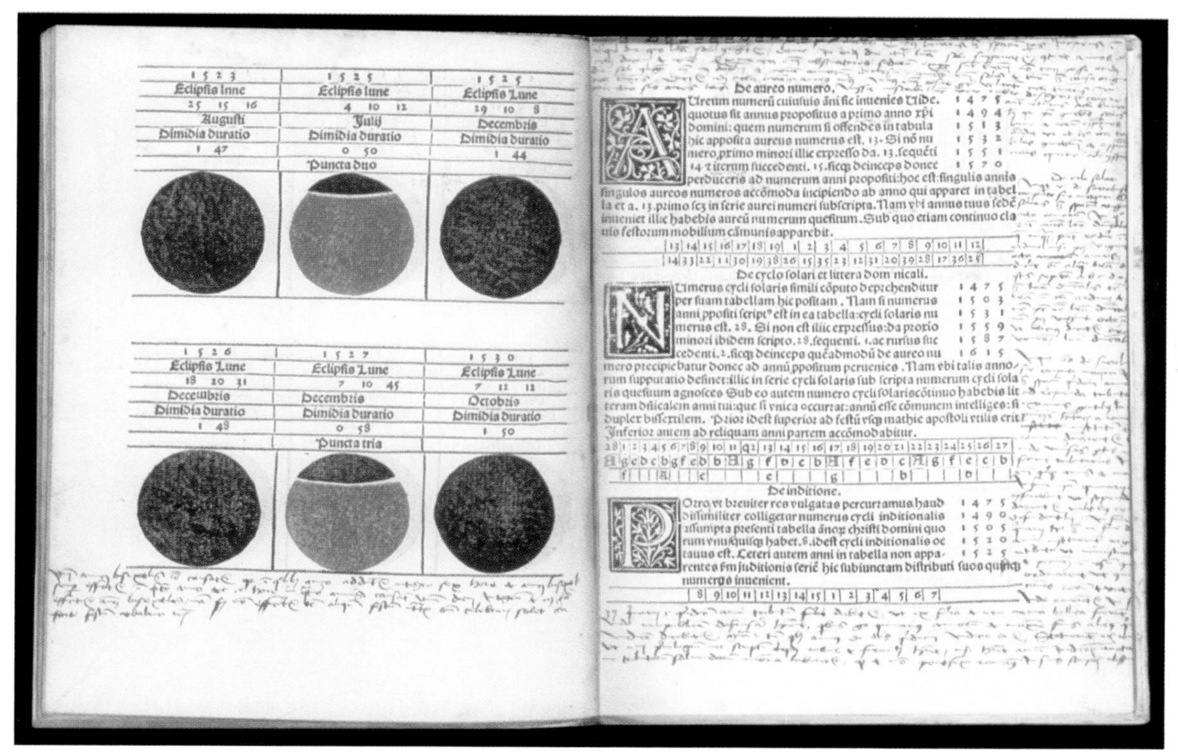

Regiomontanus published an almanac (an annual factbook) showing the position of the planets each year—a handy reference for astronomers. This 1499 edition contains eclipse predictions and other additional notations.

Toruń. At a cathedral school, he reads Regiomontanus. Toruń, a port on Poland's Vistula River, is a busy market town. Frigates and brigs from London and other faraway places tie up at riverside docks, where sailors tell adventure tales to curious boys and girls. Poland, one of Europe's largest countries, thrives with a polyglot population and bustling commerce.

Jews are a vital part of an energetic mix. When the Black Death (the bubonic plague) struck Germany in the fourteenth century, the Jews got blamed and were forced to flee. Poland provided them with a homeland. Now, in the fifteenth century, Poland's cosmopolitan empire includes them along with Poles, Germans, Ruthenians, Flemings, Armenians, Tatars, and others all living together and speaking a dozen official languages.

Polyglot ("many-tongued") means multilingual—speaking a lot of languages.

The Copernicus family is part of a merchant elite. Nicolaus's father is a copper trader, a civic leader, and a magistrate. Nicolaus is the youngest of four children. They all live in a big house on St. Anne's Street and summer in a country house with a vineyard. Theirs is a happy, prosperous

When the bubonic plague broke out in Germany in 1349, Jews were blamed and persecuted. The ones that escaped found a new home in Poland, joining Ruthenians, Tatars, Flemings, and Armenians, among many other ethnic groups. In the Middle Ages, Ukraine was called Ruthenia. So Ruthenians were Ukrainians. (*Ruthenia* is the Latin word for Russia.) Tatars were from what is today the autonomous Tatarstan in eastern Russia. Flemings were from Flanders, a country that no longer exists and is now part of Belgium, France, and the Netherlands. Armenia today is a republic, with Turkey to the west, Iran to the southeast, and Georgia to the north. The "ducky" creature (right) is a seventeenth-century doctor trying to protect himself from the plague by wearing a mask.

IT SEEMED AS IF EVERYONE WAS DYING

In 1346, Genoese sailors on a ship in the Crimea were infected with the plague. They headed home; by the time they reached Genoa, all were dead or dying. But the germs they carried were very much alive, as they were on other ships incubating a catastrophe. Some say that one-third of the population of Europe, North Africa, and the Middle East died of the Black Death (bubonic plague). In 1348 in Florence, more than half the population may have died.

Here's an excerpt from *The Florentine Chronicle* that tells of that plague year (although it is written around 1370): "When it was seen at the beginning of October that no more persons were dying of the

Above, an illuminated manuscript (a hand-painted book) from ca. 1349 shows workers burying plague victims in Tournai, a Flemish city (now in southwest Belgium).

Simple hygiene, like handwashing, can help prevent the spread of infection. But it took centuries before anyone knew about germs or made a connection between disease prevention and cleanliness. Most medieval people were filthy and stayed that way. It hadn't always been so. Until about the thirteenth century, most European villages had public bathhouses (much as in Japan today). Villagers enjoyed a hot soak. But when the forests were turned into farmland, firewood became scarce. Heating bathwater became expensive. Coal was used in some towns, but it gave off noxious fumes. By the fourteenth century, the bathhouses were gone and bathing had become a luxury for the rich. Most Europeans itched, smelled, and were breeding grounds for germs; they would have liked a good bath. That's a fifteenth-century German bathhouse below—women on one side, men on the other. Note the purple-capped furnace.

pestilence, they found that among males, females, children and adults, 96,000 died between March and October." The disease kills quickly, sometimes within 24 hours. It is death in a horrible form (boils, vomiting, intense pain, hemorrhages, uncontrollable coughing, heart failure). After 1351, the worst of the killer plague was over, but outbreaks occurred for another century and more. Italy endured a terrible epidemic in 1630.

Everywhere there were corpses, vacant houses, deserted towns, and fields left to rot. Slaves were brought from Russia to help do needed work. (Some of them died, too.) According to Isaac Asimov, "No disaster that we know of, either before or since, came so close to wiping out humanity." (Pandemics are not just a problem of the past. Today AIDS is devastating much of the continent of Africa.) Fear of the plague paralyzed the times. It left a shortage of labor, a need for technology to fill the void, and questions, lots of questions, about the meaning of life. For those who survived, the world would never be the same.

Take a closer look at the wall in this photograph. This is the schoolroom at the Collegium Maius in Kraków, where Copernicus studied geometry.

family—until an outbreak of the plague comes to Poland—and then, when Nicolaus is 10, his father dies. Uncle Lucas (the wealthy, influential bishop of Varmia) comes to the aid of the bereft family. A few years later, that good uncle sends Nicolaus off to the University of Kraków.

Kraków, with fortified walls and towers, is another vibrant river city and Poland's capital. (Later, the capital will move to Warsaw.) The university is one of Europe's best, known for religious and political tolerance, as well as for its professors of mathematics and astronomy.

It is soon clear that Nicolaus Copernicus is meant to be a scholar; he is an outstanding student. Uncle Lucas gets him a position as a canon at Frombork Cathedral. Usually canons are priests associated with a cathedral. Copernicus probably

Someone who is BEREFT is suffering from the loss of a loved one.

Why Are Universities Important? Read This and Find Out

Most historians say the modern Scientific Revolution began with Copernicus. (There's no end date; it's still with us.) Why did it happen in Europe and not in Islamic lands or in China? History professors J. R. McNeill and William McNeill, in their book *The Human Web*, say the universities spawned in Europe during the twelfth and thirteenth centuries made the difference. "The survival of universities gave European scientists a supportive community not quite paralleled elsewhere in the world. Europe had more than one hundred universities by 1500, and by 1551 new ones had also sprouted in European colonies in Mexico City and Lima."

Salamanca University in Spain dates from the thirteenth century. It's one of the oldest university buildings still in operation.

The division of Europe into many nations and the breakup of Christianity (into Catholic and Protestant) were also factors, according to the McNeills (who happen to be father and son). If you weren't free to express your ideas in one nation or one church, you could try another. "The Scientific Revolution in general required the combination of a political landscape that gave protected space to thinkers and broader circumstances that favored the long-distance flow of ideas and information." Universities provided protected space for thinkers; the printing press made ideas easy to spread; and seagoing caravels and far-ranging caravans took ideas to distant places.

A Certain Kind of Science

What exactly is the Scientific Revolution? It's the changeover from the idea that science should be based mostly on deep thinking (the Greek way) to the idea that only through observation, prediction, experimentation, and exact measurement can scientific truth be determined—that's the scientific method. As measurement techniques got better and better, belief in the scientific method came to govern science. That concept dominates this book. It's known as classical science. Scientists, following the scientific method, will make much of the universe understandable.

But, just to shake up your mind, here's a thought from advanced physics: In the twentieth century, scientists found that subatomic particles play games with classical science. The act of observing the particles changes their behavior. The more precisely scientists measure a particle's position, the more uncertain the information about its momentum becomes. These behavior changes are summed up by what is known as the Uncertainty Principle.

So can experimentation and observation lead to scientific truth? Is there really any certainty in science? These are among today's unanswered questions.

doesn't serve as a priest, but he does have a few churchly administrative duties, and that gives him a share of the revenues (which means "money") from the lands owned by the church. Thanks to his uncle's nepotism, the job helps pay for his studies. If he wants to go away he can do so and still collect income. (This kind of job is called a sinecure.)

Today, NEPOTISM is about giving favors to any family member, but it originally meant favoritism to a nephew. So Uncle Lucas was literally engaging in nepotism when he gave his nephew a cushy job.

His uncle arranges for him to continue with his education, this time in Italy. In 1496 he heads over the Alps—on foot—and is soon learning about medicine, church matters, philosophy, and mathematics at universities in Bologna, Rome, Padua, and Ferrara. He studies ancient Greek and reads Aristotle in his original language. When he finishes his schooling and returns to Poland, Copernicus is one of the most learned men of his day.

This is the Renaissance (REN-i-sahns), and many widely accepted ideas are being blown away. Before this time, artists and philosophers focused almost solely on God and otherworldly concerns. Most of the new thinkers are still deeply religious, but their attention is on *this* world and on human beings.

Renaissance artists are painting and sculpting nature and the human form as the eye sees them. Renaissance intellectuals are sometimes called humanists. Perhaps most important, they have confidence in themselves, which the generations before them lacked. They are inspired by the ancient Greeks but not intimidated. They understand that the Greeks were human and creative; so are they. Jean Bodin (ca. 1530–1596), a French political philosopher and the author of an enlightened plea for religious tolerance, compares his time with that of the ancients and says, "The age which they call golden, . . . if it be compared with ours, would seem but iron." A brash statement? You bet. That defines the age.

Since we have minds that can act as time machines, let's use them to visit self-confident Renaissance Italy. If we zoom in near the end of the fifteenth century, we'll find the artist Michelangelo—just 21 in 1496—in Rome completing his first major statue; it is of a young man and a sleeping cupid. The multitalented 44-year-old Leonardo da Vinci is in Milan,

Leonardo da Vinci's *Last Supper* (in a convent in Milan) depicts a Passover Seder, a holiday celebrating the birth of the Jewish nation. Jesus Christ is flanked by his 12 apostles (followers picked to preach the Gospel). The artist chose to use bold, geometric windows and doors in contrast to the fluid grouping of people. He also played with perspective as no one had before. Christ, larger than any other figure, commands both the painting's vanishing point and its foreground. The whole fresco, painted over a door, is designed to look like a deep, upper room.

where he lives well as court painter, architectural advisor, theatrical producer, and military engineer to the local prince, Ludovico Sforza. Leonardo is working on a fresco called *The Last Supper*. He has been experimenting with paints, and, unfortunately, those he has chosen will not hold up. Still, this painting will be one of the world's greatest. In his burgeoning (rapidly growing) notebooks, Leonardo is sketching intriguing machines, including a flying invention that uses revolving paddles (see pages 32–35).

In Venice, glassmakers have found a way to make pure crystal glass, and in fancy homes, wine is now poured from glass beakers. In the Tuscan village of Pienza, we can visit a "new town" built by Pope Pius II and laid out according to ideas on urban planning that go back to and improve on the first-century Roman architect Marcus Vitruvius. In Florence, the awesome dome on the cathedral—its technology and beauty—has turned architecture into an accepted art form and architects into popular heroes.

Our sojourn here will be pleasant as long as we know the right people. Italy's city-states are constantly at war. Now and then, armies come from France and Spain, adding intensity

In 1492, Martin Behaim made the first known map of Earth as a globe. It did not include the soon-to-be-discovered Americas or the Pacific Ocean.

to the fights. For the officers in their splendid armor, war is a way of life, thought to be valorous and not terribly dangerous. For peasants, it is something else. Crops and towns are burned, and villagers raped and killed. (The life expectancy for a peasant woman is 24.)

Meanwhile, the rich merchants and nobles live in opulent palaces with landscaped courtyards, elegant ballrooms, gorgeous wall hangings, fine furniture, handsome murals, and sometimes hundreds of servants.

In this energetic environment, Ptolemy's *Geographia* is a best-seller. That ancient geography text uses lines of longitude and latitude to plot out land and sea on a spherical Earth. It is a vast improvement over the maps of the Middle Ages, where the "Ocean Sea" was drawn as an edge circling the flat discs thought to represent the world. Ptolemy is inspiring a generation of adventurers. One, a sailor from Genoa whose brother is a mapmaker, is convinced he can sail west—across the Atlantic—reach Asia,

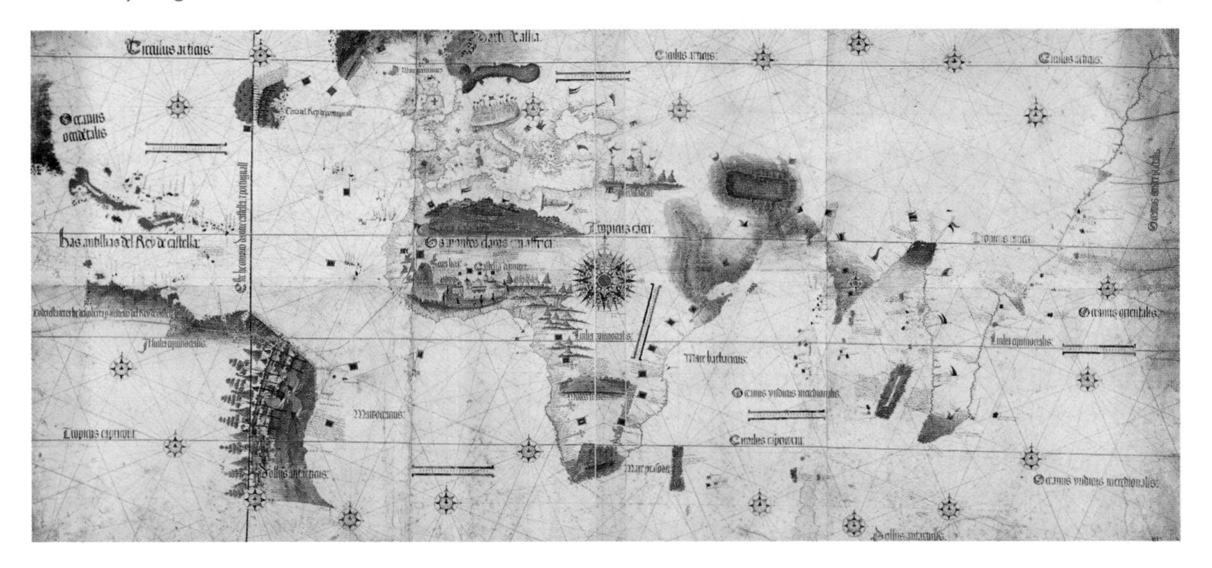

This world map was drawn in 1502, after Vasco da Gama discovered a sea route from Europe to Asia. It shows India with a degree of truth, but it later became important because of that line drawn through South America. The Portuguese claimed all the land east of the line; the Spaniards took everything to the west.

and actually prove, empirically and beyond question, that the Earth is a sphere. He gets the backing of the king and queen of Spain and, in 1492, sets sail. When he sights land, he believes he has done it—arrived in Asia's Indies. (The sailor—you know his name—also believes he has found the biblical Garden of Eden.) He, of course, is wrong. He has bumped into a continent that is unknown to Europeans. He has found a world that is new to them. Among other things, that "new" world will add potatoes, corn, and peppers to the "old" world's diet; those foods will help sustain a growing worldwide population.

In Italy, the Polish student Nicolaus Copernicus learns of Columbus's voyage and of others that follow. He studies some of the new maps. He also studies an astonishing new drawing technique called perspective; it uses principles of geometry to show three dimensions on a two-dimensional piece of paper. Drawings no longer have to look flat; they can show depth. They can mirror the real world. Many who see the new paintings think magic is involved. But it is not magic. It is the legacy of our friend Filippo Brunelleschi, the architect who designed Florence's cathedral dome (see page 7). He combined his knowledge of mathematics and surveying to create a new art form. His disciple, Leon Battista Alberti (1404–1472), who is an architect, musician, poet, and author, explains the principles of perspective in a well-respected treatise on architecture. Others take the idea and run with it.

Art is soon used as a tool of investigation—to plot, plan, and visualize. It is no coincidence that Renaissance artists, like Leonardo da Vinci, are also scientific experimenters.

Perspective is not the only technological breakthrough in the arts. Some artists are using polished lenses focused through a tiny peephole to project images of scenes and people onto canvases. It helps make their paintings exquisitely precise. Painting is now far more than decoration. It illuminates and enlightens. It adds to the excitement of the times.

In Italy, rival city-states strive to outdo each other in wealth and originality. Florence, Milan, and Venice are the

To make a pinhole projection, you just need a pinhole—and a sunny day. Prick a tiny hole in an index card, hold another card behind it, and line up the cards so that a shadow falls on the second card. (Note: Never look directly at the Sun; use the shadow to line up the cards.) A small, upside-down image of the Sun will appear on the second card. If the image is blurry, adjust the distance between the cards. Renaissance astronomers used this same technique to project images of the Sun on cathedral floors. You can sometimes see Sun images on forest floors too. Small gaps between tree leaves are natural pinholes that project dapples of light—lots of round Suns—on the ground.

PERSPECTIVE was more science than art, originally. The word is from the Latin *per-* ("through") and *specere* ("look"), which describe the science of optics—light, vision, and lenses. You can have a perspective (a viewpoint), put things in perspective (see the big picture), or gain perspective (an overall understanding), but if you have a lack of perspective, your prospects (a related word) are limited.

PUTTING IT IN PERSPECTIVE

The ancients didn't realize that there are mathematical laws of perspective that can make two-dimensional drawings appear to have depth. They did understand diminution: that objects appear smaller when they are farther away, and that an artist can suggest distance through size. Some understood that curved lines can imply shape and volume (check the map on page 13) and that, as parallel lines go from close to distant, they seem to come together and meet in a vanishing point. But they didn't make a science of it.

Renaissance artists did. Two Florentines, Filippo Brunelleschi and Leon Battista Alberti, discovered that a precise geometry of grids and lines of vision can create the illusion of depth with exactitude. With Alberti's mathematical framework—laws that fix the degree at which objects appear to get smaller as they get more distant—Western art was transformed.

The first great painting using perspective is *The Holy Trinity* (right) by Masaccio (muh-SAH-chee-oh). It is a fresco in a church in Florence, and in Masaccio's day, people lined up to see it. They couldn't believe that an artist could actually draw three-dimensional depth. Most thought there must be a hole in the wall. In a society that didn't see many new ideas, perspective was an astonishing technological breakthrough; it showed the power of creativity.

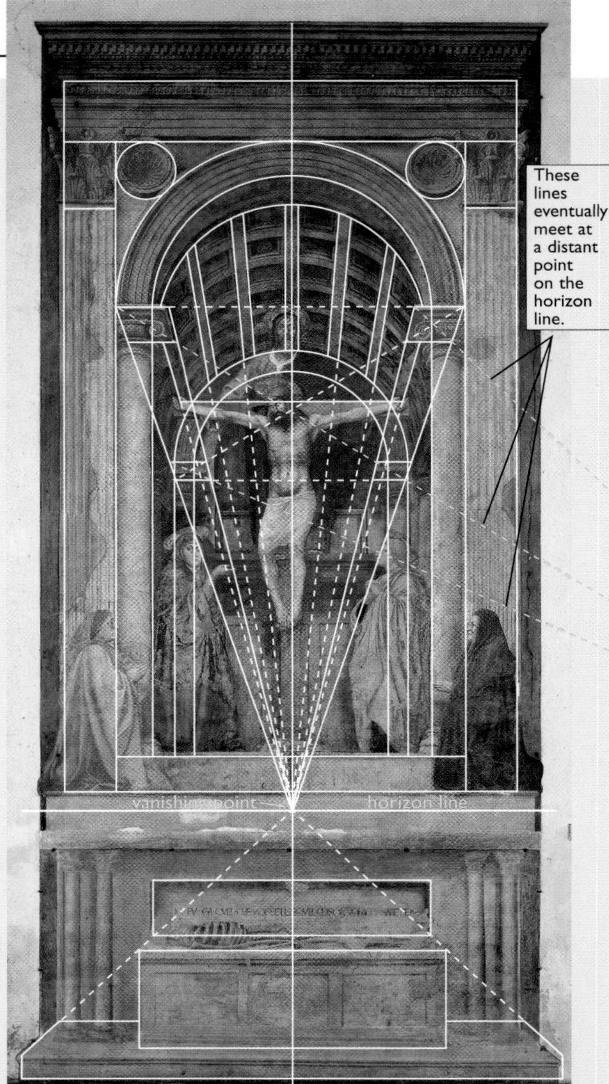

Masaccio's real name was Tommaso Guidi, which means "clumsy Thomas." (If that were my name, I'd change it, too.) He died at age 27. Very few of his paintings have survived. All are masterpieces.

In *St. John the Baptist Retiring to the Desert* (left), Giovanni di Paolo painted his key figure large in both foreground and background. That's medieval thinking (it's also the way small children draw). But then, being an early Renaissance man, di Paolo added geometric lines and some depth and perspective. You can call this a transitional painting (made ca. 1454).

Look what Raphael (1483–1520) has done in this wonderful wedding painting (below)! He's made a flat canvas seem to have depth. Note the rectangles on the plaza angling into the background, and making it appear distant, just as Alberti decreed. The focal (or vanishing) point is behind the door in the middle of the canvas. This is Renaissance perspective at its best. For more details, find a book on this fascinating subject.

Andrea Mantegna (1431–1506), a northern Italian, took perspective and filled a room with its illusions at the Gonzaga Palace in Mantua. In his ceiling painting (above), insiders seem to be laughing at the spectators below. The figures of babies are brilliantly foreshortened.

Germany's Albrecht Dürer (1471–1528) went to Italy to learn about "secret perspective." Later, he wrote a treatise on measurement, which described perspective and illustrated some devices to help artists accomplish it. In Dürer's woodcut (below), an artist is using an aid to draw a lute (an early guitar) that is foreshortened (as the eye sees it).

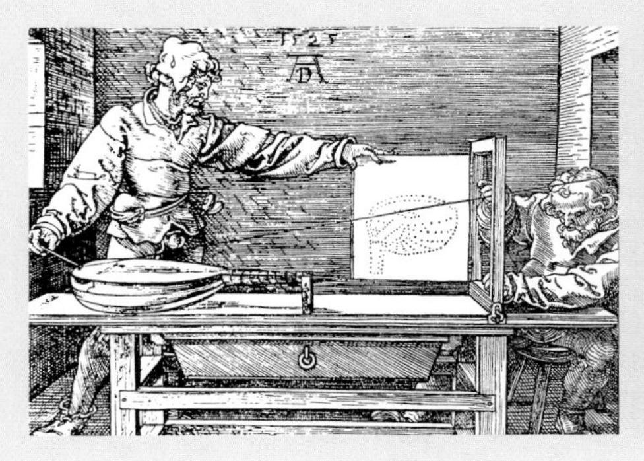

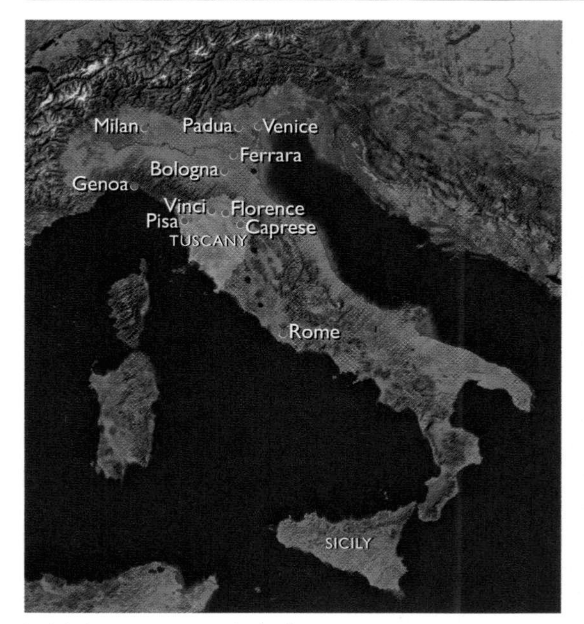

Italy's boot seems to kick the island of Sicily, as seen from a satellite.

three largest. Prosperous Florence (near where Leonardo was born in Vinci and Michelangelo in Caprese) is in Tuscany and nominally a republic, but Lorenzo de' Medici, his family, and some 800 other families rule. (They are mostly bankers and textile merchants.) Lorenzo, called the Magnificent, is a patron of artists and intellectuals. He has helped make this a golden time for the arts. But Lorenzo dies in 1492 (at age 43), and his son Piero attempts to take his place. He thinks little of the idea of republican government, chooses the wrong friends, and will be known as Piero the Unfortunate.

An eloquent and determined preacher, Savonarola, is disgusted: he is trying to save mankind from sin. There is plenty of sin to worry about—especially in Renaissance Italy. The pope plays favorites; there is much corruption, and violence and murder are everyday occurrences. Savonarola helps overthrow Medici rule (in 1494) and brings a more democratic government to Florence. He has a vision of a Christian republic that will be a light to guide all of Italy and perhaps the rest of the world. But his very success dooms him. The old families are alarmed. Besides, he is a zealot who has burned books and works of art. His enemies (including the pope) strike back. In 1498, Savonarola is tortured, hanged, and burned. (What must young Copernicus think of all this?)

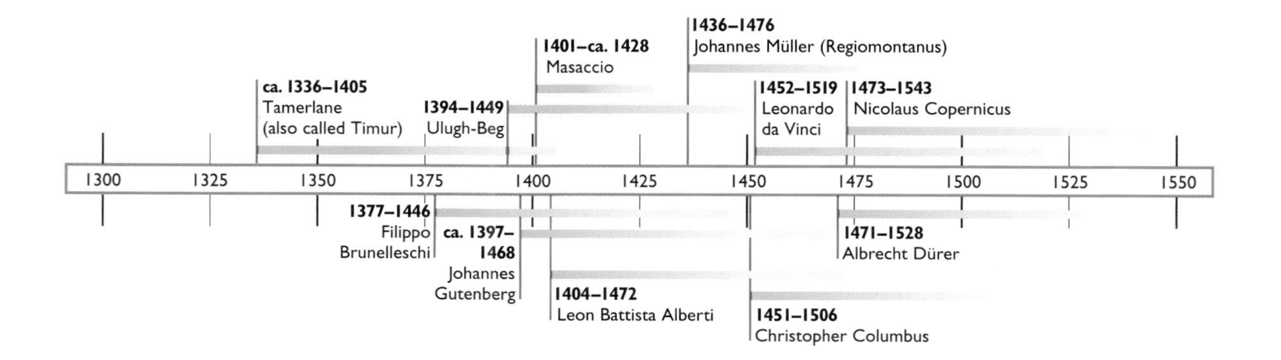

The oval portrait is Dominican monk Girolamo Savonarola (1452–1498) at the height of his influence. Soon after, the rebellious spiritual leader was accused of heresy, hanged, and burned in the Piazza Signoria, the main town square in Florence, as shown in the painting *The Martydom of Savonarola*.

Two years later, in 1500, Christian pilgrims (and others just looking for a good time) flock to Rome for a year-long feast of celebrations. (It is called the Year of the Jubilee by the Catholic Church.) Our scholarly Pole is now in Rome giving lectures on mathematics. Do you think he might be having some fun too? (Read a biography of Copernicus to find out.)

Nicolaus, who has the soul of a poet, paints portraits (including his own) and writes verse. He is especially interested in stargazing. (Telescopes will not be invented for 100 years; he must gaze with bare eyes.) He studies the works of the experts on astronomy; they are those masters of old—Aristotle, Hipparchus, and Ptolemy. In all the centuries since their time, no one seems to have improved on their astronomical (pun intended) thinking.

So, of course, Copernicus reads Aristotle. As to Ptolemy, Copernicus owns two copies of Ptolemy's masterpiece, *Almagest*. One of them is Regiomontanus's new, up-to-date version. In it, Regiomontanus has gone out of his way to make fun of the "foolish notion" that the Earth moves.

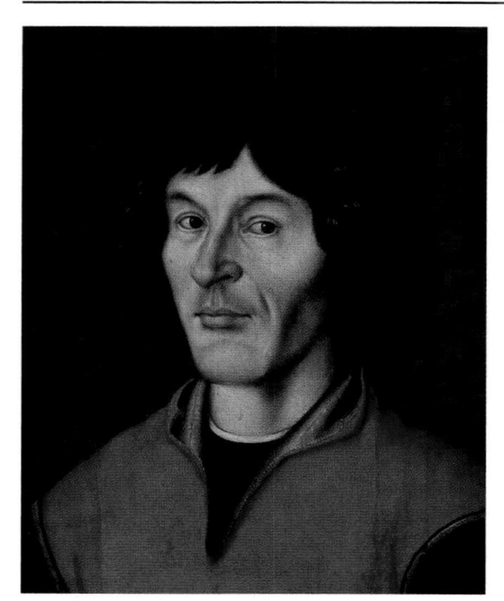

Like many medieval and Renaissance thinkers, Copernicus took a Latin pen name. His Polish name was Zepernik.

Regiomontanus agrees with Ptolemy's basic idea of an Earth-centered universe. But Regiomontanus's calculations do point out inconsistencies in Ptolemy's details.

In Ptolemy's model, the planets are a real problem. Their paths are imperfect. To try to make sense of that, Ptolemy has them orbiting focal points just beyond Earth. The planets loop about, creating circles within circles—called epicycles—while the stars move in unison. It's very complicated. Besides, if you follow Ptolemy's math, the Moon's orbit should get closer to Earth than it does. And why do Mercury and Venus behave so differently from Mars, Jupiter, and Saturn? All of this is bothersome.

Copernicus is sure that God's universe is elegant. Ptolemy's model is awkward and seems to have flaws. Copernicus begins searching for what he calls, "the principal thing . . . the shape of the universe."

He has another reason to search. The pope's secretary has asked him to do it. The calendar bothers Pope Leo X, who, in 1514, sends a letter to all the kings and rulers in the Christian world. He asks them to consult their wisest astronomers. Leo hopes to reform the calendar; it has a serious flaw.

The calendar everyone uses was developed by the ancient Egyptians and adopted by Julius Caesar in 45 B.C.E. (and called the Julian calendar). It is based on a year that is just about 365¼ days long. (Leap years handle that one-quarter-day problem—well, almost.)

POET'S CORNER

When they come to model Heaven,
And calculate the stars; how they will wield
The mighty frame; how build, unbuild, contrive
To save appearances; how gird the Sphere
With Centric and Eccentric scribbled o'er,
Cycle and Epicycle, orb in orb.

—John Milton (1608–1674), *Paradise Lost*

The poet John Milton was also a political thinker and, like others of his day, fascinated by science. (If you want to be a great poet—as Milton was—it helps to tune in to all the arts and sciences.) Notice here, "gird the Sphere" (which means "to encircle or bind [it] with a belt"), "Cycle and Epicycle," "orb in orb." Whose system is he describing?

Actually, Earth takes 365 days, 5 hours, 48 minutes, and 46 seconds to orbit the Sun, which is 11 minutes and 14 seconds less than the Julian calendar's year. Besides that, our not-quite-symmetrical globe has a slight wobble, throwing

INFERIORITY ISN'T COMPLEX

Why aren't Mercury and Venus like the other planets in the sky? Mars, Jupiter, and Saturn (the only other planets known to Renaissance astronomers) sometimes appear all night, high in the sky at midnight. Mercury and Venus are visible only in the west around sunset or in the east around sunrise—and never very high above the horizon.

Saturn can take three years to pass through one zodiac constellation (Sagittarius, say) and enter the next (Capricorn). Jupiter takes about a year, give or take a few months. Venus can zip through constellations—Taurus to Gemini to Cancer to Leo to Virgo—roughly one per month. Mercury—named after the speedy messenger god—is even zippier.

Why are Mercury and Venus special? If you believe that Earth is the center of the universe, you're clueless. But if you buy the idea that the *Sun* is at the center of a *solar system*, then the answer is simple. Mercury and Venus are the only two inferior planets—planets that orbit closer to the Sun than Earth does. Planets beyond Earth are called superior (but only in a positional sense, not because they're better).

Imagine yourself on Earth in the diagram. You can't see Mercury and Venus when they're behind the Sun. You can't even see them when they're in front of it. (Here, Mercury's shadowed half is facing us, like a new moon.) So every time you do see these planets, they're beside the Sun. Can't see the Sun in the middle of the night? That's why you can't see Mercury or Venus then, either—only at sunrise and sunset.

Why the high-speed pace of Mercury and Venus? That too is because they're near the Sun. Both planets travel through space faster than Earth, while the outer planets, Mars through Pluto (the farthest planet), travel progressively slower. (How and why this happens is in chapter 11.) Also, both planets are close to Earth—and closer objects appear to move faster than distant ones. (You can easily see this for yourself next time you're riding in a car and looking out the window.)

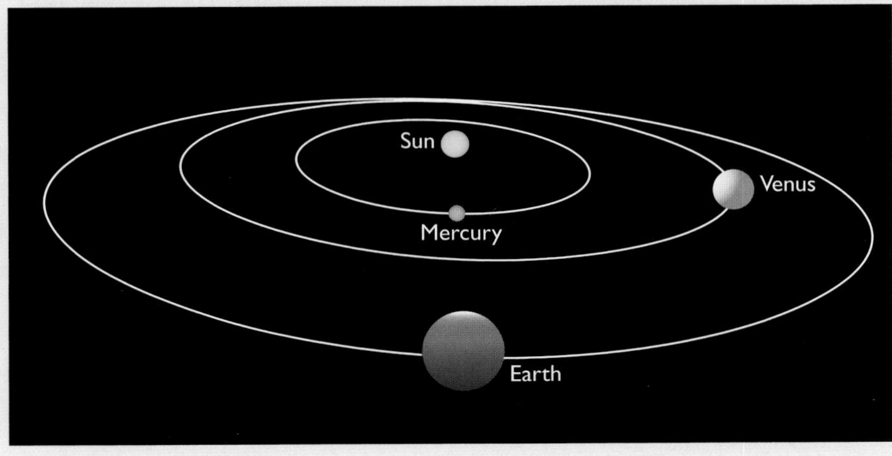

In this diagram (left) Mercury's dark side is facing Earth, making the planet invisible. When Venus is close to Earth, that planet appears in our sky as a crescent (see pages 114–115).

The Anasazi people carved this petroglyph (*petra* means "stone") 1,000 years ago in what is now the Petrified Forest National Park in Arizona. Sunlight strikes the spiral in a precise way during the equinox. Similar petroglyphs were probably the work of shamans, or religious leaders.

SOME DAYS ARE MORE EQUAL THAN OTHERS

The equinox is the time twice a year, in March and September, when the Sun is directly over the equator. Modern astronomers can pinpoint that moment down to the minute. *Equinox* means "equal night" in Latin, but don't believe it. Day and night aren't exactly equal on either equinox, anywhere on the globe. (Though they *are* equal on certain days before and after the equinox.) Near the equator, an equinox day (sunrise to sunset) is about 7 minutes longer than night. The reason: At sunrise and sunset, the Sun appears higher above the horizon than it actually is, an illusion caused by the atmosphere refracting (bending) the Sun's rays. We see the Sun before it rises and after it sets—stretching out the day.

(exaggerated scale)

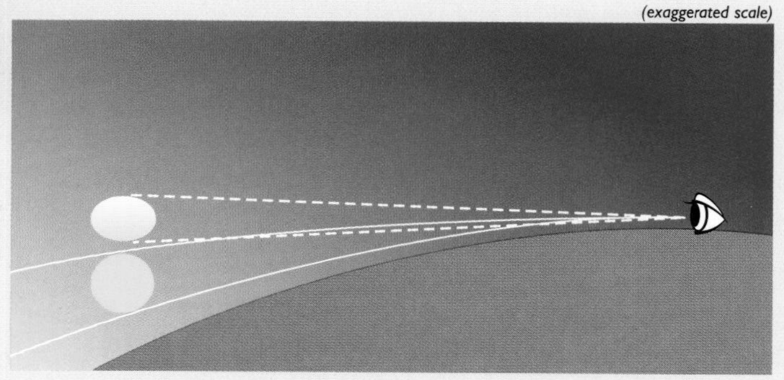

things off another hard-to-predict second or two.

By Copernicus's day, 15 centuries after Julius Caesar, the calendar no longer agrees with the seasons. That messes up the celebration of Easter. Easter is meant to come on the first Sunday after the first full moon that happens on or after the spring equinox. In Europe, the spring equinox is supposed to fall on about March 21, but, because of the faulty calendar, it is 10 days away from that date. If nothing is done, Christmas could end up in midsummer. This is a problem that has been growing over the centuries. Something needs to change.

So Pope Leo is looking for an astronomer to right the calendar. He sends four letters to England's King Henry VIII asking for help. Henry, who has been a church scholar, doesn't respond; he is having marriage problems.

Back in 1475, Regiomontanus was summoned to Rome by Pope Sixtus IV and asked to fix the calendar, just as an outbreak of the plague swept across the city. The famous astronomer was soon dead, and no one else could deal with the calendar. (Some say it wasn't the plague that killed Regiomontanus, but that he was poisoned by a rival. In Renaissance Italy, anything was possible.)

In 525, a tiny monk, Dionysius Exiguus ("Dennis the Little"), set the dates that define our calendar. The calendar in use at that time was called "Julian," after Julius Caesar. Its starting date coincided with the founding of Rome. But the Roman world had become a Christian world. Dennis assigned year 1 to the time he thought Jesus Christ was born. He declared it *anno Domini nostri Jesu Christi* ("the year of our Lord Jesus Christ"), or A.D. for short. Dennis labeled earlier years B.C. ("before Christ"). Today, we use "the common era," or C.E. in place of A.D., and B.C.E. ("before the common era") for B.C. Instead of Julian (or Dennisian!), our calendar is Gregorian, after Pope Gregory XIII—seated on the far left, above. An advisor is pointing out an error in the Julian calendar while the pope's men discuss reforms.

Copernicus (now in his early 40s) does reply. He says the calendar crisis can't be solved without a better understanding of the relationship between the Earth, the Sun, and the Moon. "The movements of the sun and moon had not yet been measured with sufficient accuracy," he says. He is soon hard at work trying to do that measuring. "From that time on I gave attention to making more exact observations of these things," he writes.

What he discovers will challenge everyone's ideas about the heavens and the Earth.

[Copernicus's] great idea was purely that—an idea, or what is today sometimes called a "thought experiment," which presented a new and simpler way of explaining the same pattern of behavior of heavenly bodies that was explained by the more complicated system devised (or publicized) by Ptolemy. If a modern scientist has a bright idea about the way the Universe works, his or her first objective is to find a way to test the idea by experiment or observation, to find out how good it is as a description of the world. But this key step in developing the scientific method had not been taken in the fifteenth century, and it never occurred to Copernicus to test his idea.

—John Gribbin (1946–), British author, *The Scientists*

What's to Be Made of Leonardo?

What's to be made of Leonardo da Vinci? Almost everyone agrees he is the outstanding, all-around genius of the Renaissance. Many believe his paintings have never been surpassed; his sculptures (though few) are breathtaking; his architectural and landscape design farsighted. As to science, he was way ahead of his time in anatomy, botany, hydrodynamics (water flow), mechanics, optics, and aeronautics (flying). He experimented and saw for himself; he used the scientific method. While most doctors of his day knew almost nothing of human anatomy, Leonardo dissected (cut up) cadavers (the dead bodies were usually those of criminals) and learned the way the eyes and muscles and much of the body works.

Leonardo da Vinci had a long, productive life. Here's how he portrayed himself as an old man.

Everyone who met him spoke of his beauty; he was strikingly handsome. A country boy from Vinci (a hamlet between Pisa and Florence), he was born in 1452, which was 40 years before Columbus's first voyage and 21 years before the birth of Copernicus. Leonardo wasn't educated as the scholars of his day were. Country boys

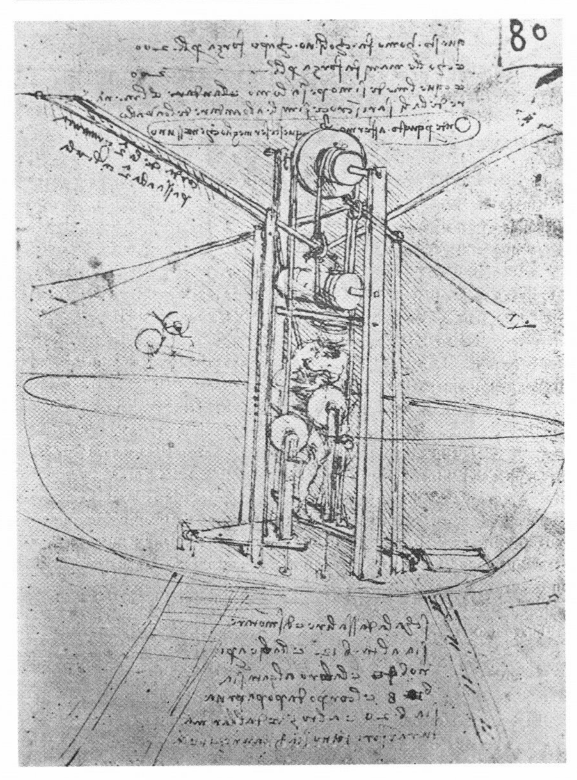

Leonard's notebooks are filled with drawings of birds and designs for fanciful flying machines. In this sketch (above), a pilot is meant to sit in the lower part of the four-winged device. He called his machines "ships of the air." He wasn't the first to dream of flying. Icarus of Greek mythology tried to fly with wax wings. In Andalusia (southern Spain) in the ninth century, Islamic scientist Abul Ibn Firnas worked hard on a flying machine; he didn't fly, but he tried and tried. In the thirteenth century, Roger Bacon was sure humans would one day take to the skies. He described "a flying machine in the middle of which a man could be seated." An Italian engineer, Giovanni Battista Danti, built a flying device around the early sixteenth century. He crashed into a church roof. Lots of dreamers look at birds and wonder: How can humans lift off, too?

weren't taught much math or to read Latin or Greek. But his curiosity was tireless and his mind energetic, so he taught himself and ended up knowing more than almost anyone of his time.

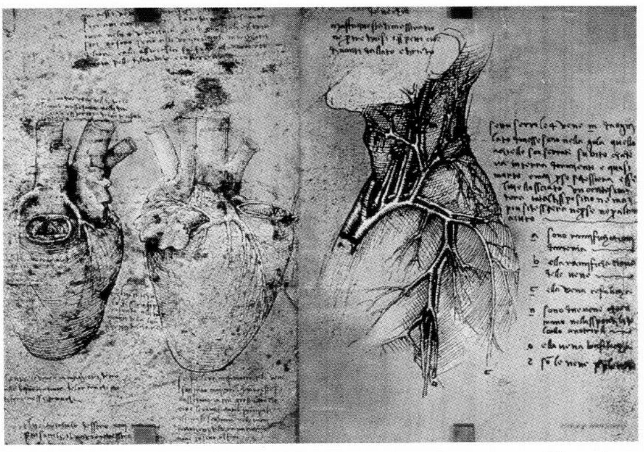

Leonardo's medical drawings were very precise, like this one of the blood vessels and muscles of the neck and heart. He drew the human body from the point of view of an engineer, thinking of it as a machine with moving, functioning parts.

The mathematician Luca Pacioli became a friend; Leonardo studied with him and illustrated his popular math book *On Divine Proportions*. Pacioli seems to have been a great teacher. Here's what Leonardo wrote:

> *No human investigation can be called real knowledge if it does not pass through Mathematical demonstrations and if you say that the kinds of knowledge that begin and end in the mind have any value as truth this cannot be conceded . . . first of all because in such mental discussions there is no experimentation, without which nothing provides certainty of itself.*

But Leonardo had problems. He had a habit of not finishing his work, and that left his contemporaries, and it leaves us, frustrated. Besides that, he didn't trust

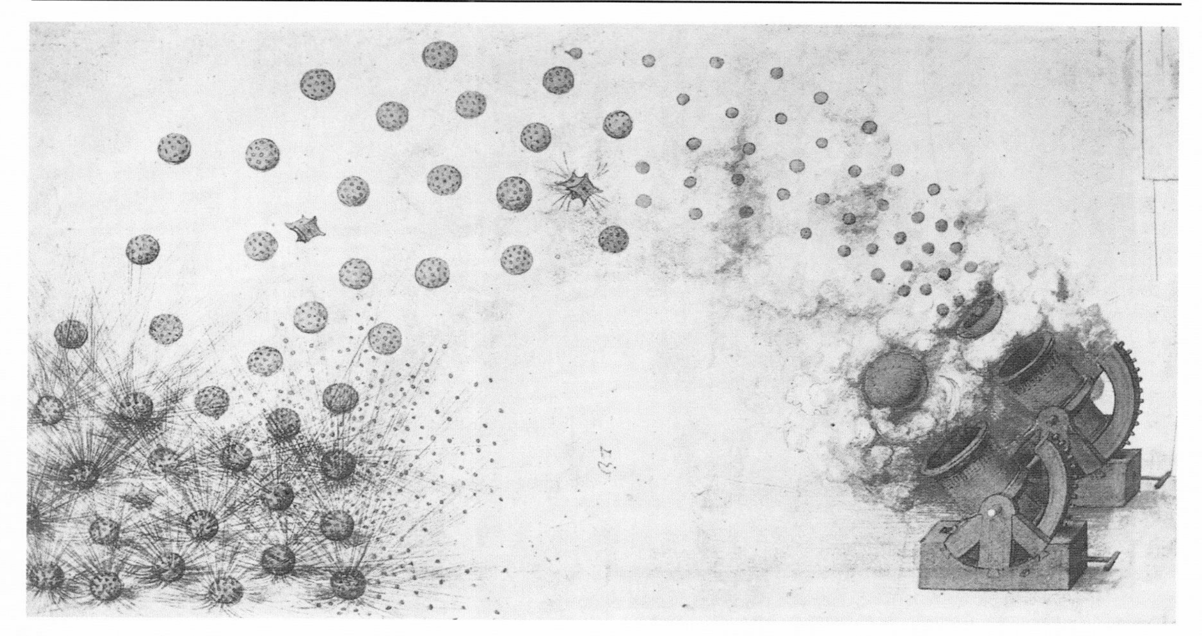

In a letter to Ludovico il Moro, Leonardo wrote, "If need shall arise, I can make cannon, mortars, and light ordnance of very beautiful and useful shapes." (*Mortar* and *ordnance* refer to cannons that fire shells at low velocity and high trajectory.) Here's his "very beautiful and useful" sketch. The drawing at right is based on the writings of Vitruvius, a Roman who described proportions of the body mathematically. Find the circle, square, and triangle. Generations of art students have copied this work.

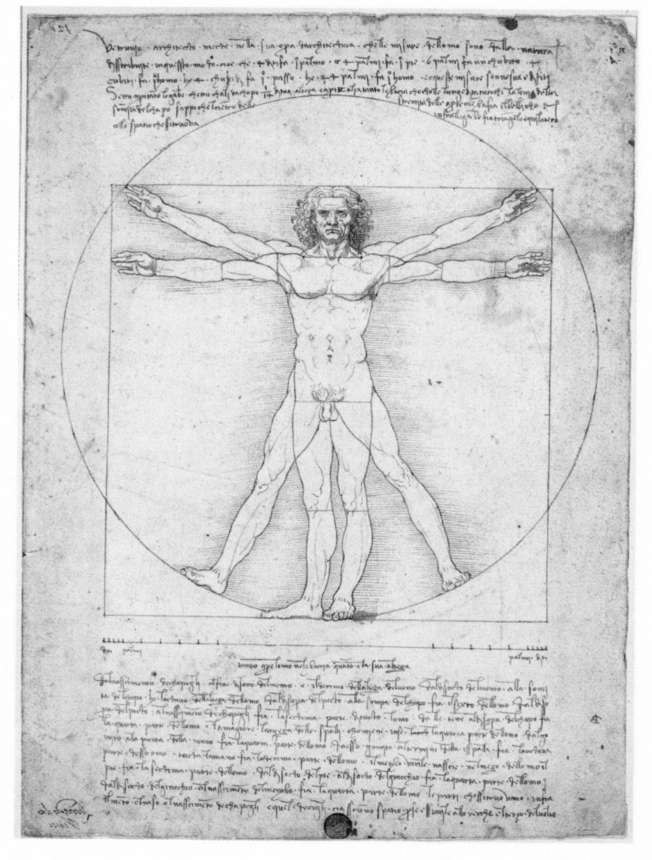

most people and he worried about plagiarism (copying of his work). So he kept his ideas in notebooks that were written in code or in a script that could only be read if reflected in a mirror—which means his scientific theories didn't influence others as they might otherwise. It is only in recent times that we have been able to read those notebooks and learn how astonishing and wide-ranging his accomplishments were.

Lady with an Ermine is a revolutionary work of art that suggests motion and animation using a calm, poised model and a wild animal. Note that her body turns one way and her head another.

Why is the sky blue? Leonardo got it right 300 years before John William Strutt (Lord Rayleigh) explained it in 1871 as the scattering of sunlight by particles in the air. Here's Leonardo in the *Codex Leicester* (one of his notebooks): "I say that the blue which is seen in the atmosphere is not its own color but is caused by... minute and imperceptible atoms on which the solar rays fall rendering them luminous against the immense darkness of the region of fire that forms a covering above them." (Imperceptible? It means "so small it's impossible to see." Luminous? It means "glowing.")

Leonardo developed a plan to control the water supply along the Arno River. His plan called for canals, aqueducts, dams, and artificial lakes. He thought it would help end the fighting between Florence and Pisa, which was mostly over water. He imagined an industrial link between the two cities, with river barges that carried goods and people to the mills and factories along the Arno. But it didn't happen. Below is his sketch of northwest Tuscany and the Arno.

3 On Revolutions and Fools

As though seated on a royal throne, the sun governs the family of planets revolving around it.

—Nicolaus Copernicus (1473–1543), Polish astronomer, *On the Revolutions of the Heavenly Spheres*

Once it was recognized that the earth was not the center of the world, but only one of the smaller planets, the illusion of the central significance of man himself became untenable. Hence, Nicolaus Copernicus, through his work and the greatness of his personality, taught man to be modest.

—Albert Einstein (1879–1955), German-born American physicist, "Message on the 410th Anniversary of the Death of Copernicus"

Although Copernicus himself saw his discovery as giving rise to even greater amazement at the Creator of the world and the power of human reason, many people took it as a means of setting reason against faith.

—Pope John Paul II (1920–2005), address at Copernicus University in Toruń, Poland

Copernicus is back in Poland and has taken up his duties as canon of Frombork's magnificent cathedral. Around 1514, he buys 800 stones and a barrel of lime and builds a roofless tower next to his house to use as an observatory. It is taller than the cathedral's towers. But often Copernicus has to put up with overcast skies. It is hard to stargaze. He is jealous of Ptolemy, who lived in the clear sunshine of Egypt, where, Copernicus writes, "the skies are more cheerful,

The cathedral at Frombork was surrounded by the residences of seven canons. The whole complex was circled by red brick walls with bastions and two towers.

where the Nile does not breathe fogs as does our Vistula." (The Vistula is Poland's longest river.)

When the Polish fog lifts, he looks. And the more Copernicus gazes at the skies, the more he is convinced that Ptolemy's model of the universe is wrong. It doesn't agree with what he sees in the heavens. The paths of the Moon and planets are especially bothersome.

A grandson of the Mongol conqueror Tamerlane (ca. 1336–1405), Ulugh-Beg took the throne at Samarkand (now Uzbekistan) in 1447. He built an observatory (above), the ruins of which you can still visit today. Ulugh-Beg was an outstanding astronomer, and the observatory was probably the best in the world at the time. But his work was unknown in western Europe, and when he died (murdered by his son), Mongol astronomy died with him.

How can he solve that problem? Perhaps he does a thought experiment—imagining himself on the Sun. Seen from the Sun, the planets' orbits are regular. When Copernicus figures that out, he decides to toss the revered Ptolemy aside and make his own model of the universe.

He has the Earth and Sun switch places! He puts the Sun in the center of the universe. In his model, it is the Sun that stands still. And Earth moves. Copernicus believes that **Earth makes a full turn on its axis every day—and that causes night and day**.

That isn't all. He finds that:

- **Earth completes one orbit of the Sun each year.**
- **The universe is much larger than anyone has previously believed.**
- **Our Earth is a planet among the others.**

Once he centers the Sun, he can figure out the order of the planets. He tracks and times their orbits and is able to line them up. **Mercury is closest to the Sun, then Venus, then Earth, then Mars, Jupiter, and Saturn.** Before, no one could put them in sequence. (Like the Greeks, he is sure that all the celestial bodies move in perfect, circular orbits. And he keeps the accepted idea of a finite universe that ends with a sphere of fixed stars. Still, he has taken a huge leap.)

What was the "Copernican Revolution"? When Copernicus moved Earth from the center of the universe, it was profoundly shocking. Earth became just another planet. What did that do to us humans? Was the universe created just for us? Or is there more to it than that? The Copernican Revolution changed ways of thinking; it raised new questions; it was troubling. It was also exciting.

Copernicus knows that these ideas of his are earthshaking! He wants to be as sure of them as possible. He begins

Martin Luther (1483–1546), a German professor of theology (religion), holds on tight to the Bible in this painting by his friend and colleague, Lucas Cranach the Elder. Luther preached that salvation comes from faith, not good works. Then he drew up a list of 95 points, including the idea that the pope and the priests have no right to forgive sins. In 1517 he nailed his list on the church door in Wittenberg. Out of that act came Protestantism.

As to Copernicus, Martin Luther said, "People gave ear to an upstart astrologer who strove to show that the earth revolves, not the heavens or the firmament, the sun and the moon. . . . But sacred Scripture tells us that Joshua commanded the sun to stand still, not the earth."

working on the geometry to prove his theories.

Copernicus has read the ancient philosophers, so he knows that Pythagoras, way back in the sixth century B.C.E., believed Earth rotates around a central fire in the heavens. He knows that Aristarchus, three centuries later, put the Sun in the center of the universe, with Earth and the planets moving around it! Copernicus mentions Aristarchus in a manuscript he is writing, and then he crosses out the reference.

If Aristarchus was right, then Aristotle was wrong. Christian church leaders, Jewish scholars, Muslim scientists, and the major university faculties in Europe all teach Aristotle's ideas. Moving the Sun and the Earth won't be easy; overturning Aristotle may be even harder.

In May of 1514, Copernicus shows a few people his manuscript. In it, the Polish canon has described a heliocentric (Sun-centered) universe. Copernicus is a good Christian, and he is aware that his ideas will challenge the whole system of medieval thought that keeps faith and science in a tight embrace. Because if Earth is not the center of the universe, perhaps it was not the sole focus of creation. Perhaps the whole concept of the human place in the universe may have to be reconsidered. Perhaps creation is more grand and complex than anything anyone has imagined.

Copernicus understands how upsetting his ideas will be to people who believe the universe revolves around us. Can Earth be shifted from center stage?

In Europe in the early sixteenth century, there seem to be problems enough without moving the universe about. Martin Luther has tried to reform the Catholic Church; unexpectedly, his ideas have led to breakaway religions called Protestant, and that starts religious wars.

Catholics and Protestants are fighting each other with terrible fury. Neither side will find this a good time to question traditional church teaching on the heavens and creation—it is one thing they both agree on. But Copernicus has a scientist's passion for understanding, and he can't stop following where his mind leads him.

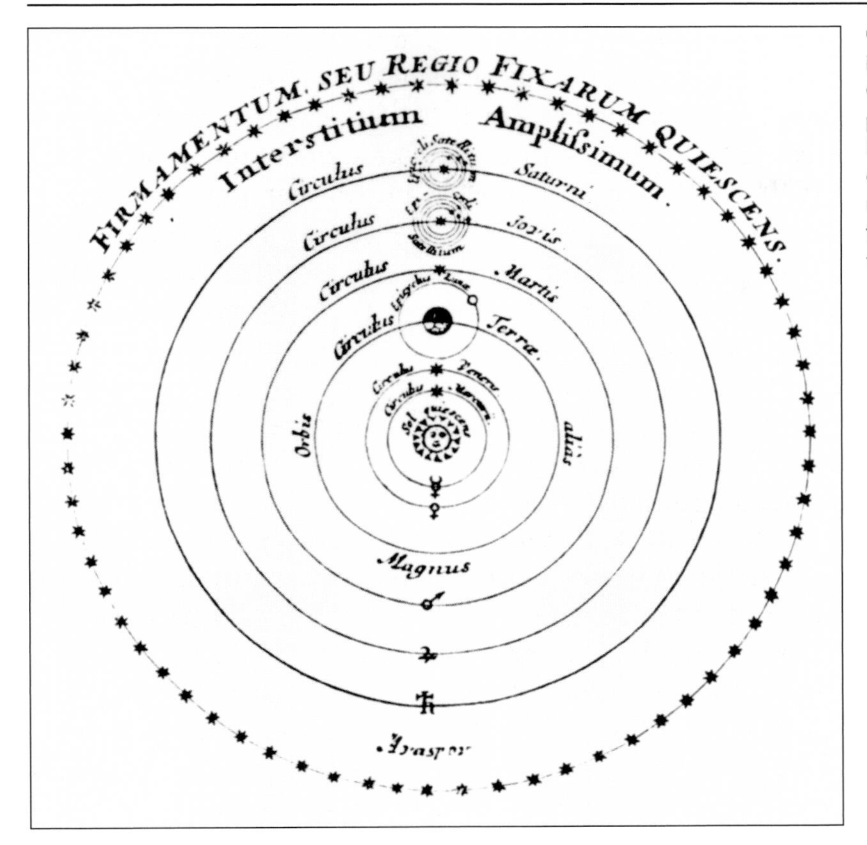

Our view of the solar system is an ongoing work of science. Copernicus improved it by putting the Sun, rather than Earth, in the center (left). That outer ring of fixed stars is now a three-dimensional field—ever-deepening as telescopes improve.

He comes to the conclusion that Earth does indeed move around the Sun. Aristarchus never did the mathematics to back up his ideas; Copernicus does it. He knows, if he is right, that most schoolbooks will have to be thrown out. The Aristotelians, who dominate the universities in partnership with church leaders, will have to admit they are wrong. Are they likely to do that? Copernicus doesn't think so. Is he afraid they will make fun of him? Most people are content with an Earth-centered universe. The Aristotle/Ptolemy model has worked well enough for centuries.

So Copernicus writes quietly. He goes on proving his ideas and expanding them into

If you are a long-distance traveler...you can get your bearings by landmarks—mountains and rivers, buildings...but...the emptiness and homogeneity of the sea...drove sailors to seek their bearings in the heavens, in the sun and moon and stars.... They sought skymarks to serve for seamarks. It is no wonder that astronomy became the handmaiden of the sailor; that the Age of Columbus ushered in the Age of Copernicus.
—Daniel J. Boorstin (1914–2004), American historian, *The Discoverers*

EARTH A PLANET? SOUNDS WEIRD. IMPOSSIBLE. AND THREE MOTIONS? ALL AT ONCE? WAS COPERNICUS LOONY?

Englishman John Milton (1608–1674) thought about that and, in a great poem titled *Paradise Lost*, he wondered:

> *...What if the Sun*
> *Be Centre to the World, and...*
> *The planet Earth, so steadfast though she seem,*
> *Insensibly three different motions move?*

If Copernicus was right, then Earth must behave like a planet. Milton seemed to be buying that idea, or at least suggesting it is possible, by referring to "three... motions." In other words, Earth moves around the Sun, it spins on its axis, and has a third motion—called precessional—caused by the planet wobbling slightly as it spins.

But take note: Poets and all thinking people, then as now, keep up with scientific progress.

Blindness doesn't stop John Milton from writing his greatest poem, as shown in this nineteenth-century painting by Hungarian artist Mihály Munkácsy. He's dictating *Paradise Lost* to his three daughters. Milton probably lost his vision due to cataracts, cloudy films over the lenses of the eyes.

Like many scholars of the times, Copernicus was often called on to practice medicine. Among his prescriptions, says scientist Rocky Kolb, are "medicines typical of his age: lemon rind, deer's heart, calf's gall, earthworms washed in wine, lizards boiled in olive oil, and that widely prescribed all-purpose magical medicinal elixir of 1510, donkey urine."

a book—but he won't publish that book even though some people are curious. And the more he holds on to it, the more curious they get.

In the book that he won't let anyone see, he writes, "At rest, however, in the middle of everything is the sun. For in this most beautiful temple, who would place this lamp in another or better position than that from which it can light up the whole thing at the same time? For, the sun is not inappropriately called by some people the lantern of the universe, its mind by others, and its ruler by others."

Meanwhile, Copernicus continues to perform his ecclesiastical (i-klee-zee-AS-ti-kuhl—it means "churchly") duties, practice medicine, write about money reform, and

Copernicus's book was finally published in 1543. That same year, Andreas Vesalius, a 29-year-old professor of anatomy at Padua University (near Venice), published his seven-volume *On the Fabric of the Human Body*, with more than 300 detailed illustrations. He also prepared charts of the human organs. This book too was an affront to the traditional thinkers. The Aristotelians taught that the heart is the origin of the nerves; Vesalius traced the nerves up the spine and neck to the brain. The Bible talked of Eve being formed from Adam's rib. And so medieval people believed that men have one fewer rib than women. But Vesalius showed that is not so.

Vesalius dissected human cadavers. Those before him based their knowledge mostly on nonhuman animal dissection. Vesalius claimed he could learn more from a butcher than from the old medical books. But was it right to dissect human remains? There was a moral issue here. Vesalius spent a year defending his work from critics, then he had enough; he retired from teaching and anatomical research. His book transformed the field of medicine.

think about mathematics and astronomy. He is a recluse (someone who keeps to himself)—timid, pious, and stubborn. He isn't looking for attention; and he doesn't want to upset church authorities.

Copernicus doesn't see himself as a revolutionary. Like Aristotle and Plato and most Christian thinkers, he believes

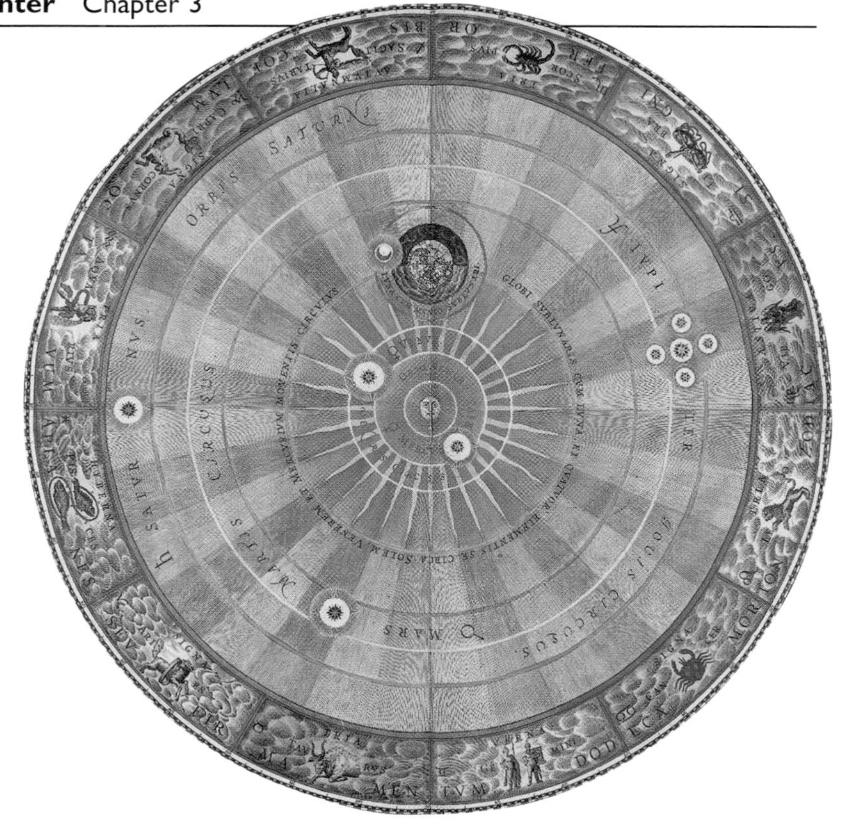

It's easy to tell that this diagram of a Copernican solar system was made a century after the one on page 39. The planetary orbits are still evenly spaced circles, but there's a new feature. Jupiter (far right) has four moons, which Galileo discovered in 1610. This *Planisphere of Copernicus* appeared in a gorgeous 1661 atlas called *Harmonia Macrocosmica*, by Andreas Cellarius.

God's universe must be perfect. That's what bothers him. He knows that neither Aristotle nor Ptolemy came up with a perfect system. Copernicus thinks that with his Sun-centered model he has found the path to the perfect universe.

This Polish church canon has no desire for controversy. "The scorn which I had to fear on account of the newness and absurdity of my opinion almost drove me to abandon a work already undertaken," he writes in the book. But when a young admirer insists that he publish it and says he will do the necessary work, Copernicus finally agrees. In 1543, as he lies dying, the first printed copy of his great work, *On the Revolutions of the Heavenly Spheres*, is placed on his bed. His mind has already left his body. (Or so the story goes.)

Does the book cause a great stir? Copernicus

Nicholas Copernicus set the Earth in motion, dislodged it from the center of the Universe, and hurled it into the skies. He undermined the foundations of the traditional cosmic order: the human species was no longer at the center of the world, and the cosmos was no longer organized around us; we had lost our most reliable bearings.

—Dominique Lecourt (1944–), French philosopher, *Dictionary of History and Philosophy of Sciences*

IT ALL COMES DOWN TO PHYSICS

Can a book change the whole world? Copernicus's tome, *On the Revolutions of the Heavenly Spheres*, is said to have started the modern era. One English edition has 330 pages, 143 diagrams, 100 pages of tables, and more than 20,000 numbers (astronomical data), which gives you an idea of its complexity. This book wasn't just about Sun-centering. It was a massive and impressive work of science and mathematics. Copernicus spent almost 40 years working on it.

Few people in the sixteenth century actually read the shy Polish canon's written words, but the few who did were enough. Copernicus's ideas soon spread like the plague's germs, right through the medieval walls. They led to the modern idea that the world is understandable—which brings us to the purpose of this book.

Just what are we trying to find out here? It's not a small thing. We're trying to discover how the universe works. It's the scientific quest.

Will we find the answer? Of course not. Science is an ongoing thing. There are always new questions and new answers. But, you'll see, what we'll learn will be staggering.

To follow this story, keep in mind that the search to make sense of the universe is two-headed. One head deals with *matter*—what the universe is made of. The other head deals with the phenomena that power the universe and make things happen: *motion, energy, and universal forces*. So think matter and think action.

was right. Most people laugh. Or yawn.

A few readers are furious. One of them is Martin Luther, who is fascinated by science but not ready for Copernicus. He says, "This fool wishes to reverse the entire science of astronomy." Which is exactly what Copernicus will do. It is called the Copernican Revolution.

The religious beliefs of John Calvin were as serious as he looks here.

John Calvin, another Protestant leader, joins the attack. He quotes Psalm 93: "The world also is stablished [meaning 'established'], that it cannot be moved."

As to the calendar, 39 years after Copernicus is in his grave, it will be reformed. Gregory XIII will then be pope, so the new calendar will be called Gregorian.

What about Copernicus's main idea, that the Earth circles the Sun? His mathematics works better than Ptolemy's, the professors admit, but what proof is that? Just look at the sky and you can see the Sun in motion.

Would you have believed in an orbiting Earth if you had lived in Copernicus's day? Hardly anyone then does.

The few who do will move the Earth.

4 Tycho Brahe: Taking Heaven's Measure

On the 11th day of November in the evening after sunset, I was contemplating the stars in a clear sky. I noticed that a new and unusual star, surpassing the other stars in brilliancy, was shining almost directly above my head; and since I had, from boyhood, known all the stars of the heavens perfectly, it was quite evident to me that there had never been any star in that place of the sky, even the smallest, to say nothing of a star so conspicuous and bright as this. I was so astonished of this sight that I was not ashamed to doubt the trustworthyness of my own eyes.

—Tycho Brahe (1546–1601), Danish astronomer, *De Nova Stella* ("About the New Star")

Then felt I like some watcher of the skies
When a new planet swims into his ken.

—John Keats (1795–1821), British poet, from "On First Looking into Chapman's Homer"

The movement of the planets is no little matter. Why are their paths different from those of the stars? What about the Sun? Does it make a daily journey around the Earth (as almost everyone believes)? Or is Earth traveling around the Sun? If so, why don't we feel it? Copernicus has brought up some troublesome questions. And the truth is, no one in the sixteenth century can answer those questions.

The Danes pronounce the name *Tycho Brahe* "TOO-koh BRAH"; most English speakers say "TY-koh BRAH-hee." Astronomy.com says "TEE-koh." Take your pick.

The ancient Greeks believed that pure thought was the way to solve problems. But pure thought hasn't solved these problems.

A Danish astronomer named Tycho Brahe has a different idea, a common-sense idea. Tycho (born three years after Copernicus dies) decides that if you want to understand why

Someone had to transform the practice of astronomical observation into the science of astronomical observation. In the latter half of the sixteenth century, there was only one person in all of Europe who seemed up to the task, an irascible [hot-tempered], red-bearded Danish nobleman named Tycho Brahe. Tycho was prepared to devote his entire life to improving astronomy.

—Alan W. Hirshfeld, astronomy professor, *Parallax*

the planets move as they do, the first thing you need to do is look at them carefully and measure their movements as accurately as possible.

This is "a tremendous idea—that to find something out, it is better to perform some careful experiments than to carry on deep philosophical arguments," says physicist Richard Feynman (FINE-muhn), commenting in the twentieth century.

It happens that Tycho is the right person to do the looking. The first telescopes are yet to be built; he is probably the best naked-eye astronomer the world has known.

But when you hear Tycho's story, you may be surprised that he turns out as well as he does. His parents, Otto and Beate, promise their firstborn son to Otto's brother, Joergen,

In the sixteenth century, Denmark-Norway was a powerful kingdom consisting of what is today Denmark, Norway, and parts of Sweden. Any ship that wanted to get into the Baltic Sea and reach its affluent ports had to pass through a narrow strait controlled by King Frederick II. The king had a castle on one side of the strait (now Denmark), and another castle on the other side (now Sweden). The king's men wouldn't let ships through unless they paid a toll—a big toll. With all the money he collected, Frederick rebuilt one of those castles (the Kronborg castle in Helsingor, shown above) and made it into something spectacular. But the king still had riches to spend. He was generous with his astronomer, Tycho Brahe.

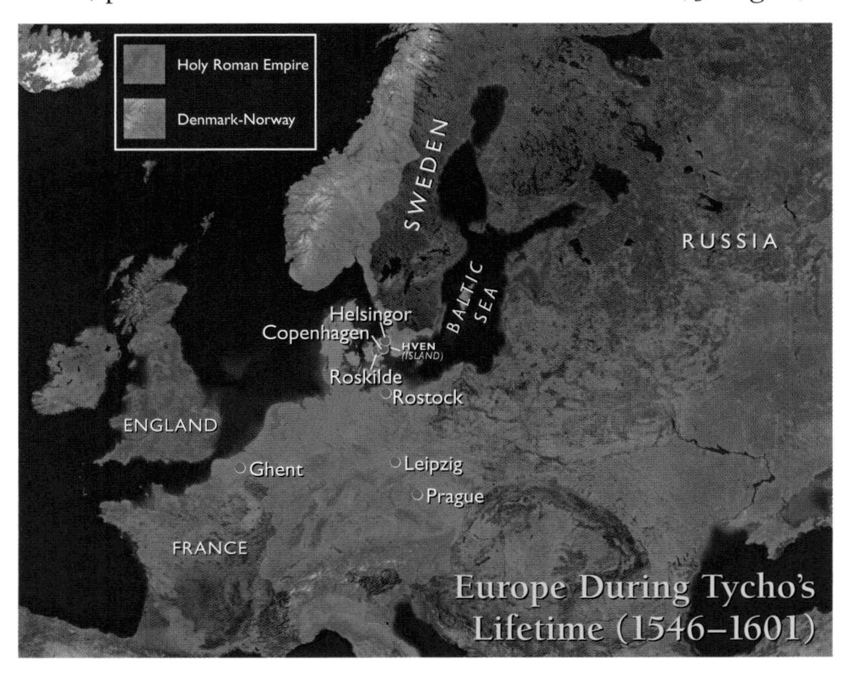

The Brahe baby was given the Danish name Tyge. When he was 15, he started using its Latin form, Tycho.

an admiral who is childless. Then they change their minds. The admiral is infuriated. He kidnaps the infant and hides him in his castle.

Otto and Beate eventually have nine other children and forgive Joergen. So Tycho Brahe is raised by an uncle and aunt as if he were their son. They send Tycho, at age 12, to the University of Copenhagen (not an unusual age for a wealthy sixteenth-century college student). Joergen expects Tycho to enter political life as part of the privileged elite surrounding the king of Denmark, Frederick II.

A SOLAR ECLIPSE happens when the Moon blocks our view of the Sun. When Earth's shadow hides the Moon, it's a LUNAR ECLIPSE.

Perhaps he might have done that but, during his first year in college, an eclipse is predicted. And it happens. That seems to Tycho, "something divine that men could know the motions of the stars so accurately that they could long before foretell their places and relative positions." Tycho buys a copy of Ptolemy's *Almagest* (his greatest work), and notes on the cover that he paid two thalers for it.

Here's a famous story about a lunar eclipse:

In 1504 Christopher Columbus was stuck on the island of Jamaica. His ships were in bad shape. While Columbus waited to be rescued, his sailors—who were rude, crude, and lazy—sponged food off the Jamaicans. After six months, the islanders threatened to kill them all.

Columbus checked his astronomical charts and saw that a total lunar eclipse was imminent (about to happen). He told the islanders that his God had power over the Moon, and if they didn't treat him well, he'd see that God blackened the orb. So when the Earth's shadow covered the Moon on the night of February 29, the Jamaicans were awed and terrified. After that, Columbus and his men got everything they wanted.

Joergen thinks Tycho's stargazing is just a teenage phase and sends him to the University of Leipzig to study law. Then, to be sure that Tycho sticks to the law, Joergen hires a companion to watch his nephew. But when the companion goes to bed, Tycho studies the stars. Tycho enrolls in a mathematics class, and the companion, doing his job, tells him that Copenhagen's chief mathematician has gone insane from studying numbers. Tycho studies them anyway. On his own, he is reading Ptolemy—and Copernicus too.

At 18, he builds his own astronomical instruments. When Tycho realizes that the existing star charts are often wildly inaccurate, he finds his life's work. His wealthy, prominent family isn't happy about this. (A stargazer? It seems a foolish career choice.) But Tycho is determined.

In 1563, war breaks out between Sweden and Denmark, and two years later Tycho's uncle calls

Tycho's looks and actions left no doubt that he was an aristocrat, an upper-class nobleman. Bright red and gold colors were off-limits to peasants, who wore only dull earth tones. So were high-collared cloaks, fancy jewelry, white gloves, and ruffs—those stiff white ruffled collars and cuffs worn around necks and wrists. Even Tycho's fake metal nose had a rich sheen to it.

him home. They hardly have time to see each other when Joergen goes off with King Frederick II (who drinks too much). Joergen and the king are riding horses across a bridge. The king falls in the water. Joergen heroically rescues him—and dies of pneumonia.

Tycho is now 19, restless, on his own, and still determined to be an astronomer. So he makes a tour of some northern universities looking for experts to teach him. He's at the University of Rostock (on the Baltic Sea in Germany) when a student taunts him at a Christmas party. Tycho, who is known for his fierce temper, has little patience with those who don't think as well as he does. He demands that they duel. They use swords, and a chunk of Tycho's nose gets sliced off. For the rest of his life, he will go about with a patch on his nose made of silver, gold, and wax; it glistens above a big handlebar mustache.

Two years later, King Frederick makes Tycho a canon at the Roskilde Cathedral. It is an easy job: he gets paid but has no churchly duties at all. He falls in a category that includes "men of learning." How did he get that sinecure? Young Tycho knows more about astronomy than anyone in Denmark, and perhaps the king feels

Tycho did a lot of nothing at Roskilde Cathedral (below). He was a canon, a religious figure attached to the church, but the job was a sinecure—all pay and no work.

When I Heard the Learn'd Astronomer

When I heard the learn'd astronomer;
When the proofs, the figures, were ranged in columns before me;
When I was shown the charts and diagrams, to add, divide, and measure them;
When I, sitting, heard the astronomer, where he lectured with much applause in the lecture-room,
How soon, unaccountable, I became tired and sick;
Till rising and gliding out, I wander'd off by myself,
In the mystical moist night-air, and from time to time,
Look'd up in perfect silence at the stars.

—Walt Whitman (1819–1892), American poet, *Leaves of Grass*

Perplexity means puzzlement. If you're perplexed, you're confused.

remorse over Uncle Joergen's untimely death on his behalf.

Tycho now has the means do what he wants with his life. And what he wants to do is look at the stars. When it comes to love, he also does what he wants, breaking society's rules. He falls in love with the daughter of a peasant and follows his heart; they will have eight children.

As to astronomy, he intends to do star charts better than any done before. No one knows how far away the stars are, but the relative distances between them, as they appear in our sky, can be plotted. Using equipment of his own design, Tycho's measurements are far more precise than others of his time.

Then something astonishing happens. A bright star lights up the sky where no star has been seen before, and no one can explain it. Tycho Brahe first sees it on November 11, 1572, and writes, "When I had satisfied myself that no star of that kind had ever shone forth before, I was led into such perplexity by the unbelievability of the thing that I began to doubt the faith of my own eyes."

Tycho knows the details of the sky as well as you know the clutter in your bedroom. And he can't believe his eyes. This star is brighter than any other in the heavens. It can even be seen in the daytime. Never before has he seen anything where this brilliant orb now shines. He runs for his servants. He wants them to look. He thinks maybe he is having visions.

He isn't. But how can it be? Aristotle said the heavens are fixed, forever. The Christian church fathers say the same thing. Just about everyone believes the heavens are God's realm and therefore perfect—and unchanging. Tycho writes, ". . . all the philosophers agree, and facts clearly prove it to be the case, that in the ethereal region of the celestial world no change . . . takes place." So what is going on?

Only once before, in Western recorded history, has a "new" star been cited. "Nor do we read that it was ever before noted by any one of the founders [the Greeks] that a new star had appeared in the celestial world, except only by Hipparchus" (ca. second century B.C.E.), writes Tycho. But no one has ever been able to account for the star that Hipparchus saw.

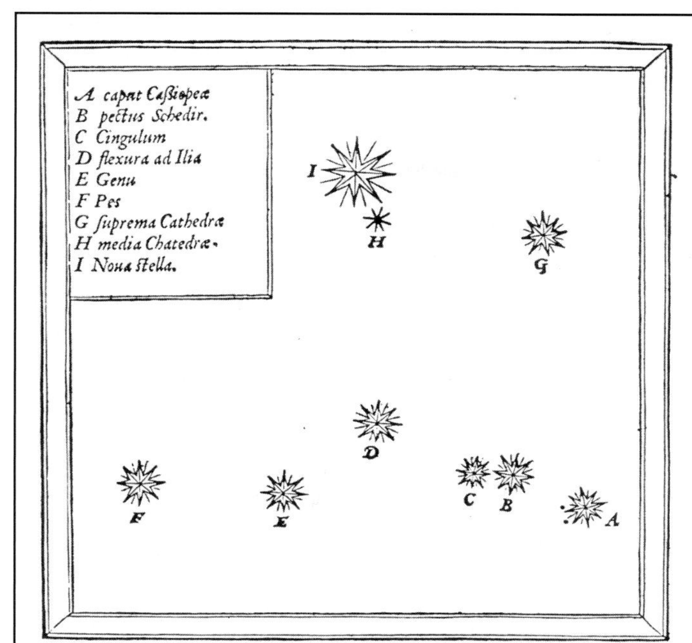

A caput Cassiopeæ
B pectus Schedir.
C Cingulum
D flexura ad Ilia
E Genu
F Pes
G suprema Cathedræ
H media Chatedræ.
I Noua stella.

Distantiam verò huius stellæ à fixis aliquibus in hac Cassiopeiæ constellatione, exquisito instrumento, & omnium minutorum capacj, aliquoties obseruaui. Inueni autem eam distare ab ea, quæ est in pectore, Schedir appellata B, 7. partibus & 55. minutis: à superiori verò

Above, the comet of 1528 is depicted as a fearful attack accompanied by a cloud of swords, daggers, and heads without bodies. The woodcut is from Ambroise Paré's *Dix Livres de la Chirurgie* ("Ten Books of Surgery"). The illustration at left is in the first edition of Tycho Brahe's *De Nova Stella*, published in 1573. This is his earliest work announcing and documenting a "new" star in the constellation Cassiopeia. The oversized star at the top, labeled I, is the supernova.

As to this dazzling orb, it must be a comet, most viewers insist. In Germany, painter Georg Busch is sure of it. He writes that it was "formed by the ascending from earth of human sins and wickedness, formed into a kind of gas, and ignited by the anger of God." He predicts horrors to come. No wonder people fear comets.

Tycho is exasperated. He knows this is no comet. Comets are supposed to be in the imperfect and changeable area between the Earth and Moon. (That's what Aristotle said.) The fiery Dane looks and measures and writes, "This star is not some kind of comet . . . it is a star shining in the firmament itself, one that has never been seen before our time."

He watches it week after week through his sighting tubes (they look like telescopes but have no lenses.) Unlike the planets, the star doesn't budge. He concludes:

THE STAR ISN'T NEW—IT'S OLD AND WORN OUT

Before, there seemed to be nothing in the sky where Tycho's "new" star now shone. In fact, the star had been there all along; it was just too faint to be seen. Then it exploded.

What Tycho saw was a rare *supernova*—a massive star that explodes with an intensity so enormous that, very briefly, it may shine as brightly as an entire galaxy. Weeks after that, it can still give off as much light as several hundred suns.

For 485 days, Tycho watched the supernova of 1572 in the W-shaped constellation Cassiopeia. Then, finally, it faded from his naked-eye view, though remnants of gas and dust remain (see opposite page). (A supernova explosion destroys the original star. A nova explosion happens on the surface of a white dwarf star, an old and super-compact object that eventually goes back to the way it was.)

In 1573, Tycho wrote a book about his observations titled *De Nova Stella* ("About the New Star"—which is really an old star, keep in mind). That book made him the most famous astronomer in Europe.

We'd love to show you a supernova explosion, but we can't. This is an artist's view. The event is too rare and the blast too fleeting to capture in a photo. In a split second a massive star runs out of fuel. No longer opposed by the outward pressure of nuclear fusion, gravity collapses the core. Atoms squeeze so tightly that their nuclei repel each other and—wow! The star explodes in a spectacular way.

> *That it is neither in the orbit of Saturn . . . nor in that of any one of the other planets, is hence evident, since after the lapse of several months it has not advanced by its own motion a single minute from the place in which I first saw it . . . Hence this new star is located neither below the Moon, nor in the orbits of the seven wandering stars but in the eighth sphere, among the other fixed stars.*

If it were nearby, he could measure its parallax (see pages 58–61), which would give him a clue as to its distance. But it's too far away. "It is a difficult matter," he writes, "to try to determine the distances of the stars from us, because they are so incredibly far removed from the Earth." His instruments are just not good enough. He guesses that the stars are "at

The first supernova of which we know was recorded in 1054 in Japan, China, and the Middle East. Judging by this 1,000-year-old petroglyph (rock carving, above) in Petrified Forest National Park, Arizona, maybe the Anasazi knew about it, too. It must have been dazzlingly bright. But it was not noted in Europe (so Tycho didn't know about it). How did Europeans miss it? Scholars think that because Europeans were taught the heavens never change, those who must have seen that supernova just didn't believe their eyes.

Below is what's left of Tycho's 1572 supernova. This X-ray image taken by the Chandra telescope in 2000 reveals a remnant, a cloud of dust and gas that's expanding into space. There's no trace of a stellar core, which suggests that the original star was a white dwarf in a binary (two-star) system rather than a red giant. If a core does survive, it can become a neutron star or a black hole.

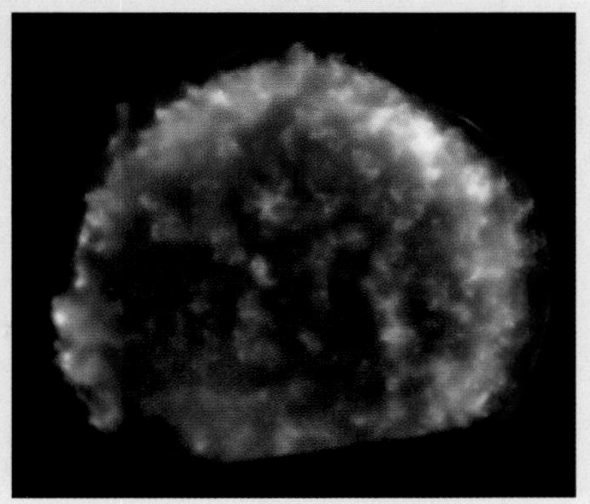

least 700 times further than Saturn." He is wrong about that; the nearest star is more than 20,000 times farther than Saturn.

But he is right that this is not a comet. He is actually seeing ancient history. This is a supernova, an exploding star that is so far away it has taken 20,000 years for its light to reach the Earth. Eventually, the supernova will fade and disappear and its atoms will be recycled in the universe. Knowledge of what has happened, and the understanding that bits of that star (nuclei from the atoms) are being scattered throughout the universe, will come in future generations. But in 1572, 25-year-old Tycho knows what he can see—it isn't a comet—and he isn't going to listen to those who tell him not to trust his eyes.

American astronomer Rocky Kolb will write of Tycho's reaction. "Impatient with those who had minds but would not think, and frustrated by those who possessed eyes but could not see, Tycho lashed out at them in the preface of a book he wrote about the new star:

> *O crassa ingenia. O caecos coeli spectatores.*
> Oh thick wits. Oh blind watchers of the sky."

But not everyone is blind. Tycho's book *De Nova Stella* brings him fame. He decides to travel through Europe to promote the book (and himself) and to look for a patron (someone who will finance his stargazing in a big way). He writes, "An astronomer, more than the student of other branches of knowledge, has to be a citizen of the world."

Tycho gives great lectures, and he easily attracts attention. In Germany, Wilhelm IV of Hesse is impressed. He sends an urgent note to Denmark's King Frederick II, "Your majesty must on no account permit Tycho to leave, for Denmark would lose its greatest ornament."

Frederick dispatches a rider. He is to bring Tycho Brahe to the royal hunting lodge near Copenhagen. The king will be Tycho's patron. He wants to be known as a monarch who

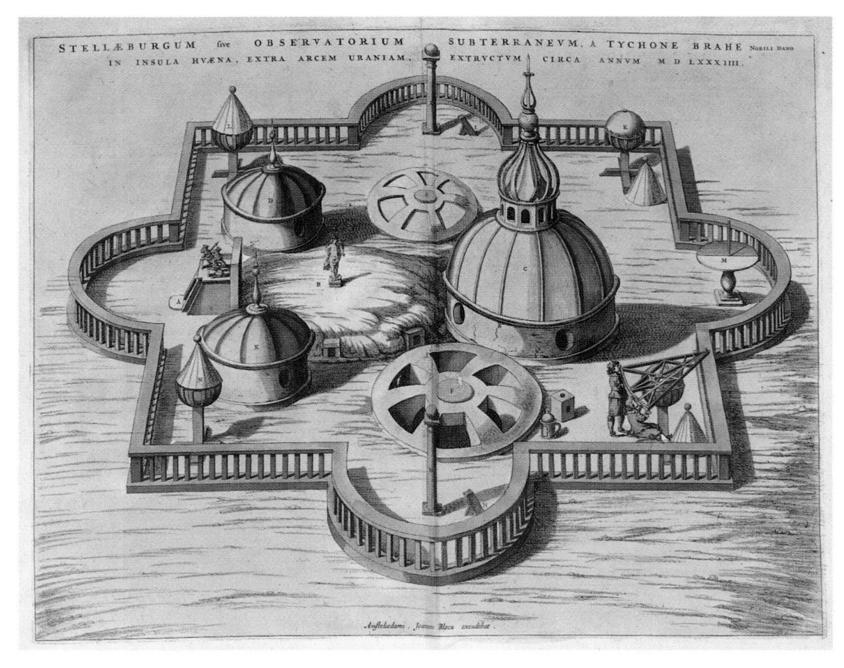

Tycho had two observatories on Hven: Uraniborg (which means "Castle of the Heavens") and Stjerneborg (which means "Castle of the Stars," shown at right). Stjerneborg was built mostly underground so no breezes could disturb the delicate instruments, which included giant quadrants for measuring angular distances between stars, and a mobile sextant.

supports the arts and sciences. Frederick grants Tycho an island—Hven (in the sound between Denmark and Sweden)—and throws in an extraordinarily generous pension. Tycho is able to live like royalty. (Rocky Kolb tells me it sounds like the right way to treat astronomers.)

Tycho builds a grand castle, Uraniborg, on Hven. It is the fanciest observatory in Europe; he calls it a "sky palace." The castle has flush toilets, an intercom, its own chemical lab, a paper mill, a printing press, a private jail, and, almost surely, the best astronomical instruments in the world. He turns the green island of Hven into the scientific think tank of his time.

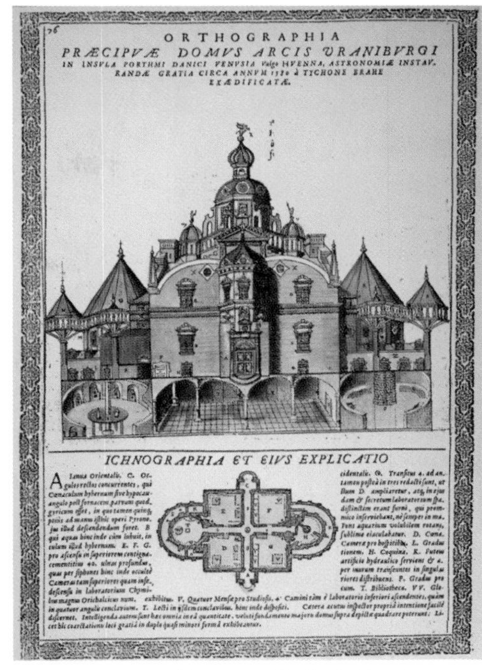

Above are Tycho's Uraniborg observatory and its floor plan.

Tycho spends years at Hven studying the skies and making exact astronomical charts and tables. To give you an idea of what is involved in observing: It takes a full year to track the Sun's movements in relation to the swath of zodiac constellations that appear in the sky in an annual, repeating cycle. It takes almost 12 years to track Jupiter's orbit, and just about 30 years to follow Saturn's circuit around the Sun. Tycho works for years, creating charts that are better than any done before.

Here's how long it takes the visible planets to orbit the Sun:

Mercury . . . 88 Earth days
Venus . . . . . 224.7 Earth days
Mars . . . . . 687 Earth days
Jupiter . . . . 11.86 Earth years
Saturn . . . . 29.46 Earth years

For him, astronomy is "heavenly chemistry. " He calls medicine and alchemy, "earthly astronomy." He studies them

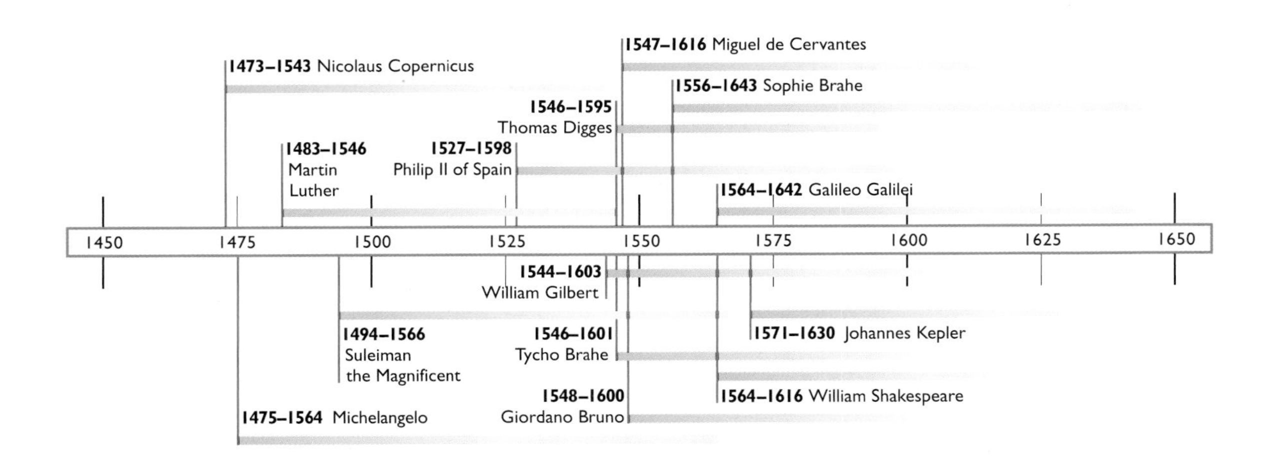

Alchemy is sometimes magic, sometimes early science. (More on that topic in chapter 19.) A lot of rich, smart Renaissance folks truly believed in it. Tycho and Sophie Brahe were among them. That's Sophie, above. Sophie's second husband, Erik Lange, lost all his money (and hers) attempting to make gold from common metals. Sophie tried to find a cure-all for all diseases. We know there is no elixir to cure all diseases, but don't laugh at the alchemists. We have hindsight; we know how things came out. That makes us think we are smart. When future generations look at us, what might make them chuckle?

all, along with meteorology (weather) and astrology (foretelling the future through the stars). Alchemy and astrology will turn out to be false sciences, but their study sometimes leads to real science. To Renaissance thinkers, all the arts and sciences are connected.

Tycho's sister Sophie Brahe often comes to Hven to join him. He calls Sophie his "learned sister" and thinks of her as a colleague. She has a castle of her own where, according to Tycho, "she designed a wonderfully beautiful garden, which is unparalleled in these northern parts of the world." When Sophie's husband dies, she is distraught. Tycho tells us,

> *She sought diversion in chemistry, with the intention of preparing certain . . . medicines. . . . Soon she was not just supplying these preparations to her friends and to the upper class . . . but also free of charge to the poor. But she found that not even this road led to the fulfillment of her intellectual ambitions, which continuously aimed further and higher.*

Sophie provides her brother with inspired help as well as intelligent conversation. It's too bad we can't listen when they talk of Copernicus. Tycho knows that Ptolemy's geocentric

A colored woodcut shows the 1577 comet as a streak across the sky above Nuremburg, Germany. Comets are impressive, but they don't zoom overhead like a jet plane. They appear to inch forward, night after night, over a period of days and sometimes weeks. Tycho Brahe was able to observe the 1577 comet (on clear nights) from November 13 to January 26 of the following year. He tracked half a dozen smaller comets in the next two decades.

(Earth-centered) system doesn't work well, but he can't believe that Earth is not the focus of the universe. Yet Copernicus's math does seem to have solved the problem of the planets. What should he believe? Tycho comes up with his own system—the Tychonian model—a compromise. Here it is:

The Earth is still the center of the universe with the Sun circling it—while the planets circle the Sun. It is an attempt to deal with the wandering planets and get rid of the epicycles, without throwing out Aristotle and Ptolemy. It seems a promising idea. Lots of people—ordinary people as well as scholars—argue over this dilemma: Ptolemy, Copernicus, or Tycho—who has it right? Do any of them? No one knows.

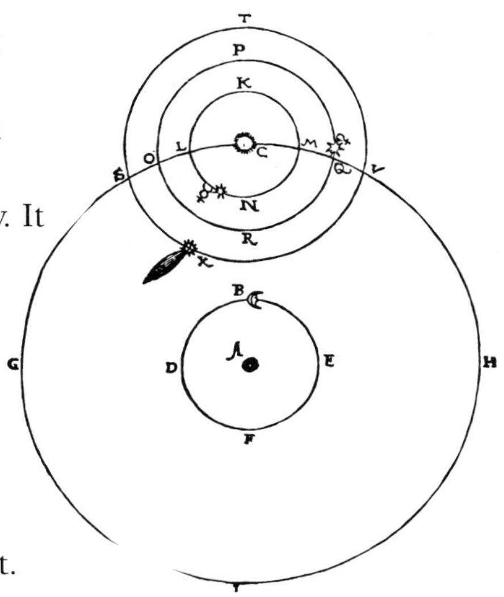

Tycho included the 1577 comet (the torpedo-shaped object) in his geocentric solar system model. Earth is in the middle, circled by the Moon and the Sun (the center of the bull's-eye at the top). Mercury, Venus, and Mars circle the Sun.

Then, in November 1577, a bold, long-tailed comet travels through the skies. For two months, Tycho follows it nightly, as clear skies allow, using the measuring instruments he has devised and his extraordinary eyesight. He plots its course. Because he is unable to detect any parallax, it is clear to Tycho that this comet is beyond the Moon, in the region of the supposed spheres. That means either Aristotle's crystal spheres don't exist or the comets have crashed through them.

PARALLAX is the way an object seems to move in relation to its background when you change your line of sight. (For a lot more on this, see pages 58–61.)

"There are no solid spheres in the heavens," Tycho concludes in his next book, thereby smashing the heavenly model that Aristotle and Ptolemy had both proclaimed. Those transparent spheres are just symbols, Tycho says they don't actually exist. He puts the celestial bodies in the heavens without any concrete support, which is an achievement that takes imagination and intellect. But if there are no crystal spheres, what does keep the stars and planets in their places? That is a question that no one yet can answer.

Tycho is without question the greatest astronomer of his time—and one of the greatest of all time. But when King Frederick dies, his 11-year-old son becomes King Christian IV, and he doesn't like Tycho Brahe's attitude. (Actually, Tycho is arrogant, and the new king has uppity advisors. It isn't a good situation.) A few years later, the boy-king writes this to the great astronomer: "[How dare you] not blush to act as if you were my equal. . . . I expect from this day to be respected by you in a different manner if you are to find me a gracious lord and king. . . . Your correspondence is somewhat peculiarly styled and not without great audacity and want of sense."

An astronomer must be cosmopolitan, because ignorant statesmen cannot be trusted to value their services.
—Tycho Brahe

King Christian takes away Tycho's island and his financial support. He hires Tycho's assistant to build an observatory in Copenhagen (called the Round Tower, shown on page 195). Eventually, Uraniborg is torn down. Tycho leaves Denmark, but he is a celebrity astronomer and in demand. He soon has a job as mathematician, astronomer, and astrologer to Europe's mightiest monarch, the Holy Roman Emperor Rudolf II of Austria.

Emperor Rudolf expects Tycho to predict the future with the stars. (You can see that, if it were possible, it would be very helpful.) Astrology is having a heyday, and Tycho's reputation as an astronomer/astrologer helps get him another fabulous observatory—this time in Prague (PRAHG). But now he is getting on in years.

No one had ever made celestial maps as splendid as his. Even Tycho himself doesn't know how to interpret all he has detailed. He needs a helper who knows how to use maps of the sky. He finds the perfect partner—someone who can take his charts and do something with them (even after he is gone). His name is Johannes Kepler. When they meet, Kepler is 28; Brahe is 53.

History has had many famous partners: Rome's Mark Antony and Cleopatra, Spain's Ferdinand and Isabella, and America's Lewis and Clark, to name a few. Tycho Brahe and

WHAT HAPPENED IN THE SIXTEENTH CENTURY?

Like it or not, we're obsessed with history. We argue about it and depend on it to justify and empower the present. Check out this diverse period of energy and achievement. Consider: The year 1600 is often cited as the beginning of the modern world. But the ideas for modernism were incubating in the 1500s.

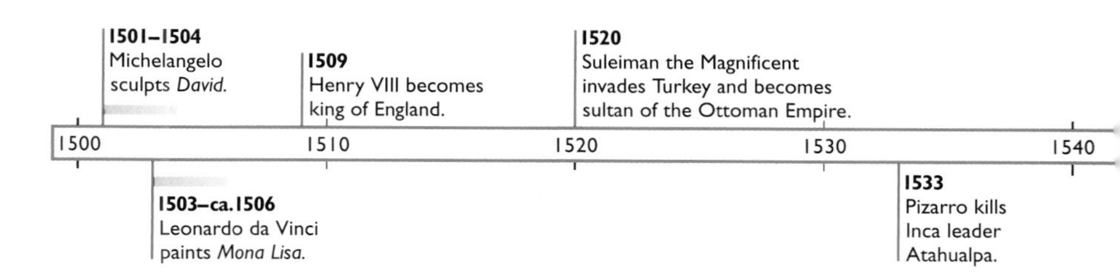

The Noble Dane, an 1855 painting by Austrian Edouard Ender, shows Tycho Brahe (left) standing in the shadow. He's demonstrating a celestial globe (a three-dimensional star map) to Emperor Rudolf II, the colorful center of attention.

Johannes Kepler are right up there with the most important partners of all time. It is too bad that they don't much like each other, but it doesn't matter. Tycho is a terrific astronomer; Kepler is a terrific thinker. Their skills complement each other.

Keep that name, Johannes Kepler, in your head. You're going to hear more about him and what he will do with Tycho's charts (especially after Tycho is dead). But first we need to head south, to Italy, where seeds from Copernicus's book are sending up sprouts.

1547
Ivan the Terrible is named czar of Russia.

1556
Akbar becomes Mogul emporer in India.

1558
Elizabeth I crowned queen of England.

1564
Spanish begin to settle Philippine Islands.

1572
Supernova blazes in constellation Cassiopeia.

1585 & 1587
English colony is planted on Roanoke Island (now North Carolina). The settlers all disappear.

1588
England defeats Philip II's Spanish Armada.

1550 1560 1570 1580 1590 1600

HOLDING A RULER TO THE SKY

When Tycho Brahe looked at the Moon from different viewing points, he could see it shift position in relation to the stars in the background. That apparent shift is called parallax. The greater the parallax, the closer the object. Tycho tracked the supernova of 1572 and the comet of 1577; they showed no parallax. That meant they couldn't be nearby. If that was true, then Aristotle made a whopper of a mistake.

What exactly is parallax? It is easy to understand if you hold a finger in front of your nose. Close one eye, then open it and close the other. What you see is parallax. It's the *apparent* change in position of an object in relation to the background when viewed from different vantage points. Your eyes are separated by an inch or so; that makes for different viewing points. So your finger appears to move in relation to whatever is in the background. The Moon does the same thing in relation to the background stars. (See diagrams at left.)

Now do the one-eye trick with your arm stretched as far as it will go. Your finger will still seem to move, but not as much.

Without moving your finger, close one eye and then the other. The finger appears to shift to the left or right in relation to the background—a parallax shift.

right eye closed

left eye closed

The farther an object is from the viewing point, the less its parallax shift.

Tycho and others tried and tried to measure the parallax of stars. They saw none. The stars are just too far away for parallax shifts to be observed without a high-powered telescope. The parallax dilemma was a challenge to Renaissance astronomers, who were eager to measure cosmic distances.

When Tycho attempted to determine the parallax of the supernova, he found it was like the stars: it didn't budge. That told him it was very far away. When it came to the comet, he had a more complicated job. Comets are in rapid motion and, because they're much closer than stars, their motion is more apparent. Tycho did a number of detailed sightings and was perplexed: was he seeing parallax or comet movement? Finally he came to the conclusion that the movement he was observing was not related to parallax. That meant neither the "new" star nor the moving comet could be near the Moon (whose parallax shift is obvious). So Aristotle's idea—that the distant heavens are unchanging—was wrong. Here were two examples of change.

"The shock dealt to the Aristotelian world view could not have been greater had the stars bent down and whispered in

Tycho designed this instrument (shown here on a book cover) to measure the parallax shift of stars, but the apparent movement was too tiny to detect.

the astronomers' ears," writes author Timothy Ferris. "Clearly there was something new, not only under the sun but beyond it."

Looking for the comet's parallax, Tycho learned something else. He watched

(continued on page 61)

THE SKINNY ON MOON PARALLAX

Here are the Earth and the Moon in scale—perhaps a surprise, since many of us picture the Moon much bigger and closer to Earth. Look how skinny that parallax angle is. That's as big as it gets from Earth's viewpoint. The Moon is our closest neighbor and so has the largest parallax angle of any celestial object.

This diagram uses the Earth's diameter as a base—the biggest possible base on the planet. (Your parallax base is the distance between your eyes.) The bigger the base, the greater the parallax angle. Yet, even with an Earth-sized base and a neighborly distance, the Moon's parallax angle is still less than 2°.

The angles for neighboring planets (and passing comets) are far smaller, less than 1°. The angle for the nearest star (Proxima Centauri)— less than 1/3600 of a degree!—is undetectable without an extremely powerful kind of telescope, which wouldn't be invented until the nineteenth century.

Parallax is more than an eye trick. It's a useful tool for measuring distances from Earth to space objects. Notice that the parallax sighting of the Moon (right) forms a triangle—a long, skinny triangle, but it has three sides and three angles like any other triangle. Using trigonometry—the mathematics of triangles—astronomers can measure the parallax angle and, knowing the length of the base, calculate the distance from the Earth to the Moon.

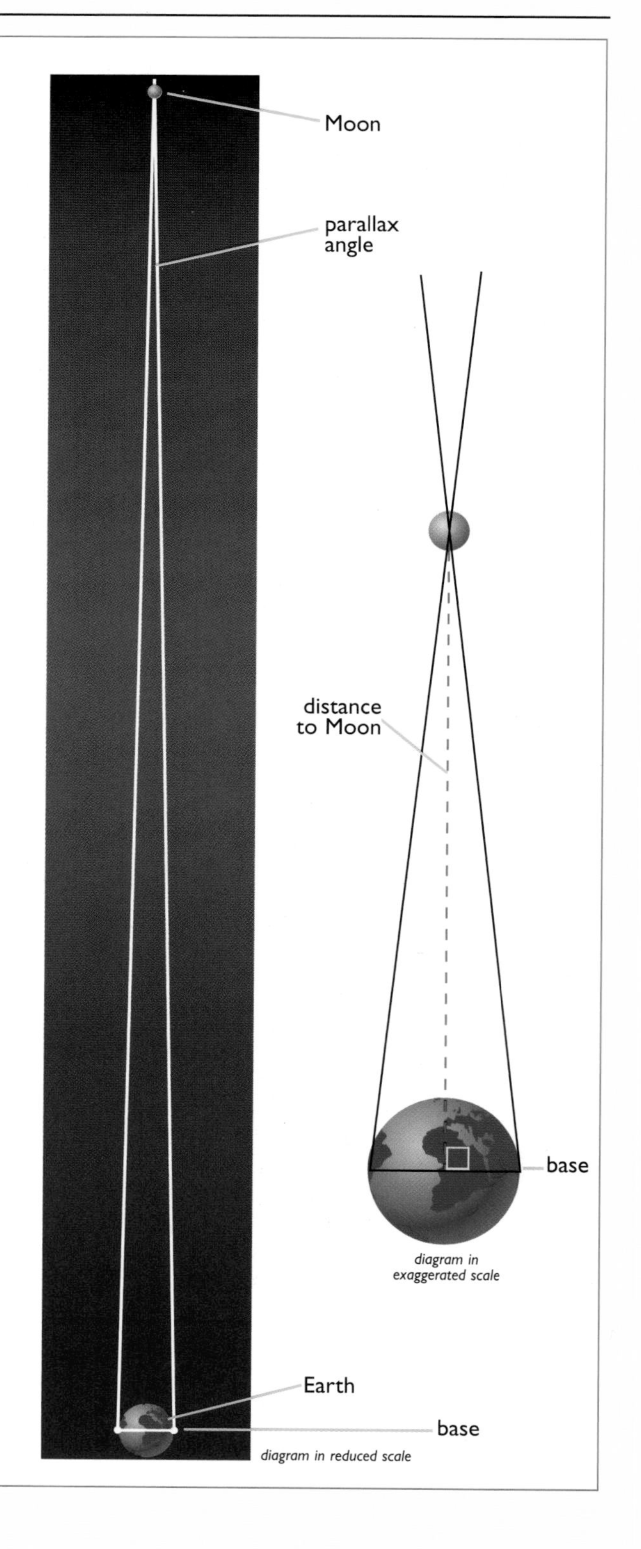

diagram in exaggerated scale

diagram in reduced scale

At the same time on November 9, 2003, astronomy fans around the globe took pictures of the Moon. When combined, the images show a clear parallax shift. The top image compares the two positions of the Moon as seen from Maldon, England, and from Colorado in the United States, which is a long base that spans two continents. In the bottom image, Moon positions were viewed from a much smaller base separation—between Maldon and Glasgow, Scotland. The longer the base, the greater the parallax shift.

the comet move in front of the Sun, then get dimmer and seem to move away from Earth, then appear back near the Sun again. The reason was clear to Tycho: the comet was orbiting the Sun! That helped him come to the conclusion that the planets orbit the Sun. But when it came to the Earth, he couldn't believe it actually moves. His observations helped other thinkers accept the Copernican model, even if Tycho himself couldn't quite make the leap all the way to a solar system.

5 Renaissance Men

The heavens themselves, the planets, and this centre
Observe degree, priority, and place,
Insisture, course, proportion, season, form,
Office, and custom, in all line of order.
—William Shakespeare (1564–1616), English playwright and poet, *Troilus and Cressida*

Truth is great and will prevail if left to herself.
—Thomas Jefferson (1743–1826), U.S. president, "Virginia Statute for Religious Freedom"

Twenty-one years after Nicolaus Copernicus's death in church quarters in Poland, a first child is born in Italy to a spirited musician, Vincenzo Galilei (vin-CHEN-zoh gal-ih-LAY), and his wife Giulia (Italian for "Julia"). They name their son Galileo.

Galileo's family coat of arms includes a ladder with three steps, which means "good help." This one crowns Galileo's tomb in Santa Croce church in Florence.

Three days later, the renowned artist Michelangelo dies at age 89. (It was he who painted the awesome mural of God's creation on the ceiling of the Sistine Chapel at the Vatican in Rome.) Two months after that, in England, John Shakespeare (who makes gloves and also sells wool) and his wife, Mary (a woman from prosperous farming stock), give birth to a child they christen William. It is February of 1564, and if you believe that the stars tell of earthly events—as the astrologers do—then you might look to the heavens for signs of genius. For these three—Galileo, Michelangelo, and Shakespeare—have between them as much genius as any time has seen.

Michelangelo the artist, Shakespeare the poet and playwright, and Galileo the scientist—each shatters ideas and ways that once seemed good enough. After them there is no going back.

Today, we know there's no scientific basis for astrology.

148151

Dialogo
DI VINCENTIO
GALILEI NOBILE
FIORENTINO
DELLA MUSICA ANTICA,
ET DELLA MODERNA.

IN FIORENZA M. D. LXXXI.
Appresso Giorgio Marescotti.

It appears to me that they who in proof of any assertion rely simply on the weight of authority, without adducing any argument in support of it, act very absurdly. I, on the contrary, wish to be allowed freely to question and freely to answer without any sort of adulation, as well becomes those who are in search of truth.

—Vincenzo Galilei, *Dialogue of Ancient and Modern Music*

After Galileo, the cosmos will dance to a new tune. (Galileo, like Tycho, is usually referred to by his first name. By the way, Tycho is a teenager when Galileo is born.)

It is Galileo's father who does much to mold his gifted son. Vincenzo Galilei is a composer-musician-performer who is not content writing madrigals (short love songs) and the expected church music. He longs to recapture the Greek ideal of popular music—music built on myth and poetry. So he experiments with new musical forms, setting speech and poetry to music. (Opera will evolve from that experimenting.)

To the left is the title page of Vincenzo Galilei's book *Dialogue on Ancient and Modern Music.* He experimented with the pitches of notes made by musical strings of different lengths. Maybe his scientific work led his son, Galileo, to experiment, too.

OF KNOWLEDGE INFINITE

Christopher Marlowe was yet another creative genius who was born in 1564 (in addition to Shakespeare and Galileo). But he died young, at 29, supposedly in a tavern brawl over who would pay for supper and ale (it may have been murder). Too bad. He was a playwright and poet—vital and unconventional. He broke down doors that Shakespeare would walk through. Here are some lines from his play *Tamburlaine the Great*:

> *Our souls, whose faculties can comprehend*
> *The wondrous Architecture of the world:*
> *And measure every wandering planet's course,*
> *Still climbing after knowledge infinite.*

Marlowe was intrigued by the new science. (Read those lines of his a few times; they mean more with each reading.)

This portrait is believed to be Christopher Marlowe.

A SOUND INTERVAL is the difference in pitch between two tones. A SOUNDING BOARD is a thin board built into the upper portion of some musical instruments, like a piano or violin, to increase resonance (the strength and depth of the sound). Sometimes a person or a group of people can be used as a sounding board for a new idea. Does the idea resonate with that group?

He also experiments with sound intervals, using weighted strings and sounding boards. Musicians are expected to compose serious music following rules set down by Pythagoras. Vincenzo breaks those rules, which annoys some of the old guard. He is a freethinker with a sharp wit and little tolerance for big-shot authorities. But revolutionary ideas don't buy bread. Although he is a respected composer and performer, Vincenzo has to work as a wool merchant in order to feed his growing family.

This map of Italy (right) is from the first modern atlas, compiled in 1603 by Abraham Ortelius.

They live in Tuscany—in Italy's west central region—in the city of Pisa and then in Florence. Italy is a collection of independent states—each with its own ruler, its own laws, and its own personality, although all share a common language and religion.

Vincenzo is a devoted father who tutors his son in Greek and Latin and classic literature, as well as in music. Young Galileo learns to play the lute and the organ and plays them well. He shows talent in drawing; he writes poetry and, while still a boy, a comic play. A strapping red-haired youth, he has his father's energy and self-confidence. When he is 11, his parents send him to a Jesuit monastery, where the scholarly monks teach him more Latin and Greek, natural philosophy (the term for science), mathematics, and religion. But when

Galileo's key contribution to the birth of science lay precisely in emphasizing the need for accurate, repeated experiments to test hypotheses, and not to rely on the old "philosophical" approach of trying to understand the workings of the world by pure logic and reason—the approach that had led people to believe that a heavier stone will fall faster than a lighter stone, without anyone bothering to test the hypothesis by actually dropping pairs of stones to see what happened.
—John Gribbin, *The Scientists*

Galileo decides to become a monk himself, Vincenzo dashes up the mountainside with an appointment for his son to see an eye doctor; he takes the boy home. Galileo never goes back. Vincenzo doesn't want his son to spend his life in a monastery. Still, that early training makes an impact; Galileo will always be devout.

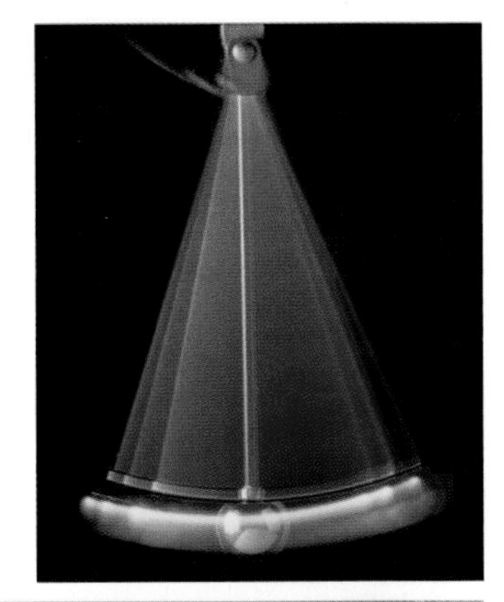

Galileo combines piety with his father's combative nature and a fierce determination to think for himself. It is a personality mix that brings him admirers—and enemies.

But he still has schooling ahead of him. The Galilei family hopes their bright son will become a doctor and bring some prosperity to the family, so they send him to medical school at the University of Pisa.

He doesn't like medical studies. He cuts classes and his professors complain. Secretly, he is attending a mathematics class. There he has found a mentor, a splendid teacher who takes a special interest in the outspoken, inquisitive young man and gives him private instruction.

One morning, when he is about 18, Galileo is sitting in the exquisitely ornate Cathedral of Pisa; he is there for early mass (or so the story goes). A monk pulls a heavy chandelier to the side to light its wicks, then he lets it go. At first the chandelier swings in large arcs, but as it slows, the arcs get smaller and smaller. Galileo's mind wanders from his prayers; he uses his medical training and times the swings of the lamp with the beat of his pulse. He notices something astonishing: Although the swings get shorter and shorter as the pendulum winds down, the period—the time of each swing—remains exactly the same.

A chandelier still swings in the Cathedral of Pisa (above). The period of a pendulum (the timing of each swing) depends only on the length of its suspension (a rod or chain, for instance). The mass of the bob (the object on the end) doesn't count. Jo Ellen Barnett explains in *Time's Pendulum*: "When a pendulum swings in a wide arc, the [bob] falls with greater velocity, and the longer distance it has to travel is made up for by its increased speed."

Galileo rushes home from church and begins to do some experimenting with strings and weights. There is something important here, and he seems to be the first to take note of it. A pendulum can be used to measure time accurately.

No matter how heavy or light the bob (the weight at end of the pendulum), it always takes the same time to complete

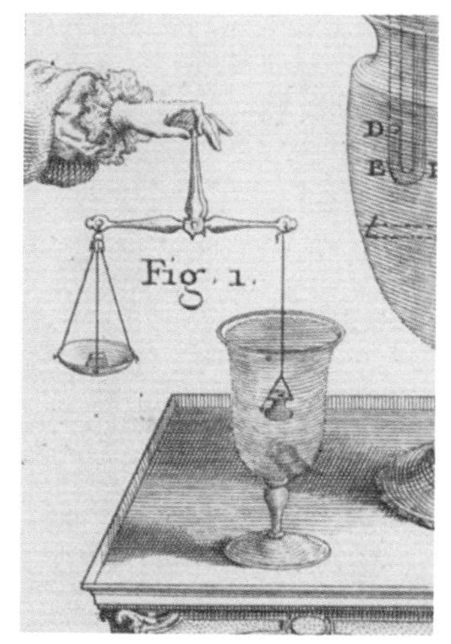

A hydrostatic balance measures the density of an object by comparing it to the standard density of water—a measurement called specific gravity. An object is dunked in a glass of water. If it's denser than the water, it sinks and weights are added to the other side of the scale to balance it. The denser the object, the more weights are required.

a swing. But if you change the length of the pendulum, the period will change. (Keep in mind, we're ignoring air resistance here.) At first, Galileo doesn't think of using a pendulum as a clock (that will come later); instead he comes up with a device for measuring the rate and variation of a person's pulse.

His pendulum is developed into a useful doctor's tool and brings him some acclaim, but not enough to get a scholarship. With family finances at a low point, he has to drop out of college. For the next few years he lives by his wits—tutoring students in mathematics while also using his active mind to experiment, measure, investigate, and invent. He studies on his own—especially the works of the ancient mathematicians Euclid and Archimedes. Then he invents a hydrostatic (water-based) balance. (It comes out of his study of Archimedes' ideas.) He realizes that without precise tools of measurement, science can't get far.

What he really wants is to be a professor of mathematics—but every time he tries to get a job he is turned down. Finally, when he is close to despair and thinking of leaving Italy, a junior position turns up. It is at the University of Pisa; it pays very little, but at last he is a professor—and soon a popular one.

Because he has an exuberant (lively) nature and knows and loves literature, music, and all the arts—as well as math and science—his lectures range broadly. Students are eager to hear what he has to say. Unfortunately, he has a habit of taunting, or worse, belittling other professors—especially the Aristotelians. That makes him enemies. And he goes too far. The Aristotelians aren't all alike. At Pisa, a few professors are beginning to think in modern terms and test ideas with experimentation. Though he hates to admit it, Galileo will learn from their examples. He is especially interested in experimenting with motion. "Ignorance of motion is ignorance of nature" is a maxim heard in scientific circles.

Today we use the term *fluid mechanics* for the study of fluids in motion or at rest. Fluids can be liquids or gases. Fluid mechanics is important in the study of the weather, in the design of airplanes and automobiles, and in the chemical industry, where liquids or gases are in flow. Hydrostatics is a branch of physics that deals with fluids at rest.

It is in Pisa that Galileo may have conducted an experiment that changes ideas about motion. Aristotle said that heavy objects fall faster than light ones. He said a 2-pound ball falls twice as fast as a 1-pound ball. Galileo thinks Aristotle was wrong—and that all objects fall at the same rate (if there is

A TRUE STORY?

Vincenzo Viviani was Galileo's assistant; he was also his first biographer. He wrote about Galileo's famous experiment—dropping iron balls of different sizes and weights from the Leaning Tower of Pisa. Which reached the ground first—the heavy or light ones?

Supposedly they crashed at about the same time. The idea was to prove that objects all fall at the same rate—which they do when there is no air resistance, as on the Moon.

If a man jumps out of a 20-story window and, at the same time, drops his briefcase, the briefcase will remain at hand level through the fall. (Don't try this experiment!)

But a feather and an iron weight dropped together from a tower will not hit the ground together. Outside a vacuum, air resistance makes a difference. (The less dense the object, the more it will be resisted by air.)

Did Galileo actually perform that experiment? No one is sure. But he told Viviani he did, and that's good enough for me. Besides, he wasn't the only one doing this kind of experiment, as you'll see if you keep reading.

The Tower of Pisa in northern Italy was built from 1173 to 1360. It began leaning after the third story was finished because the ground beneath was unstable. After reviewing hundreds of plans, engineers in the 1990s added a counterweight underground to keep it from falling over.

no air resistance). He notices that hailstones—big and small—often hit the ground at the same time. He says he doesn't think that God drops the small ones from a lower height!

It isn't hard to prove his theory. All he has to do is experiment. So he takes a cannonball and a musket ball and climbs the steps inside Pisa's Leaning Tower. If he drops those balls from the top of the tower at the same time, will they hit the ground simultaneously? They do (well, close to the same

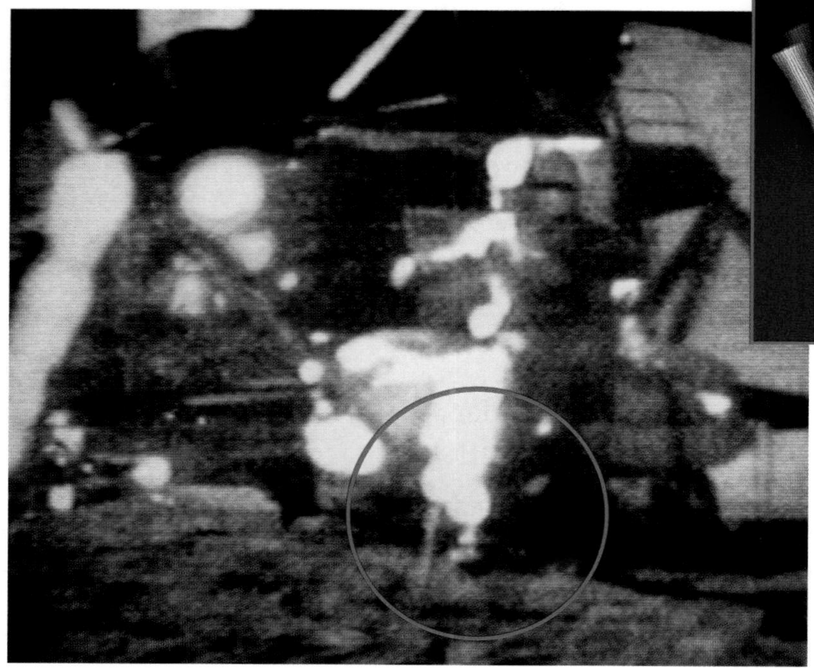

In 1971, astronaut David Scott dropped a hammer and a feather while standing on the moon—where there is almost no atmosphere. He watched them hit the ground simultaneously and announced to the world, "How about that! Mr. Galileo was correct in his findings."

If you do this experiment, make sure you drop the objects at exactly the same time and from exactly the same height. Science historian Thomas Settle tried it with a heavy and a light ball while an experimental psychologist watched. They found that a heavy object makes the hand more fatigued, which causes the experimenter to release that object more slowly.

time), smashing some of Aristotle's reputation while preparing for laws of motion that, almost 400 years later, will take us to the moon.

Today most authorities don't believe Galileo really did that experiment, but he was dropping weights from somewhere; maybe he leaned out a window. Here are Galileo's words on the subject:

> *Aristotle says that a hundred-pound ball falling from a height of a hundred cubits hits the ground before a one-pound ball has fallen one cubit. I say they arrive at the same time. You find, on making the test, that the larger ball beats the smaller one by two inches. Now, behind those two inches you want to hide Aristotle's ninety-nine cubits and, speaking only of my tiny error, remain silent about his enormous mistake.* [We know that the "tiny error" was in the measuring or dropping or in air resistance.]

After three years in Pisa—despite his achievements—Galileo's impertinence (we call it "attitude") and outspokenness have offended most of the faculty. He won't wear the long black gown professors are expected to wear. Then, to make things worse for himself, he writes a poem making fun of

GIVE SIMON SOME CREDIT

A Dutch mathematician, Simon Stevin, took two lead balls, one 10 times more massive than the other, and dropped them simultaneously onto a board (so he could hear the sounds as they hit). "It will be found," said Stevin, "that the lighter will not be ten times longer on its way than the heavier but that they fall together onto the board so simultaneously that their two sounds seem to be one and the same rap." Stevin didn't make much of this phenomenon; a few years later, Galileo did. Stevin published his results in 1586, but he wrote in Latin, so only scholars read of his work.

those gowns. His contract to teach is not renewed.

Galileo realizes that Tuscany is *not* a great place for an independent thinker. So at age 28 he heads northeast, to the bustling city of Padua in the freewheeling Republic of Venice. There he uses all his charm and wit and every contact he can find and lands a prestigious job as chair of the department of mathematics at the University of Padua. It is one of Europe's great universities.

The INQUISITION was a tribunal established in the Middle Ages to find and punish heresy. It got out of hand. By the sixteenth century, trials were held in secret, and those charged were not told who had accused them.

Padua isn't perfect. Not long after Galileo settles in, Giordano Bruno (who had hoped for the mathematics post given to Galileo) is arrested by the authorities for voicing heretical ideas. In 1593 he is taken to Rome and kept in the prison of the Inquisition. Bruno refuses to give up his ideas even though the church fathers find them outrageous. He has gone much further than Copernicus. Bruno not only says that the Earth orbits the Sun, he says the Sun itself moves. Nothing in the universe is at rest. The stars, scattered through space, are themselves suns with orbiting planets. And our planetary system is not the center of the universe, because the

Galileo was not a tall man. His students at the University of Padua couldn't see much of him when he lectured from behind a stand, so they built this rostrum with steps.

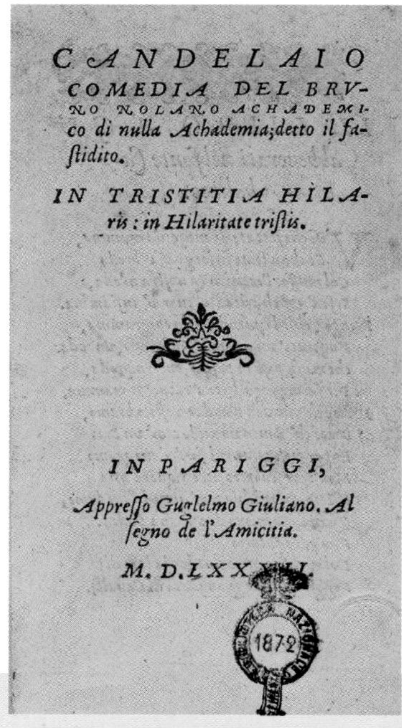

CANDELAIO
COMEDIA DEL BRVNO NOLANO ACHADEMICO di nulla Achademia; detto il fastidito.

IN TRISTITIA HILARIS: in Hilaritate tristis.

IN PARIGGI,
Appresso Guglelmo Giuliano. Al segno de l'Amicitia.
M. D. LXXXII.

Giordano Bruno was educated as a Dominican priest but eventually came to believe that religion was incompatible with experience. He found that everything he had been taught about the universe was wrong, and he knew he would be executed for his beliefs. His play *Candelaio* ("Candlemaker") is a comedy that protests corruption. The title page is at far left.

universe has no center. (Given hindsight; we know he had it right.)

Bruno has yet another thought, this one beyond consideration. He says, "There is a common soul

Queen Elizabeth I reigned over a growing British Empire. In case anyone doubted her power, she was painted standing on a map, with the whole empire at her feet.

THE VIEW FROM ENGLAND

Henry VIII split with the Catholic Church over divorce—he wanted one, the church wouldn't grant it—so he formed his own church—the Church of England. In 1536, an Act of Parliament declared the authority of the pope void in England. That rift left England with a degree of freedom in the field of religion that other European nations didn't have.

Is Earth the center of the universe? That was the big intellectual question a generation after King Henry VIII, in the England of Queen Elizabeth and William Shakespeare. Shakespeare's friend Thomas Digges (1546–1595), a well-known scientist, took the Copernican view: The Earth is not the center of the universe. Digges said that the universe is infinite—not bounded by Aristotle's spheres. He also said that the stars are suns and that there are probably other planets and that some of them might hold intelligent life. Perhaps he was influenced by Giordano Bruno (1548–1600). Or maybe they influenced each other. Bruno visited England and within the whole to which it gives being and at the same time is individual and yet is in all and every part." What he is saying is that the divine soul (God) is in every part of the universe. *He is a heretic*. According to church teaching of the time, the creator is separate from the created world. Bruno refuses to recant. In 1600 he is burned alive. But where did Bruno find his ideas?

Not from Copernicus, who was moderate by comparison. (Actually, some ideas came from Nicholas of Cusa.) But Bruno's heresy makes the authorities take a new, hard look at Copernicus, whose idea of the Earth moving around the Sun now seems contrary to scripture. Bruno has shown the danger of doubting the wisdom of the past. The heat of Bruno's immolation is meant to chill scientific questioning.

HERESY—thoughts and actions contrary to religious belief—was a crime punishable by death, which is why some accused heretics RECANTED (took back) their heretical thoughts.

An IMMOLATION is a sacrifice of life, often by fire.

Thomas Digges favored the Copernican system with the Sun at the center and the five visible planets in circular orbits (left). The English words in the outer ring describe the beauty of the stars: "perpetual shining glorious lights innumerable." Digges's friend, the great writer William Shakespeare (above), kept up with the changing views of the cosmos.

published his work both there and in France (he couldn't do it in Italy).

Queen Elizabeth's doctor, William Gilbert (1544–1603), who was investigating magnetism, believed in Copernicus's ideas, and said so. In England, as long as he taught those ideas as theory and not truth, he could do that.

So English thinkers and writers considered the ideas of Copernicus and Tycho and Bruno and Digges without fear. (By the way, Digges's father, Thomas, invented an early telescope, although he didn't publicize it, so he didn't get credit for it.) The English were fascinated by the new science: The poet John Milton visited Galileo in Italy; the poet John Donne visited Johannes Kepler in Germany. They had no problem talking to each other; they all spoke Latin.

William Shakespeare followed the controversy; all thinkers of the time did. Who was right—Copernicus, Tycho, or Aristotle? There was much disagreement on that. In his play *Hamlet*, Shakespeare named two characters for Tycho's relatives. They are Rosencrantz and Guildenstern. Was it an "in" joke for those who were scientifically savvy?

6 Gazing at a Star Named Galileo

The true method of experience...first lights the candle [with a hypothesis] and then by means of the candle shows the way,... with experience duly ordered...and from it educing axioms, and from established axioms again new experiments.
—Francis Bacon (1561–1626), English philosopher, *The New Organon*

Flout 'em and scout 'em
And scout 'em and flout 'em;
Thought is free.
—William Shakespeare (1564–1616), English playwright, *The Tempest*

Galileo isn't scared. He is a great speaker—robust and witty—who has so much to say and says it so well that his lectures soon attract pupils from all over Europe. Students are known to complain about boring professors—"paper doctors," they are called. The senate in Venice takes the charge seriously and levies a fine against those who read their lectures. Galileo is not a paper doctor. He is a star. He says he doesn't enjoy lecturing, that he prefers the one-on-one of tutoring, but he may be fooling himself. His lectures draw crowds. He is a gifted showman and an even more gifted researcher and thinker. He is able to take ideas, consider them with imagination, and come up with something original.

Galileo stands on a historic cusp between the ancient and modern worlds. He understands that for science to progress, it needs to become exact—and that mathematics can provide it with a language that is certain.

The first quotation (above) is from Francis Bacon's *The New Organon*. The original *Organum*, by Aristotle, laid down rules of logic. Bacon argued that science begins instead with experiments and observations.

A cusp can be a point or edge of something, like a tooth or the tip of a crescent moon. In architecture, it's the place where two curves meet. In astrology, it's the time when two zodiac signs overlap. The cusp in the text at left is a transition time from one era to another.

Galileo's son Vincenzo wrote that his father "was of jovial aspect, in particular in old age, of proper and square stature, of robust and strong complexion, as such that is necessary to support the really Atlantic efforts he endured in the endless celestial observations." Justus Sustermans, a Flemish artist in the Tuscan court, painted the elderly Galileo twice—including this oil portrait.

Aristotle had asked the right questions, but he hadn't answered them with mathematical proofs. Ptolemy attempted to do so. Copernicus took that idea further and in a new direction. But it is Galileo who helps make physics mathematical. With math and measurements, science can become precise. Here's his famous statement of that idea:

> *Philosophy is written in this grand book of the universe, which stands continually open to our gaze. But the book cannot be understood unless one first learns to comprehend the language and to read the alphabet in which it is composed. It is written in the language of mathematics, and its characters are triangles, circles and other geometric figures... without these, one wanders in a dark labyrinth.*

Galileo used a compass like this one (another kind of compass points north) to measure the altitude of stars and planets above the horizon.

WHO KNEW LEVERS HAD LAWS?

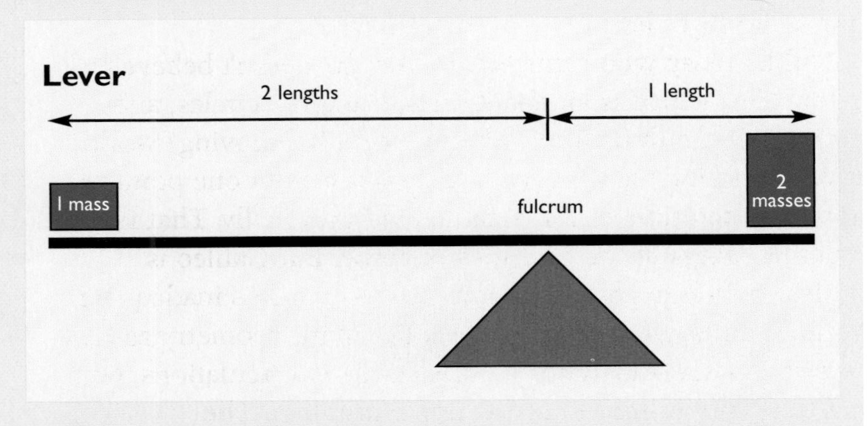

Archimedes found that a seesaw (which is a lever) reaches balance (equilibrium) when the ratio of the mass at either end is equal to the *inverse ratio* of the length of the board from the fulcrum (center support). That's a fancy way of saying: To get a seesaw to balance with a heavy person on one end and a skinny kid on the other, you'll need a short distance between the heavyweight and the fulcrum. And vice versa. That's the **Law of the Lever**. Using this law and a jack (a lever tool), a child might be able to lift a heavy truck.

More than 1,800 years before Galileo, another great mathematician, Archimedes, discovered the **Law of the Lever**. What Archimedes did—and what almost all great scientific achievers do—is find a relationship between seemingly unrelated things—in his case, mass and distance. Then he expressed that concept mathematically, so it could be used by others. Because he did that and wrote about it, it soon seemed easy and obvious. But likenesses often stay hidden until someone "discovers" them.

Galileo's experiments using balls help him find a hidden likeness that links speed to distance and time. When we say a car is going 100

SPEED measures how fast something is moving. It is always measured in terms of distance covered during a unit of time—miles per hour, for example:

$$\text{speed} = \frac{\text{distance}}{\text{time}}$$

VELOCITY requires both speed and direction: 300 kilometers (about 185 miles) per hour, west.

ACCELERATION is a change in velocity. This word is tricky because it has one meaning in everyday speech ("speeding up") and a more precise meaning to scientists. To a scientist, acceleration means "any change of velocity—speeding up or slowing down or changing direction." The key thought is change. To accelerate is to change the motion over a period of time.

$$\text{acceleration} = \frac{\text{change of velocity}}{\text{time interval}}$$

Shortly before his death, Galileo thought about how to make a self-running pendulum clock. His son Vincenzo improved the design and created this model (right).

kilometers (about 60 miles) per hour, we're comparing measurements of different things—distance and time. Galileo put us on that path.

Unlike those who came before him, he doesn't believe that geometry—which is all about lines, triangles, circles, and numbers—has to be limited to the static (unmoving) world. If you can measure the *time* it takes to go from one point to another, then time can be treated mathematically. That idea will find wings in the centuries to come. But Galileo is limited by the mathematics of his day—and it is inadequate. Yes, he can measure speed and motion with geometry as he knows it—but sometimes it takes so many calculations, he has to spend months doing a simple problem. The mathematics of calculus, which can measure motion efficiently, is crying to be invented. (Hold on, it's coming.)

Technology will have to improve, too. Galileo is stuck with measuring instruments that are not very accurate, like the clocks of his time. When Galileo drops balls of different masses, he can note that they hit the ground at about the same moment—but he can't measure the time of the drop exactly—or the intervals of time along the way. There are no clocks or measuring instruments precise enough. Is there a measurable pattern or mathematical law governing free fall? Galileo wants to know.

Like Leonardo da Vinci, nearly a century earlier, Galileo suspects that falling objects increase their velocity (pick up speed) as they fall. A rock dropped from a high place seems to fall faster and

FREE FALL is the movement of objects under the influence of only gravitation. Acceleration in a free fall equals the acceleration due to gravity (g). Balls rolling down a ramp aren't in free fall because friction acts on them, but Galileo had no choice. He didn't have the instruments to measure free-fall acceleration precisely. Since heavy balls have a harder time overcoming friction than light ones, make sure you use balls of identical mass if you try Galileo's experiment.

Modern strobe photography captures a falling ball and feather in one-second intervals. Air resistance would normally cause the feather to flit around, but this image shows what an airless, free-fall descent looks like for any object on Earth—heavy, light, or in between. The distance between images increases from top to bottom because the objects are accelerating—dropping a little faster and farther each second.

FRICTION is often defined as "rubbing or roughness." A better way to think about it is "stickiness and slipperiness." Friction is the resistance to slipping and sliding when two objects come in contact. If there's a lot of resistance, they stick. If not, they slide. Whichever words you use, friction slows things down.

faster as it heads for the ground. What can he do to find out if that is true? He needs a stopwatch or fast photography, and neither have been invented. He wants to understand the implications of free fall (which is another term for gravity), but he's limited by the instruments of his day. What can he do?

Galileo decides to adapt the dropped-balls experiment by slowing the fall. He does that by rolling brass balls down a groove on a slanted board (an inclined plane). He's introducing friction, and that affects the experiment. Still, he has set up a way to measure acceleration. He finds: The more inclined the slant, the faster the fall.

He begins with a moderate slant, carefully measuring the length of the board (the *distance*) and the *time* balls take to travel it. Then he makes the experiment more complicated: He measures sections of the board to see if the balls keep a uniform speed or if they accelerate (change speed) as they descend the plane.

Galileo designed this wooden ramp to measure acceleration. As a ball rolled down, it rang small bells dangling above the track. The time between rings was the same, even though the bells were farther and farther apart. That meant the ball was speeding up at the rate indicated by the intervals between bells.

The balls do not roll the length of the board at a constant speed. Their roll gets faster and faster—by a percentage that never varies—no matter how many times he tries the experiment or how much he tilts the ramp. The **acceleration is constant**, which means the speed always changes by the same amount during equal periods of time.

Even more startling, he finds that the changes of speed follow the list of odd numbers. In other words, if a ball falls past one notch on the board in one second, it will pass three notches during the next second, five notches during the third, and so on. We call that **Galileo's Law of Free Fall**. He guesses (rightly) that what happens on an inclined plane is the same thing that happens in a free fall from a tower.

MATH, THE LANGUAGE OF SCIENCE

Galileo's **Law of Uniformly Accelerated Motion** is expressed in a simple equation:

velocity = accleration × time

This equation stated another way is:

distance = ½ accleration × time^2

Abbreviated, it looks like this:

$d=\frac{1}{2}at^2$

You can use this equation to figure out the distance of a falling (or rolling, in Galileo's case) object from its starting point. After one second (or one heartbeat, or one drip from a bottle), the distance is 1 notch; after two seconds, it is 1 notch plus 3 more, totaling 4; after three seconds it is 4 plus 5 more, a total of 9; and so on. The pattern of numbers is—astonishingly—the same as the sequence of squares of integers: 1, 4 (2^2), 9 (3^2), 16 (4^2), etc. The proportions are always the same. I find that amazing.

Now this is a little tricky, so be sure you understand: Free fall (or gravity) is a measure that doesn't consider friction. Friction changes things. So the inclined-plane experiment isn't an exact measure of free fall. (Actually, neither is a fall from a tower; the friction there is air resistance.) But it does reveal a law or pattern of acceleration that doesn't change—it's a constant. That free-fall law is part of a general law with a fancier name—the **Law of Uniformly Accelerated Motion**. The inclined-plane experiment gives a measure of gravity. And knowing how to measure gravity will soon help explain things about the way the universe works.

The LAW OF UNIFORMLY ACCELERATED MOTION says that the distance traveled is in squared proportion to the time taken. If the time is twice as long, the distance is four times as long. If the time is three times as long, the distance is nine times as long. Time gets squared. You'll see that happen a lot in nature. This law is the basis for the science of dynamics, the study of motion and forces. We couldn't design cars, planes, or rockets without it.

Just how does Galileo do his measuring? He needs to measure time to figure out velocity and acceleration. But how can you measure time when there are no reliable clocks? Galileo builds a water clock:

> *For the measurement of time, we employed a large vessel of water placed in an elevated position; to the bottom of this vessel we soldered a pipe of small diameter giving a thin jet of water which we collected in a small glass during the time of each descent; . . . the water*

thus collected was weighed; . . . these weights gave us the differences and ratios of the times, and this with such accuracy that although the operation was repeated many, many times, there was no appreciable discrepancy in the results.

He is measuring time by weighing it. Even more important, he is introducing the practice of accurate measurement by repeating his trials.

Galileo isn't finished with motion. Everyone believes that moving objects have a built-in tendency to stop moving, to come to rest. That's what Aristotle said, and that's what objects seem to do. But Galileo wants to be sure of it. So he does more experiments measuring motion. He finds out that what seems to be true isn't necessarily so.

Galileo slides some blocks on a piece of polished wood. They move. Then he polishes the wood some more and slides the blocks again. They move farther, then they stop. His head starts churning. What makes the wood blocks stop? Can it be

MESSING AROUND WITH MOTION

Do heavy balls roll farther than light balls? Do they start out slower? Don't take my word for it; see for yourself. You can make a simple ramp with a cookie sheet or a grooved ruler, and then roll marbles down it—ones of equal mass, different masses, different sizes, and so on.

If you already tried dropping balls of different mass (see page 75), you know that the balls hit the floor at the same time. That's the **Law of Free Fall.** But what about two objects of different mass that are thrown with different amounts of force? Try it: Line up two coins on the edge of a table. Position a ruler next to them as shown. Then pull back and, holding the ruler at one end, pivot it quickly. One coin will travel farther than the other because the ruler hits it with more force; but listen for the thud of the coins as they hit the floor. The two coins land at the same time because they fall vertically at the same rate (thanks to gravity), but the coin that travels farther travels faster horizontally.

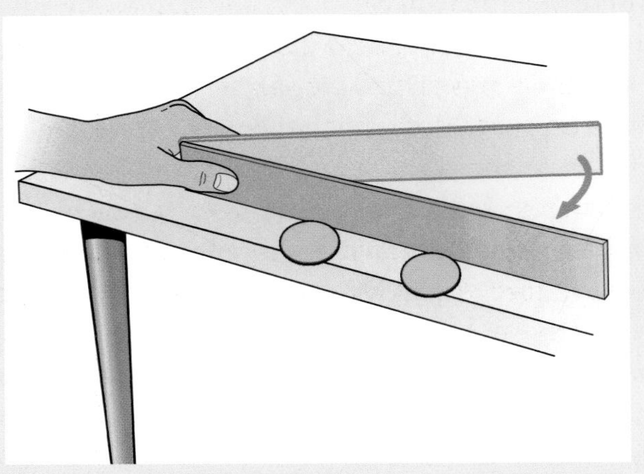

Which coin will hit the ground first?

This nineteenth-century mural by Giuseppe Bezzuoli is a three-ring circus of science. In the center, a kneeling monk uses his pulse to time an object rolling down a ramp. Galileo (standing, in the gown) shows the experiment results to a philosopher, while eager students watch. Skeptical scholars (left) search Aristotle's book for explanations but don't find any. The colorful man seated at right is Giovanni de' Medici, the son of Cosimo I. He's unhappy because Galileo demonstrated that one of the prince's inventions didn't work. Don't overlook the setting—that's the cathedral and Tower of Pisa in the background.

the rubbing of the pieces of wood against each other? What if he could make a track so smooth that there is no friction? Would the blocks keep moving? Galileo thinks so, but he can't prove it. He can't build a frictionless track.

But he is right. It is friction that causes the motion of a block on a straight horizontal plane to stop. If there is no friction, a moving object will continue moving forever. That tendency to keep moving (or to keep still) is called **inertia**. **An object that is moving will continue moving at the same velocity forever; an object that is at rest will stay at rest**—if there is no net force acting on the object. That's the **Law of Inertia**. It's a very important law, but it will take more minds and more time for it to be recognized and accepted. (The idea of inertia will help explain why the Earth stays in motion and why the Moon keeps in orbit. We wouldn't have space travel without it.)

Galileo deals Aristotle a serious blow with this concept. Although, at the time, only a few people understand that.

7 Moving Relatively or Relatively Moving?

There is perhaps nothing in Nature older than MOTION, about which volumes neither few nor small have been written by philosophers; nevertheless, I have discovered by experiment some properties of it which are worth knowing and which have not before been either observed or demonstrated.
—Galileo Galilei (1564–1642), *Dialogues Concerning Two New Sciences*

Pure logical thinking cannot yield us any knowledge of the empirical [experimental] world; all knowledge of reality starts from experience and ends in it.... Because Galileo saw this, and particularly because he drummed it into the scientific world, he is the father of modern physics—indeed, of modern science altogether.
—Albert Einstein (1879–1955), German-born American physicist, *Ideas and Opinions*

Motion. How is that for a subject? Boring? You might think so, but with his special genius, Galileo makes it anything but dull. Understand motion, and you understand how the Earth turns and why we don't feel it happening. Understand motion, and you understand why a bullet doesn't fly straight. So read this and take your mind where Galileo suggests:

> *Shut yourself up with a friend in a large room below decks on a big ship, and bring with you there some flies, butterflies, and other small flying creatures. Have a large bowl of water with some fish in it; also hang up a bottle that empties drop by drop into a wide-necked bottle beneath it. Then, with the ship standing still, observe carefully how those small winged animals* [the flies,

> butterflies, etc.] *fly with equal speed to all parts of the room; how the fish swim indifferently in all directions; how the drops all fall into the bottle placed underneath; and how, in throwing something to your friend, you don't have to throw harder in one direction than another, as long as the distances are equal. Jumping with your feet together, you will jump as far one way as another.*

This is a thought experiment; it is something that good scientists usually do well. Just follow Galileo's directions and imagine yourself in a big cabin below the deck on a ship tied up at the dock. There is no porthole, so you can't see outside. Inside the handsome wood-paneled cabin, fish are swimming in a fishbowl, flies are buzzing about, water is dripping from one bottle to another, and you're throwing a ball and playing hopscotch with a friend—all as you might do if you were at home.

> *When you have observed all of these things carefully (though there is no doubt that when the ship is standing still everything must happen this way), have the ship proceed with any speed you like, so long as the motion is uniform and not varying this way and that. You will discover not the least change in all the things going on in the cabin, nor could you tell from any of them whether the ship was moving or standing still.*

Are you still doing the thought experiment? As Galileo says, the ship is now speeding, but the water is calm, and you can't feel any ship motion.

> *In jumping, you will jump over the same spaces as before; your jumps won't be larger when you head toward the ship's stern* [back of the ship] *than toward the prow* [front, or bow, of the ship], *even though the ship is moving quite rapidly, or even though during the time that you are in the air the floor under you will be going in a direction opposite to your jump.*

> *In throwing something to your friend, you will need no more force to get it to him whether he is in the bow of the ship or the stern.*
>
> *The droplets will fall just as before into the bottle beneath without flying toward the stern of the ship, although while the drops are in the air the ship runs many spans.* [A span is an old unit of measurement.]

Has anything changed for you and the flies and the fish in that room below deck? Do you have to work hard to walk or jump? The answer is no. If you don't believe Galileo or me, get on a train, plane, or ship and you will be able to walk about, jump, or throw a ball—all normally. (None of this works if the vehicle sways or hits bumps, because that causes acceleration. Galileo is talking about smooth, steady motion.)

Read Galileo's passage a few times to be sure you get it. Even better, do what I often do: write it in your own words.

Why is this idea about motion so special? To begin, it answers one of the questions that keep the Aristotelians from accepting Copernicus's idea. They say that if Earth were moving, we would all feel it. We would get blown over; birds would be left behind by moving trees; and if we jumped forward, we would land behind.

Galileo realizes that none of that happens on a moving ship. Why don't we observe motion in a closed room on a ship? He can't explain it—he doesn't know why it is so—but he can describe what his experiments tell him.

His observation—that if you're in a closed vehicle or cabin, no mechanical experiment can tell you if the vehicle is moving or not—comes to be called **The Principle of Relativity**. Here it is in Galileo's own words (well, his words translated from Italian): "**All steady motion is relative and cannot be detected without reference to an outside point.**"

It means that you can't know if you're moving unless you look at something out the window.

Galileo's sentence will set off charges of dynamite in the scientific world.

***Relative, relativity*—what do those words mean? Well, your Aunt Tillie is a RELATIVE, but physicists may not have her in mind when they use the word. For scientists, *relative* means "something dependent on or connected to something else." A dictionary definition of RELATIVITY is "a condition in which the existence or importance of one entity depends on another."**

But is it true? Can you tell you're moving in a car or ship or plane without looking at something through the window? Try it and see for yourself. Keep in mind, if the ride is bumpy, you will know something is going on. Galileo is talking about smooth, steady motion. (I've repeated this because it's important.)

What Galileo does next will change science. It's the thing geniuses do. He connects two seemingly unconnected things—the ship's motion and the Earth's motion. It takes some imagination to do it; no one has done it before.

Galileo knows that we don't feel the effects of motion on a smooth-sailing ship. Could Earth be like a ship in space? Galileo thinks so.

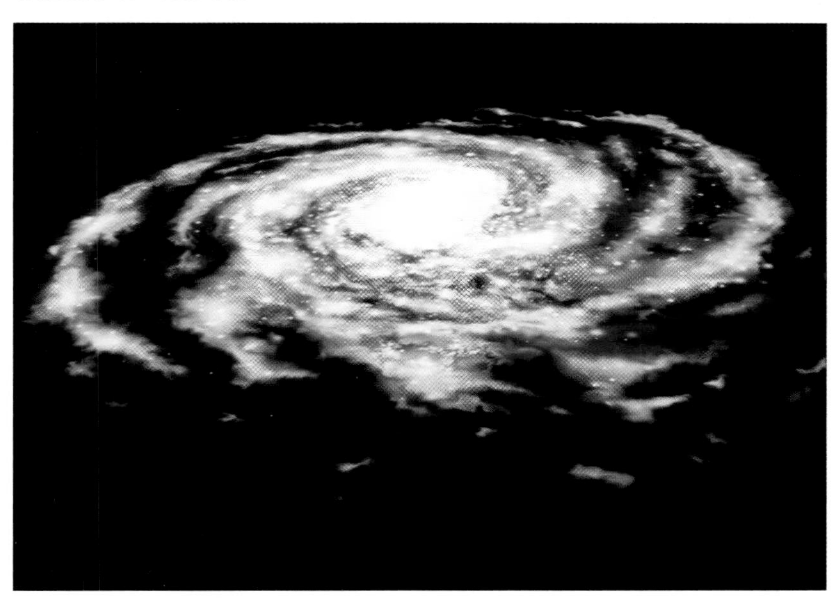

The Milky Way is a spiral galaxy that might look something like this to a creature in another galaxy. We think it has six arms (our galaxy, not the creature) spinning around a center. Our solar system is whirling inside one of the moving arms at the same time Earth is orbiting the Sun and spinning on its axis. Feeling dizzy? You shouldn't be, according to Galileo's thought experiment.

Right now you and the chair you're sitting in are moving—and very fast—on this Earth of ours. How fast? If you stand at the equator, the Earth will be carrying you along as it spins on its axis at about 1,670 kilometers (1,038 miles) per hour. The whole twirling Earth is zooming around the Sun at just under 30 kilometers (18.5 miles) *per second*. Our solar system is orbiting the center of the galaxy at about 217 kilometers (135 miles) *per second*. That's not all. Our Milky Way Galaxy is speeding toward our nearest neighbor, the

The SOLAR SYSTEM is the Sun and all the matter influenced by its gravitational pull—nine planets, their moons, countless asteroids and comets (many of which are beyond Pluto), and dust.

A GALAXY, like our Milky Way, consists of billions of stars orbiting a dense center. We now know that some of these stars have orbiting planets.

WHY DON'T WE FEEL EARTH'S MOTION?

That was a sensible question; lots of people asked it. Here's an experiment that a French philosopher did to try to find an answer.

In 1640 (two years before Galileo died), Pierre Gassendi (1592–1655) got the fastest vehicle of his time—a naval galley. He had it rowed across the Mediterranean while a series of balls was dropped from the top of the mast to the deck. All the balls fell at the foot of the mast. Not a single one got left behind by the motion of the boat. Did that explain why things are not left behind by a moving Earth and why we can't feel Earth move? Gassendi, Galileo, and now some others thought so. But not everyone agreed.

Pierre Gassendi was a skeptic; he tested ideas scientifically before accepting them.

Andromeda Galaxy, at about 130 kilometers (81 miles) per second. The Milky Way and Andromeda are both part of the Local Group of galaxies, which is speeding toward the constellation Hydra at roughly 600 kilometers (373 miles) per second. Have you ever felt any of that motion?

To repeat: "All steady motion is relative and cannot be detected without reference to an outside point." But there *are* outside points; we don't live in a closed box. Why can't we see our motion when we stand here on Earth and look at the sky?

We can. It's Earth's movement that we're "seeing" when we watch the planets, stars, and Sun move across the evening sky. (It's like riding in a car and seeing trees move by outside the window. Think of Earth as a vehicle—which it is—and the stars as trees. The picture is a bit more complicated because stars do move, while trees are rooted.)

Galileo had this to say about the phenomenon:

> *I think that anyone who considers it more reasonable for the whole universe to move in order to let the earth remain Fixed, would be more irrational than one who climbs to the top of a cupola to get a view of the city...and then demands that the whole countryside revolve around him so that he would not have to take the trouble to turn his head.*

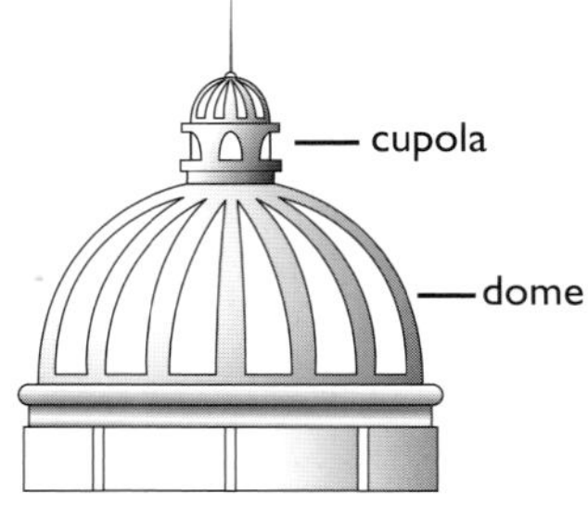

A CUPOLA is a little round room atop a dome or other roof. A church belfry is a cupola. The word comes from the Latin for "little barrel or tub."

A sailor on top of the mast of a ship drops a ball. That's the same ship, same sailor, and same dropped ball in both diagrams. So why does the ball's path look different—one path going straight down (left) and one with an arc (right)? The diagram on the left shows what a sailor on the deck sees. The diagram on the right shows what an observer standing on the shore sees. Two people are watching the same event, and yet seeing it differently. That's Galilean relativity. In this case, the reason for the difference is that the sailor on deck is moving along with the ship, but the observer on the shore is fixed in place.

There's still more to this awesome Principle of Relativity. So catch your breath, relax, and be prepared to stretch your mind with another thought experiment.

Picture a moving ship and a man high on the ship's mast. That man drops a ball down to a sailor on the deck right below him (see left diagram, above). If you're standing on the ship watching, the ball seems to go straight down. But if you're on shore looking at the moving ship and the ball, you see something different (see right diagram, above). To your eyes the ball doesn't go straight down. Because the ship is moving, from the shore you see the ball making an arc. (But for both observers the ball lands at the same distance from the mast.)

Hold on. People on the ship see one thing (the ball going down), and those on land see something else (the ball making an arc)? Who is seeing what really happens?

Both of them. In this case the visual truth is relative. The ball does go down *and* it does make an arc. What you see depends on where you are. Galileo doesn't explain why; he just says this is what happens. And he is right. (Albert Einstein, in the twentieth century, will help with the explaining.)

We're still not finished. There is one more idea to put in your head. (This is a tough chapter; stick with it; the ideas here are important.)

> It has been observed that missiles and projectiles describe a curved path of some sort; however, no one has pointed out the fact that this path is a parabola.... This...I have succeeded in proving....My work is merely the beginning,...other minds more acute than mine will explore its remote corners.
>
> —Galileo Galilei, *Dialogues Concerning Two New Sciences*

According to Aristotle, a cannonball shoots out in a straight line; then, when its motion peters out, it falls vertically to Earth, as shown here. Of course, this only happens in the mind (and Wile E. Coyote cartoons). So why didn't Aristotle (and a thousand years' worth of his followers) test the idea?

Galileo's experiments with motion lead him to understand the flight of balls and bullets and other projectiles. That path of a moving object through space is called its trajectory (truh-JEK-tuh-ree).

The Aristotelians said that horizontal motion (e.g., a pitched ball) is straight. They assumed that when a ball runs out of force, it falls vertically, which is straight down. They thought the path was a random thing: some balls drop one way, others another. It was typical of the Aristotelians that they wrote learned papers on the subject, but they didn't get out and toss balls and try to observe and measure. Galileo does. He finds that all trajectories take the same basic shape.

The first serious use of a cannon was in 1346 at Crécy, France, during the Hundred Years' War. The cannons then were primitive things; it was the English longbow archers who carried the day. That would change.

According to the Law of Inertia, a ball thrown horizontally will move horizontally with a speed that doesn't change, because no horizontal force acts on it.

But the Law of Free Fall (gravity) says that the vertical motion toward Earth is one that gets faster and faster.

Galileo figures out that all trajectories are combinations

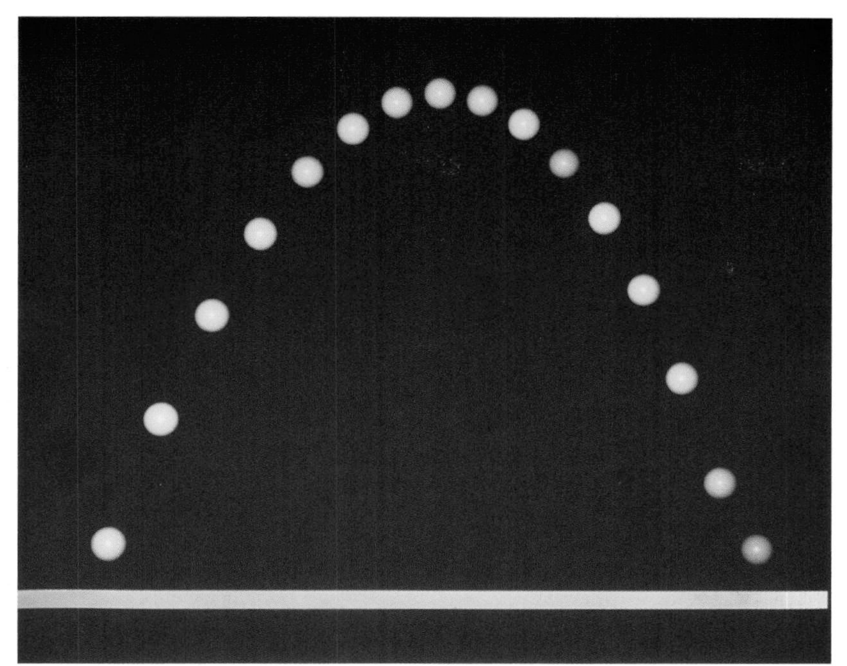

A projectile is any object thrown into the air—like this ball. You too can be a projectile by throwing yourself into the air (in other words, jumping). All projectiles on Earth follow the same predictable curved paths because they're all under the same influence of gravity. They slow down as they rise, reach a maximum height, and then speed up as they fall.

of steady horizontal motion with changing vertical speed, and he shows that the result is a parabola. Since he has worked out the geometry of those laws, he has no problem combining them. He discovers that the path through space of any thrown or fired object is always in the shape of a parabola (see photo above). From the stream that comes out of a drinking fountain to a ballplayer's pitch, each is parabolic. (Air resistance, wind, etc., change things; they can make that not exactly true.)

To repeat: Galileo's research shows him that trajectories have two independent motions. One is horizontal; it carries a constant (unchanging) velocity that is initiated by a propelling force (the explosive in the gun, the power of the

If an astronaut coasting in space pushes a ball off a table, the ball travels in a straight horizontal line at a constant speed (first picture). If you let a ball drop on Earth, without pushing it, gravity pulls it down with increasing speed in a straight vertical line (second picture). The third and fourth pictures show a ball moving steadily forward and at the same time increasing its downward speed due to gravity.

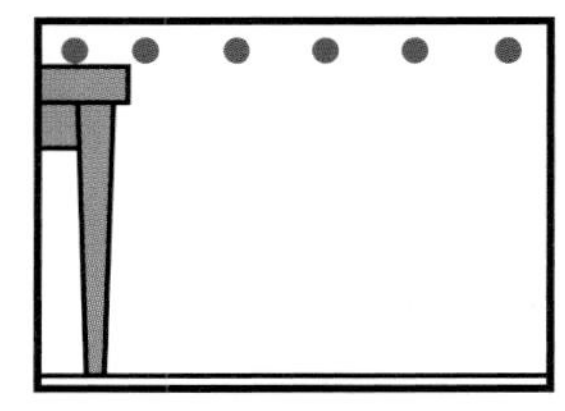

horizontal motion

vertical motion (free fall)

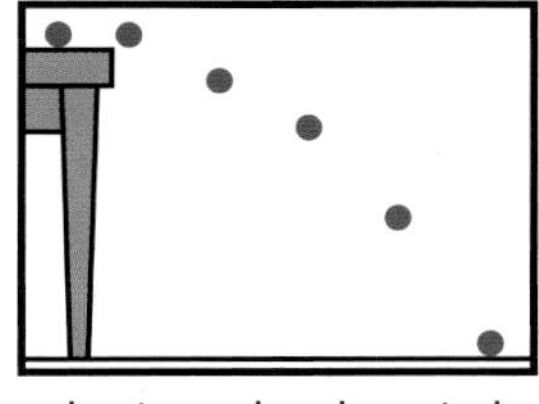

horizontal and vertical motions combined

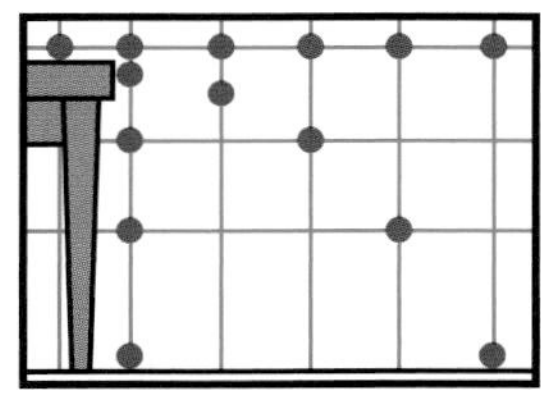

comparison of motions

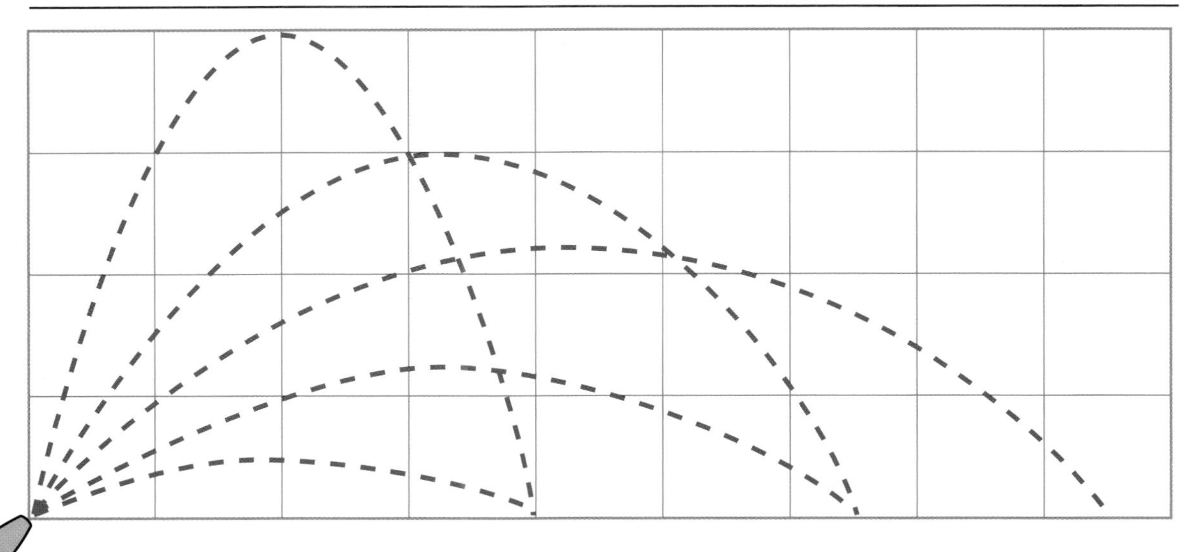

To launch a cannonball (or baseball or shot put) as far as possible, what's the best angle? If the cannon points almost straight up, at a steep angle, the cannonball gets more height than distance. But if the cannon points nearly parallel to the ground, the ball falls short quickly.

pitcher's arm). The other motion is vertical and changes according to Galileo's formula of free fall (the acceleration due to gravity). Combining both paths produces a predictable parabolic curve. (If there is air resistance, as there often is, the path will fall short of a parabola.)

Now that information *is* immediately useful. It makes it possible to figure out where a cannonball will fall when fired with a given velocity and a given angle of elevation. It leads to the science of ballistics. It changes the design and placement of guns and cannons. Those in power pay attention to this scientific breakthrough. Societies almost always respond when there is a military advantage.

As to the Principle of Relativity, it doesn't seem very important in Galileo's time. Hardly anyone makes mention

A ballista was an ancient military machine for hurling stones. The word originated from the Greek verb *ballein*, "to throw." BALLISTICS is the modern science (and geometry!) of projectiles in motion.

Shine a flashlight directly at a wall, and the light forms a circle. Angle the flashlight a little to one side, and the circle (left) stretches into an ellipse (middle). Hold the flashlight parallel to the wall, and you have a parabola (right).

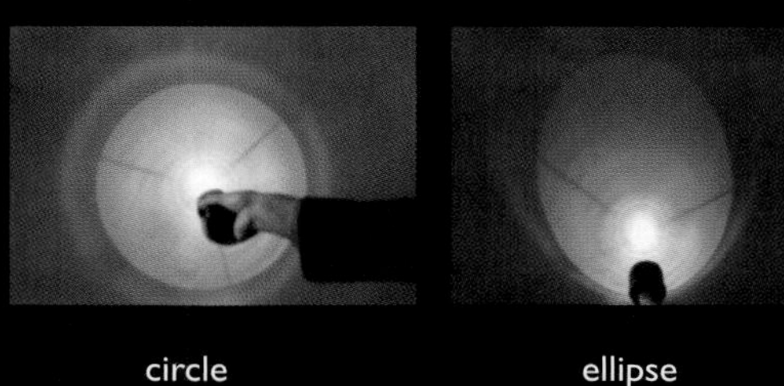

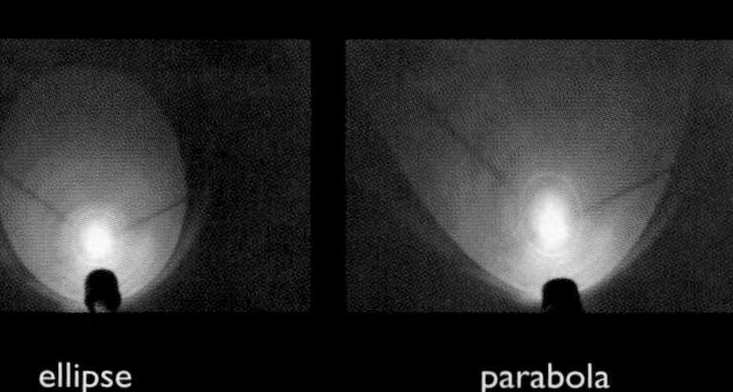

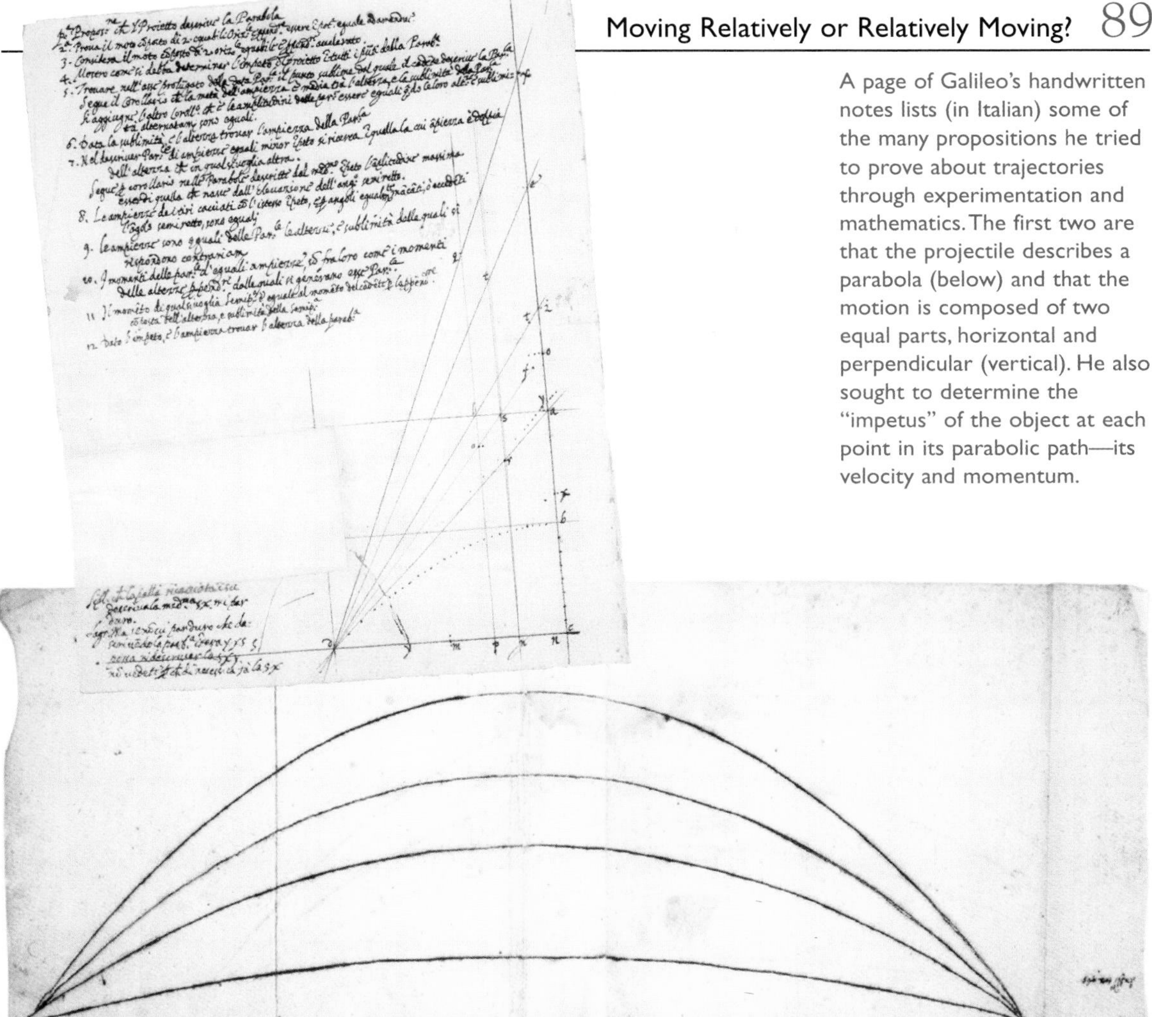

A page of Galileo's handwritten notes lists (in Italian) some of the many propositions he tried to prove about trajectories through experimentation and mathematics. The first two are that the projectile describes a parabola (below) and that the motion is composed of two equal parts, horizontal and perpendicular (vertical). He also sought to determine the "impetus" of the object at each point in its parabolic path—its velocity and momentum.

of it before the twentieth century. Science is like that: You never know when or where ideas will germinate.

Relativity will have to wait a while; when its time comes, it will turn out to be *very* significant. Common sense makes us want to believe that science has absolute, always-the-same rules. Relativity says: Hold on; two viewers (one on ship, one on shore) can experience reality differently. So which experience is real? Just what is reality? Three centuries after Galileo those questions and the search for their answers will help create a new scientific age. An expanded Theory of Relativity will be at its core.

8 Are Novas Really "New" Stars? As to Supernovas—Wow!

A supernova can liberate more energy in one minute than is released by all the normal stars in the observable universe during the same amount of time. Only a fraction of this energy—as little as one one-hundredth of one percent, in some cases—is emitted as visible light, but that is enough for the supernova to outshine the entire galaxy it inhabits.

—Timothy Ferris (1942–), American science writer, *The Whole Shebang*

Twinkle, twinkle, little star,
How I wonder what you are.

—Ann Taylor (1782–1866) and Jane Taylor (1783–1824), British writers, "The Star"

Despite his enemies—who are often jealous professors—the times are right for Galileo's talents. Even the heavens help his cause. Galileo was eight when the nova that Tycho wrote about brightened the sky. Of course he saw it and heard his elders talk of it with wonder. In 1604 another nova disturbs the heavens! This time Galileo is 40 and living in Padua, Italy. The star is eye-blinking brilliant—but no one understands or even guesses that it is a supernova, the last gasp of a huge dying star. They do know that they are

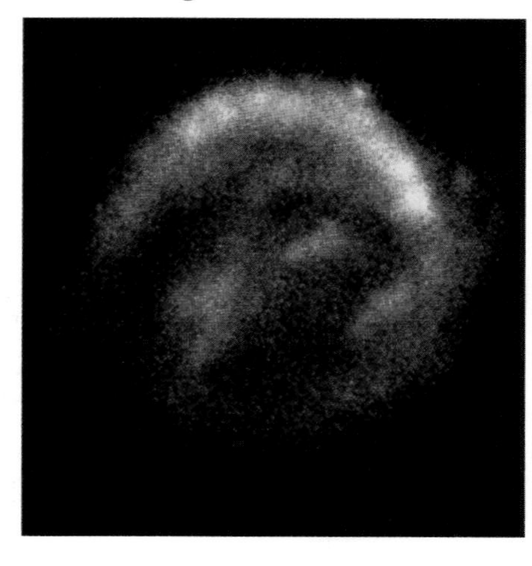

A modern X-ray image shows what's left of the supernova Galileo saw in 1604—a loose cloud of hot gas. Galileo and his contemporaries used the term *nova*, which means "new," to describe what they thought they were seeing—new stars. We now know that novas and supernovas are old, dying stars that explode and thus become bright enough to be newly visible.

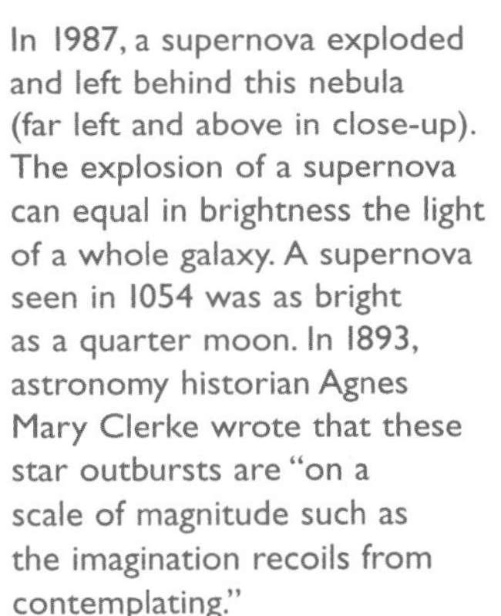

In 1987, a supernova exploded and left behind this nebula (far left and above in close-up). The explosion of a supernova can equal in brightness the light of a whole galaxy. A supernova seen in 1054 was as bright as a quarter moon. In 1893, astronomy historian Agnes Mary Clerke wrote that these star outbursts are "on a scale of magnitude such as the imagination recoils from contemplating."

seeing something momentous. It is a "new star" to them, and it doesn't fit with the idea of an unchanging universe. Tycho's nova made him realize that Aristotle's model of the universe was wrong; this nova turns Galileo into a serious astronomer. Galileo notes that the new star shows no parallax (see pages 58–61). He realizes that means it is very, very far away. He even writes a poem about it:

No lower than the other stars it lies
And does not move in other ways around
Than all fixed stars—nor change in sign or size.
All this is proved on purest reason's ground;
It has no parallax for us on Earth
By reason of the sky's enormous girth.

The heavens haven't finished providing delights for astronomers. In 1607 a brilliant comet makes fireworks in the deep sky. Since there are no electric lights to compete, it's hard to miss any unusual illumination in the inky black heavens. So this comet is not only noticed, it is big news. (Imagine a world without television; might you spend time looking at the skies?)

Think of an ASTRONOMER as the one who practices the science of observing, measuring, and interpreting the light that comes to us from stars, planets, comets, and moons. What we see in the night sky is light that has made the journey to Earth. Actually, there's more to this story than those visible light rays. Modern instruments can help us "see" the invisible part of the electromagnetic spectrum—infrared, X rays, and other forms of light.

Almost 100 years after Galileo, Edmond Halley discovered that comets take long but predictable orbits. In 1705, he wrote that the comet seen by Galileo in 1607 would reappear in 1758. And it did! This was no miracle; it was the work of a person's brain. We now know comets are made of mostly dirty ice. Instead of terror when one appears, we get out our telescopes, we take notes, and some of us even have comet parties.

Petrus Apianus was the first western astronomer to notice that comet tails always point away from the Sun. In his 1540 book, *Astronomicum Caesareum* ("The Emperor's Astronomy"), he describes a comet seen in 1531, later known as Halley's comet. His book included hand-painted adjustable star charts like the one below. The paper disks could be rotated by pulling strings.

The comet is going in the opposite direction from the planets' orbits, and speedily. Pamphlets and broadsides tie the comet to bloodshed, famine, and the death of princes. People quote the popular playwright William Shakespeare, who wrote in his play *Julius Caesar* in 1599, "When beggars die, there are no comets seen; The heavens themselves blaze forth the death of kings." As to the astrologers, they say comets are omens—and not omens of good.

So when the comet etches the sky, most people run to their priests, say prayers, or hide. They don't look for explanations in the laws of the universe, because they don't believe there are any universal laws—at least not any that men and women can comprehend.

But Galileo is different. He wants to understand. "I do not . . . believe that the same God who gave us our senses, our speech, our intellect, would have us put aside the use of these, in order to teach us instead such things that, with their help, we could figure out for ourselves." When he says "we could figure out for ourselves," he's pushing for scientific thinking (rather than looking for explanations in the Bible). Of astronomy he writes, "so little notice is taken [in the Bible] that the

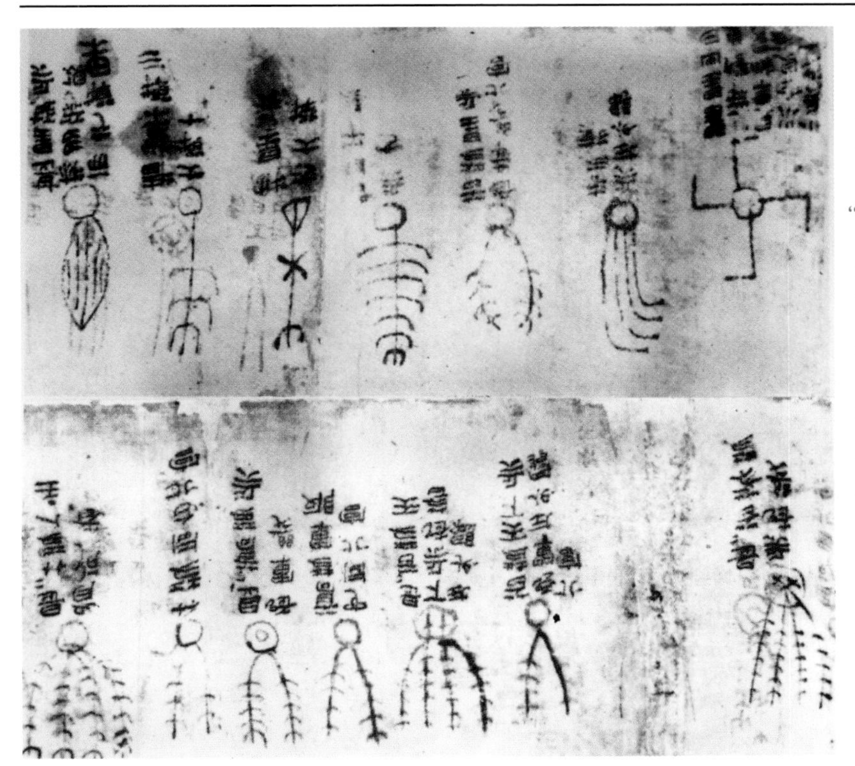

This drawing (left) is from a Chinese atlas of comets dating from the 400s B.C.E. Discovered in northern China in 1973, the atlas predicted wars and crop yields.

"There are many comets that were observed in China, Japan, and Korea from ancient times through most of the seventeenth century that were never reported in Europe," says Gary Kronk, author of *Cometography.* "The strong astrological beliefs of the Chinese prompted them to have teams of astronomers who scrutinized the sky on every clear night, one for each compass direction (north, south, east, and west).... They thought comets foretold the future, so the earlier they spotted them, the better they could prepare themselves. For the same period of time, the Europeans only seem to have observed the very bright comets."

names of none of the planets are mentioned."

Galileo is determined not to put aside his senses or his intellect. The celestial wonders that come at the beginning of the seventeenth century—the supernova (that he called a "nova") and the comet—focus his mind's direction. He studies the heavens with even more fervor than before. He is convinced that it is God's wish that he do so.

But the schools and universities teach that nothing new can appear in the skies. The universe is said to be exactly as it was when it was created—perfect (except for the imperfect Earth). There can be no such thing as change in the perfect heavens. Why would God change perfection? "The movement of that which is divine must be eternal," said Aristotle.

For Galileo, Science Is Measurement

Copernicus, like most of those who came before him, worked in his head. He wasn't much of an experimenter. Galileo was. He thought in terms of number, weight, size, and motion. He measured, tested, and helped establish the scientific method.

Galileo described his approach:

> *No sooner do I form a conception of a material or... substance than I feel the need of conceiving that it has boundaries and shape, that relative to others it is great or small; that it is in this place or that; that it is moving or still; that it touches or does not touch another body; that it is unique, rare, or common.*

Upon seeing a comet in 1607, astrologers predicted the death of King Henry IV of France (left). He didn't buy it. Henry said, "One of these days, their prediction will come true, and the single occasion when the prediction was fulfilled will be better remembered than all of the other occasions when it was not." In 1610 a religious fanatic assassinated the popular king; maybe the astrologers thought their prediction was fulfilled.

What are thinking men and women to believe? Aristotle's astronomy—based on the idea that the universe is unvarying and eternal—works in harmony with church philosophy that is centered on eternal truths. The new astronomy—based on observation and measurement—is showing a changing universe. It challenges what almost everyone has been taught for thousands of years.

This is dangerous—especially to those in authority. If people question Aristotle's thoughts on the stars, might they also question church ideas about God? The power of churches and rulers and teachers are all braided together. Yank one strand, and the other two will feel the pull.

To the Greeks, experimentation seemed irrelevant.... It was Galileo who overthrew the Greek view and effected the revolution. He was a convincing logician and a genius as a publicist. He described his experiments and his point of view so clearly and so dramatically that he won over the European learned community.

—Isaac Asimov (1920–1992), Russian-born American writer, *Asimov's Guide to Science*

ABOUT SUNS AND STUFF (GALILEO DIDN'T KNOW ANY OF THIS)

Our Sun—it's a medium-sized yellow star—was born about 5 billion years ago in a swirl of gas and dust. That cloud of gas and dust was cold, just a few degrees above absolute zero, until matter started to clump together and molecules rubbed against each other and things began heating up. Gravitation pulled more particles together, which led to more heat, creating a glowing protostar ringed by a shell of gas and dust. After 50 million years (no time at all, cosmically) the protostar was hot enough to start nuclear fusion, which means it was millions of

Giants and dwarfs? These fairy-tale names for stars have a familiar ring, but keep in mind that all stars are unimaginably massive. The terms are relative: (1) A neutron star has a larger diameter than the smallest observed black hole, a massive star that's super-compacted. (2) A white dwarf star, which is roughly the size of Earth, has a diameter 700 times that of a neutron star. (3) The Sun's diameter is 100 times that of Earth, yet it looks like a yellow pea next to a beach-ball-size red giant (4).

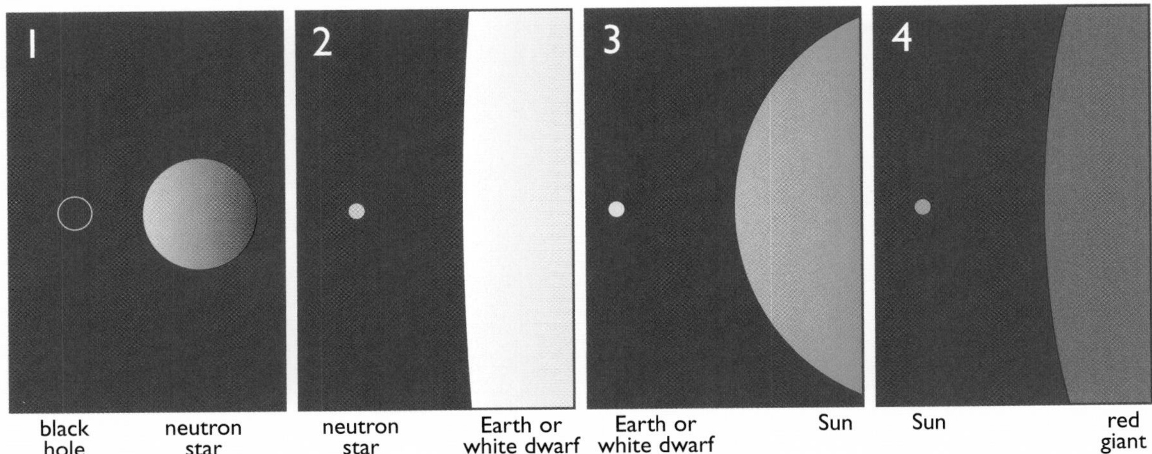

The Life and Death of a Sun-Sized Star

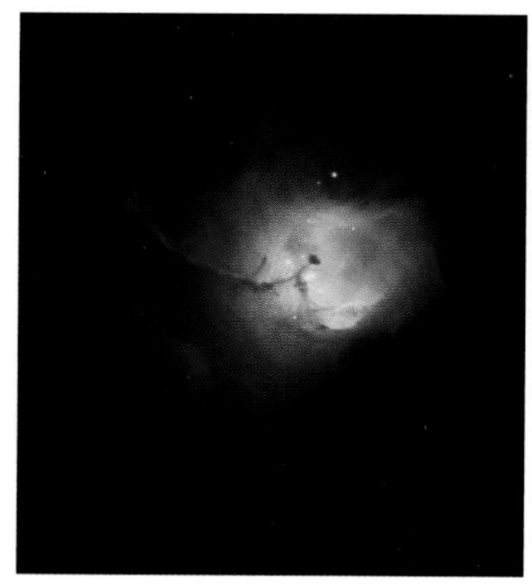

1. A protostar is wrapped in a cocoon of gas and dust. Gravitation causes the particles to condense. Heat and pressure increase, and nuclear reactions begin.

2. A yellow main-sequence star like our Sun gradually converts hydrogen atoms (H) into helium atoms (He) through nuclear fusion.

degrees hot. The nuclei of hydrogen atoms began fusing and converting into helium, giving off energy in the process, which made the Sun shine. (It's still happening.)

Things haven't changed a whole lot since that stellar birth (at least as far as we can see), and if you can stick around for the next 5 billion years, the Sun should still be shining and still fusing hydrogen. But then old age will begin to claim our Sun. The Sun will swell up—maybe to 55 times its present size—and turn from yellow to red as its surface cools. It will give off 400 times more energy, vaporizing Mercury. I wouldn't suggest staying on Earth. The oceans will boil away, and the land will bake.

The Sun—having become a "red giant"—will hang about for another 100 million years or so, burning itself up and shrinking until it turns into a small "white dwarf" (about the size of the Earth)—insignificant in the vastness of the heavens. Helium will be fueling most of this phase, not hydrogen. But the story hasn't ended. That white dwarf will be so compact and so dense that a teaspoon full of it will weigh as much as an elephant (but there won't be any more elephants in this solar system). Our now-aged Sun will continue to shine, losing heat in doing so. And it will keep shrinking. What about us humans? We'll be inhabiting another solar system or maybe another galaxy or another universe. Otherwise we'll be ashes.

3. The hydrogen runs out, and the star swells to a red giant, 10 to 150 times bigger. That's an enhanced image of Betelgeuse (opposite, near left), an even larger red giant, a supergiant that forms the shoulder of the constellation Orion. A supergiant has more luminosity (it shines brighter) and volume than a giant.

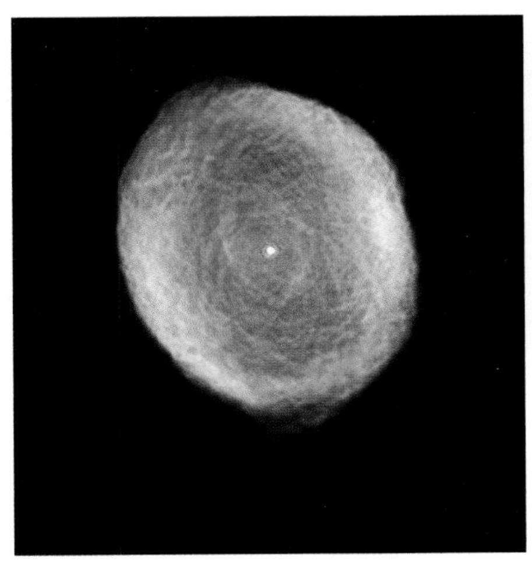

4. A dying stellar core pushes away gas and dust to form a planetary nebula. The Spirograph Nebula (above) once looked like our Sun.

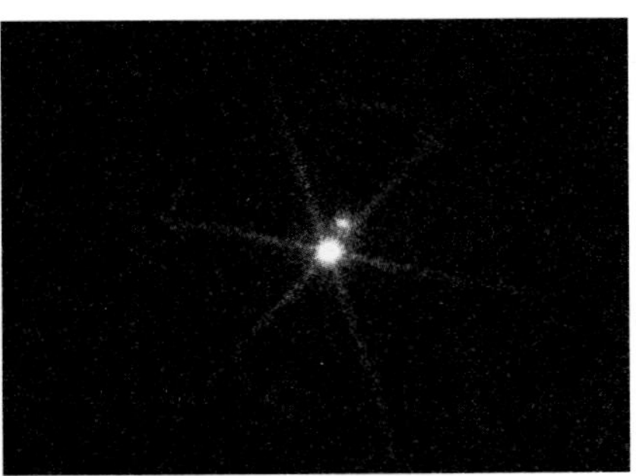

5. When the core finally dies, a white dwarf forms. This star is hot, dim, and dense: at its center, a spoonful weighs as much as an elephant. Sirius B, revealed in an X-ray image (above), has as much mass as the Sun, and yet is a little smaller than Earth. (That dot just above it is Sirius A, the Dog Star.)

How do scientists know what's going to happen to stars (including ours) millions of years from now? The laws of physics can predict how energy and matter will interact; the amount of mass is the key to a star's life cycle. But there's evidence in the skies too. This stunning image of nebula NGC 3603 (right) shows the entire life cycle of stars in one glimpse. Start with the small, dark dust-and-gas clouds in the upper right corner—the Bok globules—and the giant pillar of gas near the bottom. These are early formations of stars being born. The bright blue-and-white blob in the middle is a cluster of young superhot stars. To the upper left of it is an aged star, Sher 25, which is a blue supergiant. It is spewing a ring of pale-colored gas as its life winds down.

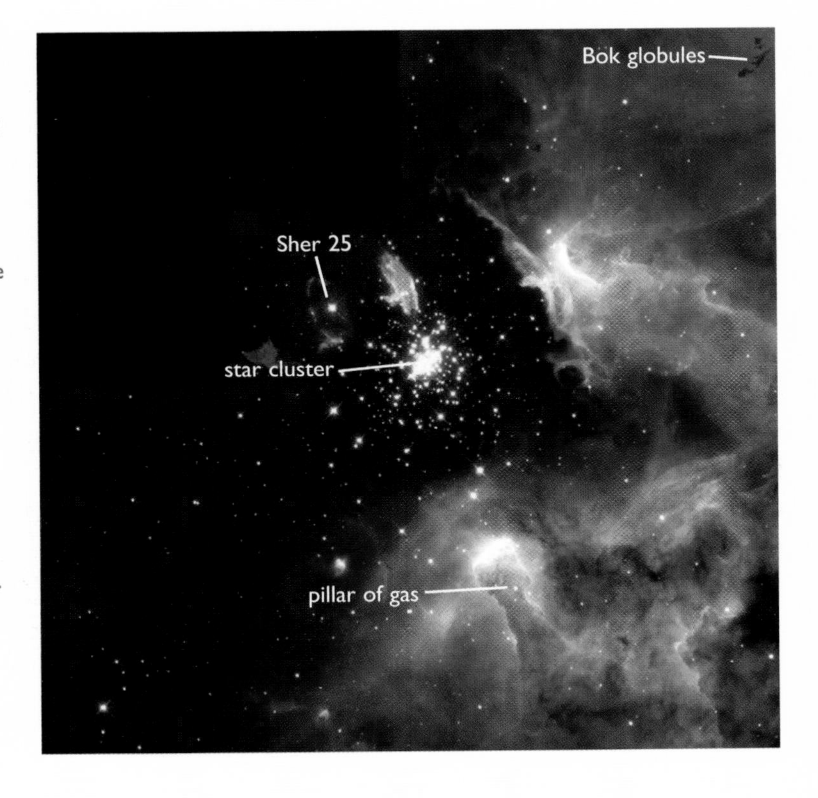

9 Moving the Sun and the Earth

> Who better than Galileo to propound the most stunning reversal in perception ever to have jarred intelligent thought: We are not the center of the universe. The immobility of our world is an illusion. We spin. We speed through space. We circle the sun. We live on a wandering star.
>
> —Dava Sobel (1937–), American author, *Galileo's Daughter*

> To conceive of the moon not as a polished mirror but as a dry, dirty stone without water or atmosphere would be a shock to all humankind but a sacrilege to the church. If the moon had earthlike features, they must exist for human beings, for had not God made the heavens only to please and to benefit man? Was he saying that there were men on the moon? If so, how could they have descended from Adam? Or escaped the Flood?
>
> —James Reston Jr. (1941–), American author, *Galileo: A Life*

Galileo is a bit arrogant, and often impatient with people who aren't as smart as he is. But he is such a good teacher and he has so many fans that he manages to poke fun at the Aristotelians and get away with it—especially after he takes a group of senators to St. Mark's Square, the water-lapped *piazza* with its gorgeous cathedral that is the pride of the city of Venice. Galileo and the senators go up into the Tower of St. Mark's, the highest spot in the city. There, Galileo lets everyone look through a viewing tube he has just built. It is 1609.

A few months earlier, Galileo heard a rumor about a miraculous instrument invented in the Netherlands that made far things seem near. He was told that if you looked through that Dutch spyglass, a man a couple miles away would seem almost at your elbow. Things seen through the tube not only appeared closer but bigger too.

The Dutch are treating this scope as a kind of fun toy or novelty and selling copies madly. The Dutch models (there are more than one) use two concave lenses, which give an upside-down image. They magnify about 3 times.

Working quickly, without ever seeing one of the telescopes, Galileo comes up with something far better. He puts one concave lens and one convex lens in a tube, and the convex lens turns the image upright. He grinds his lenses to exact measurements. His first telescope has a magnifying power of about 5 or 6. He quickly improves that model.

The tube he takes to Venice magnifies 9 or 10 times. The senators are amazed when they look through its eyepiece. They can see ships more than 55 kilometers (about 35 miles) away. Later Galileo writes, "I have made a telescope, a thing for every maritime and terrestrial affair and an undertaking of inestimable worth. One is able to discover

THE TUSCAN ARTIST

In his epic poem *Paradise Lost* (1667), John Milton compares Satan's shield to the giant Moon that Galileo ("the Tuscan artist") saw with his telescope:

...The broad circumference
Hung on his shoulders like the moon, whose orb
Through optic glass the Tuscan artist views
At evening, from the top of Fesolè,
Or in Valdarno, to descry new lands,
Rivers, or mountains, in her spotty globe.

Milton doesn't accept or reject the new astronomy. He pictures God chuckling over the meager efforts of humans to understand:

...he his fabric of the Heavens
Hath left to their disputes—perhaps to move
His laughter at their quaint opinions wide.

The Venetian senate and doge (city ruler) got their first glimpse through Galileo's telescope on a balcony above St. Mark's Square in Venice, Italy. The response, Galileo reported, was "the infinite amazement of all." The setting of this historic moment in science was restaged for a photograph (left).

enemy sails and fleets at a greater distance than customary, so that we can discover him two hours or more before he discovers us, and by distinguishing the number and quality of his vessels judge whether to chase him, fight, or run away."

For a city that depends on its sea trade—and worries about pirates, as Venice does—this invention seems miraculous and very valuable. Galileo is treated like a hero. He is given a lifelong appointment to the University of Padua (today that's called tenure)—and a raise. He needs the money. He has two daughters and a son who depend on him for support, as do other relatives, and he always seems to be broke.

As a young man, Galileo had said that he intended "to win some fame." He now has it. His telescopes are marveled at throughout Europe. (He keeps improving the magnification.) But when he points his telescope at the sky, he is on his way to changing the whole scientific world. He will be first to use a telescope as an aid to astronomy and then write about what he sees.

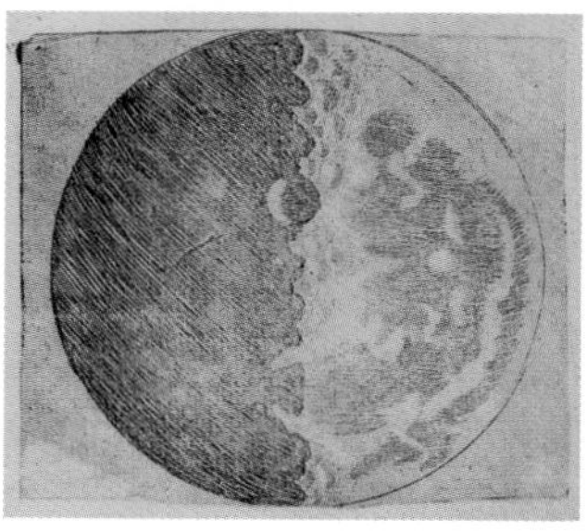

During its first-quarter phase, the Moon appears half lit and half dark (photo, above). Galileo chose that phase to make a sketch for a good reason: In the area where light meets dark, sunlight strikes the surface at a sharp angle, creating shadows that bring out features—mountains, chasms, craters, and maria (dark, smooth "seas" of hardened lava).

To begin, he can see the Moon clearly, and it doesn't look like a smooth disk (as the Aristotelians believed). The Moon has mountains and valleys—Earth-type landforms. Is that possible?

When Galileo looks at Jupiter, he can't believe what his eyes tell him. Jupiter has four "little stars" nearby! Are they stars or planets? Five planets have been known since the days of ancient Babylonia (Mercury, Venus, Mars, Jupiter, Saturn), and now, all at once, Galileo seems to have found four more! He can't explain them with Ptolemy's model or Aristotle's theories.

Galileo keeps a nightly log of what he sees. Three of the "stars" are to the west and one is to the east of Jupiter. Then "only three stars are present." Then all four are in a row. They keep changing their positions. What's going on up there? The only explanation of this behavior that makes sense is that they are orbiting Jupiter. They must be moons. But only Earth is supposed to have a moon. If the planet Jupiter has moons, then maybe Earth with its Moon is a planet. This is brain-shaking.

Galileo is just getting started with his observing. For the next few years he will have an eye to the eyepiece. He sees spots on the Sun—they appear to move across its surface and change size and shape as they do so. Watching those spots leads him to believe that the Sun rotates. But how can the Sun have spots? According to Aristotle, the Sun is perfect. Galileo has much to consider.

When he turns his telescope to the Milky Way, which had seemed just a hazy cloud, he realizes it is made up of stars. As for stars, with the increased power of the telescope, he can see so many that they seemed like grains of sand on a beach. (Actually there are more stars in the sky than sand grains on all of Earth's beaches; it will take a few centuries and huge telescopes for us to realize that.)

Galileo describes what he sees in a slim book called *The Starry Messenger* (written in Latin, as is customary). It is immensely popular in Europe and even has readers in China. "He hath first overthrown all former astronomy. . . and next all astrology," writes Sir Henry Wotton, the English ambassador at Venice. That book turns Galileo into a celebrity.

He intends to widen his audience, so his next books are in Italian, not scholarly Latin. "I wrote in the colloquial tongue because I must have everyone able to read it," he says. He wants ordinary people, many of whom come to his lectures, to know of the new ideas. "I want [them] to see that just as Nature has given to them, as

SIDEREVS
NVNCIVS
MAGNA, LONGEQVE ADMIRABILIA
Spectacula pandens, suspiciendaque proponens
vnicuique, præsertim verò
PHILOSOPHIS, atq; ASTRONOMIS, quæ à
GALILEO GALILEO
PATRITIO FLORENTINO
Patauini Gymnasij Publico Mathematico
PERSPICILLI
[illegible]
[illegible]
QVATVOR PLANETIS
[illegible]
MEDICEA SIDERA
NVNCVPANDOS DECREVIT.
VENETIIS, Apud Thomam Baglionum. M DC X.
[illegible]

Never look directly at the Sun, as Galileo did. Use a pinhole viewer (see page 23) to project an image of the Sun. If your timing is right, you might even see sunspots—tiny black dots.

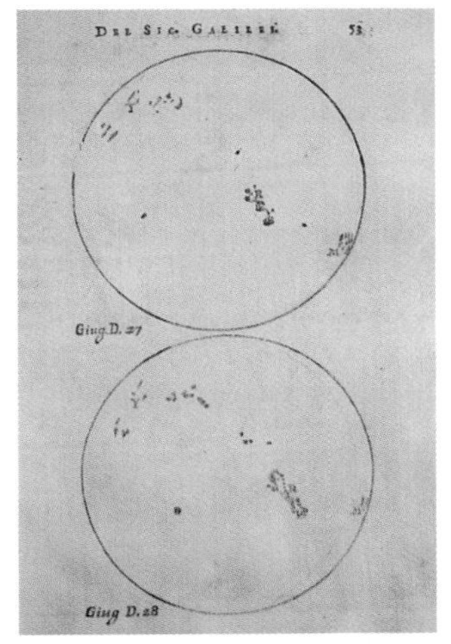

Chinese astronomers identified sunspots in the fourth century B.C.E. In the West, the heavens were supposed to be perfect, so sunspots were thought impossible. As telescope technology raced across Europe in 1610, several scientists observed the dark blotches. That's Galileo's sketch, above. He thought the spots were clouds stuck to the Sun's rotating surface. Other scientists had other ideas. Galileo was right about the rotation, but we now know that sunspots are relatively cool areas that give off less energy (that's why they're dark). The photo of the Sun (above left) was taken in 2003.

Galileo's *The Starry Messenger*, or *Sidereus Nuncius* in Latin (left), describes his first telescopic discoveries, including Jupiter's moons.

JUPITER'S SPACE-TRAVELING COMPANIONS

On January 7, 1610, Galileo aimed his new telescope at Jupiter for the first time. Out of the inky darkness appeared three points of light very, very close to the planet and, days later, a fourth one. Galileo called them stars and kept a nightly vigil. He sketched and recorded his observations (below) and soon reached a stunning new conclusion.

The evidence was overwhelming. He reported that these objects weren't stars at all. Besides being unusually close to Jupiter, the tiny dots were neatly lined up with the planet's equator in an east-west row. They shuffled positions nightly—sometimes hourly. But what really threw Galileo for a loop was that *they never left Jupiter's side*. They traveled with the planet as it slowly orbited the Sun. And they weren't just along for the ride. Galileo was amazed to discover that they were orbiting Jupiter "as Venus and Mercury about the sun."

Galileo called the flighty foursome the "Medicean stars" in honor of the ruling Medici family. These days, they're called the Galilean moons—Ganymede, Callisto, Io, and Europa. Modern telescopes and spacecraft have revealed that the seemingly identical points of light, are, in fact, as different as fire and ice (opposite page).

—LJH

January 7, 1610
[T]wo stars were near [Jupiter] on the east and one on the west....I was not the least concerned with their distances from Jupiter, for...at first I believed them to be fixed stars.

January 8
[O]n the eighth...I found a very different arrangement. For all three little stars were to the west of Jupiter and closer to each other than the previous night....I was aroused by the question of how Jupiter could be to the east of all the said fixed stars when the day before he had been to the west of them.

January 10
[O]n the tenth...[o]nly two stars were near him, both to the east.

January 13
[F]or the first time four little stars were seen by me....They formed a very nearly straight line, but the middle star of the western ones was displaced a little to the north from the straight line.

NOTE: The fourth "star" was Ganymede, Jupiter's largest moon, which had been traveling behind the planet but then emerged into view.

January 15
[T]he four stars...were all to the west and arranged very nearly in a straight line, except that the third one from Jupiter was raised a little bit to the north....They were very brilliant and did not twinkle, as indeed was always the case.

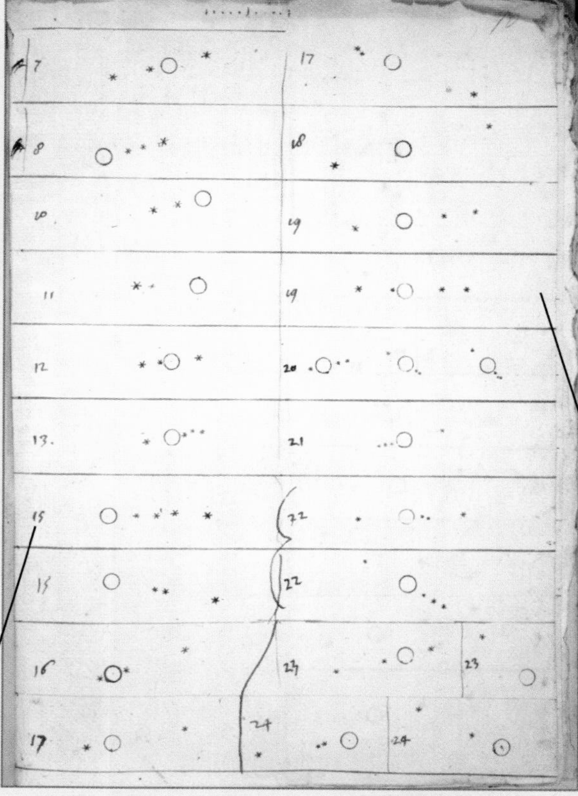

The Galilean moons appear to play musical chairs because they're orbiting Jupiter at different speeds. Like planets, they follow Kepler's laws—closer means faster. Io, the closest moon, zips around the planet in less than two days. Callisto, the farthest moon, takes almost 17 days. Kepler published his first two laws in 1609, the year before Galileo began his Jupiter vigil.

January 19
There were three stars exactly on a straight line through Jupiter....At this time I was uncertain whether between the eastern star and Jupiter there was a little star, very close to Jupiter, so that it almost touched him.

NOTE: Galileo's record isn't perfect. His telescope had a tiny field of view that sometimes missed far-flung Callisto. Also, Jupiter's bright reflection sometimes blotted out little Io, the closest moon. And, of course, on some nights, Earth's clouds hid everything from view.

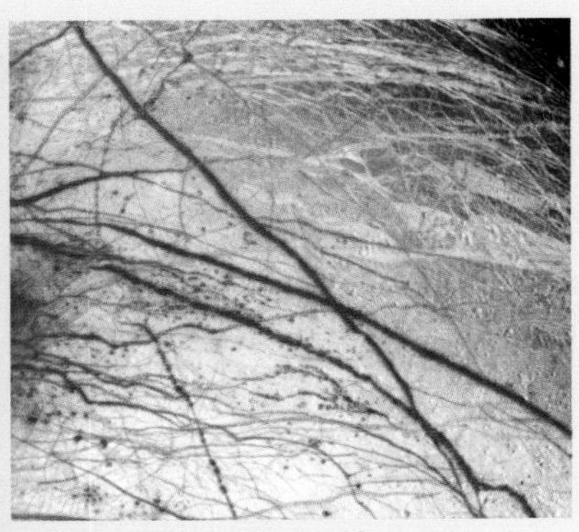

Jupiter (left) has some 60 space-traveling companions, it turns out. Most moons are tiny potato-shaped asteroids snagged by the planet's gravitation while passing by. The Galilean moons (top) are rounder and far bigger than all the others. Ganymede outsizes both Pluto and Mercury. Callisto is the most crater-pocked object we know, with a surface almost as old as the solar system. Little Io has a molten core, the result of being yanked this way and that by the gravity of Jupiter and the larger moons. Volcanic outbursts constantly resurface the young, colorful crust. Io's fire is countered by Europa's ice, shown in a false-color (or computer-colored) image above right. The smooth, frozen crust is streaked with brownish-red minerals that may have risen through cracks. Indirect evidence suggests that a slushy ocean flows below the crust.

An OPPORTUNIST takes advantage of any opportunity or prospect that comes along in order to profit from it. Sometimes profiting means sacrificing principles and ideals, so the word isn't necessarily a compliment.

well as to philosophers, eyes with which to see her works, so she has also given them brains capable of penetrating and understanding them all."

Galileo loves the acclaim and makes the most of it. You could call him an opportunist and not be wrong. Maybe he has to be: those relatives of his keep nagging him for money. His two daughters are nuns; their monastery needs help, and he gives money generously, even though he has little to spare.

This is the philosophical idea of Aristotle (from *On the Heavens*) that Copernicus and Galileo were challenging: "The movement of that which is divine must be eternal.... Heaven [is] a divine body, and for that reason to it is given the circular body [the Sun] whose nature it is to move always in a circle. Why, then, is not the whole body of the heaven of the same character as that part? Because there must be something at rest at the centre of the revolving body.... Earth then has to exist; for it is earth which is at rest at the center."

Galileo sets out to impress the Grand Duke of Tuscany, Cosimo II de' Medici. When Cosimo offers him a position as his "philosopher and chief mathematician," Galileo is quick to take it. The pay is good, and he will have freedom to write and study and experiment. Besides, court positions are more prestigious than university chairs. It means he will move from Venice to Florence (the chief city in Tuscany).

That is a bad move politically. The Venetians, who have just given him that lifetime appointment, are miffed. And Tuscany isn't as open-minded a place as the Venetian state.

Galileo doesn't worry. He is at the height of his powers—consulted in fields that range from military engineering to astronomy. He has lots of friends, no tolerance for stupidity, and stubborn self-confidence. So he concentrates on science and doesn't bother about his enemies.

The wealthy Medici, a family of bankers, ruled the old Republic of Florence during the Renaissance. Under their patronage, the arts and literature flourished. By Galileo's prime (the early seventeenth century), they had assumed royal titles and authority. In 1609, when Cosimo de' Medici was a pupil of Galileo, Grand Duke Ferdinand I died and his son, not yet 19, was suddenly enthroned as His Serene Highness the Grand Duke Cosimo II, sovereign of all Tuscany. A year later, Galileo was appointed his philosopher and mathematician. The painting at left depicts Cosimo II (he died at age 30) with his wife Maria Magdalena and his son Ferdinand II (shown here in later life).

Each time Galileo looks through his telescope, he makes notes and calculations and thinks hard about what he is seeing. It isn't long before he is sure that Copernicus *was* right—the Earth does revolve around the Sun. That means Aristotle was wrong. Galileo says that if

This is Florence, the influential and beautiful center of the Renaissance, in a painting by Giorgio Vasari (1511–1574). Many of Italy's geniuses lived here: the writers Dante, Petrarch, and Machiavelli; the artists Michelangelo, Botticelli, and da Vinci; and the architect Brunelleschi, who designed the magnificent Duomo cathedral (just left of center).

Aristotle could look through his telescope, he'd change his ideas. Aristotle *can't* look through a telescope—he is long gone—and the Aristotelians (the followers of Aristotle's ideas) *won't* look.

"My dear Kepler," Galileo writes to the German astronomer who was Tycho's partner, "what would you say of the learned here, who afflicted with the stubbornness of the mule, steadfastly have refused to cast a glance through the telescope? What shall we make of all this? Shall we laugh, or shall we cry?"

It is enough to make anyone cry. The scholarly Aristotelians aren't about to look through a new invention or change their minds. Neither are the church leaders—and they have authority. Slowly, they begin to think about using their power. To accept Galileo's ideas would disrupt everything they know and teach.

Galileo is undaunted. Quoting a Catholic cardinal of an earlier generation, he says, "the Bible shows the way to go to heaven, not the way the heavens go."

Galileo didn't have it quite right. The actual quote is: "The purpose of scripture is to teach how one goes to heaven, not how heaven goes." It was spoken by Caesar Baronius (1538–1607), a church historian, librarian of the Vatican, and Catholic cardinal.

WHO DID INVENT THE TELESCOPE?

The word TELESCOPE comes from two Greek words: *tēle-*, meaning "far," and *skópos*, meaning "seeing." How many English words derived from *tēle-* do you know? Hint: What about the big box that you often look at? How about other *-scope* words?

Leonard Digges, who is said to have invented the telescope in the mid-1500s, never got a chance to tell people of his discovery. Digges took part in a plot to overthrow England's Queen Mary. The plot was discovered, and all Digges's property was taken from him; he was lucky to stay alive. Later, his son Thomas Digges (a well-known scientist) wrote this: "My father... was able, and sundry times hath, by proportional glasses duely situated in convenient angles, not only discovered things far off... but also seven miles off declared what hath been done at that instant."

It's Hans Lippershey (LIP-er-shee) who usually gets credit for the invention. He was a Dutch lens grinder who made eyeglasses. One of his apprentices, playing with two lenses, put one in front of the other and noticed that faraway things seemed to come closer. He showed his boss, who mounted two lenses in a tube and, in 1608, applied to the

Galileo's telescope was made of wood, paper, and copper. It was 1.3 meters long (a little more than 4 feet). It had a micrometer, a tool for measuring angular distances—how far apart objects appear to be in the sky.

On March 13, 1610, Sir Henry Wotton, the English ambassador to Venice, wrote to King James of England with big news of a major technological breakthrough:

> *Now touching the occurents of the present, I send herewith unto His Majesty the strangest piece of news (as I may justly call it) that he hath ever yet received from any part of the world; which is the annexed book (come abroad this very day) of the Mathematical Professor at Padua, who by the help of an optical instrument (which both enlargeth and approximateth the object) invented first in Flanders and bettered by himself, hath discovered four new planets rolling about the sphere of Jupiter, besides many other unknown fixed stars.*

Dutch government for a patent on a "certain instrument for seeing far." Within two weeks, two other patent applications were filed for similar devices. (The grapevine was active in the Netherlands.) The States-General (the Dutch ruling body) decided it was so easy to duplicate the "instrument" that there was no point in awarding a patent.

Galileo heard a rumor about the device, used his knowledge of mathematics and optics to make a cutting-edge model, and was soon studying the stars. Here are Galileo's words on the making of his telescope, from *The Starry Messenger*:

> *About ten months ago a report reached my ears that a Dutchman had constructed a spyglass, by the aid of which visible objects, although at a great distance from the eye of the observer, were seen distinctly as if near. . . . A few days after, I received confirmation of the report in a letter written from Paris . . . which caused me to apply myself wholeheartedly to consider the means by which I might invent a similar instrument. This I did a little while after, through deep study of the theory of refraction. First I prepared a tube of lead, at the ends of which I fitted two glass lenses, both plane on one side, but on the other side, one was spherically convex, and the other concave. Then placing my eye to the concave lens I saw objects satisfactorily large and near, for they appeared*

TWO KINDS OF TELESCOPES

One of Galileo's biggest problems was field of vision—how wide a view he had while looking through the tube. The greater the magnification, the smaller the field of vision. Galileo could take in only about a quarter of the Moon's surface at a time with his telescope.

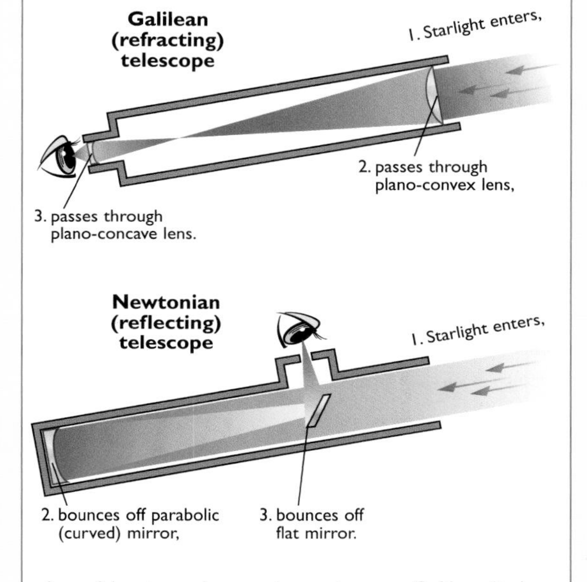

Isaac Newton, who was born the year Galileo died (1642), improved telescope technology by using curved mirrors instead of lenses.

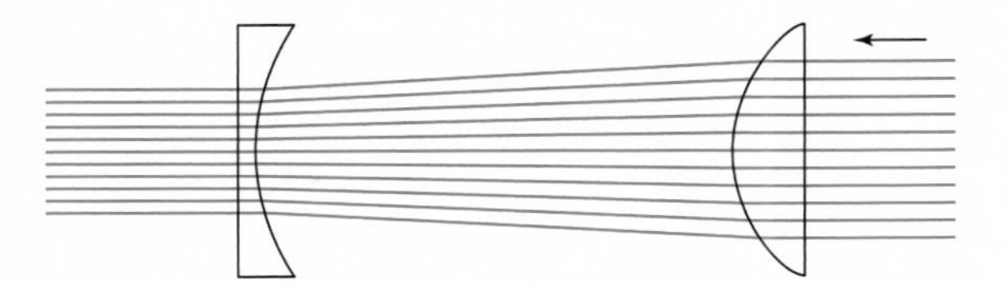

Here's what happens to light rays from a distant object that pass through a Galilean telescope. The rays are moving from right to left in this diagram. They start out parallel to each other in a wide beam (right). The plano-convex lens (on the right) bends them so that the beam is narrower and the rays are no longer parallel. The plano-concave lens (left), which is near the eyepiece, straightens the rays so that they are parallel again, but the beam stays narrow.

> *three times closer and nine times larger than when they are seen with the natural eye alone.*

Later, both Lippershey and Galileo found that when they reversed the lenses, they had a microscope.

The Andromeda Galaxy is a mind-boggling 2.2 million light years away (roughly). Yet you can see it with your own two eyes, unaided by lenses or mirrors. It's that smudge in the upper right corner of the image above. That smudge becomes a glowing whirlpool of white (right) through a high-power telescope. The two white fuzz balls are smaller galaxies.

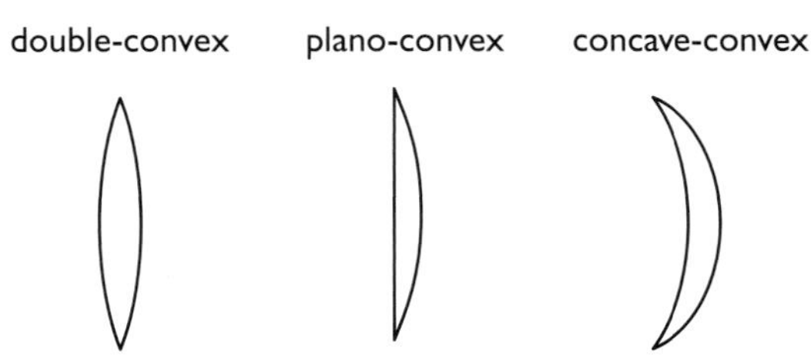

Convex lenses curve outward on one or both sides. Light rays pass through the lens, bend, and converge (meet) at a focal point. An image appears upside down after the rays pass through that focal point. A straight side is called a plane, so a plano-convex lens (middle) has one straight side and one outwardly curved side. In *The Starry Messenger*, Galileo wrote that both his lenses were "plane on one side."

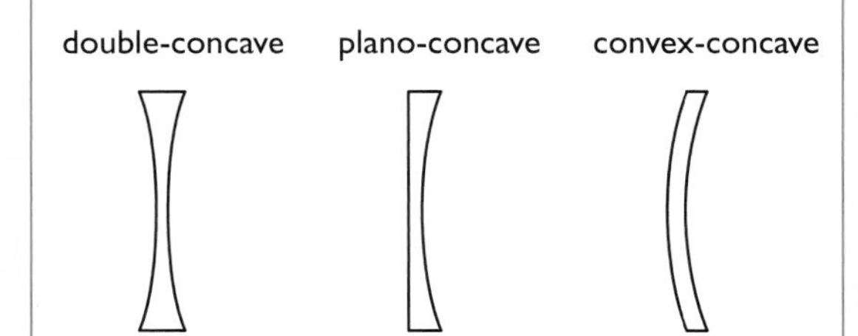

Concave lenses curve inward on one or both sides. They diverge (spread apart) light rays. (Concave lenses are also called diverging lenses.)

TODAY'S TELESCOPES

The first telescopes collected visible light, but that's just a sliver of the energy that comes to us from space objects (see page 369). NASA's Chandra X-ray telescope (above) orbits Earth, where it has an atmosphere-free view of the universe. It's tuned to "see" X-ray energy, like these glowing auroras at Jupiter's north and south poles. X-ray telescopes have revealed an extreme universe with explosive stellar gases and energetic newborn stars.

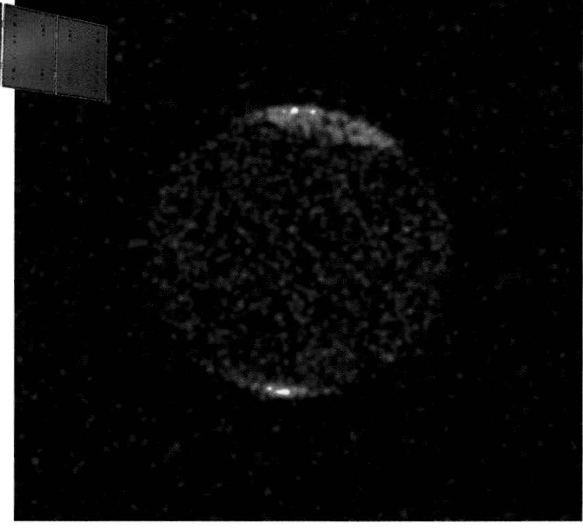

Fast-spinning stars called pulsars and other space objects give off radio waves, another form of electromagnetic (EM) energy. Towering, dish-shaped antennae, like this Very Large Array (VLA) near Socorro, New Mexico, capture the waves. You'd feel like an ant standing next to one. The giant dishes (27 in all) are spread over an area bigger than most cities, but they play well together. They pool their collected data to act like one huge radio telescope. To see how it works, watch the 1997 movie *Contact*. The VLA costarred with Jodie Foster.

Energy patterns (light rays, X rays, radio waves, and so on) can be recorded as computer data. Computers turn the patterns into graphs and maps that reveal interesting information. This map of Mercury (right) is based on radio energy patterns collected by the VLA. Mercury is the closest planet to the Sun, and the colors stand for temperature ranges just below its surface. (Radio waves, including radar, can penetrate dense objects, just as X rays zip through skin to reveal bones.) The red-hot spot is facing the Sun, and the freezing-cold blue areas are in the shadows.

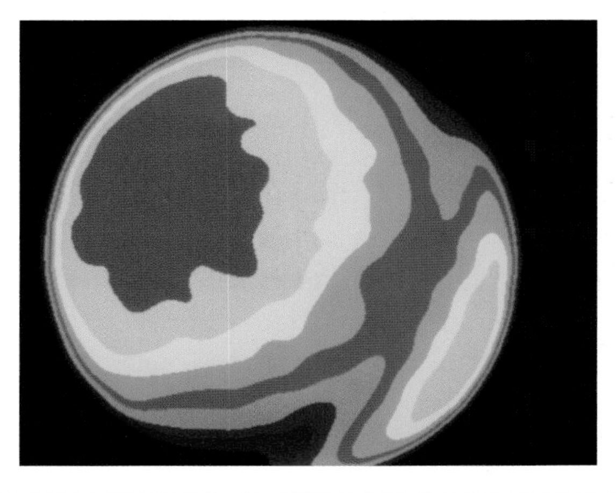

10

Do You Think You Have Troubles?

Let me speak first of the surface of the Moon.... I feel sure that the surface of the Moon is not perfectly smooth, free from inequalities and exactly spherical, as a large school of philosophers considers with regard to the Moon and the other heavenly bodies, but that, on the contrary, it is full of inequalities, uneven, full of hollows and protuberances, just like the surface of the Earth itself, which is varied everywhere by lofty mountains and deep valleys.

—Galileo Galilei, *The Starry Messenger*

The history of science as a whole is the record of a select group of men and women who have dared to be wrong.... A thousand pages of adverse criticism have been written about them by men who were themselves incapable of being wrong because they would never think of exposing themselves to criticism, let alone failure.

—Lloyd A. Brown (1907–1966), American historian and cartographer, *The Story of Maps*

This 1510 painting is packed with Christian symbols. The hand of God holds the universe (not literally, Galileo said, but metaphorically). The wounded Jesus Christ stands under a lamb. St. Catherine, under the Bible, holds a lily and wears a crown of thorns. The eye of God hovers above all.

Galileo doesn't help himself with the church authorities when he writes a letter to a friend saying that the Bible, though certainly true, is not meant to be taken literally when, for example, it talks of the "hand of God," or the "tent of heaven." Those phrases are metaphors, he says, meant to convey ideas. He even argues that scientific fact can be used to understand and interpret the Bible.

Why doesn't he just keep his ideas to himself,

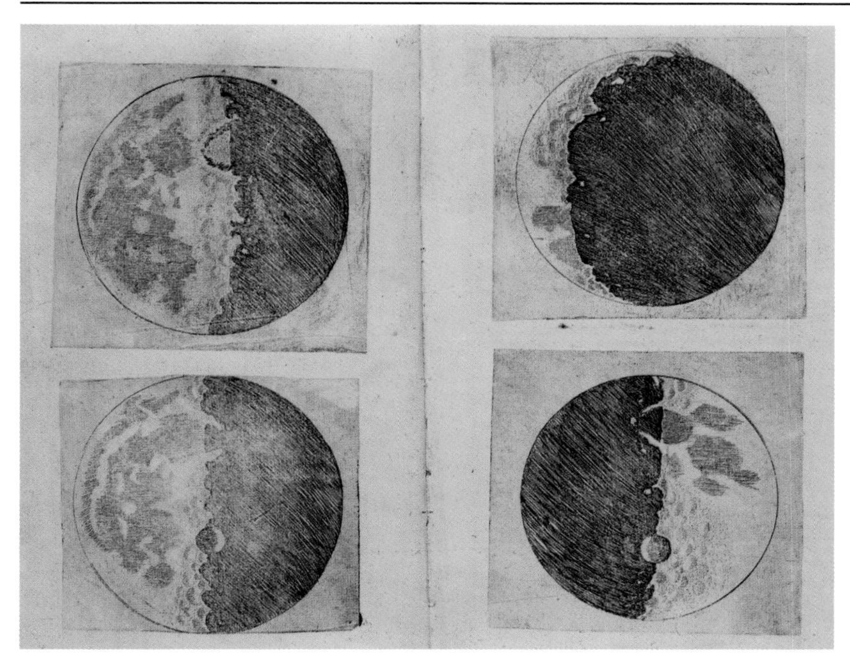

Anyone can see that the Moon looks blotchy, not smooth. Galileo sketched every blemish he observed during various phases (left). People who believed in heavenly perfection claimed that those dark spots were light tricks, not deformities. The surface was uniform, they insisted, but light reflected differently off some parts due to varying colors. Galileo's evidence to the contrary included the unmistakable fact that the mountains cast shadows—which he used to compare their heights.

the church leaders ask? But Galileo can't seem to keep quiet, even though he knows speaking out may put him in danger. What makes things worse (to the leaders) is that because Galileo now writes in Italian, ordinary people are reading what he has to say. The educated elite write in Latin, which keeps learning a scholarly thing. Galileo is making knowledge democratic—available to all.

Earth moves on its axis, and it also moves around the Sun. That's what Galileo writes. Should he keep quiet? When he describes the sunspots he sees through his telescope, a critic writes, "He defiles the dwelling place of the angels by seeing spots on Sun and Moon, and lessens our hope of Heaven." Some people are really upset.

But then he writes a book full of wit and ideas that seem sound to his critics. In *The Assayer* (*Il Saggiatore*), he describes nature in terms of mathematics and atoms—and he does it with beautiful language and images. Galileo is a superb

Anyone who doubts the inevitable triumph of a correct idea, forcibly stated and passionately defended, need only look to the case of Galileo.

—Rocky Kolb, *Blind Watchers of the Sky*

The cover of Galileo's *Dialogue Concerning the Two Chief World Systems* (1641 edition, left) shows an imaginary conversation between Aristotle, Ptolemy, and Copernicus. Posing in the book as Simplicio, Sagredo, and Salviati (which also speaks as Galileo's alter ego), they discuss Earth's motion, the organization of the planets, and the tides of the sea.

writer. The book is widely read. He talks to the pope, Urban VIII, about his idea for a book on the new ideas in astronomy. In order for it to be published, the pope says the book must also present the systems of Ptolemy and Aristotle. What Galileo does, in *Dialogue Concerning the Two Chief World Systems* (1632), is take three characters—Sagredo, Salviati, and Simplicio—and have them debate the issue of the place of the Sun and Earth. Simplicio speaks for Aristotle. (That name of his—*Simplicio*—gives you a clue as to what Galileo thinks of his ideas.) It's a good book but not much of a debate. Galileo is sure his sound, logical arguments will convince his enemies. Instead, he enrages them. He also makes it clear beyond doubt that he is a Copernican. And he gets Pope Urban angry, too. (The pope mistakenly thinks he is being called Simplicio.)

But since Galileo turned his telescope to the sky, he has been convinced that Copernicus was right. One of the arguments against Copernicus is that since the Moon orbits the Earth, the Earth can't possibly orbit the Sun at the same

We now remember Pope Urban VIII (right) for his participation in the case of heresy against Galileo. Yet he was a learned man who first supported Galileo's theories against church officials stuck on the Aristotelian view. When Galileo criticized Aristotle's ideas in the voice of Simplicio, the pope could no longer defend Galileo.

ABOUT POWER, FAITH, AND CENSORSHIP

It's easy to fault the Catholic Church for its condemnation of Galileo, but new ideas often don't fare well. Back in ancient Greece, Aristarchus came up with the idea of a Sun-centered universe. That seemed crazy. Just look at the sky, said his critics, and you can see the Sun go around the Earth. Besides, Zeus and the other Greek gods were supposed to be in charge of the heavens. Aristarchus's ideas were seen as a threat by Greek authorities.

Fast-forward to Renaissance Europe. The church fathers felt responsible for the faith of Catholics, so they banned books they thought were dangerous or immoral. They made an official list in 1559, called the *Index of Prohibited Books*. It was a sin to read or even own any book on the list. In 1616 (when Paul V was pope), Copernicus's book was put on the Index.

As to Galileo, when it became clear that he had embraced Copernicanism, his critics used that against him. Urban VIII (who became pope in 1623) said, "May God forgive Mr. Galilei for meddling with these subjects.... We are dealing with new doctrines in relation to Holy Scripture. The best course is to follow the common opinion." In a book titled *Galileo: A Life*, James Reston Jr. says, "[The pope] saw it as an issue not of truth or falsity, but of obedience and disobedience."

When Galileo was charged with heresy (in 1633), he was nearly 70, a devout Catholic, and he remembered Bruno's fate. Galileo recanted: "Because I have been enjoined, by this Holy Office, altogether to abandon the false opinion which maintains that the sun is the center and immovable ... I abjure, curse, and detest the said errors and heresies, and generally every other error and sect contrary to the said Holy Church; and I swear that I will never more in future say, or assert anything, verbally or in writing, which may give rise to a similar suspicion of me." Galileo's writings were put on the *Index*. They stayed there until 1835.

Finally, in 1966, publication of the *Index* ceased. Then in 1979, Pope John Paul II decided it was time to set up a commission and revisit the Galileo issue. "The Church's experience, during the Galileo affair and after it, has led to a more mature attitude and a more accurate grasp of the authority proper to her," he said. Four years later, John Paul said this: "I hope that theologians, scientists, and historians, animated by a spirit of sincere collaboration, will more deeply examine Galileo's case, and by recognizing the wrongs... will dispel the mistrust that this affair still raises in many minds, against a fruitful harmony between science and faith." On October 31, 1992, 350 years after Galileo's death, Pope John Paul II formally acknowledged that the church had made an error regarding Galileo.

Is this just an issue from the past, interesting only as a historical curiosity? Perhaps not. Evolution, cloning, and stem-cell research are three of today's ideas that sometimes divide science and religion. Should they? And how about censorship? Is it ever justified?

Accused of heresy, Galileo recanted his ideas before the Inquisition in 1633.

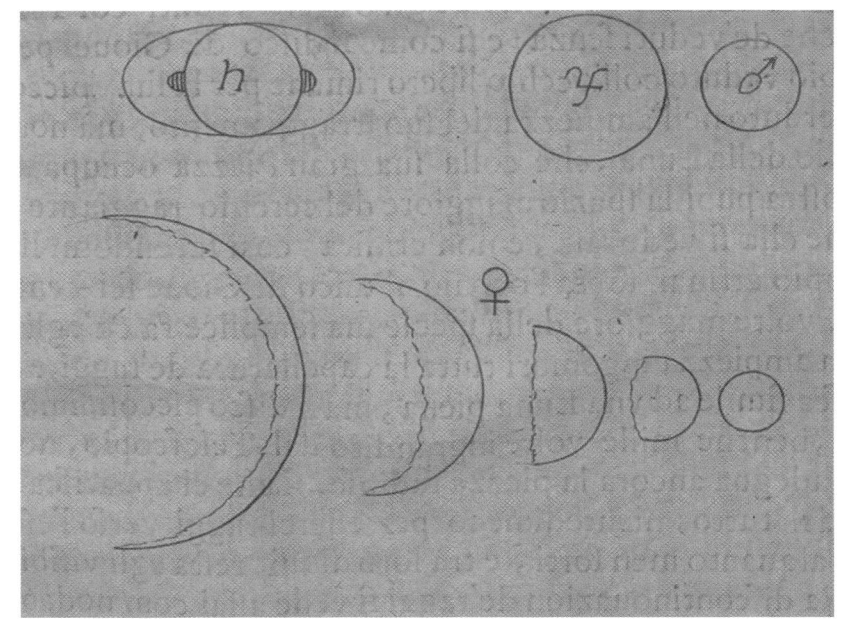

In 1610 Galileo's telescope revealed that just after Venus zipped past Earth in its orbit, it appeared as a thin sickle or crescent (first drawing in second row). Then, as the faster-traveling Venus moved ahead of and away from our planet, the sickle filled in. Venus became a full-phased circle (last drawing) at the point farthest from Earth, directly opposite the Sun. Galileo argued that this pattern of phases could happen only if Venus was orbiting the Sun, not the Earth.

time; Earth and Moon would get separated from each other (or so almost everyone believes).

When Galileo discovered four moons in orbit around Jupiter, he knew it knocked out that argument. If Jupiter can move and bring along four moons, it should be possible for the Earth to do something similar with its one moon.

Benedetto Castelli, a Benedictine monk and gifted scientist, studied with Galileo in Padua and became extremely devoted to him. Throughout his life, he assisted and supported his former teacher. When Galileo was accused of heresy, Castelli defended him, speaking both as a religious official and scientific ally.

Galileo gets a letter from a former student, Benedetto Castelli, who says that if Copernicus was right and the planets do circle the Sun, then Venus must show phases (like the Moon's phases). Galileo studies Venus with his telescope, and there they are—phases! Castelli has a hypothesis; Galileo tests it through observation and proves it correct. This is the scientific method in action! And it offers more evidence that Copernicus had it right.

But none of that matters to the church leaders, including Pope Urban. It took the Catholic Church 73 years to ban Copernicus's

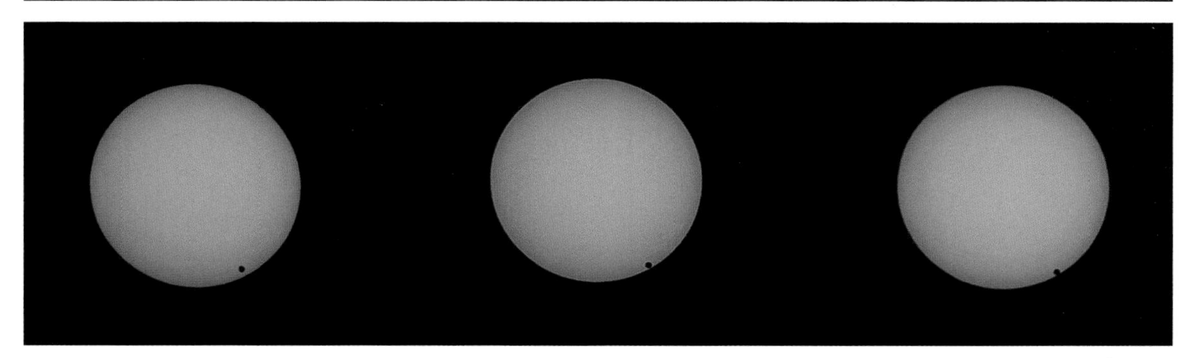

In June 2004, Venus made headlines by crossing in front of the Sun. (The planet is that tiny black dot in the photo above.) Look for a second showing in June 2012, but the next pair of transits won't be until 2117 and 2125. Nineteenth-century astronomers relied on transits of Venus to measure the distance between the Earth and Sun (using trigonometry).

book *On the Revolutions of the Heavenly Spheres*. Less than a year after the publication of Galileo's *Dialogue*, he is summoned to the Inquisition, and in 1633 he is accused of heresy. For someone who is sincerely devout, as Galileo is, this is personally awful. At a time when the Catholic Church dominates the European world, it is also serious—very serious. The church takes action. Read a biography of Galileo to find out what happens. It's one of history's big stories, and it doesn't reach its conclusion until many centuries later. But don't just blame the Catholic Church; it is more complicated than that. Change is wrenching—and Galileo is challenging the scientific and religious thinking of his time.

The big commotion that the Catholic Church made over Galileo and his ideas was hard on Galileo but, in the long run, good for science. Before, everyone had just accepted what the church had to say about science. Afterward, everyone wanted to know more. What was going on? Why all the fuss? Suddenly lots of people had opinions about science.

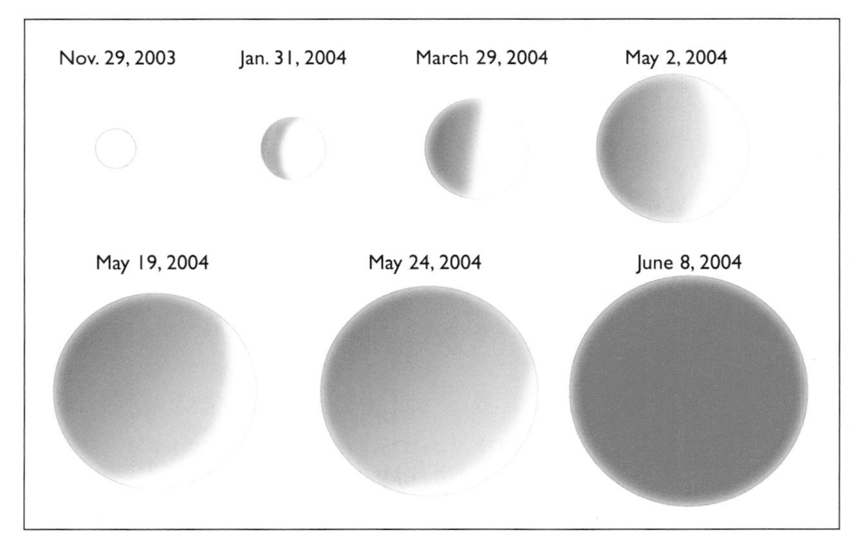

Though Venus has phases, like the Moon, these modern images and Galileo's sketches (opposite page) show one big difference: Unlike the Moon, Venus appears to grow larger as it passes from full phase to crescent phase. That's because it's getting closer to Earth. At full phase, Venus is on the opposite side of the Sun—as far away as it can be.

Getting Along by Staying Apart: Religion and Science

Religion is the search for truth; science is the search for understanding. The difference is subtle, but it's important. In earliest times, there was no separation between the two. Priests and rulers, who were the keepers of knowledge, controlled both faith and scientific learning. Europe's Holy Roman Emperor was both a political and a religious leader; so was England's king.

By Galileo's day, science was getting into high gear. Kings and prelates could no longer keep up with scientific thinking. But it isn't easy to give up power, and control of science was a powerful thing. That created a serious problem, because science can only exist in a free atmosphere. Scientists need to consider all ideas, even those that seem heretical. So controls don't work. Religious authorities weren't used to that kind of intellectual freedom. That was the heart of the conflict over Galileo.

In 1644, the English poet John Milton protested Parliament's decision to require government approval of all published books. His pamphlet, the *Areopagitica* (ah-ree-oh-puh-JIT-i-kuh), is an impassioned defense of freedom of speech. In it, Milton wrote of an incredible meeting (imagined here by the nineteenth-century painter Annibale Gatti): "There it was that I found and visited the famous Galileo, grown old, a prisoner to the Inquisition, for thinking in astronomy otherwise than the Franciscan and Dominican licensers thought."

Like Galileo, most scientists of his time were religious. Johannes Kepler said, "There is nothing I want to find out and long to know with greater urgency than this. Can I find God, whom I can almost grasp with my own hands in looking at the universe, also in myself?" And devout Isaac Newton said, "This most beautiful system of the sun, planets, and comets could only proceed from the counsel and dominion of an intelligent and powerful Being." For them, there was no conflict between religion and science.

Not so for many others. After more than a century of religious wars, Thomas Jefferson saw the need for a law that would keep religion, science, and politics apart. He intended to create a society where free thought would flourish, a society where no church could control the mind's thoughts. That idea was first put into a legal document near the end of the eighteenth century, when Jefferson wrote the *Virginia Statute for Religious Freedom*. Separation of church and state is an American-born concept. Later, James Madison added that separation idea to the United States Constitution in the First Amendment. It says, "Congress shall make no law respecting an establishment of religion, or prohibiting the free exercise thereof."

At right are some other thoughts on the issue. You can find many more if you search a bit.

"Truth is compared in Scripture to a streaming fountain; if her waters flow not in a perpetual progression, they sicken into a muddy pool of conformity and tradition."

—John Milton, (1608–1674), English poet, *Areopagitica*

"Truth is great and will prevail if left to herself."

—Thomas Jefferson (1743–1826), American president, *Virginia Statute for Religious Freedom*

"The religion that is afraid of science dishonors God and commits suicide."

—Ralph Waldo Emerson (1803–1882), American essayist and poet, *Journal*

Faith is a fine invention
When gentlemen can see.
But microscopes are prudent
In an emergency.

—Emily Dickinson (1830–1886), American poet, *Second Series*

"Science without religion is lame, religion without science is blind."

—Albert Einstein (1879–1955), German-born American physicist, *Out of My Later Years*

"Science gives man knowledge, which is power; religion gives man wisdom, which is control."

—Martin Luther King Jr. (1929–1968), American Civil Rights leader, *Strength to Love*

"Profound harmony... can exist between the truths of science and the truths of faith."

—Pope John Paul II (1920–2005), 1979 Address at the Einstein Session of the Pontifical Academy of Sciences

11 Poor Kepler

[Tycho] Brahe had many superb assistants. The most brilliant of all was a strange, mystical mathematician-astronomer named Johannes Kepler...who would discern simple and profound truths in Brahe's mountain of observational data....He was among the first to promote the idea that a force was needed to make sense of the solar system.

—Leon Lederman (1922–), American Nobel Prize winner in physics, *The God Particle*

As regards movement, the sun is the first cause of movement of the planets and the first mover of the universe.

—Johannes Kepler (1571–1630), German astronomer, *Epitome of Copernican Astronomy*

Do you think you have a hard life? Just listen to this:

> *At age four I nearly died of smallpox.... My hands were badly crippled.... During the age of 14–15, I suffered continually from skin ailments, severe sores, scabs, putrid wounds on my feet which wouldn't heal.... On the middle finger of my right hand I had a worm, on the left hand a huge sore.... When 16, I nearly died of a fever.... When 19, I suffered terribly from headaches and disturbances of my limbs.... I continually suffered from the mange and the dry disease.*

That is Johannes Kepler telling his story. Kepler is born in Germany in 1571 to the daughter of an innkeeper. Later he will describe his mother as "small, thin, swarthy, gossiping, and quarrelsome, of a bad disposition." (There may be another side to this tale. When he is six, his mother takes him outside and points out the "great comet." That helps trigger Kepler's interest in astronomy.) Kepler will call his

Sometimes you will see Kepler's name as Johann, sometimes Johannes. Which is right? In Germany, where he was born, he was Johann. But scientists used to communicate in Latin. In Latin, he was Johannes—that's the name he seemed to have preferred, so it's the one I decided to use.

father "an immoral, rough and quarrelsome soldier" who is "vicious, inflexible."

That father, who is hardly ever around, deserts the family permanently when Johannes is 17.

Does the boy run to his grandparents for help and love? Not likely. His grandfather is "arrogant . . . short-tempered and obstinate." (Granddad, a fur merchant, is mayor of their little village.) His grandmother is "lying . . . of a fiery nature . . . an inveterate troublemaker, jealous, extreme in her hatreds, violent, and a bearer of grudges."

But Kepler does have some luck: The local duke sends the brightest sons of his poor subjects to the university. Johannes Kepler is very bright. He has more luck when he studies with a rare professor who believes in Copernicus and his ideas. Kepler never doubts Copernicus.

Johannes Kepler, shown here in a painting by Hans von Aachen, would appear to be a serious man. At a party, his idea of fun was to find a mathematical method for measuring how full the wine barrels were. The party, believe it or not, was for his second wedding.

It's his personal life that seems to go from awful to more awful. At age 26 he writes a description of himself using the third person: "His appearance is that of a little lap dog. . . . He hates many people exceedingly and they avoid him. . . . He has a dog-like horror of baths."

When he tries to teach mathematics at a seminary (religious school), he mumbles, so his students make fun of him and stop coming to his classes. Too bad for them. They are missing out on one of the best scientific minds of all time. That becomes clear when Kepler publishes his first book; it places astronomy in a mathematical web. Kepler sends a copy of his book to the most renowned astronomer in all of Europe—Tycho Brahe.

Tycho is impressed. He invites young Kepler to visit him at his castle in Prague and later hires him as his assistant. (Which is doubly fortunate for Kepler, who otherwise might have been arrested for being Protestant in Catholic Austria.)

There is no way Kepler can study the stars on his own. Because of a childhood illness, he has poor eyesight. Kepler aches to get his hands on all of Tycho's calculations; he knows he can do something with them. But Tycho is

ARROGANT describes one who has obnoxious self-importance, especially where it isn't appropriate. OBSTINATE? It means stubborn, headstrong, or pigheaded. Mules can be obstinate; so could Kepler's grandfather. As to INVETERATE, it means long-established and deep-rooted. (Unfortunately, some people are *inveterate* liars; they keep lying, again and again.)

It was no secret that Emperor Rudolf II suffered from depression, fits of rage, and paranoia (unfounded fear of people plotting against him). In this sixteenth-century painting by Lucas van Valckenborch, Rudolf appears to be seeking a cure, but his mental state grew worse, not better. In 1611, he was forced to sign over his kingdom to his younger brother, Matthias. The monkey on the attendant's shoulder symbolizes vice, folly, and vanity.

unwilling to hand over his life's work to someone he hardly knows. He doles out his findings a bit at a time.

Things don't go well, and then, just a year after Kepler's arrival, Tycho becomes very ill. He is delirious and keeps saying, "Let me not have lived in vain." In a moment of lucidity, before witnesses, Tycho turns his papers and charts over to Kepler.

Kepler uses them well, but not as Tycho intended. Tycho expected him to show that Copernicus was wrong and that his Tychonian Earth-centered system is right. But Kepler keeps an open mind. He will study those charts and measurements for years and years before coming to conclusions.

After Tycho dies in 1601, Kepler, who is just 30, succeeds him as Imperial Mathematician to the Holy Roman Empire—but his personal troubles aren't over. Emperor Rudolf II, who has his own problems (he will be deposed in 1611), often forgets to give his mathematician/astronomer a salary. That isn't fair, but it doesn't stop Kepler's mind from working.

Kepler is fascinated with light and wants to understand how it behaves and how we see with it. That leads him to consider the eye. The ancients, especially Ptolemy, said that vision comes when rays from inside the eye shoot out, hitting objects and bouncing back. They had it backward. We see because light rays bounce off objects *into* our eyes.

By Kepler's time, eyeglasses have been in use for a few hundred years, but no one knows how or why they work. Kepler, who wears specs, explains that in the normal eye, light rays meet at a single point on the retina, where they compose upside-down images (which the brain inverts). But in some eyes, the light rays come to a point in front of or

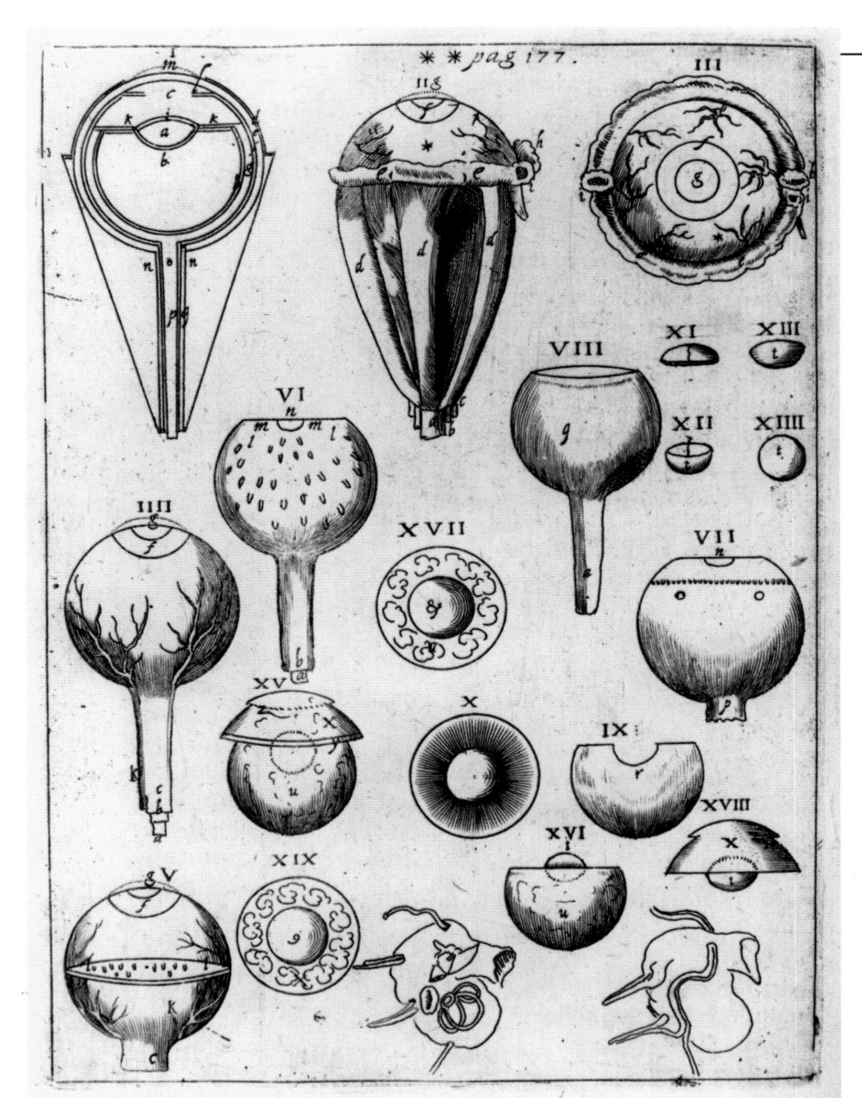

Since the curved lenses of eyeglasses bring images in focus, Kepler wondered if the lens of the human eye worked the same way. He made these scientific drawings of eyeballs (left) for his 1604 book on optics. Kepler's study revealed that the lens projects an image on the retina (the back wall of the eyeball) and that the image is blurry if the focal point falls short or long (below).

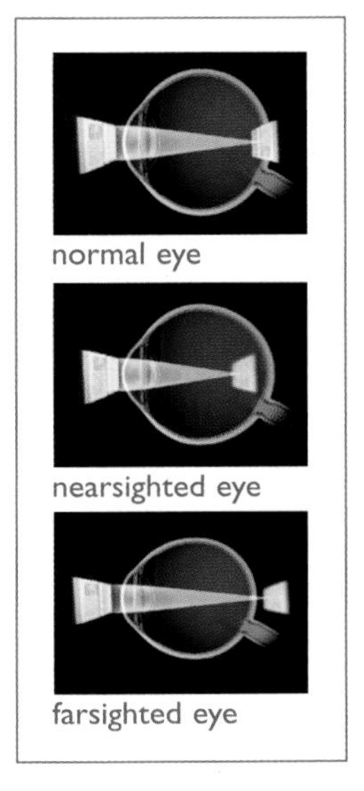

behind the retina—and vision is fuzzy. (We call it nearsighted or farsighted.) The curved lens of an eyeglass can correct that problem.

Kepler writes a book (in 1604) that discusses lenses, optical instruments, the light from the Sun, and eyesight. It is the start of the modern science of optics.

Then he hears about Galileo's telescope and writes to the Tuscan astronomer asking for one, "so that at last I too can enjoy, like yourself, the spectacle of the skies." When Kepler finally gets a telescope, he says it is "more precious than any

THE MYSTERY OF THE SIX-ARMED STARS

It's amazing how something as simple as a snowflake can set a scientist wondering. Kepler couldn't get snowflakes out of his head: "Our question is, why snowflakes . . . fall with six corners and with six rods, tufted like feathers. . . . For if it happens by chance, why do they not fall just as well with five corners or with seven?"

Kepler had a patron, a rich supporter, who was interested in philosophical problems and language too. So he shared his musings over snowflakes. "I am well aware how fond you are of Nothing," Kepler wrote, knowing his patron would chuckle over his pun on *Nothing*, because snow is "white nothing" in poetic German.

But how *does* water vapor transform itself into beautiful, white six-armed crystals? Why not five arms? Experience with science told Kepler that in searching for an answer he was likely to head down wrong paths. He wrote, "I will have my say, on the chance of coaxing the truth from the comparison of many false trails."

Kepler was right. He did take false avenues. In part, it was because he didn't know about molecules, and so he didn't look for an answer in the makeup of water. After observing that water vapor is invisible and water droplets are round, not six-sided, he concluded, "The cause was not to be looked for in the material, but in an agent."

He searched for that "agent"—an external cause—by examining other six-sided shapes in nature, especially the honeycombs in beehives. He observed that each hexagonal cell in the honeycomb shares a "party-wall" (common barrier) with multiple others—an efficient way to store honey without using a lot of wax. Kepler wrote, "The bee, therefore, by nature has this instinct . . . to build in this shape rather than others."

But snowflakes don't store honey and don't have instincts. There had to be a physical reason for their six-sidedness. What was it? Kepler thought about cold—an absence of heat. Cold made water vapor condense and freeze. But, he wrote, if anything, "it is more plausible that condensation should assume a quite flat shape" or that round water droplets should stay round after freezing. To him, even four-sided shapes, like cubes, made better sense because their diameters cross at neat, right angles.

And, of course, Kepler examined the structure of snowflakes—lots and lots of them—in endless, baffling six-armed variations. He deduced that whatever created the six-cornered structure, it began in the center but grew outward from the corners in the form of "rods." But he still couldn't answer Why? And so

A SCEPTER (SEP-ter) is usually reserved for kings and queens. It's a wand or staff that is a symbol of their regal power.

scepter." As to Galileo's *The Starry Messenger*, that book inspires Kepler to write three papers: One describes the way a telescope works; the other two confirm and support Galileo's discoveries. Galileo writes to Kepler, "I thank you because you were the first one, and practically the only one . . . to have

Wilson "Snowflake" Bentley, born in 1865 in Vermont, was the first to use photography to show the unique beauty of snowflakes. Through much trial and error, he perfected a way to capture snowflakes on black velvet and to position a camera nearby without melting them. At left are several of Bentley's photographs mounted together. A modern camera captured more detail of a six-armed crystal (opposite page). Why six arms? Water molecules freeze in a hexagonal pattern, as in the illustration below, bottom.

he finally said it was their nature to be so, and knew it wasn't a good answer. "I have not yet got to the bottom of this," he admitted.

No one got to the bottom of it for some 300 years. With X-ray machines and an understanding of atomic structure, early twentieth-century scientists finally solved the mystery. Here's part of it: All snowflakes are made of water molecules—namely H_2O, a trio of atoms (top right). Frozen together, a lattice framework of three-atom molecules can form only hexagons (bottom right). The hexagons have six corners, and each corner has a hydrogen atom sticking out. The hydrogen atom tends to hook up with an oxygen atom of another water molecule, which explains why snow crystals grow six arms, one at each corner.

Why does one snowflake differ from every other? Kepler was on the right track when he thought about cold. At very cold temperatures, ice crystals have sharp tips and snowflakes form faster—into those feathery, six-armed stars. At warmer temperatures, slow-formed snowflakes are smoother-edged and simpler—hexagonal columns, needles, or plates.

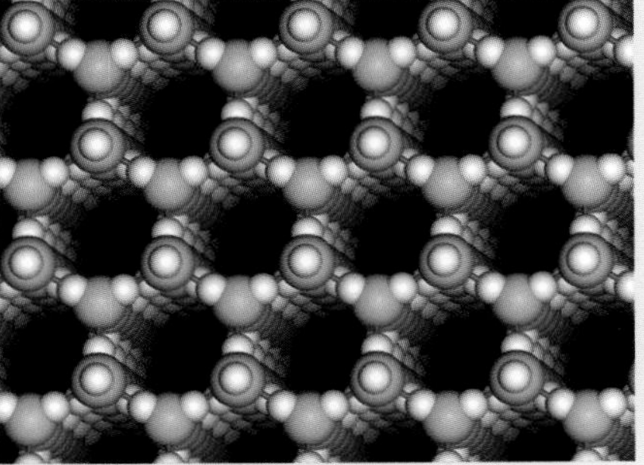

complete faith in my assertions."

Crossing a stone bridge in Prague on a winter day, Kepler notices the snowflakes covering his jacket. They are all six-cornered and perfect. "I do not think that even in snowflakes these ordered figures occur by chance." That thought leads him

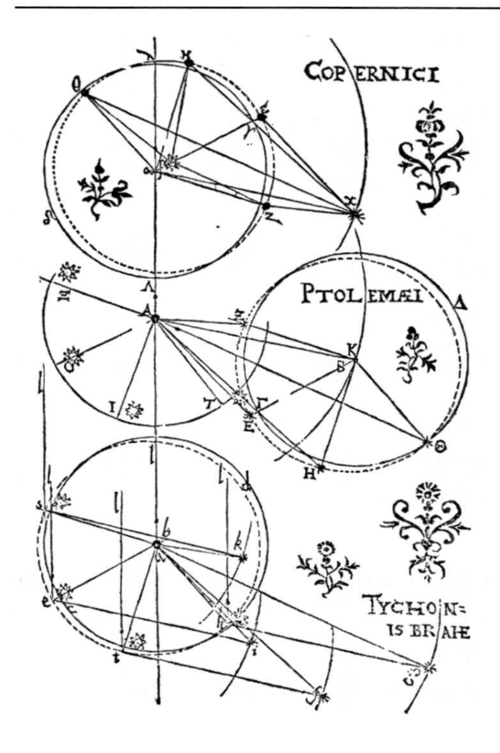

Copernicus, Ptolemy, and Tycho Brahe pictured different ways to explain the irregular orbits of planets. Ptolemy's Earth-centered scheme required epicycles—orbits within orbits (middle drawing). Kepler used the geometry of triangles to show that, no matter whose system you used, the orbits must be ellipses, not circles. This drawing is from his 1609 book *Astronomia Nova* ("New Astronomy").

to write a book on the geometry of snowflakes (published in 1611).

Meanwhile, he keeps studying Tycho's maps of the skies. Those charts convince Kepler that Copernicus was right (even if Tycho never understood that). Kepler writes books that go further than Copernicus. He not only sees Earth and the other planets revolving around the Sun, but he believes that a force from the Sun accounts for the planets' motion. The farther a planet is from the Sun, the weaker the force. That's a far-out thought. Some traditionalists say that angels push the planets. Most are sticking with those moving transparent spheres.

But if Aristotle's heavenly spheres don't exist, something is needed to replace them. What is keeping the stars and planets in their orbits? Could the Sun hold the answer to that question? Copernicus's Sun is passive and stationary. Kepler believes the Sun has some kind of power (he calls it "vigor"), but what is it? He isn't sure, but he is heading in a productive direction.

In 1957, a German musician and composer, Paul Hindemith, wrote an opera called *The Harmony of the World*, based on the life of Johannes Kepler. Hindemith ran away from home at age 11 because he wanted to be a musician and his parents didn't want him to be one. Later, Adolf Hitler threw him out of Germany and banned his music. He was welcomed in Britain, Switzerland, and the United States (where he became a Yale professor).

Hardly anyone is interested. With fierce religious wars decimating much of Europe, who has time to read books of astronomy? *The Harmonies of the World* is the title of one of Kepler's great works. Another is titled—unwisely—*Epitome of Copernican Astronomy*.

DECIMATING originally meant "killing every tenth person." Today it means killing a large part of a group.

Kepler says it is the hand of God that has guided him to astronomy. He writes to his old teacher, "I wanted to become a theologian [religious scholar], and for a long time I was restless. Now, however, observe how through my efforts God is being celebrated in astronomy." And: "I am satisfied to guard the gates of the temple in which Copernicus makes sacrifices at the high altar." Kepler believes that God has made the heavens into a book in which an orderly plan is written, and that the human mind is designed to understand it. Copernicus glimpsed that plan; he, Kepler, is attempting to describe it in further detail.

He does a thought experiment. Imagining himself on the Moon, then on Mars, and then on the Sun, Kepler thinks and calculates and works—using Tycho's charts—until he comes up with three landmark laws that will be known as **Kepler's Laws of Planetary Motion**. They are science-changing whoppers (details on page 130).

Kepler wrote a book about a man who traveled to the Moon in a dream. That book, *Dream or Astronomy of the Moon*, with descriptions of the Moon's surface, is thought to be the first book of science fiction, written long before Jules Verne came along and made science fiction popular.

His big idea is that planets travel around the Sun in a path that is an ellipse, not a circle. And the closer a planet is to the Sun, the faster it will move, according to a rule that can be calculated.

The Greeks believed the heavens were perfect; they thought the circle was the perfect form, so they deduced that the planets' orbits must be circular. *Through all of world history, no one seems to have questioned that idea.* Everyone just accepted the idea that the planets travel in perfectly round orbits.

But Kepler has Tycho's charts and measurements, especially those of Mars. (Tycho paid a lot of attention to Mars.) What Tycho saw and noted doesn't work with the mathematics of circular orbits. Something is wrong. When finally Kepler tries elliptical orbits, Tycho's observations work. He realizes that **the planets, including the Earth, travel along paths that are *slightly* elliptical.** In addition, he discovers that the Sun is always located at one of the two foci of those ellipses.

In one sense, the Greeks were right. There is a basic form in the heavens; it just happens to be *elliptical*. (You can read more about ellipses on pages 129–131.)

Kepler writes, "I thought and searched, until I went nearly mad, for a reason why the planet preferred an elliptical orbit. . . . Ah, what a foolish bird I have been!" (He may have thought himself foolish, but science historian Charles Gillispie calls Kepler's insight "one of the great elastic feats of the human mind.")

Kepler is so amazed by his discovery that he falls on his knees and exclaims, "O God, I am thinking Thy thoughts after Thee."

And new Philosophy calls all in doubt,
The Element of fire is quite put out;
The Sun is lost, and th'earth, and no man's wit
Can well direct him where to look for it.

—John Donne (1572–1631), English poet (one of my favorites), *The First Anniversary*

Donne visited Kepler. Too bad we can't hear their conversation. What does he mean by "The Sun is lost"?

Johannes Kepler was born in Weil, Germany in 1571. Throughout his life, he was on the move, living in German and Austrian states that were part of the Holy Roman Empire. The Thirty Years' War (1618–1648) devastated the European continent and took many millions of lives. When the Peace of Westphalia (1648) ended the war, Europe was no longer dominated by the Holy Roman Empire and new borders were established. It was the end of a European world with shared values and the beginning of the rise of separate national states.

The discovery that planetary orbits are not round is one more hit—a really hard one—to Greek thinking.

"The sun will melt all this Ptolemaic apparatus like butter, and the followers of Ptolemy will disperse," writes Kepler, who has fully accepted the Copernican view. "My aim is to show that the heavenly machine is not a kind of divine, living being, but a kind of clockwork."

Kepler realizes that if Earth, which is not perfect, is a planet, then maybe the other heavenly bodies are imperfect, too. He is the first important thinker to say that **the planets are like the Earth in being made of matter.** That idea prepares the way for the study of space.

Everyone knows Kepler is brilliant, but no one wants to pay him for thinking—especially since he holds the unpopular opinion that Earth is *not* the center of the universe. To make things worse, the horrors of the time keep multiplying. In much of Europe, but especially in Austria and Germany, Protestants and Catholics are killing each other in what comes

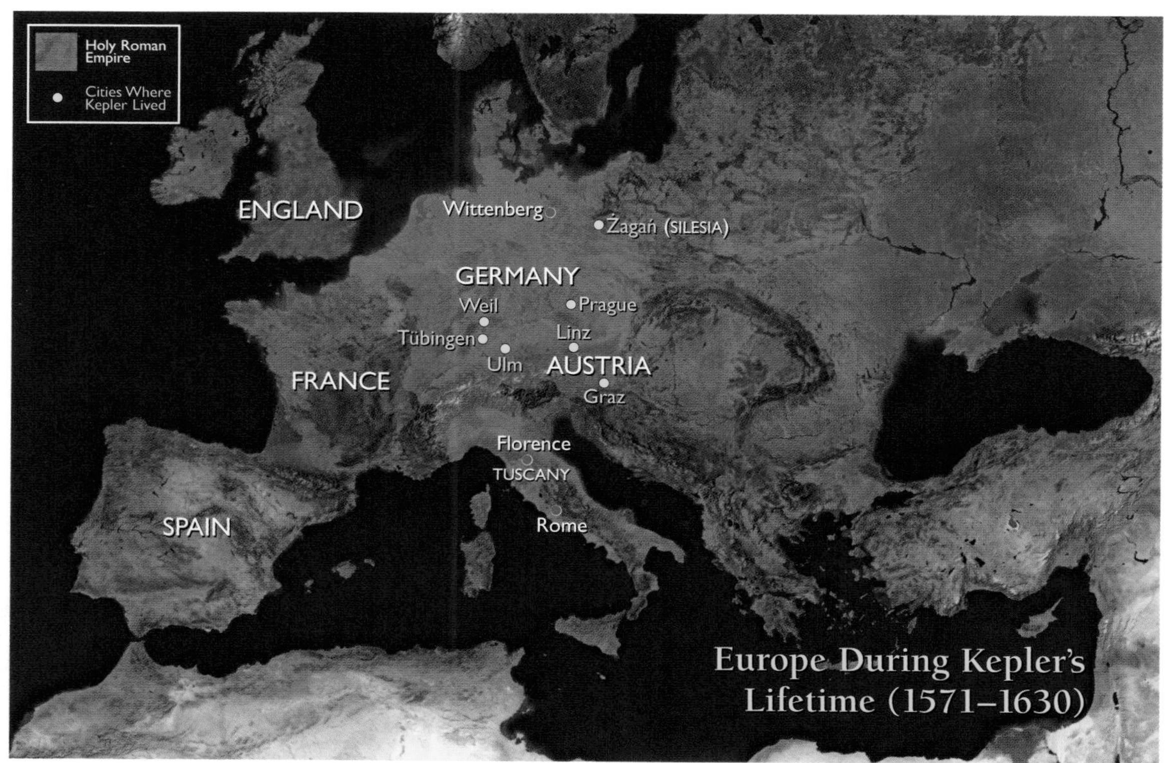

Europe During Kepler's Lifetime (1571–1630)

to be called the Thirty Years' War. It is all about religion.

Kepler is Lutheran, but his children are forced to attend Catholic mass. His mother is accused of being a witch—no joke in the seventeenth century; witches are burned at the stake. Kepler speaks out on her behalf (it takes courage) and helps save her from a fiery death.

His son and then his wife die of the bubonic plague, which travels with the warring soldiers. In part because of his peaceful ideas (he believes Catholics and Protestants can live harmoniously), Kepler is excommunicated by (thrown out of) the Lutheran Church. He is very devout, so this is terribly upsetting; excommunication makes him an outcast. Hardly anyone wants to have anything to do with him—or his scientific ideas—and his mousy personality doesn't help. To earn income, he casts horoscopes. That is astrology, and he knows it is hokum. In letters to friends, he calls his clients "fatheads." He describes astrology as "silly and empty." His last patron, an imperial general, asks for an astrological forecast. Kepler sees disaster in the man's future (he is right). The general loses his job and then his life. But no one appreciates forecasts like that.

Kepler remarries and has three new children, and then is again without income. He is on a journey to try to collect funds he is owed, when he dies of a fever. He is not quite 59. Before he dies, he writes his own epitaph, expecting it to be placed on his grave:

> *I used to measure the heavens, now I shall*
> *measure the shadows of the Earth.*
> *Although my soul was from Heaven, the*
> *shadow of my body lies here.*

Today, no one knows where his body lies.

Tycho Brahe smashed the idea of round crystal spheres with his observations of

Martin Luther's ideas led to the Reformation—a time when new Protestant religions were established and many Europeans became Protestant. Then came the Counter-Reformation, in which many became Catholic again. Some of the conversions were forced—one way or the other; your religion was apt to shift with state power. Those who wouldn't change their beliefs often faced death, imprisonment, or exile. Many fled to America.

Kepler tried to picture planetary orbits as perfect circles, nested inside each other like the five Platonic solids (cube, tetrahedron, octahedron, icosahedron, dodecahedron). He ditched this idea when he discovered the orbits were ellipses.

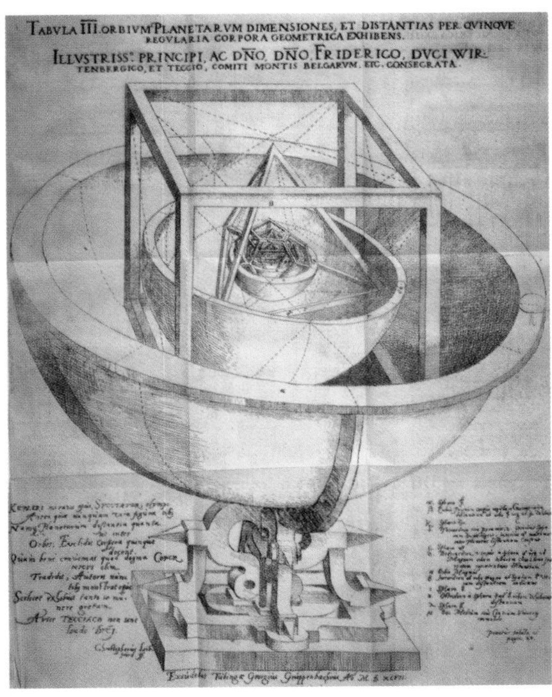

comets. Johannes Kepler went even further with elliptical orbits that can be verified scientifically.

But what keeps the planets from flying away into space if there are no crystal spheres? Kepler thought it must be a force from the Sun, maybe magnetism, which was a sensible theory; it just happened to be wrong.

In Italy, Giovanni Borelli (1608–1679), who got Galileo's old job as math professor at the University of Pisa, helped spread the word of Kepler's ellipses. Borelli suggested that Jupiter exerts an influence on its moons much as the Sun influences the planets in their elliptical orbits. What kind of influence could it be? Borelli, who is best known for his study of the structure of the human body, doesn't figure that out.

Someone needs to do so. If that someone can take Galileo's thoughts on motion, Kepler's laws on the planets and their orbits, Borelli's hints on cosmic influences, and put them together into a grand theory that can be proved mathematically—well, that will change the world of science.

A boy, soon to be born in England, will grow up and do just that.

WHERE ARE WE GOING?

Despite the tragedies of his life, Kepler was given rare gifts. He was able to see and hear harmony in the heavens. He predicted that someday there would be "celestial ships with sails adapted to the winds of heaven" filled with explorers "who would not fear the vastness" of space. (Do today's spaceships qualify as celestial ships? Are astronauts the explorers he foresaw?)

THOSE ECCENTRIC ELLIPSES

An ellipse is not any oval. (Egg shaped won't do.) To draw one, first stick two pushpins or tacks in a piece of heavy cardboard and put a loop of string around them, as shown at left, bottom. With the tip of a pencil, pull the string tight and run the pencil along the inside of the loop to trace an ellipse. Move the two tacks farther or closer apart and draw another ellipse; it will be flatter or rounder. If you could put the tacks on top of each other, you would trace a circle.

An ellipse has two fixed points inside (represented by the tacks), called foci (plural of focus). Mark any spot on the perimeter of the ellipse, measure the length to each focus, and add the two measurements. Start in another spot on the perimeter and do it again; the total will be the same. It will always be the same, no matter which spot you choose.

(continued on next page)

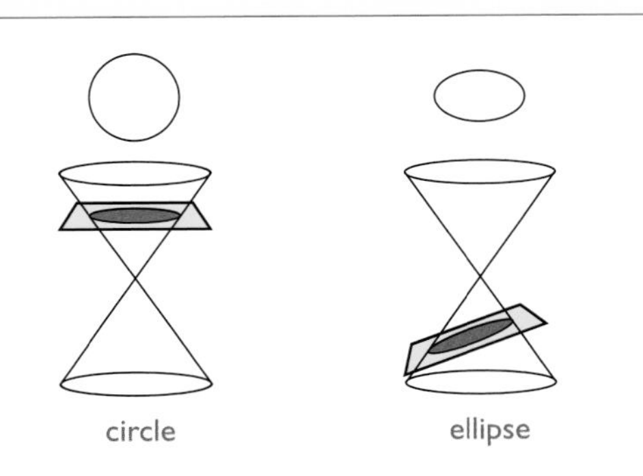

An **ellipse** is an elongated circle. To make one, you can slice through a cone at an angle that is not parallel to the base. (A parallel slice will produce a circle.)

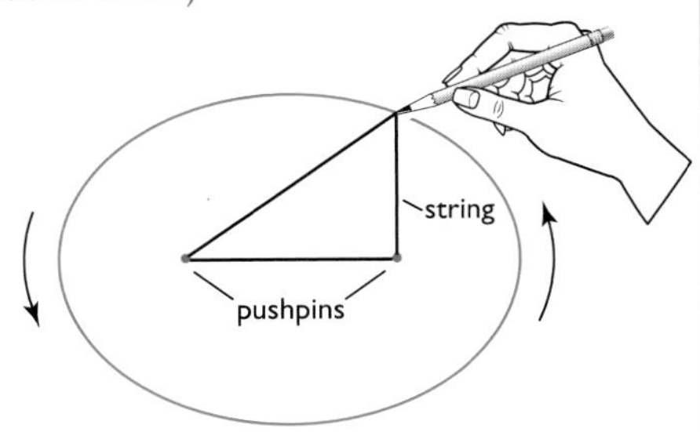

How to Draw an Ellipse (see text at right)

KEPLER'S THREE LAWS DESCRIBE THE WAY PLANETS MOVE

1. Each planet travels around the Sun in a path that is an ellipse. The Sun is off-center inside that ellipse, so the planet's distance from the Sun changes as it orbits.

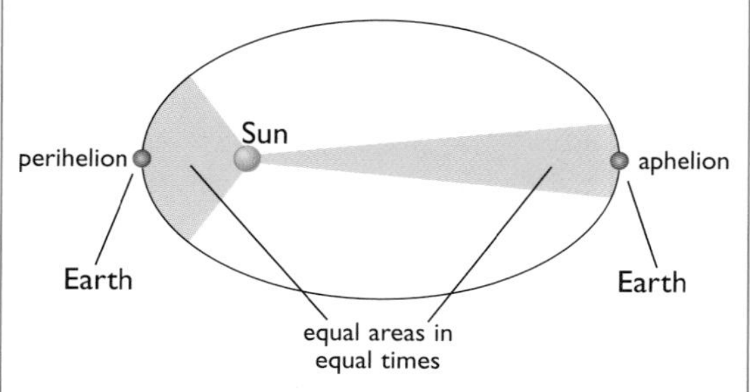

A greatly exaggerated ellipse shows Earth at perihelion (its closest point) and aphelion (its farthest point) from the Sun. The true orbit is almost a circle.

2. The speed of a planet's orbit depends on its distance from the Sun: closer is faster, and farther is slower. A planet sweeps out equal areas of its ellipse in equal intervals of time.

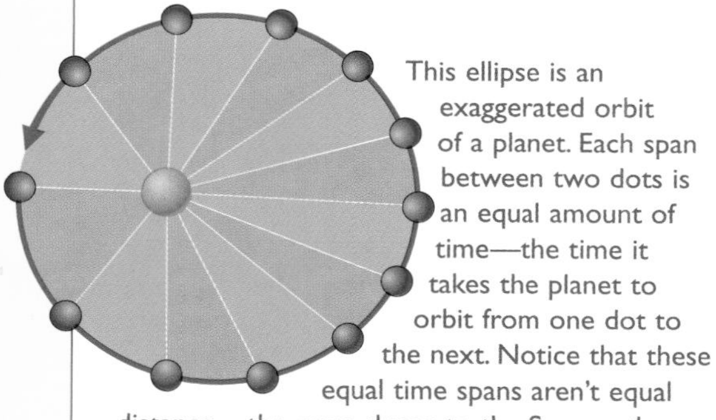

This ellipse is an exaggerated orbit of a planet. Each span between two dots is an equal amount of time—the time it takes the planet to orbit from one dot to the next. Notice that these equal time spans aren't equal distance—the ones closer to the Sun, on the left, are longer. To cover that longer distance in the same amount of time, the planet must travel faster. Each slice—blue or green—has exactly the same area.

3. The farther a planet is from the Sun, the longer it will take to go around the Sun, and the longer its year will last.

In the case of the Earth's elliptical orbit of the Sun, the Sun isn't in the center—it's at a focus. (Nothing is at the other focus.) It was Kepler who found that that was true of every planet's orbit.

Apollonius, way back in the ancient Greek world, did the first serious study of ellipses. He probably thought it was just an exercise in pure mathematics. Most mathematicians will tell you they never know where their work will lead.

Here's more on ellipses, which are central to understanding astronomy. The diameter of a circle is the same no matter where you measure it. An ellipse's diameter (a straight line drawn through its center) varies in length. The longest diameter is called the major axis; the shortest is the minor axis. (A circle has only one axis—the diameter.) The flatter the ellipse, the greater the difference in length between its major and minor axes and the greater its eccentricity. A circle has zero eccentricity.

There's still more to learn about ellipses. The two foci are always on its major axis, and each is at an equal distance from the ellipse's center. (Again, a circle has just one focus, which is dead center.)

EARTH IS JUST A LITTLE ECCENTRIC

We now know, thanks to Kepler, that planetary orbits are ellipses. They aren't ellipses by much, but they are definitely not circular. The flatness of an ellipse—how far it is from being a perfect circle—is called its eccentricity.

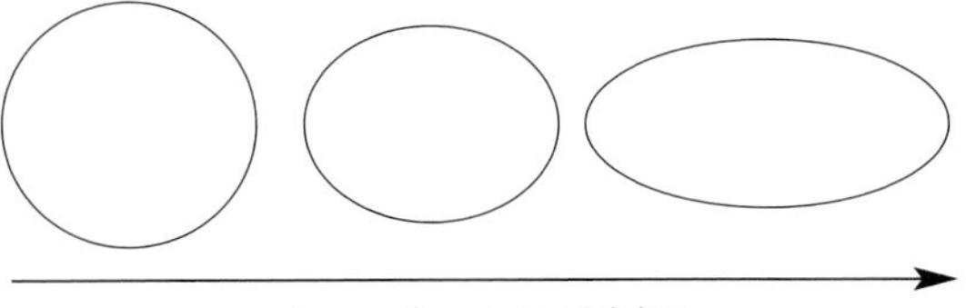

Earth's eccentricity is .017, which means our planet is about 5 million kilometers (roughly 3 million miles) closer to the Sun at perihelion (its nearest point) than at aphelion (its farthest point). That difference sounds like a lot, but it's just a tiny fraction of the size of Earth's orbit. It's so minor that if I drew a picture of Earth's orbit, it would look pretty much like a circle. You'd need precise measuring tools to see that it's an ellipse.

Earth's eccentricity changes—from nearly circular to more elliptical and back again—in a 100,000-year cycle. In periods of greater eccentricity, Earth moves both closer and farther from the Sun, so there's more variation in the amount of solar radiation that reaches our planet. That variation affects the climate—which makes Earth's eccentricity an interesting and important thing to study. Careful, though: Don't confuse climate with seasons. Seasonal change is caused by the tilt of the Earth. Climate is affected by lots of factors, including the tilt and the changing distance between Earth and Sun.

Eccentric: An eccentric person is a weirdo. *Strange* is a synonym for *eccentric*. Scientifically, an eccentric is something that deviates from a circular form or path, like an ellipse. It can also describe something rotating on an axis that is off-center, which results in a sort of wobbly, eccentric motion.

Eccentricity: The flatness or elongation of an ellipse—or any conic section (circle, ellipse, parabola, hyperbola). The eccentricity of a circle is 0; the eccentricity of an ellipse is a decimal between 0 and 1—so the smaller the decimal, the more circular the shape is.

Orbit: The path of a celestial body as it revolves around another body. *Orbit* comes from a Latin root meaning "circle." When Kepler found that planetary orbits are not circular, the word, like those orbits, got stretched a bit.

The red line is Earth's current eccentricity (.017). Compared to 225,000 years ago, when it was at a peak of about .05, Earth's current orbit is closer to circular. The eccentricity changes in 400,000-year cycles (roughly), and we're nearing the end of the second cycle shown on this graph. (Read the graph from right to left, or past to present.) Within each larger cycle are four shorter cycles lasting about 100,000 years.

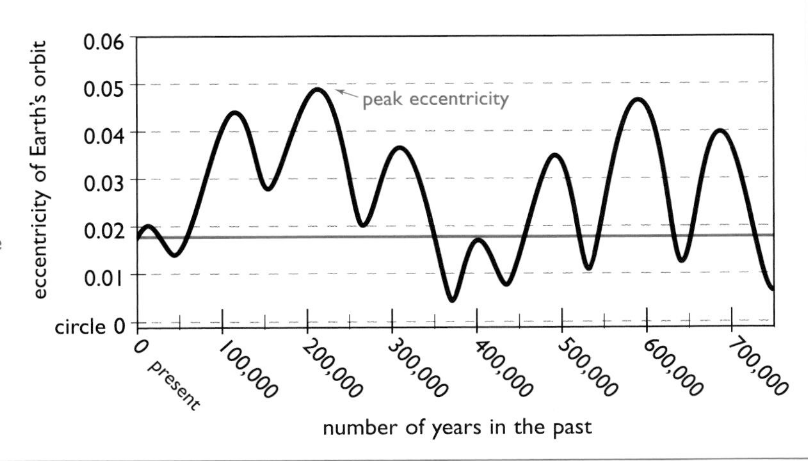

12 Descartes and His Coordinates

I think, therefore I am.

—René Descartes (1596–1650), French mathematician and philosopher, *Discourse on the Method...*

Descartes had likened the world to a giant machine, but many viewed such reductions as threats to the power of God.

—Alan Lightman (1948–), American author and professor, *A Modern Day Yankee in a Connecticut Court*

[Descartes's] years of education convinced him primarily of his own ignorance and the ignorance of his teachers, and he resolved to ignore the textbooks and work out his own philosophy and science by studying himself and the world about him.

—John Gribbin (1946–), British science writer and astronomy professor, *The Scientists*

Galileo is 32 and Kepler is 24 when René Descartes (ruh-NAY day-KART) is born in France in 1596 into a moderately well-to-do family. His mother dies when he is just a baby, and he is a sickly kid. So when he is sent off to a religious boarding school (run by Catholic Jesuits), the combination of his poor health (he has a chronic cough) and his brilliant mind get him special privileges. He doesn't have to leave his bed in the morning if he doesn't want to. Descartes gets into the habit of doing most of his work in bed and stays with that for the rest of his life. He never marries, so he never has to worry about anyone else's schedule. He does go to law school and he serves in the army (where he is a soldier/engineer), but even there he seems to be allowed to work in bed. He's not lazy; he works hard at what interests him, and he understands that he has a talent for deep thinking.

Descartes was a rationalist, which means he relied mainly on reason as a source of knowledge. Philosophers known as empiricists rely mainly on observation and experiment to find knowledge. Those two viewpoints keep turning up through the ages, with a balance between the two often being the best course.

Descartes is a philosopher (some will call him the Father

of Modern Philosophy), and he has an idea he wants to share with as many people as possible. So he writes it in simple, everyday French in a book whose full title is: *Discourse on the Method of Rightly Conducting the Reason and Seeking the Truth in the Sciences*. His idea, a common-sense idea, is that if you have a problem or a difficult concept, the way to attack it is to break it into parts and then deal with each of those parts separately. It's called analysis, which is from the Greek word for "breaking up or dissolving." It's a method that assumes that, although the world is complex, its parts can be studied individually to understand the whole. Some people have always worked that way, but Descartes makes it an acceptable scientific approach.

Descartes was a slim, dapper man with long dark hair and a mustache. As befitted a gentleman, he wore clothes of fashionable taffeta fabric and strapped a sword to his side. Although he never married, he had a daughter named Francine. He was devastated when she died at the age of five. Those close to him found him generous with his time. He helped his valet become a skilled mathematician and his shoemaker become an astronomer.

While most of his contemporaries think that supernatural (unknowable) forces guide much of nature, Descartes believes in a world where almost everything can be known if humans use reason and mathematical proof. (He distrusts the senses.) Descartes compares the universe to a clock that can be studied part by part. He sees all of nature as open to scientific and mathematical analysis. It is a unifying idea.

Some of Descartes's most important work is in mathematics.

One of his greatest ideas comes to him when he is in bed (no surprise). He is watching a fly and he wonders how to chart its position as it moves. He figures that at every moment, the fly is located at the intersection of three lines (one going north–south, one east–west, and one up–down). He knows that points on a map can be located exactly by two lines: longitude (north–south) and latitude (east–west). The Greeks were the old pros of that invention. Descartes wants to follow his fly on a mathematical grid that introduces a third dimension

The reading of good books is like a conversation with the best men of past centuries—in fact like a prepared conversation, in which they reveal only the best of their thoughts.

—René Descartes, *Discourse on the Method* . . .

Algebra is the mathematics of using letters and symbols to stand for numbers in an equation: $x + 3 = 8$. It comes from the Arabic *al-jabr*, meaning "mending broken bones." The idea was that when you solve an algebraic equation, you take all the "broken" parts (x, 3, 8) and meld them into a whole (5), like this:

$$x + 3 = 8$$
$$x = 8 - 3$$
$$x = 5$$

Algebra gets far more interesting when the letters can change value: $s = d/t$ (speed equals distance divided by time). For example, doubling the distance you travel in the same amount of time means you have to double your speed. If it takes you twice the time to travel the same distance, you're going half as fast. With algebra, you can compute ratios like these quickly.

(up–down). He starts with a horizontal plane, which has a grid made from east–west lines cutting through north–south lines. (Think of the way some cities are laid out, with avenues going in one direction and numbered streets in the other.) Then Descartes *adds a third family of lines at right angles to the horizontal plane*. Those new lines represent up and down. He labels points on the lines with letters—thus combining algebra with geometry.

Descartes picks three representative lines, one N–S, one E–W, one U–D (up–down) and marks them with numbers just like a number line (these numbers will be called "Cartesian coordinates"). The lines themselves are called "axes" (plural of *axis*), and three together (labeled x, y, and z on math graphs) make a "Cartesian coordinate frame" (like the three-dimensional cloud graph on the opposite page).

While the idea of longitude and latitude lines goes back to the Greeks, Descartes is first to develop this concept into a system that can be used in abstract mathematical constructions.

All of this allows mathematicians to plot and analyze variable patterns—like the movement of a fly, your heartbeats per minute as you run, or the air pressure in a tire as the temperature changes—to name just a few relationships. Do you want to see if there is a national pattern between family income and crime, or between locality and disease? Descartes's coordinates allow you to graph and compare different variables. His mathematical ideas are called "Cartesian," because the name *Descartes* in Latin is *Cartesius*.

Descartes's Cartesian coordinate system frees algebra from the straitjacket of the one-dimensional number line and gives it three dimensions. He enhances the voice that geometry has in the language of numbers. He provides stepping stones for the calculus that is to come. Perhaps most important, he links algebra and geometry so that we can use algebra to compute

I could not possibly exist with the nature I actually have, that is, one endowed with the idea of God, unless there really is a God; the very God, I mean, of whom I have an idea.

—René Descartes, *Discourse on the Method*...

geometrical quantities like area or volume. Today, Cartesian math is often called "analytical geometry," and analytical geometry has become essential in physics and engineering.

Labeling those Cartesian coordinates leads Descartes to establish the notation that we use in algebra today. You can thank him for the familiar symbols in the equation $ax + by = c$, which describes a straight line in the plane (example above). In Descartes's notation, a, b, and c are unchanging quantities (called "constants"), and x, y are unknown quantities (called "variables"), which vary subject to the equation.

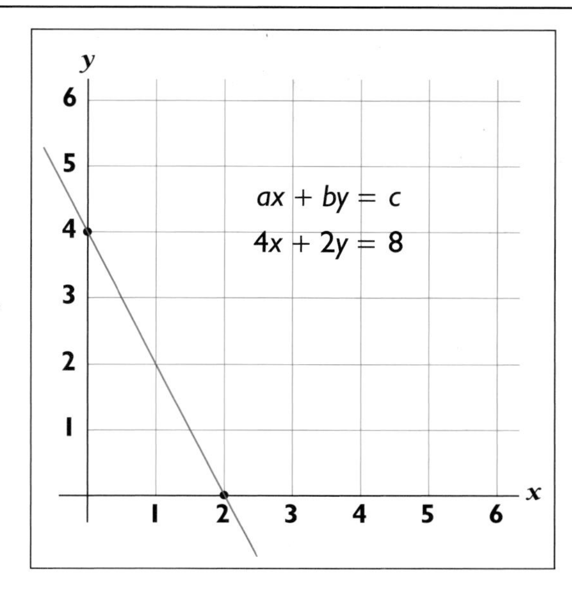

In an equation for a straight line, $ax + by = c$, the letters a, b, and c are constants because they each represent one number that doesn't change. So for $a = 4$, $b = 2$, $c = 8$, you have $4x + 2y = 8$. You can put in any value for a variable, x or y: If you put in 0 for x, then $y = 4$. If you put in 0 for y, then $x = 2$.

What if you add a third dimension, a z coordinate? You gain a depth of information, as in this computer model of a moving and growing thundercloud (below). The grid records the changing length and width of the cloud and its shadow; the vertical axis records its mounting height. The cloud is moving to the northeast. The picture graph along the bottom indicates how the cloud grows and changes at intervals of time. Each picture represents a dot along the red scale of minutes.

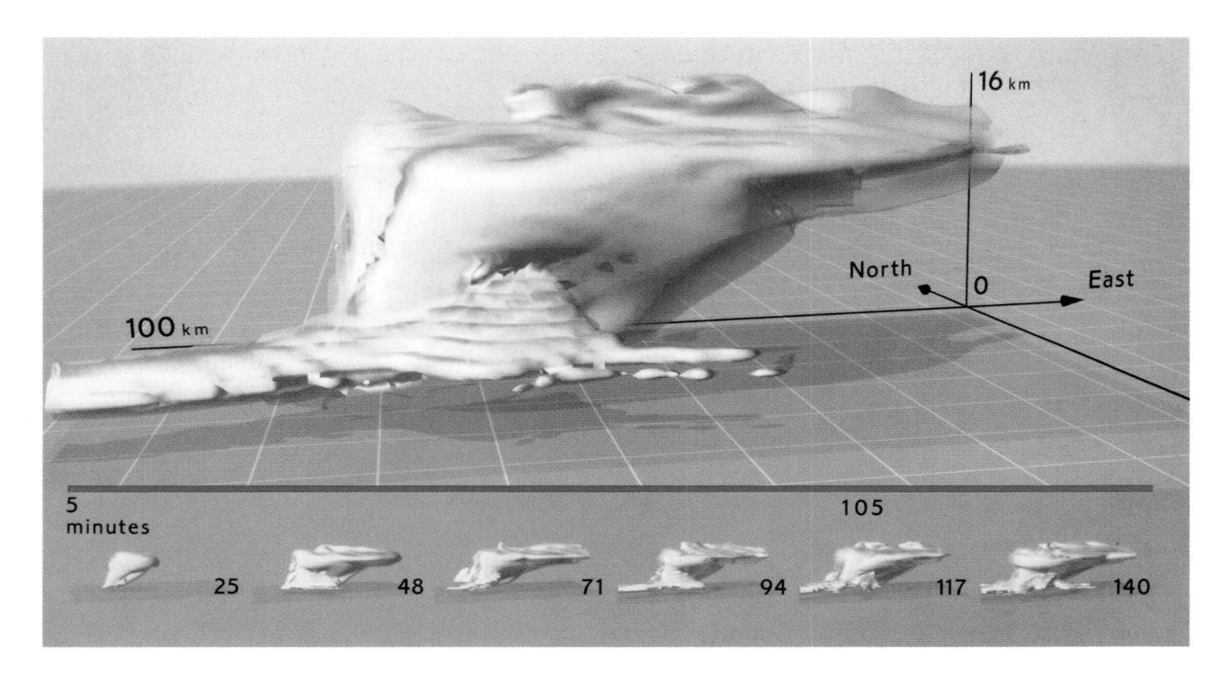

XX—FROM CODES TO EQUATIONS

France was fighting a civil war when Spain got involved. The French intercepted a coded message meant for the Spanish king, Philip II. No one could read it. (This was in 1589.) The king of France, Henry IV, decided to give the message to François Viète (1540–1603), a French lawyer who did math for fun. Viète worked on it for a month and broke the Spanish code. When Spain's king realized that his secrets were being found out, he complained to the pope that the French were using black magic. King Philip couldn't believe that any human could decipher his complicated code.

In 1591, Viète wrote a math book that had equations with letter symbols for unknown numbers. It is said to be the first book a modern student would recognize as algebra. He used vowels for unknowns and consonants for constants (known quantities). He didn't use the term *algebra*. The word has an Arabic root, and Viète didn't want to get involved in the Christian-Muslim rivalry. So he called it "analysis." Thinkers borrow from each other. René Descartes took that word "analysis" and popularized it. He changed Viète's vowels into the *x*'s and *y*'s that you'll find in today's algebra books.

Reminder: An exponent (like the tiny 3 in 4^3) indicates how many times to multiply a number (like the 4) by itself. Any number with the exponent 0 is equal to 1: $x^0 = 1$ and $55^0 = 1$.

In physics, a COLLISION is any interaction between particles or bodies of matter in which there's mutual influence (usually an exchange of energy). The objects don't have to touch; magnets collide when they repel each other, for example. Collisions are very important in the study of gases: The bangs and whams of gas molecules change kinetic energy (the energy of motion) into thermal energy (heat). You'll see why in chapter 34.

Action at a distance means objects influence each other without coming into contact—like the Earth's gravity keeping the Moon in orbit, or two magnets repelling each other.

Descartes also introduces the use of exponents (5^2) and the square-root sign ($\sqrt{}$).

Some later scholars will say that what Descartes has done is almost as important as going from Roman numerals to Hindu-Arabic numbers. But others won't agree at all. They will see symbols as, well, just symbols and not a sign of great genius.

Actually, Descartes is kind of arrogant. He claims to be setting all of knowledge on a rock-solid foundation, but he is often wrong. For instance, Descartes takes two billiard (pool) balls of varying sizes and writes down laws for what happens when they smash together and fly apart. Some 50 years later, Gottfried Leibniz will come along and graph the collisions of Descartes's billiard balls using Cartesian coordinates. The graphs turn out to be wild, with impossible jumps and gaps. Descartes got it wrong on collisions, and Leibniz will shoot him with his own coordinates, which shows the power of those coordinates.

Like him or not, Descartes is very influential. His coordinates extend and transform mathematics. In physics, he makes an important contribution by building on Galileo. Galileo realized that objects in motion continue in motion unless stopped by a force (like friction). But Galileo thought that natural motion was curved. (He was wrong about that.) Descartes understands that the natural motion of objects and of planets is in a straight line at a constant velocity. (He is right.) That's a concept called inertia.

But if objects naturally take a straight path, what makes planets follow curved orbits? Descartes rejects the possibility of any kind of force projecting through the universe and leading to *action at a distance*. That sounds like hocus-pocus to him. (He's missing a good idea here.) As to space, Descartes doesn't think there is any such thing as a void or vacuum. He says space can't be empty. So what about those curved orbits? Descartes believes the universe is filled with invisible matter that swirls in vortices (whirling masses) that keep the stars and planets on curved paths.

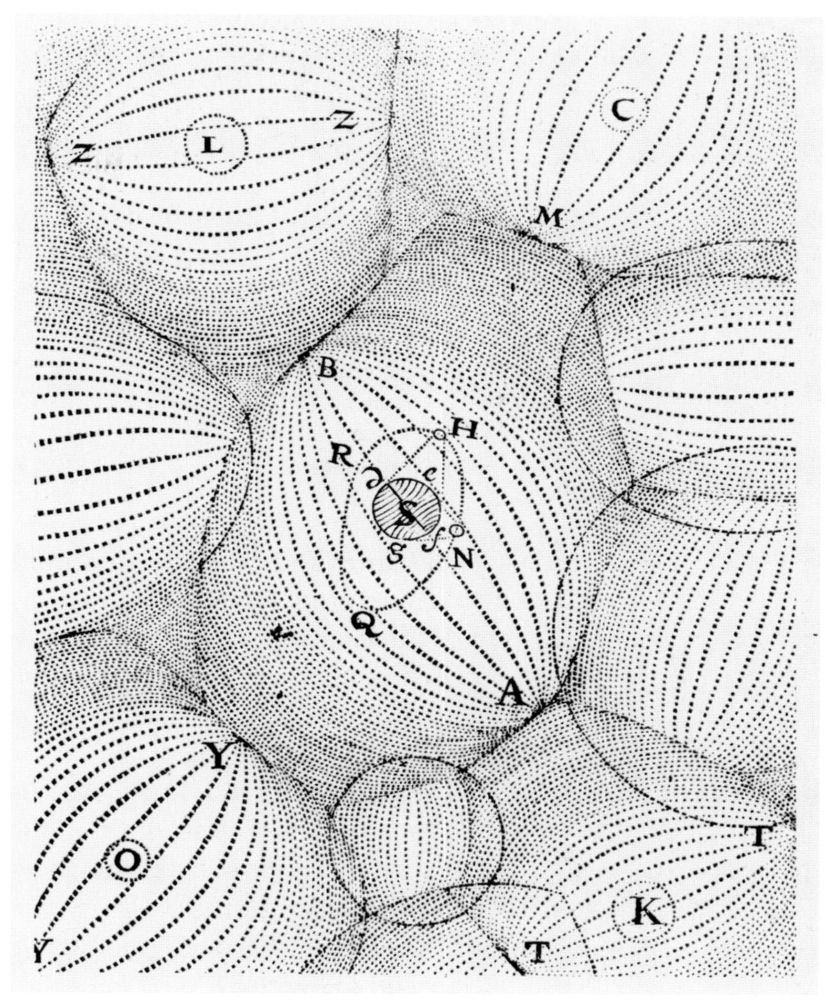

Descartes figured (wrongly) that planets must be carried along in their orbits by some sort of whirling current (above). He based that idea in part on Aristotle's belief that "nature abhors a vacuum." By that, the Greek philosopher meant that no space could remain completely empty; something had to fill it. The drawing (above) is from *Principia Philosophiae* ("Principles of Philosophy"), a 1644 book by Descartes.

It's a great idea; it makes sense; lots of thinking people believe it. It happens to be wrong. But wrong ideas can be helpful. They get people thinking and questioning and testing. When it becomes clear that there are no vortices, someone has to find the force that does keep stars and planets in their orbits. Can there be something that produces action at a distance? That is a big question that needs answering.

Descartes is pondering dangerous ideas (like inertia and the motion of planets) at a time when scientific writing is

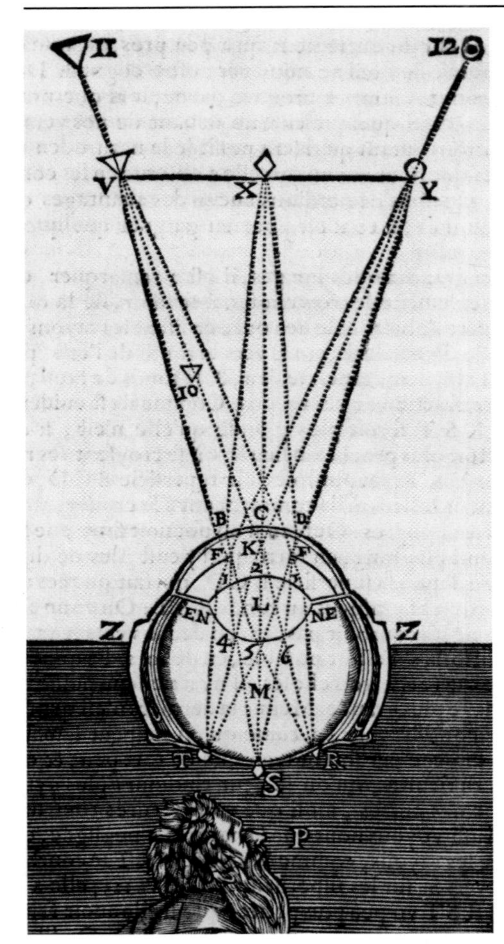

Descartes wrote an essay called "Dioptrics" that included this illustration of how light enters the eye. The essay examined reflection (light bouncing) and refraction (light bending). He wrote that light travels through an invisible medium called *the ether* that fills the universe. Does the ether exist? Lots of scientists thought so. More on this to come.

An ATHEIST is someone who doesn't believe in the existence of any god. It comes from the Greek word *átheos*—*á* ("without") + *theos* ("god"). ATHEISTIC is the adjective form.

being closely watched by church authorities. He tries not to offend the Catholic Church. In a letter to Father Marin Mersenne, a priest and lifelong friend, Descartes says:

> *No doubt you know that Galileo had been convicted not long ago by the Inquisition, and that his opinion on the movement of the Earth had been condemned as heresy. Now I will tell you that all the things I explain in my treatise, among which is also that same opinion* [as Galileo] *about the movement of the Earth, all depend on one another, and all are based on certain evident truths. Nevertheless, I will not for the world stand up against the authority of the Church. . . . I have the desire to live in peace and to continue on the road on which I have started.*

But in one book he does include technical descriptions of cannonballs and how their trajectories (paths) are influenced by the rotation of the Earth. The rotation of the Earth? According to the church, the Earth does not rotate! That's a Copernican idea, and Copernicus has been condemned in Italy, France, Germany, and Belgium (after a Catholic Church edict of 1616).

Descartes's books are put on the church *Index of Prohibited Books*. He is in danger of arrest in Catholic nations. He has been living in liberal Holland, where he has been safe, and then some Dutch Protestants find his books "atheistic." Life in Holland begins to get uncomfortable. Still, he is not eager to leave.

Meanwhile, Queen Christina of Sweden decides to add glory to her court by inviting the famous philosopher/mathematician to instruct her. It's a solution to his problems with church authorities, and he is flattered by attention from

Math historian E. B. Bell describes Queen Christina of Sweden as "a tough morsel of femininity who was as hardened to cold as a Swedish lumberjack." In this eighteenth-century painting by Louis-Michel Dumesnil, the queen and her court listen to Descartes (on her left) giving a lecture on geometry.

royalty. Unfortunately for Descartes, the queen is not a stay-a-bed, and she expects him to arrive at her chambers at 5:00 A.M. This is during a cold Swedish winter, and the royal castle is drafty. Descartes soon has pneumonia; he doesn't make it through the winter (he dies in 1650). The grisly part of the story is to come (see next page): Descartes's body is sent back to France, but his head is chopped off and kept in Sweden. Finally, in the nineteenth century, his skull is sent home to rest.

A Grave Story

Descartes complained during his lifetime that people wouldn't leave him alone. "I desire only tranquility and repose," he said. He longed for quiet time to think and work; but he didn't find much tranquility in his very active life. Even in death, everyone seemed to want a piece of him.

After his burial in Sweden, the French people demanded that the great philosopher's body be entombed in France. So 16 years later, it was sent off. The French ambassador, who arranged for the corpse's return, removed the right forefinger to keep as a souvenir for himself. One of the Swedish guards (Israel Planstrom) went still further: he cut off the head and engraved in the bone, "Descartes's skull." (He substituted someone else's head in the coffin.) The corpse's journey from Sweden to France took eight months; this was during 1666–1667. Many thought Descartes might eventually be named a saint, so according to the *Internet Encyclopedia of Philosophy*, "anxious relic collectors along the path removed pieces of the body." In Paris, thousands turned out for the reburial, although King Louis XIV forbade any funeral oration.

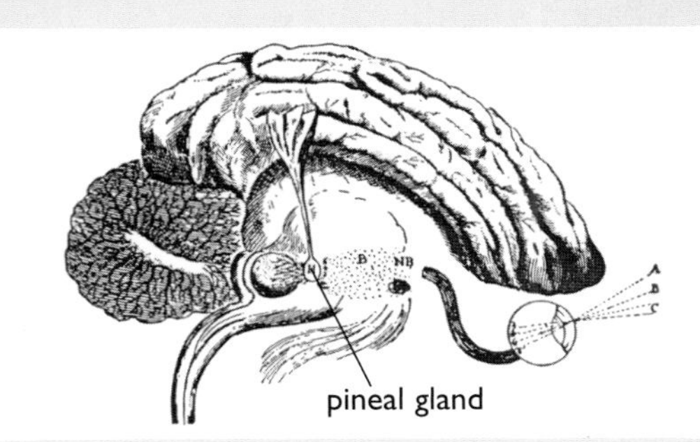

Descartes thought the mind controlled the body through the pineal gland, in the center of the brain. He was partly right: This gland makes you sleepy. The engraving is from his *Treatise on Man*.

Back in Sweden, Planstrom died and the skull was sold with his other possessions. For the next 150 years, it passed to a succession of owners who engraved their names on the bone. Meanwhile, in France, the coffin was moved from one place to another, then reburied in 1819. That's when the skull was found missing. A Swedish chemist, Jöns Jakob Berzelius, was upset at the blame laid on his country. According to a biographer of Descartes, Emile Aron, when Berzelius "read an 1821 gazette report of the sale of the skull at a Stockholm auction, he . . . bought the skull . . . and gave it to the French Academy of Sciences."

Morbid science souvenirs aren't limited to Descartes's body parts. This ornate jar contains Galileo's middle finger from his right hand. It was removed in 1737, just before the rest of the scientist was reburied in a fancier tomb. You can see the finger in person at the science-history museum of Florence.

Still, even today, head and body are not joined in tranquil rest. The body is at the Church of Saint-Germain-des-Prés while the head is at the Musée de l'Homme (both in Paris). As a philosopher, Descartes thought a lot about the separation of mind and body. Some people see some irony here.

If you think this could only have happened in the past, consider this: After Albert Einstein died in the twentieth century, someone took his eyes, someone took his brain, someone else took a slice of that brain. They all claimed they wanted the pieces for scientific research. There's no evidence that was true.

AT LAST, PROOF OF THE LAST

On June 21, 1993, a shy 40-year-old mathematician, Andrew Wiles, began a presentation to some 20 mathematicians at the Isaac Newton Institute for Mathematical Sciences at Cambridge University in England. No one knew where he was heading, but the math was exciting. Wiles's lecture was scheduled to be in three parts. A buzz made its way through the conference.

We say Pierre de Fermat (above) did not publish any of his work, but there is more to the story. The scientific community kept in touch through middlemen, like the philosopher and theologian Marin Mersenne. Mersenne worked in mathematics as well, but he was chiefly known as the best-connected scholar in France. Everyone corresponded through him, including Fermat and Descartes.

The second day, the room was filled. He hadn't said what he was trying to prove, but the hints were enticing. The guess was that he might be about to solve a 350-year-old puzzle posed by Pierre de Fermat (1601–1665), a contemporary of Descartes and one of the most brilliant mathematicians of all time. Fermat, who was the king's councilor to the parliament of Toulouse (a city in southern France), never published his work in mathematics, so during his lifetime only a few people knew of his accomplishments. Today we recognize him as the originator of modern number theory (look it up). We know of his achievements in math because of letters he wrote to friends and also because of notes he put in

TRIANGULAR TRIPLETS

Diophantus's problem can be written as the algebraic equation $x^2 + y^2 = z^2$. Any three positive integers that satisfy that equation are called Pythagorean triples (3, 4, 5 is a Pythagorean triple; so is 5, 12, 13). Pythagorean triples are the lengths of the three sides of a triangle, and they always produce a right triangle (one with a 90-degree angle).

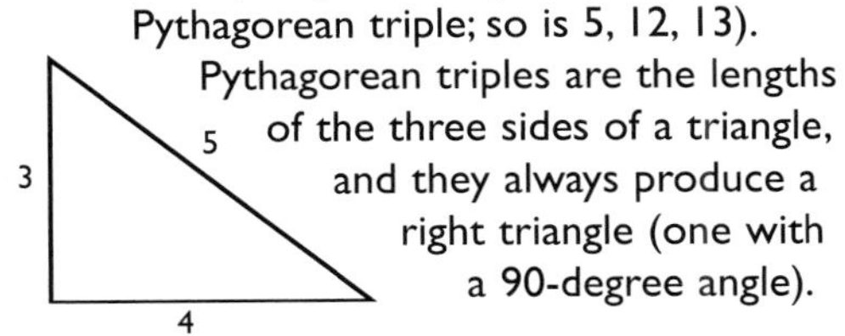

the margins of his books. His son published them after he died.

Fermat pored over the works of the ancient Greeks, which were being translated in his time. "I have found a great number of exceedingly beautiful theorems," he said in a letter to a friend about a book written by the Greek mathematician Diophantus (ca. 210–290 C.E.). Diophantus restated the Pythagorean Theorem—*In a right triangle, the square of the hypotenuse is equal to the sum of the squares of the other two sides.* Put in an equation, it can be: $3^2 + 4^2 = 5^2$. (Check this out for yourself; both sides of the equation equal 25.)

Here's the rub: according to Diophantus you can't do the same thing with any powers greater than 2. That means, for instance, you can't cube a number and then find two smaller cubed numbers that added up will give you the same answer. But was he right? Here's the problem stated mathematically: $x^n + y^n \neq z^n$ has no integer solutions for $n > 2$ and $x, y, z \neq 0$.

"I have discovered a truly marvelous proof of this, which, however, the margin is not large enough to contain," Fermat wrote in the margin of Diophantus's book. In other words, he said he had proved this impossibility, but he didn't have space enough to show his proof. After he died, that tantalizing note became famous. The missing proof was called Fermat's Last Theorem.

But did he actually have a proof? Or was he mistaken? It seems as if it would be an easy thing to prove, until you try it. Just how do you prove an impossibility for all possible numbers? For 350 years, one mathematician after another attempted to prove Fermat's Last Theorem. No one could do it.

On June 23, the third and final day of his presentation, Andrew Wiles wrote equation after equation on the blackboard. The room was electric with anticipation. Finally the quiet professor said, "And this proves Fermat's Last Theorem." The audience exploded with applause, cameras flashed, E-mails spread the news worldwide.

The next day an article on the front page of *The New York Times* announced, "At Last, Shout of 'Eureka!' in Age-Old Math Mystery." *The Washington Post* called Wiles "the Math Dragon Slayer."

As a boy growing up in Cambridge,

"Perhaps I could best describe my experience of doing mathematics in terms of entering a dark mansion. You go into the first room and it's dark, completely dark. You stumble around, bumping into the furniture. Gradually, you learn where each piece of furniture is. And finally, after six months or so, you find the light switch and turn it on. Suddenly, it's all illuminated and you can see exactly where you were. Then you enter the next dark room."
—Andrew Wiles (at left), as quoted in *Fermat's Last Theorem*

England, Wiles had heard about the mystery of Fermat's Last Theorem. He was living in Princeton, New Jersey, when he decided to try to find the missing proof. He didn't tell his colleagues; he just set up a table in his attic and went to work. For seven years he hardly saw his family. It was worth it. His proof was filled with mathematical ideas so deep, complex, and exciting that no one was sure where they would lead.

Other mathematicians began checking the proof. And, to everyone's dismay, an error was found. In mathematics there is no such thing as "almost right." Wiles was humiliated, crestfallen. Maybe Fermat's Last Theorem really had no proof.

Wiles went back to his attic. Others tried to help, but no one could find the solution. Months passed. He worked and reworked the proof. On September 19, 1994, he decided to give up; he took a final look at the papers piled on his desk; and then, Wiles said, "Suddenly, totally unexpectedly, I had this incredible revelation . . . it was so simple and so elegant, and I just stared in disbelief." This time there was no error. Fermat's Last Theorem had been conquered. Wiles had not only used the most advanced math to solve the theorem, he had invented a whole lot of new math!

A twenty-first-century mathematician, Rich Schwartz, comments:

> *Most people think of math as being sort of ordained in heaven and "completely done." Actually, new mathematics is being created all the time. Probably more new math has been created in the past 100 years than in all of time before that. Mathematics has few of the stops that many others sciences have, such as the need to explain physical reality, and so it advances at a remarkable pace. I have the feeling that a lot of pure mathematics being created now will eventually find its application in other sciences, but maybe not for hundreds of years.*

13

What's the Big Attraction?

And make us as Newton was, who in his garden watching
The apple falling towards England, became aware
Between himself and her of an eternal tie.
—W. H. Auden (1907–1973), English-born American poet, "Prologue"

I do not know what I may appear to the world, but to myself I seem to have been only like a boy playing on the seashore, and diverting myself in now and then finding a smoother pebble or a prettier shell than ordinary, whilst the great ocean of truth lay all undiscovered before me.
—Isaac Newton (1642–1727), English mathematician and physicist, quoted in *Memoirs of Newton*

It turns out that Copernicus, Galileo, and Kepler *are* right about the Sun and the Earth. Still, most people don't seem to get it. They can't quite believe that Earth is *not* at the center of things. It takes Isaac Newton—born on Christmas Day in 1642 (not long after Galileo died)—to finally make gravity and Earth's place in the universe clear and beyond-doubt understandable.

Newton arrives prematurely, and so small that his mother says he could fit "in a quart pot." Like a fair number of achievers, Newton doesn't have an easy childhood. The year of his birth is the year civil war breaks out in England. It is the king and the Church of England against the Puritans. The Puritans win (for a while). The king—Charles I—loses his head (really).

A Calendar Tale

Isaac Newton was born on Christmas Day in 1642. That's true and not true too. It *was* Christmas Day in England, but in most of the rest of Europe, it was January 4, 1643. A new calendar had been adopted on the continent in 1582.

Remember Pope Gregory XIII and his calendar (page 31). England didn't want anything to do with that Gregorian calendar. The English monarchs had broken away from Catholicism and called the calendar an example of the pope's trickery. It would be 1752 before Great Britain and the United States signed on to the same calendar used on the European continent.

Woolsthorpe Manor, where Isaac Newton grew up and later performed his most important research, is located in the East Midlands of England. Its isolation in the countryside probably provided some protection against the plague in 1665.

Newton's father, an illiterate yeoman (a farmer who owns his own land) fights and dies for the king.

The family lives in the tiny hamlet of Woolsthorpe, and, like most country folk, no Newton can write—or even sign his or her name. But Isaac's mother, Hannah Ayscough, is literate; she comes from an educated family, and her brother is a Cambridge University graduate.

Hannah gets married again—this time to a prosperous 63-year-old minister, Barnabas Smith. Barnabas is a pill and a snob too. Anyway, he doesn't want 3-year-old Isaac on his premises (that is part of the marriage contract). So, after being the center of his mother's attention, the child is handed over to his grandparents to be raised. They send him to school—he might never have had an education if his grandparents hadn't been around—but he is miserable being separated from his mother. He grows up with chips on his shoulders—a bunch of them. As a young boy, in confessing his sins, Newton mentions "threatening my father and mother Smith to burn them and the house over them." When Newton is 10, Barnabas Smith dies and Hannah returns to Woolsthorpe with three young children.

Two years later, Isaac is sent to grammar school at the

EXTRATERRESTRIALS ON THE MOON? WHY NOT?

Newton's landlord, the apothecary Mr. Clark, lent him a book called *Mathematical Magick*, by John Wilkins, a master at Trinity College at Cambridge University. Young Newton cherished the volume and put thoughts from Wilkins into his notebook, especially ideas on "secret writing" or code. Wilkins also wrote a book titled, *The Discovery of a New World; or, a Discourse tending to prove, that it is probable there may be another Habitable World in the Moon.*

Why shouldn't the Moon have inhabitants? asked Wilkins. "I thinke that future ages will discover more; and our posterity, perhaps, may invent some meanes for our better acquaintance with these inhabitants." Wilkins expected the "art of flying" to be discovered. Just because ideas seem strange is no reason to reject them, he said. "How did the incredulous gaze at Columbus, when he promised to discover another part of the earth?"

nearby town of Grantham; there he boards with Mr. William Clark, an apothecary (druggist). He has a room in the attic where he carves his name into a wallboard and draws birds, ships, and some of Archimedes' diagrams on the walls. The Clarks are generous and kind, and in the apothecary, Isaac is taught to mix and use chemicals. He learns some math from Mr. Clark's brother, who is a doctor. Isaac is a quiet boy who spends most of his time reading and tinkering and making things. He builds a kite with a lantern on it, a clock run by drips of water, and a tiny windmill powered by a mouse!

These French druggists are preparing and selling medicines in an apothecary, which comes from the Latin word for "storehouse." Today's storehouses of medicines are called "pharmacies," from a Greek word meaning "drug, poison, charm, or spell" (take your pick).

At school he is taught Latin, arithmetic, and elementary surveying useful to a farmer's son. Newton isn't much interested in school subjects; he is near the bottom of his class. Alone with his thoughts, he is sometimes filled with despair. In his Latin notebook he writes of himself: "A little fellow; . . . Hee is paile; There is no room for me to sit; In the top of the house—In the bottom of hell; What imployment is he fit for? What is hee good for? He despaired. I will make an end. I cannot but weepe. I know not what to doe."

SOME ENGLISH HISTORY

Newton was born during England's civil war. King Charles I was beheaded (in 1649) and Puritan Oliver Cromwell took over as Lord Protector of a new government called the Commonwealth. Parliament became powerful. But the Puritans didn't have much of a sense of fun—they outlawed things like dancing and theater. They tried to legislate morality.

Newton was in college during the Restoration, when England got a king again (in 1660). Most English people were delighted. King Charles II, officially crowned in 1661, was known as the "Merry Monarch," and for a while, England did seem merry. Then there were two tragedies: The plague struck in 1665, killing about 70,000 people; the Great Fire of London burned for five days in 1666, destroying 13,000 houses, lots of great buildings, and perhaps the plague germs. Cambridge University had schooled many Puritans; it fell out of favor during the Restoration, and many Puritans headed for America.

Charles II parades through London on April 22, 1661, the day before his coronation. The Merry Monarch let people enjoy music and theater, much of which had been banned under Oliver Cromwell.

According to his own account, it is a bothersome bully who turns Newton into a serious student. They have a fistfight; Newton wins, but that isn't enough. Newton decides to challenge his foe in a battle of intelligence. The bully is one of the best students in the class. Newton soon surpasses him. But that doesn't make him happy or popular; he will never learn how to enjoy himself. (Many years later, Newton is asked why anyone should study Euclid. The question makes him laugh out loud. It is the only record of that ever happening.)

When he is 16, his mother decides Isaac has had enough schooling and that he should manage her farm. Newton is not meant to be a farmer. He sits and reads while the cattle

NEWTON IN COLLEGE

Newton went off to Trinity College at Cambridge University in 1661. It was a year after the Restoration of the monarchy, and England was filled with political intrigue. Oxford University was home to the Royalists; Cambridge had been home to the now-ousted Puritans. So when Newton arrived, the 400-year-old institution, despite a distinguished past, was in bad shape.

According to a German visitor, the town was "no better than a village... one of the sorriest places in the world." Fewer than a third of the students stayed to get a degree. Scholarship was at a low point, and the curriculum was out of date. What it did offer inquiring minds, like Isaac Newton's, was a terrific library.

At age 18, Newton was two years older than the average student. He was also much poorer. His mother could have helped him; she chose not to. Most students came with their own servants to wait on them. Newton, as a scholarship student, was expected to be a kind of valet (or servant) to his tutor. Happily for Isaac, the tutor didn't spend much time at Cambridge.

The Wren Library at Cambridge was built after Newton studied there. It holds the first edition of Newton's *Principia*, one of his most important books.

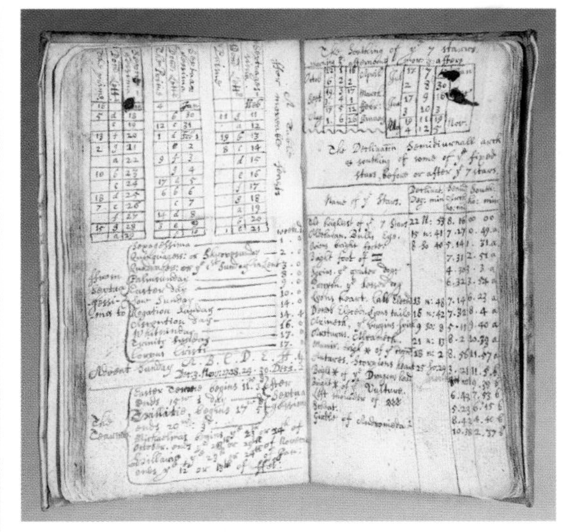

Newton casually scrawled important astronomical observations into ordinary notebooks. These scribbles made in 1659 are now a treasure of the Morgan Library in New York City.

and sheep wander off. When the animals damage a neighbor's cornfields, young Newton is hauled into court and fined four shillings and four pennies. (That makes his first public recognition a criminal conviction.)

Newton's mother realizes she has made a mistake, and when the schoolmaster and her brother urge it, she agrees to send Isaac back to school. She may also want the teenager out of her house. Later, in a personal "list of sins" he includes "peevishness with my mother," and "punching my sister."

Do you think angry kids are a new phenomenon? Not so. All his life, Newton keeps his anger and an extreme sensitivity to criticism. It doesn't seem to get in the way of his talent. At school he is always reading and thinking, and he has such a powerful mind that he easily tops his classmates on examinations, but he still doesn't have friends. When he goes to Cambridge University—as a scholarship student—those he leaves behind "rejoiced at parting from him."

Newton is lonely at college—no one likes him much—so he turns to books. *Amicus Plato amicus Aristoteles magis amica veritas*, he writes in Latin in his notebook. (It means: "Plato is my friend; Aristotle is my friend; but my greatest friend is truth.")

While Cambridge students are still being taught Aristotle's view of science, that notebook of Newton's shows that he is studying the new thinking on his own. It has pages of diagrams of the Copernican solar system. Of course Newton makes himself familiar with the work of Galileo and Kepler. And he also reads the French mathematician René Descartes. Descartes's idea—that the universe is like a great machine that can be broken down into working parts that can be measured and analyzed—will influence Newton's thinking.

One of the big scientific problems of the time deals with motion. The emerging technology has made it relevant and profitable to be able to analyze things like the flight of a cannonball. Galileo understood that and attempted to measure moving objects. (For Aristotle, there was little reason to measure motion.)

How do you measure things in motion, especially when the rate of motion keeps changing? How do you measure the

"The curriculum [at Cambridge] had grown stagnant. It followed the scholastic tradition laid down in the university's medieval beginnings. . . . The single authority in all the realms of secular knowledge was Aristotle. . . . Supplemented by ancient poets and medieval divines, it was a complete education, which scarcely changed from generation to generation," writes James Gleick in his excellent biography, *Isaac Newton*.

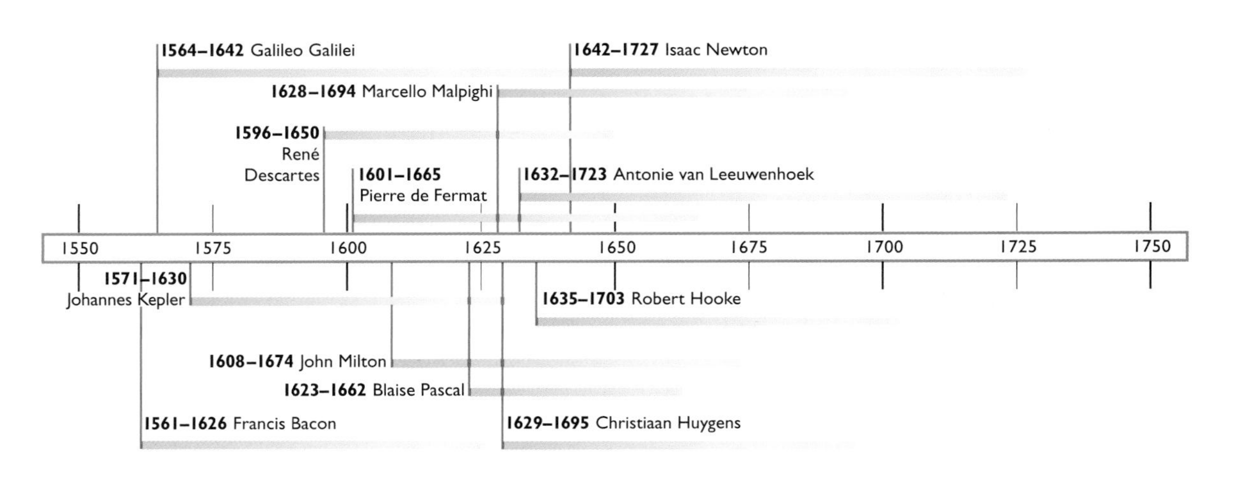

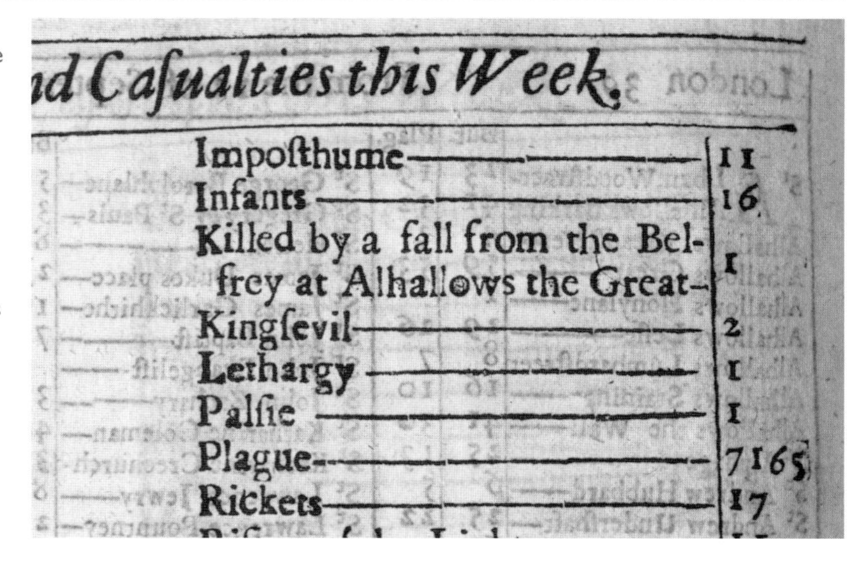

d Caſualties this Week.

Impoſthume	11
Infants	16
Killed by a fall from the Bel-frey at Alhallows the Great	1
Kingſevil	2
Lethargy	1
Palſie	1
Plague	7165
Rickets	17

How deadly was the Great Plague of London? Compare the Death Register toll for one week in September 1665: In a city of roughly 450,000 people, 1 person fell to death, 2 people gave in to the "king's evil" (tuberculosis), imposthumes (festering sores) claimed 11 people, 16 infants died for unstated reasons, and 17 victims had rickets (caused by poor diet). The plague took 7,165 souls that week—more than 1,000 per day.

time and position of a planet in a moving orbit? Newton realizes that the measuring and analyzing called for in his "modern" world requires mathematical tools that don't exist.

What is missing is calculus, the mathematics of continuously changing quantities. But in the seventeenth century, calculus hasn't been invented.

So while he is still an undergraduate at Cambridge, Newton begins to invent it. This is how calculus comes about: Newton goes to the Sturbridge Fair at age 20, and buys a book on astrology. He sees an illustration that he can't understand because he doesn't know trigonometry. So he buys a book on trigonometry. But he can't follow it, because he doesn't know geometry. So he buys Euclid's *The Elements* and studies it carefully. With that base, he will invent calculus. Actually, he does most of the thinking that leads to calculus at his mother's house. (Even when he is angry at her, he always seems to love his mother, and she him.)

Newton goes home to Woolsthorpe in 1665. Cambridge University has closed its doors because of an awful epidemic of the plague. (More than one in six Londoners will die.) At home he builds bookshelves for himself and begins filling the pages of a large blank book he has inherited from his stepfather. (Paper is scarce and valuable.) This will be the

Fifteen hundred shorn,
what comes the wool to?
. . . I cannot do 't without
counters.
—William Shakespeare,
The Winter's Tale

Shakespeare's England was a land of sheepherders and farmers—and most of them couldn't do arithmetic without something to help them count. By Newton's time, a manufacturing economy was beginning to emerge. The artisans and merchants who were coming into their own valued toolmaking abilities, measuring skills, and knowledge of numbers—in other words, mechanical and practical talents.

most productive year of his life. In this notebook, the lonely thinker poses questions, then he works hard finding answers. He plays with numbers endlessly. "Truth," he says later, is "the offspring of silence and unbroken meditation."

Aristotle taught that objects fall to their "natural place"—the center of the universe. But in Copernicus's world, Earth is no longer in the center. A new explanation was needed. Newton provided it.

Sitting in his mother's backyard, he watches the Moon shining in the evening sky and wonders why the Moon doesn't just dance off into space. What keeps it going around the Earth? There are no strings between the Earth and the Moon—but something must be holding that Moon in its orbit. Then he watches an apple drop from an apple tree, and that gets him thinking.

Newton thinks about the force that makes an apple fall to Earth (the Greeks called it gravity), and then he uses his brain (and mathematics) to make a connection between that falling apple and the Moon in the sky.

The Earth can't possibly be turning on its axis, said those who thought it a crazy idea, because an outward (centrifugal) force would make it spin into pieces. Newton's calculation of gravity showed that the force of gravity at the Earth's surface is more than strong enough to keep Earth from flying apart.

He asks himself: What if the apple tree grew into the sky until the Moon hung like an apple on its branches? Would the same force that pulls an apple toward the center of the Earth reach the Moon? Would it pull the Moon toward Earth?

The apple is about 6,400 kilometers (about 4,000 miles) from Earth's center; the Moon is about 384,000 kilometers (about 239,000 miles) away from Earth's center.

Newton figures out how much each must drop in one

This painting by Robert Hannah depicts the moment Newton came up with the Law of Universal Gravitation after observing an apple drop from a tree. Although there were no witnesses to the event, this key moment in science is a favorite of many painters—and cartoonists too.

second in relation to those distances. He comes up with a simple formula to do that (it's called the Inverse Square Law—keep reading to learn about it).

Later he writes, "I began to think of gravity extending to the orb of the Moon. . . . [Then I] compared the force requisite [necessary] to keep the Moon in her Orb with the force of gravity at the surface of the earth and found them answer pretty nearly." Indeed they do.

Scientists now have a new word for universal gravity: GRAVITATION. Use it to refer to the force of attraction between masses no matter what or where they are in the universe. The force is expressed in the Law of Universal Gravitation.

He has discovered a link between action on Earth and action in the Heavens. It is *universal gravity*, the force that keeps your feet on the ground and acts at a distance to keep the Moon and stars and planets in their orbits.

But can the Moon actually be falling toward Earth's center as apples do? (The answer is yes.) Then why doesn't it crash? (See page 177 for the surprising answer.)

There is still more to it than that. Of course Earth pulls on the apple; everyone knows that. But the apple also pulls on Earth. Every object in the universe exerts a pull on every other object! The amount of the attraction is determined by the mass of the objects and their distance from each other. (Our massive Earth is much more attractive than you are.)

When it occurred to Newton that the Law of Gravitation that works for an apple falling from a tree is the same Law of Gravitation that holds the Moon in orbit, he realized he was onto something big. Among other things, it meant experiments done on Earth could explain happenings in space.

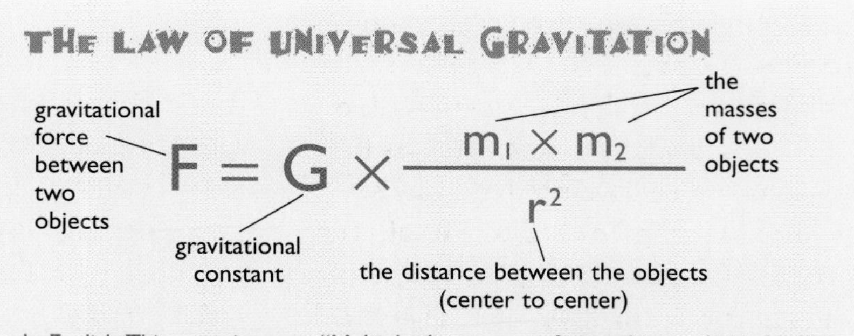

In English: This equation says, "Multiply the masses of two objects, divide by the squared distance between them, and multiply by G. Your answer is the gravitational force between the two objects (F)." What is G? It's a constant—its value doesn't change—but Newton had no idea what the number was. To find it, he would have had to measure the force between two known objects of known distance—a force much too tiny for him to detect. Now we know that G, the gravitational constant, for two 1-kilogram objects that are 1 meter apart is this: .000000000067 newton. The unit of force—the newton (N)—was named in Isaac's honor. One N is the force needed to accelerate an object with a mass of 1 kilogram by 1 meter per second per second.

Distance determines the pull of gravity? Yes, but not quite as you might imagine. Newton explains that gravity's pull follows the **Inverse Square Law**. This is the way he states it (but he uses Latin): **Every object in the Universe attracts every other object with a force proportional to the product of the masses and inversely proportional to the square of the distance between them.** If those words lose you, here is Newton's Law of Universal Gravitation in easier language:

The Inverse Square Law doesn't just apply to gravitation. Other phenomena that get much weaker with distance include electric charges, light, sound, and radiation. Their influence starts at a source, a point, and fans out in all directions. If you double your distance from the source, the influence becomes ¼ as strong.

Gravity is an attractive force. The greater the mass of an object, the greater its attraction to another object. Gravity weakens over distance, by the square of that distance. When the distance doubles, the force becomes ¼ as strong. When the distance quadruples, the force is 1/16 as strong.

The Law of Universal Gravitation applies to keys dropped from your hand, and to Earth in its orbit around the Sun. And that's what is really important about it: **It applies everywhere in the universe.**

Almost everyone in Newton's time believes that nature has laws for the heavens and other, different, laws for the Earth. Newton shows that isn't so. The fall of the apple and the orbit of the Moon are controlled by the same force. He realizes that the apple is flying through space (along with all of Earth) just as the Moon is. He ties the whole universe together with one set of rules that can be used to explain the motion of the stars and planets and of wagons and tennis balls too. This is an enormous idea.

When Newton enunciated the law of gravity he did not say that the sun or the earth had a property of attraction; he said that all bodies from the largest to the smallest have the property of attracting one another, that is, leaving aside the question of the cause of the movement of bodies, he expressed the property common to all bodies from the infinitely large to the infinitely small. The same is done by the natural sciences: leaving aside the question of cause, they seek for laws.

—Leo Tolstoy (1828–1910), Russian novelist, *War and Peace*, Second Epilogue

But Newton doesn't publish his work on gravity, because he fears it might bring him fame—or criticism—both of which seem to terrify him. "I see not what there is desirable in public esteem, were I able to acquire and maintain it. It would perhaps increase my acquaintance, the thing which I chiefly study to decline." So, for a while, no one knows exactly what he is doing.

14 Gravity—How Absurd!

In...the year 1666 (at which time I applied myself to the grinding of optick glasses of other figures than spherical) I procured me a triangular glass prism, to try therewith the celebrated phenomena of colours.

—Isaac Newton (1642–1727), English mathematician and physicist, *Opticks*

In roughly a year, without benefit of instruction, he mastered the entire achievement of seventeenth-century analysis and began to break new ground....The [unknown] young man not yet twenty-four, without benefit of formal instruction, had become the leading mathematician of Europe.

—Richard S. Westfall (1924–1996), American science historian, *Never at Rest: A Biography of Isaac Newton*

The celebrated English poet John Dryden writes a patriotic poem about what he calls "the year of wonders": 1666. The poem tells a heroic tale of the victory of the English fleet over the Dutch. It is also about London's survival of the Great Fire, which raged across the city while it was still reeling from the devastation of the plague.

When it comes to this Anglo-Dutch War, Dryden is putting a gloss on the truth. It was a shabby unnecessary affair, fueled by commercial jealousy, with lots of lives lost on both sides. No one really won.

As to the Great Fire of London, it *was* horrific, and the response of the British people *was* heroic. The fire started in a bakery on the night of September 2, and then blazed on for five days and nights, destroying 13,200 houses, 87 churches, 6 prisons, and 4 bridges—until there was almost nothing left to burn. People threw their belongings into the Thames River and then jumped in too. About four-fifths of the city was consumed.

John Dryden, like Newton, was a Cambridge graduate who resided at Trinity College. Dryden became a successful poet and dramatist during the Restoration, when the arts were allowed to flourish.

Fire, Fire Burning Bright

The Great Fire of London was an enormous tragedy, but then the city got lucky. The two men chosen to direct the rebuilding, Christopher Wren and Robert Hooke, were extraordinarily talented and public-spirited. Hardly anyone in England (except for Isaac Newton) knew more math and science than Hooke, and Wren was a mathematician with rare architectural skills. As city surveyors, they were asked to bring back a familiar London, but on a grander scale. And that they did.

Both were founding members of the Royal Society, so they planned a city that would be friendly to science. A new Royal College of Physicians building included a scientific laboratory and an up-to-the-moment anatomy theater. (These were new concepts.)

Asked to build a monument to the Great Fire, Wren and Hooke designed a Doric column topped with a flaming urn; it doubled as a telescope. Looking like a huge stone needle, it held two lenses. Otherwise its interior was clear from a basement laboratory to the urn, which when lifted, opened the shaft to the sky.

Members of the Royal Society hoped to track stars at exact, six-month intervals and record any change of position. If they could do that, they thought they could prove the rotation of the Earth. But their equipment wasn't precise enough to account for minute outside interferences.

A spiral staircase in the Great Fire Monument was used for experiments.

Other science experiments were more successful. From a high platform on the staircase that ringed the interior of the column, they could swing a pendulum or lower a thermometer. And they did both. (You can still visit the monument in London on Fish Street.)

The pride of the city had been old St. Paul's Cathedral. It burned beyond use. Christopher Wren designed a new St. Paul's (an architectural masterpiece). He incorporated a telescope in its design, although it never worked quite as intended.

While this was going on, Hooke kept his main job as curator to the Royal Society. Most members were rich; he was not. Hooke, who had engineering skills, was supposed to perform regular experiments at society meetings. He did that while doing his own scientific research and rebuilding the city.

The Great Fire of London was one of the greatest disasters that ever hit the city. Few people died, but about 80 percent of the buildings burned.

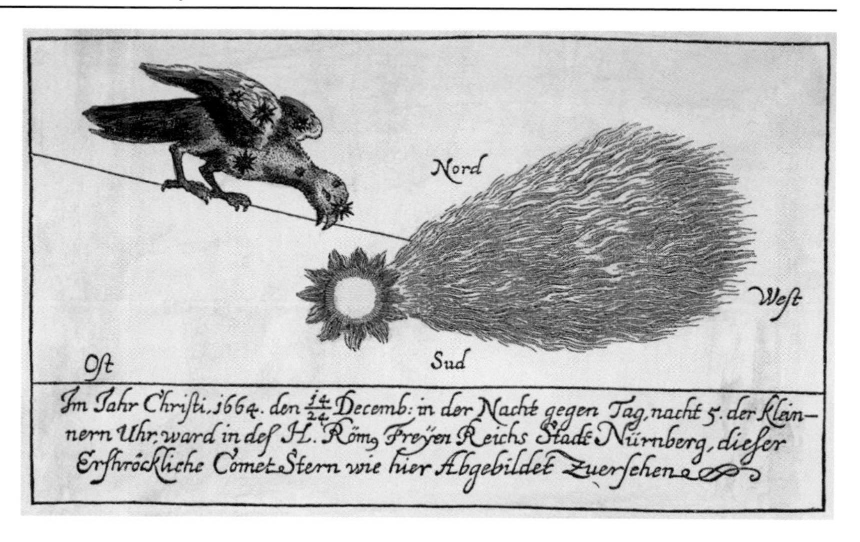

The end of the seventeenth century brought a bevy of comets, including one in December 1664, documented in this German engraving. Many people thought comets were warnings of dire events to come. So when the plague hit in 1665 and the Great Fire struck a year later, comets around those years seemed especially ominous. In *Annus Mirabilis*, Dryden wrote:

> *The utmost malice of their stars is past,*
> *And two dire comets which have scourged the town*
> *In their own plague and fire have breathed their last,*
> *Or dimly in their sinking sockets frown.*

Dryden's poem, dedicated to the king, somehow makes all this sound triumphant. Then he promises imperial glories to come. It is what most English people want to hear.

The poem is titled *Annus Mirabilis: The Year of Wonders, 1666. Annus mirabilis* (AN-uhs mi-RAB-uh-lis) means "miracle year" in Latin. In 1668, a year after the poem is published, John Dryden is named England's poet laureate.

Dryden has no idea that the title of this poem will later be used to describe the work of a then-little-known Cambridge professor. But that's exactly what happens. So put the year 1666 in your head. (All those sixes make it an easy year to remember.) It is known in the scientific world as THE *annus mirabilis*—but not because of a Dutch/British war or London's fire.

During that astonishing year, 1666, Isaac Newton develops his ideas on gravitation, lays the foundation for calculus, comes up with a theory of colors (what they are and how we perceive them), and begins working on laws of motion. He is 23. Those theories will revolutionize science and mathematics. But no one knows that for a while, because Newton doesn't publish his work.

He is still working out his ideas, and he is not ready to have them widely seen. Newton is a perfectionist, possessed by thoughts and theories, and he wants to be sure he gets things right. Much later he writes of this time in his life,

> This is the most remarkable fact about the attraction of gravitation, that at the same distance it acts equally on equal masses of substances of all kinds.
>
> —James Clerk Maxwell (1831–1879), Scottish physicist, *Matter and Motion*

saying, "In those days I was in the prime of my age for invention & minded Mathematicks & Philosophy more than at any time since."

Like all good scientists, he asks himself questions and then searches for answers. But there's a question about gravity that he can't answer. He knows that gravity exists in matter; the question is Why?

This is what he writes to the Reverend Richard Bentley (who is a friend and very interested in the new science): "That gravity should be innate, inherent, and essential to matter, so that one body may act upon another at a distance through a vacuum . . . is to me so great an absurdity, that I believe no man who has . . . a competent faculty of thinking, can ever fall into it."

A painting by Godfrey Kneller shows Newton when he was 46 years old. Newton was half that age during his "miracle year"—which was actually about 20 months long. At age 23, he created the basis for modern mathematics, mechanics, and optics.

Gravity an absurdity? Yes; Isaac Newton, that seventeenth-century marvel, that force of nature, wrote that. He can describe gravity, he can measure it, he knows it is there, but why? Why is it "inherent" in matter?

When he holds keys in his hands and releases them, they drop down toward the center of the Earth. But why? Why don't the keys just hover in place? Or why don't they fall up?

He doesn't know.

ANNUS MIRABILIS—THERE'S MORE THAN ONE

The gravity act doesn't end with Newton. In the twentieth century, Albert Einstein's General Theory of Relativity builds on Newton's ideas. (Einstein had mathematical tools that Newton didn't have.) Too bad Isaac wasn't around to learn of it. Einstein gave us a picture of the universe that helps us explore space and understand the action inside atoms. We still don't know all we'd like to about gravitation. Finding a more complete theory is one of the challenges for twenty-first-century scientists.

We have learned that the universe is expanding and, because of the Inverse Law of Gravitation, the larger the universe gets, the less effective gravitation is in slowing the expansion. Which means our universe is getting bigger and bigger, faster and faster. (More on that in book 3 of this series.)

Much of today's knowledge of the universe is based on an understanding that came from Isaac Newton: **Physical laws that work on Earth also work throughout the observable universe.**

As to the label *annus mirabilis*, science now has two of them—1666 and 1905. That second date is the year Einstein published four scientific papers that would change the world. He was 26. His paper on general relativity was yet to come.

SOME COOPERATION PLEASE? NOT US, SAID ISAAC AND ROBERT

Robert Hooke (1635–1703) was seven years older than Newton. Their rivalry, based on misunderstandings, became nasty.

How can something invisible be a force? How can it have attractive power? Descartes said there must be some medium in the atmosphere and beyond through which gravity works. Robert Hooke said there doesn't have to be anything. He was thinking about magnetism when he said there could be "action at a distance." Isaac Newton also described gravity as action at a distance and probably got the idea from Hooke.

Newton hated Hooke, but he took many of Hooke's ideas and made something of them. For a long time it was only Newton's opinion of Hooke that got into history books. Recent scholarship has shown that Hooke was a first-rate scientist and that Newton's nastiness toward him was in part an unwillingness to admit he had built on some of Hooke's work. But Hooke could be bristly, and Newton wasn't the only one he antagonized.

Newton and Hooke actually had a lot in common. Both were serious scientists who believed in experimentation and proof. "The truth is," Hooke wrote, "the Science of Nature has been already too long made only a work of the brain and the fancy: It is now high time that it should return to the plainness and soundness of observations on material and obvious things."

Isaac Newton's hates were legendary, but he could be nice to those he liked. And he loved his niece, Catherine Barton. When she caught smallpox, he worried. Here is a bit of a letter he wrote to her: "Pray let me know by your next how your [face is] and if your fevour be going. Perhaps warm milk from ye Cow may [help] to abate it. I am Your loving Unkle, Is. Newton."

FEIGN (FAYN) means "to pretend," but it comes from the Latin verb *fingere*, which means "to form or fabricate" (as in shaping clay or dough). The Latin meaning adds a twist to the expression "I feign no hypothesis."

Newton does figure out that the force of gravity works at a distance—through the void of space—which is a huge breakthrough idea. He can account for it mathematically, but why or how it works through space he doesn't know. He says, "I have not been able to discover the cause of those properties of gravity...and I feign no hypotheses."

He actually writes these words in Latin: *Hypotheses non fingo*. Newton got that line—"I feign no hypotheses"—from Galileo. It is already famous. It means he isn't going to speculate or guess. He is being modest and scientific. He knows what he can prove, and he intends to stick with that.

Here are some more of Newton's words from that letter to Reverend Bentley. "I have explained the phenomena...by the force of gravity, but I have not yet ascertained the cause of gravity itself....Pray do not ascribe the notion to me."

Well, Newton doesn't get his wish. When we think of gravity, we think of Isaac Newton!

"What an ego! One day it's Newton's laws of dynamics, then it's Newton's theory of gravitation, and Newton's law of hydrodynamic resistance, and Newton's this and Newton's that."

Isaac Newton will proclaim in 1687 (when he finally publishes his ideas) that gravity is the universal force that describes how, but not why, two objects attract each other with a force proportional to their masses and inversely proportional to the distance between them.

Time and Space? They Are Absolutely Absolute, Said Isaac

Time ticks away, one second after another, the same here, the same there. Or so it seems to our senses, and so it was to Isaac Newton: "Absolute, true and mathematical time, of itself, and from its own nature, flows equably without relation to anything external," he wrote.

As to space: "Absolute space, in its own nature, without relation to anything external, remains always similar and immovable."

When you think of Newton, keep that absolute idea in mind. Gottfried Leibniz didn't agree. And be prepared: In the twentieth century, all our sensible ideas on time and space will be challenged. Time and space aren't what they seem to be. As to time ticking uniformly on a giant cosmic clock? Albert Einstein explained that it doesn't.

CALCULUS? WHO DONE IT?

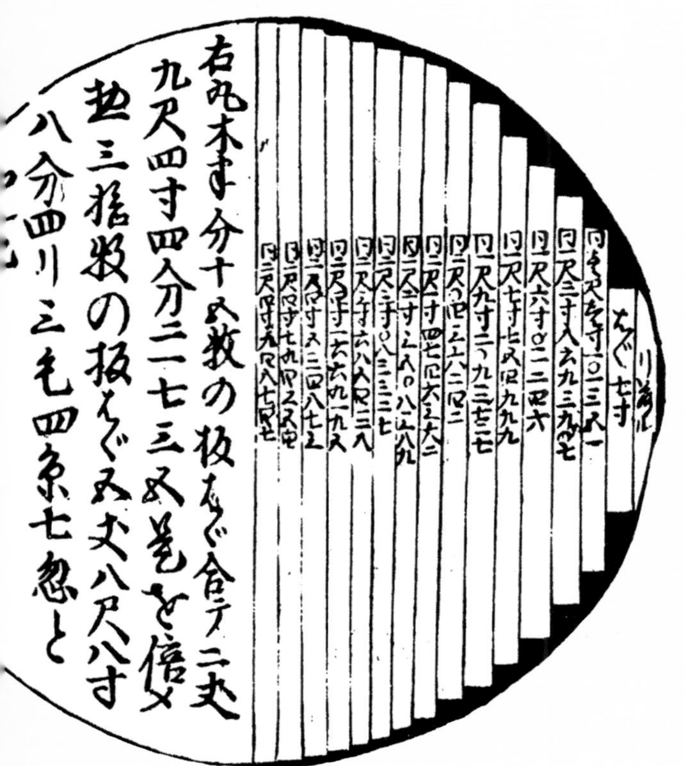

This rectangular approach to estimating the area of a circle is from a 1670 book titled *Kokon Sanpoki*, by Kazuyuki Sawaguchi, who was a student of Seki Kowa.

With arithmetic, you can compute quantities. With geometry, you can measure and compare lines and spaces. With algebra, you can manipulate ratios in an equation—how a change in speed affects the distance traveled, for example. But what if the answer you seek is a moving target? What if an object (say, a planet in an elliptical orbit) keeps speeding up and slowing down and changing direction along a path that's constantly curving? How do you measure it mathematically?

A number of mathematicians were on the trail looking for an answer to that question. François Viète (1540–1603), John Napier (1550–1617), William Oughtred (1574–1660), René Descartes (1596–1650), Pierre de Fermat (1601–1665), and Blaise Pascal (1623–1662) were some of them. They came up with logarithms (exponents like 10^1, 10^2, 10^3, etc.), a coordinate system, graphs, imaginary numbers, and other tantalizing ideas. All of the elements that would lead to calculus were in those ideas. But no one recognized the bits and pieces for what they were.

(more about calculus on page 162)

Calculus? What's That?

Mathematician Seki Kowa (1642–1708) was born the same year as Isaac Newton, but almost halfway around the globe, in Japan. The two geniuses never met or exchanged ideas. Yet Seki's geometry of *enri* ("circle principle") shares the same germ of an idea that led Newton and Leibniz to calculus.

You can't measure the area of a circle (or any curved shape) with a ruler, so Seki filled one with skinny rectangles. (The drawing, opposite, is by one of his students.) He measured each rectangle's area (that's easy—length times width), and then added them up. His answer was close to the circle's area, but it fell short. There were little black unmeasured spaces between the four-sided strips and the circle's curve. Seki demonstrated that skinnier rectangles fill in more of that space—and so give a more accurate measure of the circle's area.

With *enri*, you can keep drawing skinnier and skinnier rectangles (up to a point). But you'll always fall just short of the circle's area.

Calculus takes this simple idea infinitely further to produce an exact area. Imagine those strips as they slim down. They approach a width of 0 but, of course, they can't possibly be 0 (or they wouldn't exist). The calculus of Newton and Leibniz says, let them. Let 0 be a destiny you can never reach. Let 0 be, in the terms of modern calculus, your limit.

As the skinny rectangles head toward that unreachable limit and fill up more of the circle, their total area increases. For a circle with a radius of 1 (use any units you want), the series of rectangle areas might look like this: 3, 3.1, 3.14, 3.141, 3.1415.... Unlike the rectangles aiming at impossible 0, these numbers are heading someplace real—toward the actual area of the circle.

Now, picture rectangles filling the circle, but jutting just beyond its perimeter, as shown in red below. As these oversized strips narrow down, they do the opposite of the inner rectangles; they get decreasingly closer to the circle's area: 3.3, 3.25, 3.21, 3.17, 3.145....

The exact area of the circle is trapped between two infinite series of numbers—one increasing and one decreasing. Like a baseball runner caught between bases, you just have to "tag" it. That's what calculus does. It's a shortcut way to put your finger on a number that's otherwise out of reach. The general idea is that, by treating a limit (rectangles of 0 width) as part of an equation, you can pinpoint its relationship to a real figure (the area of the circle).

Of course, you *could* just use that familiar algebra equation: $a = \pi r^2$. Plug in any radius, square it, and then multiply by π (pi = 3.14159...), and you have the area of a circle of any size.

So why do you need calculus? Not all curves are circles. In physics, they can be ellipses, like Kepler's planetary orbits. (A big reason Newton invented calculus was to measure the constantly changing speed, direction, acceleration, and gravitational forces of orbiting planets.) Curves can be parabolas, like the trajectories of objects that Galileo plotted (see chapter 7).

In fact, physics curves can be any wiggly, snakelike, irregular shape you can imagine. Picture a graph of fluctuating temperatures on a stormy day, or the changing pressure of a tire over the lifespan of a car, or the fickle velocity of wind on a blustery day. Whatever the curve, whatever the measure of changing energy or matter, calculus can handle it.

—LJH

John Napier designed a calculator (left) for large numbers. It reduced multiplication to a series of additions ($5 \times 4 = 5 + 5 + 5 + 5$, for example) and division to simple subtractions. Leibniz's "Stepped Reckoner" (below) computed numbers, too, but also found roots (3 is the cube root of 27 because $3 \times 3 \times 3 = 27$).

Calculus is the branch of mathematics that deals with continuous change. Using calculus, you can compute the velocity or acceleration of a runner at any point during a race (see graph, page 163), the varying stresses along a bridge, and the volume of a curvy, irregular lake. The components of calculus were available in the seventeenth century, but no one had put them together. Then Newton did it but did not publish his work. Working separately in Germany, Gottfried Wilhelm Leibniz (1646–1716) also invented calculus. He published his work. That led to a famous fight between the two. Who was first? (Probably Newton.)

Newton had a hard time admitting that any of his contemporaries had ability. He was sure Leibniz had copied his idea. (That wasn't true.) Mathematicians and scientists in England lined up behind Newton; those on the European continent backed Leibniz. The fight over who should get credit for calculus became mean-spirited, counterproductive, and silly (as fights often are). Today, we mainly use Leibniz's calculus notation, which is more economical, thanks to later mathematicians who streamlined it.

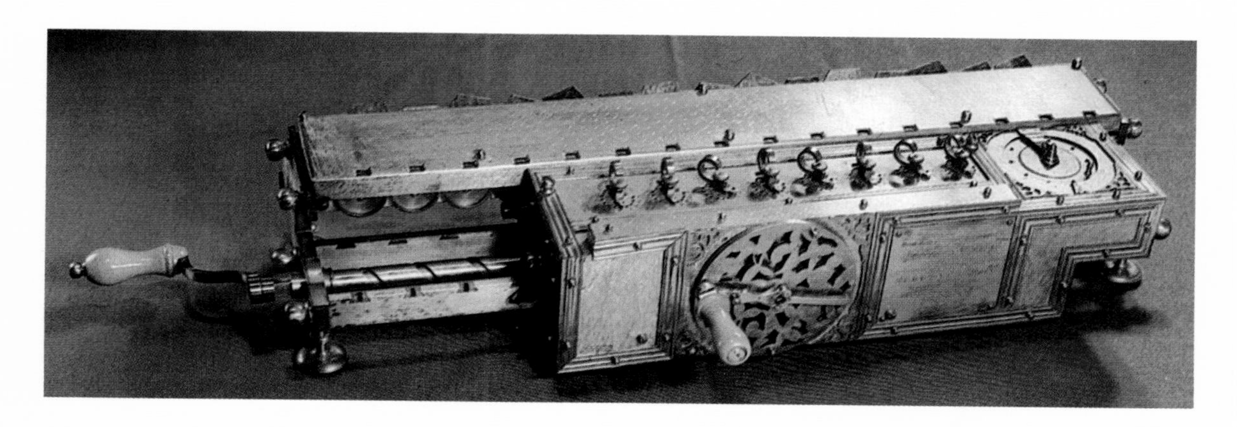

CALCULUS—RUN WITH IT!

This curve (below right) shows the changing speed of a sprinter in the first 10 seconds of a race. The higher the line (measured on the *y* axis), the faster the runner was moving at that point in time (the *x* axis). Could you outrun this runner? To answer that, you have to know how much ground was covered during those first 10 seconds. Geometry (the graph) shows you. Distance equals the shaded area under the curve. Check it out: If a faster runner comes along, the curve goes higher and the shaded area—the distance covered—gets bigger. Same thing if you increase the time (make the *x* axis longer)—the distance also increases.

To find the exact value of that curvy area, you can't measure it with a ruler or count the grid squares. Instead, you need the "skinny rectangle" technique (see page 161)—officially called *integral calculus*. Imagine rectangles just under and just above the curve, with their total areas converging toward each other as their widths shrink. An integral operation instantly sums up their changing areas and pinpoints their ultimate meeting point—the total area under the curve.

How does this amazingly powerful operation actually work? That's what you learn in calculus class—but don't let anyone tell you it's hard. It just takes a little training and practice, like running a race. As it turns out, the sprinter is almost unbeatable; the distance covered in 10 seconds is 100 meters—a world-record pace!

Another powerful calculus operation does the opposite of integral calculus (similar to subtraction undoing addition). Instead of finding an area, *differential calculus* can single out any point along that curve and reveal the runner's exact acceleration at that instant.

Try it: The dot on the graph stands for the runner's speed at the 3-second mark. You could take a ruler and draw a line from that point to the *y* axis and read the value (about 10 meters per second). Okay, but how is the runner's speed changing at that point? With geometry (the graph), you can see acceleration at a glance—just by looking at the slope of the curve. The runner started at 0, sped up really quickly (the steep part), and then leveled off near the very end. To see the acceleration at any one point, you can draw a tangent line through the point, as shown, and look at its slope. Pick another point, draw another line, and you can compare acceleration at those two points. Keep it up, and you'll soon find that the acceleration is changing from one point to the next.

The hitch is, there are infinitely many points on the curve. Differential calculus, ever helpful, uses that fact to pin down the exact acceleration at any point.

—LJH

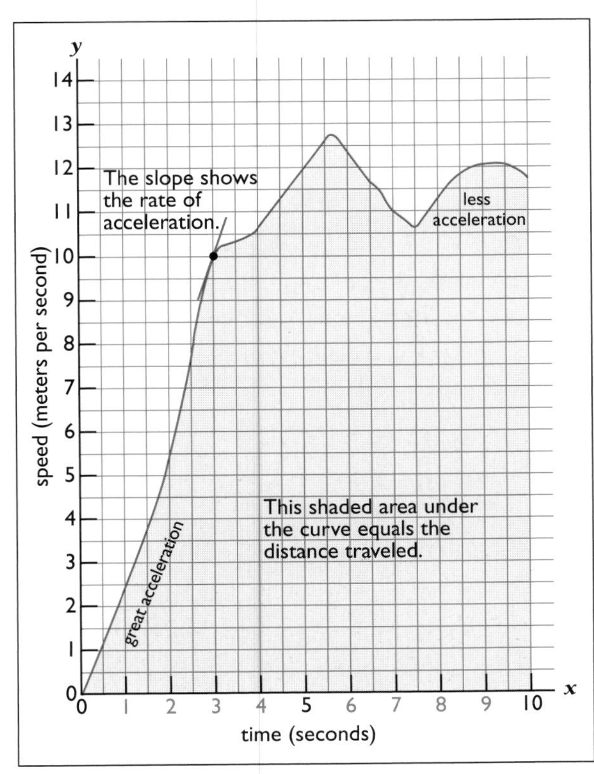

15

Newton Sees the Light

And from my pillow, looking forth by light
Of moon or favouring stars, I could behold
The antechapel where the statue stood
Of Newton with his prism and silent face,
The marble index of a mind for ever
Voyaging through strange seas of Thought, alone.
—William Wordsworth (1770–1850), English poet, *The Prelude*

Watch the stars, and from them learn.
To the Master's honor all must turn,
Each in its track, without a sound
Forever tracing Newton's ground.
—Albert Einstein (1879–1955), German-American physicist, from an interview in *The Saturday Evening Post*

Newton keeps to his peculiar ways—using his mind on everything that he finds of interest. He is now back at Cambridge as a professor, although hardly anyone goes to his lectures. His assistant, Humphrey Newton, says, "So few went to hear him, and fewer understood him, that oft times he did in a manner, for want of hearers, read to the walls."

As to food, that same assistant writes, "So intent, so serious upon his studies that he ate very sparingly, nay, oftimes he has forgot to eat at all, so that going into his chamber, I have found his mess untouched, of which, when I have reminded him, would reply—'Have I!' and then making to the table, would eat a bit or two standing."

No matter. Eating isn't important to him. He is thinking. And reading. He reads what Hooke and Kepler have to say about light and color. Then he takes a prism and observes the

Favour and *colour* are British spellings; in the United States those words are spelled *favor* and *color*.

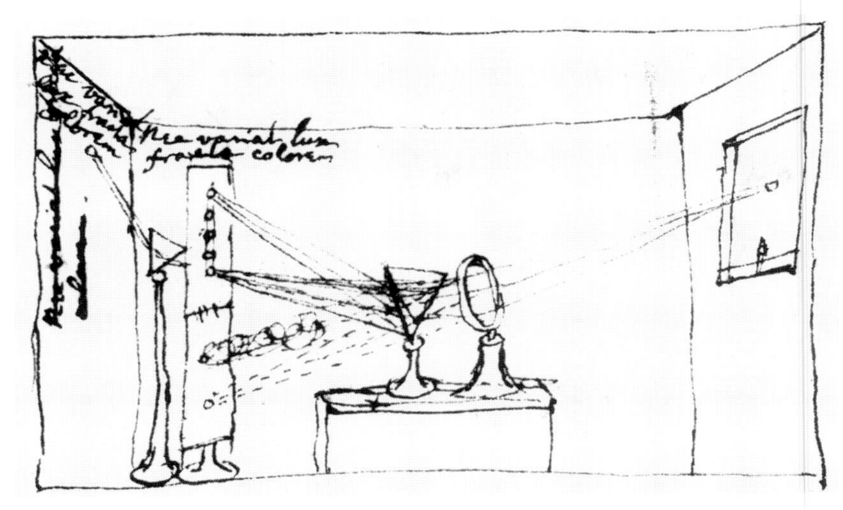

A sketch (left) from Newton's 1672 notebook shows sunlight entering through the window at right, passing through a triangular prism, and splitting into a spectrum of colors. One of the earliest known studies of optics (the science of light and vision) was done by Islamic mathematician Ibn al-Haytham (965–1040), also known as Alhazen. His sketch of lenses is below.

way light shining through the triangular glass divides itself into colors. He thinks about that process, and writes about optics, which is the study of light. He asks himself Exactly what is light made of? And how does it travel? What is color? Then he answers those questions by describing his research. Here is Newton writing of his experiments with prisms.

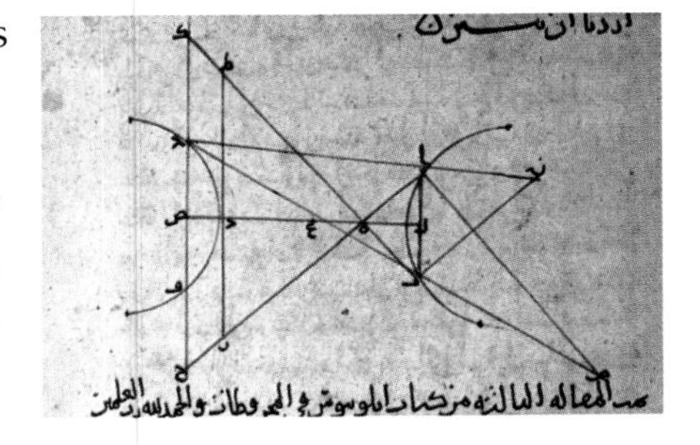

> *In a darkened Room make a hole in the* [windowshade] *of a window, whose diameter may conveniently be about* [one-third] *of an inch, to admit a convenient quantity of the Sun's light: And there place a clear and colourless Prisme, to refract the entring light towards the further part of the Room, which, as I said, will thereby be diffused into an oblong* [spectrum].

From the time of Aristotle, and perhaps before, philosophers have believed that light is a simple, homogeneous entity. But when Newton guides a beam of light through a tiny hole and a triangular glass prism, he sees a rainbow of colors—from red to violet—forming a rectangle on the opposite wall. Why isn't it white? Light doesn't seem to be either simple or homogeneous. He isn't the first to see light pass through a prism and become a rainbow. The general thinking is that

Newton was the first person to call a rainbow a spectrum of colors. To a modern scientist, a SPECTRUM is any range of entities—like the colors red, orange, yellow, green, blue, violet—appearing in order of magnitude. We now know that a rainbow is a thin slice of a much wider electromagnetic (EM) spectrum, not discovered in Newton's time.

The color spectrum spans from red, which is almost at the same angle as the white light, to the more sharply bent violet. The colors blend seemlessly, one into the next. In his early sketches, Newton labeled five of them: red, yellow, green, blue, and violet. He later added orange and indigo (deep purple), but the true number is infinite.

somehow the prism creates those colors.

Newton's experiments prove something else. He shows that white light—ordinary visible light—is the sum of light rays, each of which produces a different color. Those rays are of varying wavelengths, and each color bends at a slightly different angle, creating the rectangular lineup. That bending, known as refraction, is a kind of signature that designates a color. "If the Sun's light consisted of but one sort of rays, there would be but one colour in the whole world," he writes. A prism doesn't create colors; it separates them when each band of color refracts differently.

Newton does another experiment: He takes a long piece of cardboard and paints one half of it bright red, the other bright blue. Then he puts the paper in the sunlight, refracting the red and then the blue through a glass prism. He finds that the "blue half will be lifted higher by the refraction than its red half." (By "lifted higher," he means refracted at a greater angle.) Nothing he can do changes that refraction—or that color. There are no rainbows here: Red sent through a prism always stays red; the same with blue.

Everyone believes that colors are a mixture of white and black. Newton stands back from the pages of a book, he sees what happens when white and black blur together: they become gray. No color appears. It is an easy experiment, but no one has done it before and understood its significance. He realizes that white light contains the colors; black is just the absence of light.

Newton says that just "as sound in a Bell or musical String . . . is nothing but a trembling Motion" so color is just a tendency "to reflect this or that sort of Rays." There is no mystery to it. Color is light rays that set up a motion that, when it meets our eye, gives us the sensation of color. Color doesn't exist without light.

When he isn't writing or thinking, Newton puts on his scarlet professor's gown and lectures to the one or two students who turn out to hear him. The talking seems to help him clarify his thoughts. He is doing groundbreaking work, but the students don't know that.

Newton is thinking about the very nature of light. He suggests that light is made up of "corpuscles" (tiny particles) that are shot from their source like bullets from a gun. (Is he right? More on this to come.)

What makes a rainbow in the sky? Almost everyone has wondered about that. According to Newton, when there is both sun and rain at the same time, the raindrops become prisms and bend and separate the sunlight. An observer on Earth, with his back to the Sun and his eyes looking at clouds (through the raindrops), sees the refracted spectrum we call a rainbow.

Something important is going on here: Newton is doing experiments. Scientists of his day hardly ever do that. Mostly they debate ideas. "The proper Method for inquiring after the properties of things is to deduce them from Experiments," writes Newton.

His experiments with light prove that white light is the sum of the rainbow of separate and distinct colors. He writes a paper about his discoveries and sends it to the Royal Society in London. "But the most surprising, and wonderful composition was that of Whiteness," he writes in that report. "There is no one sort of Rays which alone can exhibit this. 'Tis ever

The knowledge of light that Newton's experiments initiated led in unexpected directions. Today we have confirmed that the Sun and other stars are made of the same elements as the Earth, because we can analyze their radiant energy, which includes light. We've learned that each element absorbs and gives off energy in a unique way. Like a fingerprint, an element's spectrum reveals distinct patterns of lines (shown in the diagram above). The lines correspond to the wavelengths the element absorbs or emits. The science of all this is called spectroscopy.

You might know CORPUSCLES as blood cells, but that meaning didn't come along until the mid-nineteenth century. The word comes from the Latin for "little bodies"—as in small particles of matter.

Newton's telescope (the photo below and Newton's sketch at right) was much shorter than Galileo's—about as long as this book is wide—and yet it was more powerful. It's called a reflecting telescope because it uses mirrors instead of lenses. His first one, built in 1668, had metal mirrors. Today's Newtonian telescopes use glass mirrors coated in silver or aluminum.

Newton had terrible bursts of temper. He imagined wrongs where they didn't exist. You could describe him as paranoid (which means he had an abnormal tendency to mistrust others). But he also felt genuine remorse when he discovered he had made a mistake, and he was quick to apologize, which is admirable.

By the time Newton's book *Opticks* was published in 1704, his rival Robert Hooke was dead.

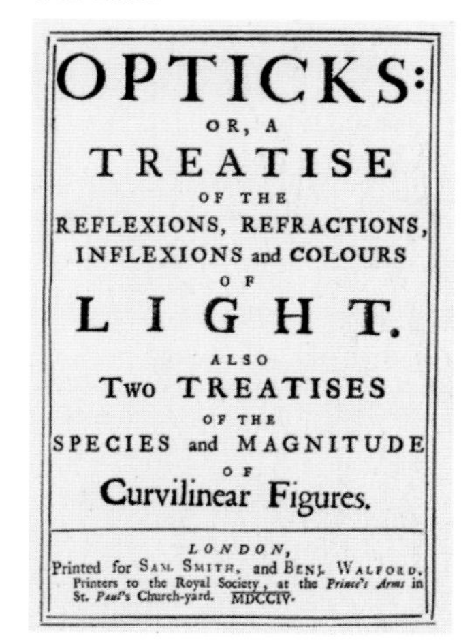

OPTICKS:
OR, A
TREATISE
OF THE
REFLEXIONS, REFRACTIONS, INFLEXIONS and COLOURS
OF
LIGHT.
ALSO
Two TREATISES
OF THE
SPECIES and MAGNITUDE
OF
Curvilinear Figures.

LONDON,
Printed for SAM. SMITH, and BENJ. WALFORD, Printers to the Royal Society, at the *Prince's Arms* in St. *Paul's* Church-yard. MDCCIV.

compounded . . . [of] all the aforesaid primary Colours."

But when his paper on light is circulated at the Royal Society, it is attacked. One of its critics is Robert Hooke (an officer of the Royal Society), who has also written about light, also believes in experimentation, and disagrees with Newton over the nature of light. Hooke thinks light is wave action, not particles.

Newton can't handle criticism. He grows to hate Hooke. He says he is being "persecuted" and resolves never to publish again. And he doesn't for many years.

But he doesn't stop working. His experiments with prisms lead him to experiment with a telescope. There is a problem with Galileo's refracting telescope. It uses two lenses. The rim of one of those lenses often acts as a prism, blurring the light and creating rainbow hues at its edges. Newton devises a *reflecting* telescope with a curved mirror in place of one lens. It solves the problem of blur and turns out to be more powerful too. In 1668, Newton puts a mirror inside a squatty telescope the length of a small hand; it magnifies 40 times. That's as good as a refracting telescope 10 times its length.

It is typical of Newton that he builds his own telescope—shaping and polishing the mirror himself. When someone asks him where he got the tools to do that, he says he made

HOOKE'D ON MICROSCOPES

Robert Hooke's greatest work, published when he was 29, is titled *Micrographia.* Written in clear, readable English (rather than scholarly Latin), the book focuses on the small-scale world as seen through his microscope (below right). Among other things, Hooke describes the compound eye of a fly, the construction of a butterfly's wing, and the structure of feathers. He identifies fossils as the remains of once-living creatures and plants. The book includes detailed drawings by his architect friend Christopher Wren. Samuel Pepys (PEEPS) wrote in his diary that he stayed up until 2:00 A.M. with "the most ingenious book that ever I read in my life."

Hooke wrote, "By the help of Microscopes, there is nothing so small, as to escape our inquiry." He looked at the point of a needle, which seemed sharp and regular, but under a microscope was blunt and ragged. He put a book under his lens and focused on a period at the end of a sentence. It too was ragged and looked "like a great splotch of London dirt." What was to be made of this wondrous tiny irregular world?

Elsewhere, other pioneers of microscopy also had their eyes to the lens. They included a Dutchman, Antonie van Leeuwenhoek (1632–1723), and an Italian, Marcello Malpighi (1628–1694).

But Hooke may have known little or nothing of their work. Working on his own, he invented a greatly improved compound microscope that used two or more lenses. When he looked at cork under the microscope, he noted its pores and called them "cells." Those pores aren't cells in the modern sense, but because of Hooke's terminology, that name was given to biological cells when they were discovered in the nineteenth century.

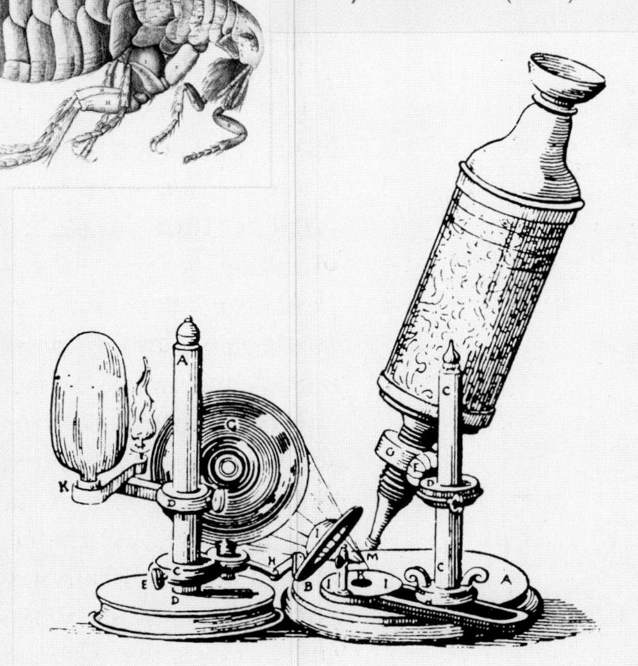

Hooke's microscopic world, illustrated by Christopher Wren, revealed astonishing structures never seen before. A flea (center) has powerful jumping legs that end in grappling hooks, a flat body for sliding between hairs, and a mouth tool for piercing skin. Hooke gave the name "cells" to the tiny holes in cork (above).

AN "ILLUMINATING" TOOL

Seventeenth-century scientists and artists were fascinated with the *camera obscura* (below), a technological marvel of the time. They let a pinpoint of light enter a dark room, and the entering rays projected an upside-down image of the scene outside the window. Johannes Kepler improved the *camera obscura* by adding a convex lens. Painters used it as a guide, which might explain the near-photographic precision of seventeenth-century Dutch art. Artist Samuel van Hoogstraten said it provided "illumination" for the young painter, using "illumination" in a double sense—light and the gain of knowledge.

Dutch painter Jan Vermeer (1632–1675) is thought to have composed several of his masterpieces with the aid of a *camera obscura*. The *Music Lesson* (above) shows a dramatically receding room with perfectly condensed objects, an effect of the fore-shortening view through a *camera obscura*.

them. "If I had [waited] for other people to make my tools and things for me, I had never made anything of it." He studies Jupiter and its moons and sees Venus through its phases.

You might think that all his scientific work would be enough to keep Newton busy, but he also spends much time writing about religion and alchemy.

His religious beliefs explain, in part, his secrecy. Newton has something to hide. Reading the Bible in its original languages, he finds he cannot believe in the Trinity (God the Father, Son, and Holy Ghost). Newton is a professor at Trinity, one of the colleges at Cambridge University. The authorities at Trinity College would not be happy to hear that he thinks their college is named after an error. When

"This most beautiful system of the sun, planets, and comets could only proceed from . . . an intelligent and powerful Being," wrote Newton, who was deeply religious (as were Copernicus, Galileo, and Kepler).

Parliament passes a bill of religious toleration, it says two groups are not included in that toleration: Catholics and "any person that shall deny, in his preaching or writing, the doctrine of the blessed Trinity." Newton is deeply religious, but in the England of his day, he lives with the terror of being exposed as a heretic.

Alchemy is something else. It is popular and seems to have a future. Newton has a furnace built near his rooms at Trinity College. There, he can melt, distill, and calcine—which means to heat something until it loses its moisture and turns to an ashy powder called calx. Calx, according to the eighth-century Arabic alchemist Geber, "is the treasure of a thing." Ever precise, Newton weighs calx carefully using a balance scale. Alchemy will prove to be a dead end, but no one knows that without treading its path. And, unlike the Aristotelians, the alchemists are experimenters. That practical training serves Newton well.

Isaac Newton stands between the old world of mystery and alchemy and a new world of logic and proof. His feet are in both those worlds; but it is he, more than anyone else, who lights the path leading into that new world.

The seventeenth-century illustration above gives some idea of how strange and intricate alchemy was. The two circles in the center spell out "hot, cold, dry, wet" (on the left) and the four elements in relation to the planets (on the right). Alchemy was sometimes tied to astrology, and the seven known metals (gold, silver, tin, copper, iron, lead, mercury) were tied to the seven heavenly bodies.

16 Newton Moves

[Newton's] first law tells us what happens to the motion of a body in the absence of force. It says that nothing happens.

—Brian L. Silver (d. 1997), physical chemist and science historian, *The Ascent of Science*

Like effects in nature are produced by like causes, as breathing in man and in beast, the fall of stones in Europe and in America, the light of the kitchen fire and of the sun, the reflection of light on the earth and on the planets.

—Isaac Newton (1642–1727), The *Principia*

We are Newtonians, fervent and devout, when we speak of forces and masses, of action and reaction; when we say that a sports team…has momentum; when we note the inertia of a tradition or bureaucracy; and when we stretch out an arm and feel the force of gravity all around, pulling earthward.

—James Gleick, modern American author and journalist, *Isaac Newton*

Is motion something inside an object? That is one of the scientific puzzles of the seventeenth century. If a wagon moves, of course the horse pulling it has a lot to do with that movement. But most people believe that some force in the wagon itself also helps make it move. As to the Moon and planets—almost everyone thinks there is something inside the Moon that makes it move around Earth. Then Newton comes along and says that isn't so—for a wagon or for the Moon. An outside force is necessary for anything, even the Moon, to change its state of motion or of rest. That may not seem like a big idea, but it is huge.

Newton proves the point with three laws of motion, laws that help make him one of the

Stargazers observe planets by eye and by telescope in part of an eight-panel painting created by Donato Creti in 1711.

FUN WORDS? WELL, IMPORTANT WORDS IS WHAT THEY ARE

Before you can deal with Newton's laws, there are some words you need to be clear about. The first of them are **inertia** and **force**.

Galileo understood that **inertia** is not motion or lack of motion; it's **resistance to change**. (Technically, it is a measure of the resistance to change.) Something with a lot of inertia is hard to push around. A rock has much more inertia than a pebble. You can move a pebble with a finger; a rock may take heavy equipment.

A **force** is a push or pull that acts on an object's inertia (like the whack of a baseball bat on a ball). A force changes a body from a state of rest or of uniform motion.

Newton wanted to find a measure that would have nothing to do with Earth's gravity. He came up with the idea of **mass**, which is the quantity of matter in a body as measured by its inertia.

The English language was well established when the concept of mass was discovered. Mostly because of old language habits, we sometimes tend to mix mass and **weight**, but they are not the same. Mass does not change with the strength or weakness of a gravitational field. Weight does.

In the physics of Aristotle, the "natural" state of motion of an earthly body is rest.... The principle of inertia, first developed by Galileo and later by René Descartes, and perfected by Newton, is that a state of motion with uniform velocity is as natural as a state of rest.

—Rocky Kolb, American physicist, *Blind Watchers of the Sky*

greatest—some say *the* greatest—scientists ever. Those laws govern the motion of bodies, from tennis balls to galaxies. Once you get them, you'll see that Newton's Laws of Motion are surprisingly simple, and yet, by using them, we humans will go to the Moon and someday probably farther.

So here they are, **Newton's Laws of Motion:**

His **first law** is sometimes called the **Law of Inertia.**

It says: **Every body remains in a state of rest, or of unchanging motion in a straight line, unless it is acted upon by a force.** In other words, the kitchen table won't get up and walk away unless you exert the force of your muscles and lift it. If something is at rest, it will stay at rest (unless a force interferes).

I find it helps to split the first law into two statements. What they both say is: Things don't change by themselves.

1. A body at rest will remain at rest unless a force interferes.
2. A body moving in a straight line will continue to do so at a constant velocity, unless a force slows it down or speeds it up or moves it out of its path.

In everyday language, INERTIA is sometimes used as a synonym for laziness: "His inertia is keeping him in his seat." It relates to the scientific meaning, which is "resistance to motion, action, or change."

That word BODY has a whole lot of meanings. To a scientist, it can be almost any tangible object, from a molecule to a planet.

OOMPH!

Sports commentators like to talk about momentum. Teams have it—then lose it—and sometimes regain it. In the end, the team with the most momentum usually wins. So what is momentum?

In Latin, the word means "movement." In English, it's been described as "moving power," "motion content," and (my favorite) the "oomph of a moving object." Oomph? That's not a scientific definition, but it fits.

Physicists define momentum in mathematical terms: mass times velocity (m × v). Notice that you need both mass and velocity (velocity is speed and direction) to have momentum. A stationary elephant can't run you over if its velocity is 0. Really massive and speedy objects, like charging bull elephants, have so much oomph they're scary. But there's an important "or" to remember: An increase in mass *or* an increase in velocity means greater oomph. A lightweight but superspeedy object, like a bullet, can have the same momentum as, say, a running house cat.

In January 2003, during liftoff of the space shuttle *Columbia*, a piece of lightweight foam came loose. It weighed less than a large loaf of bread. But because it hit the space shuttle at high velocity—877 kilometers (545 miles) per hour—the foam punched a small crack in the wing. During reentry, on February 1, the intense heat widened the crack, and the spacecraft overheated and broke up, killing all seven astronauts.

Now that isn't what we see every day. Give a wagon a shove, it will move for a while and then slow down and stop. It's logical to think as Aristotle did: Things don't keep moving forever.

Yet Newton says the wagon *will* keep moving unless something interferes. What is it that stops most wagons in our everyday world? Friction. Newton says that if we put that wagon in a frictionless vacuum, it will keep going.

(And he has been proved right. We now know that a spaceship beyond the pull of the Earth's gravity will continue moving in a straight line unless some force—like gravitation from another object—changes its path.)

On to the **second law**. (These laws are keys to much of the world of science, so stay with me.) It says: **The force acting on a body is directly proportional to, and in the same direction as, its acceleration.** Which means: Give a wagon a kick, and it will go forward rapidly. Give it a pat, and it will hardly move.

Remember: ACCELERATION is any change of velocity—either speed or direction. It includes slowing down as well as speeding up.

Something follows from the second law. It is this: The resistance to changes in motion increases as the mass increases. Sometimes this is stated as: Resistance to change in motion is inversely proportional to the mass of an object. What does that mean? Given the same horsepower, a motorcycle will leave a car in the dust.

Newton's second law is about **acceleration** and **force**. **Force is the product of mass and acceleration.** It leads to a simple but much-used formula that allows us to measure changes in motion precisely:

$F = ma$

Force equals mass times acceleration.

Put another way:

$a = F/m$

Acceleration equals force divided by mass.

Newton's **third law** completes his laws of motion. It is sometimes called **the Law of Action-Reaction**. Here is how he states it: "To every action there is always opposed an equal reaction; or the mutual actions of two bodies upon each other are always equal, and directed to contrary parts. . . ." Or, in slightly simpler language: **For every action there is an equal and opposite reaction.** What it means is for every push there's an opposing push back. The Earth pulls on an apple, and the apple also pulls on the Earth. In his words:

"Whatever draws or presses another is as much drawn or pressed by that other. If you press a stone with your finger, the finger is also pressed by the stone. If a horse draws a stone tied to a rope, the horse (if I may so say) will be equally drawn back towards the stone."

When Newton used the word DRAW, he wasn't thinking of taking pencil to paper. In his time *draw* meant "pull." Substitute *pull* for *draw*, and his third law will make more sense.

Rockets follow the third law. So do bungee jumpers. So does a bouncing ball. And so do the Earth and the Moon with their gravitational interaction.

When you hit a nail with a hammer, which one exerts a force, the hammer or the nail?

If you said "hammer" or "nail," you're not thinking as Newton did. Newton thought in terms of interactions between things. Each object interacts on the other. The hammer pushes the nail, but it is brought to rest by the nail's push back. (You can see why all this was a breakthrough. It isn't the way we normally think about things.)

Reminder: It's the second law that makes it clear that there is a difference between mass and weight. The mass of a body is its resistance to acceleration. Weight is its response to the force of gravity. If two men of equal mass are in the same place (like the surface of the Earth), they have equal weight. Put one on the Moon, and he'll weigh less.

Newton realizes that the very same laws of motion that

A FEW THOUGHTS ABOUT FALLING

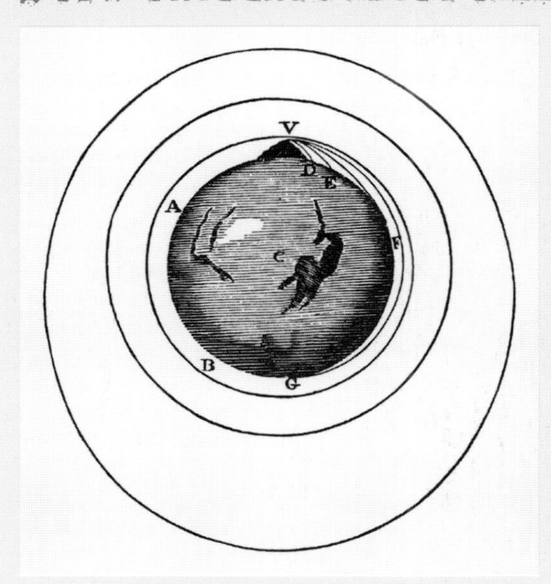

Picturing a cannon on a very tall mountain (V, above), Newton imagined, in his 1687 *Principia,* how an object might achieve orbit. The first artificial satellite, *Sputnik*, was launched 270 years later, in 1957.

To picture in your mind's eye why orbits are free falls, it helps to do a thought experiment. Here's one that Newton did. He imagined a cannon on top of an impossibly high mountain. Fire a cannonball horizontally, as shown at right, and it falls to the ground (A). Fire it with greater force, and the ball travels farther, but it still falls to the ground (B).

What if you could fire the cannonball far enough that it fell just beyond the curve of the Earth (C)? It would become a satellite—an object in orbit around the Earth. As long as the cannonball satellite maintained orbital velocity, it would fall forever, never hitting the ground.

Meanwhile, back on the mountaintop, suppose you fire a fourth cannonball with even greater force. It would zip beyond Earth's curve—and off into space (D). In other words, it would have escape velocity—enough speed and the proper direction to overcome Earth's gravity.

Keep in mind that this is all in your mind. It's a thought experiment. If cannonball C were a space shuttle, it would be orbiting just above the atmosphere at about 7.8 kilometers (4.8 miles) per second. Many satellites, higher still, travel 3 kilometers (almost 2 miles) per second. And the Moon? It averages 1 kilometer (.6 miles) per second. See the pattern? Just as Kepler said back in chapter 11, the higher the orbit, the slower the speed.

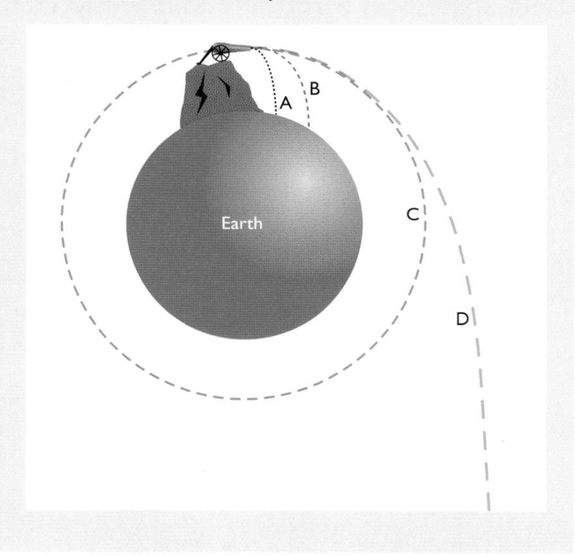

work on the Earth will also work on the Moon. A rock dropped from a tower in Italy will act the same way as a rock dropped from the Great Wall of China or from a vehicle that lands on Mars. No one told us that before Isaac Newton. He put us at home in the universe.

Newton's laws of gravitation and of motion are like brothers

NEWTON'S LAWS OF MOTION

1. Objects tend to stay at rest or move in a straight line at a constant speed unless acted upon by an outside force.
2. The force acting on a body is directly proportional to, and in the same direction as, its acceleration.
3. For every action there is an equal and opposite reaction.

Galileo looked at objects in motion and believed that their paths curved with the Earth's curve. So he thought uniform motion was curved. He was wrong, but it was a reasonable guess. It was the French mathematician René Descartes (1596–1650) who first said the natural tendency is for an object to move in a straight line. Newton asked, if Descartes is right, why doesn't the Moon shoot off into space? When he answered that question mathematically, he had his Theory of Universal Gravitation.

linked through family ties. Dealing with one seems to bring in the others. Newton's understanding of inertia (first law of motion) tells him the Moon (or any object) should keep moving in a straight line in space unless there's a force acting on it.

What happens when the force is at right angles to the motion—neither speeding it up nor slowing it down, but rather changing its direction? In that case, the object maintains its speed, but its path becomes curved.

Since the Moon's path is curved, there must be a force acting on the Moon to make it go around Earth. When Newton determines what that force has to be, and then figures out that the same force is pulling the apple from his mother's tree—he ties motion on Earth to motion in the heavens. This is the genius of Newton—realizing that the Moon's curved path is really the Moon "falling" to the Earth.

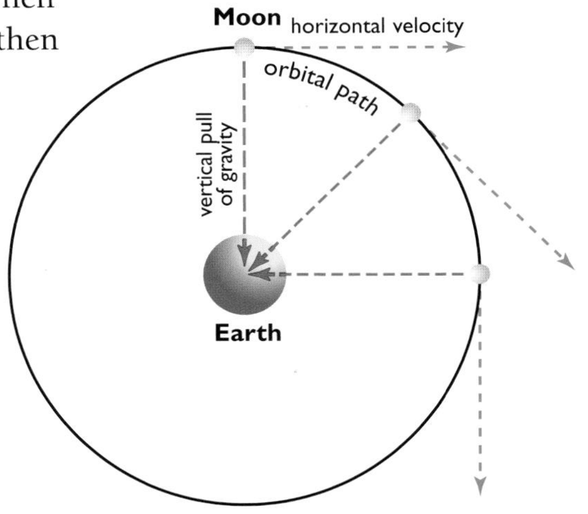

Yes, thanks to gravitation, the Moon is in a free fall toward Earth, just like an apple. But it never gets to the ground, because the Moon's inertia is maintaining its motion in a "straight" horizontal line. Newton realizes that it is the combination of forward movement (inertia) and downward pull (gravitation) that keeps the Moon moving parallel to Earth. Then he confirms it mathematically. Calculus helps.

But he doesn't tell people he is working on these ideas. "Solitude was the essential part of his genius," says James Gleick, a twenty-first-century biographer. Except for a small group of men with whom Newton exchanges letters and the very few who know him at Cambridge, hardly anyone is aware of what is going on in his head.

17 Fame Finds Newton

It is hard to exaggerate the importance...of the theory of gravitation...this fact that all the moons and planets and stars have such a simple rule to govern them, and further that man could understand it and deduce how the planets should move! This is the reason for the success of the sciences in the following years, for it gave hope that the other phenomena of the world might also have such beautifully simple laws.
—Richard P. Feynman (1918–1988), American physicist, *Six Easy Pieces*

Newtonian mechanics, and its new mathematics...represented a change in the perspective of human thinking...from a static society waiting for something to happen to a dynamic society seeking understanding, knowing that understanding implies control.
—Leon Lederman (1922–), American physicist, *The God Particle*

Vision is the art of seeing things invisible.
—Jonathan Swift (1667–1745), English author, *Thoughts on Various Subjects*

Newton has little interest in telling others of his findings, and he has no social skills.

When he does attempt to be convivial, it doesn't work. One night he actually tries to entertain some acquaintances. A bottle of wine is needed. Newton goes to his room to fetch it, sits down at his desk, gets absorbed in his work, and forgets all about the wine and his hungry guests.

Despite his solitary habits, word of his accomplishments gets around. Others are interested in the problems that Isaac Newton seems to be solving. And he actually may have an ambivalent attitude toward friendship and acclaim. Newton really seems to want them—as long as no one bothers or criticizes him. Astronomer John Flamsteed says Newton is "impatient of contradiction ... a good man at the bottom but,

In Newton's day, CONVIVIAL was a synonym for festive—anything having to do with a social feast. *Vivere* is the Latin verb for "to live," and *convivial* soon expanded to include people who like to live it up—sociable types who enjoy eating and drinking with friends.

AMBIVALENT means to consider two possibilities and be undecided about which way to go.

through his nature, suspicious." Thinkers begin to seek him out; his work is too good to ignore.

In 1684, Edmond Halley comes calling. Halley is a British astronomer who is trying to understand and explain the orbits of the planets. Are they circular or elliptical? If the Earth makes a perfect circle around the Sun, and the Moon makes a perfect circle around the Earth (as Copernicus said they do)—something is wrong: no one can make the mathematics of those orbits work exactly.

Halley knows that Kepler said the orbits *aren't* perfectly round—that the path of the planets is elliptical, like a stretched-out circle. Kepler used Tycho's charts to discover that is so, but he didn't know why. Kepler didn't have the mathematics to prove his theory. (Calculus hadn't been invented.)

It just happens that the architect Christopher Wren has offered a handsome book to Edmond Halley or Robert Hooke, whoever can explain Kepler's third law (about elliptical orbits) mathematically. Hooke says he can do it, but he doesn't. Halley goes to Cambridge to see if Newton can help him.

When Halley brings up the question of orbits and their shape, Newton says that he has already proved the point mathematically—and that Kepler was right: planetary orbits are elliptical.

Of course Newton hasn't shown his work to anyone, and in his cluttered room, he can't find the calculations. Newton

If you research Halley, you'll find both "Edmund" and "Edmond." The first one is old English, and the second is a French adaptation. After the Normans (from France) invaded England in 1066, French became the language of royalty, class, and sophistication for centuries; many English people used French names. Either way, Edmond/Edmund means "rich protector"—which is what Halley was to Newton.

THE LION'S CLAW

Swiss mathematician Johann Bernoulli had a problem he couldn't solve. He challenged Europe's leading mathematical thinkers to come up with a solution. No one was able to help him.

Newton received the puzzle at dinnertime; by four in the morning, he had it solved. The next day, he sent off the answer anonymously. Bernoulli wasn't fooled. The proof told him that it was his nemesis, Isaac Newton, who had found a mathematical answer. Bernoulli said, "I recognize the lion by his claw." Bernoulli was involved in the fight between Leibniz and Newton over who discovered calculus first.

NEMESIS is a strong word for an opponent. A nemesis usually can't be overcome.

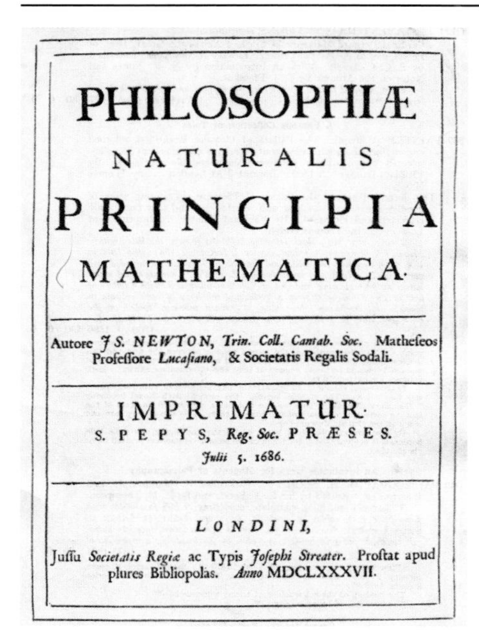

PHILOSOPHIÆ

NATURALIS

PRINCIPIA

MATHEMATICA.

Autore *J S.* NEWTON, *Trin. Coll. Cantab. Soc.* Matheſeos Profeſſore *Lucaſiano*, & Societatis Regalis Sodali.

IMPRIMATUR.

S. PEPYS, *Reg. Soc.* PRÆSES.

Julii 5. 1686.

LONDINI,

Juſſu *Societatis Regiæ* ac Typis *Joſephi Streater*. Proſtat apud plures Bibliopolas. *Anno* MDCLXXXVII.

If Newton had written the *Principia* in English, scientists who spoke French or German or Italian wouldn't have been able to read it. But in the scholarly world that was centered in Europe in the seventeenth century, every serious thinker read Latin. By writing in Latin (1687 edition, above), Newton made his words available to a wide scientific-minded audience. The average English citizen—who couldn't read Latin well—had to wait until 1729, when the book was first translated into English. Actually, the average English citizen never read it even in English; it's not easy reading.

Principium **(prin-SIP-ee-um) is Latin for "principle," a fundamental rule or law; PRINCIPIA is the plural form. It's pronounced prin-SIP-ee-uh in English, but with a hard *c* in Latin (prin-KIP-ee-uh). The root means "first part or beginning."**

says he'll do the work again. Three months later, Halley gets the proof he is looking for. Halley now realizes that Newton has proved a great many other things. He urges him to publish his work. Newton hasn't told anyone (except in some letters and conversations) about the force of gravity or calculus or his laws of motion. And he says he won't publish. Halley keeps after him, even saying he will pay the publishing costs. Still Newton hesitates, but Halley is hard to refuse. Everyone likes him, including Newton. In addition, Halley is a fine mathematician; Newton can talk to him and be understood.

Finally Newton agrees. For a year and a half he writes furiously—all in scholarly Latin—hardly stopping to eat or sleep. Halley keeps encouraging him. It isn't easy. The Royal Society agrees to publish what he is writing, but they won't put up the money to do so. Halley not only takes care of the expenses, he reads manuscript proofs, arranges for illustrations, and finds a printer. (Halley is moderately rich; he gets Robert Boyle, a scientist who is very rich, to help with the cost.)

In 1686, Newton is done, and the *Philosophiae Naturalis Principia Mathematica* ("Mathematical Principles of Natural Philosophy") is published the following year. (It's usually referred to as the *Principia*.) According to the *Encyclopedia Britannica*, "it is not only Newton's masterpiece but also the seminal work of modern science."

There aren't many who get through its 250,000 words (626 pages). A Cambridge student is supposed to have pointed to Newton and said, "There goes the man that writt a book that neither he nor anybody else understands." Newton thinks about doing a popular version of his *Principia*, and then decides against it, he says, "to avoid being baited by little smatterers in mathematicks."

But the book is an enormous success if you consider its influence, which comes very quickly. Those who do read it are the thinkers of the age. (Thomas Jefferson and John

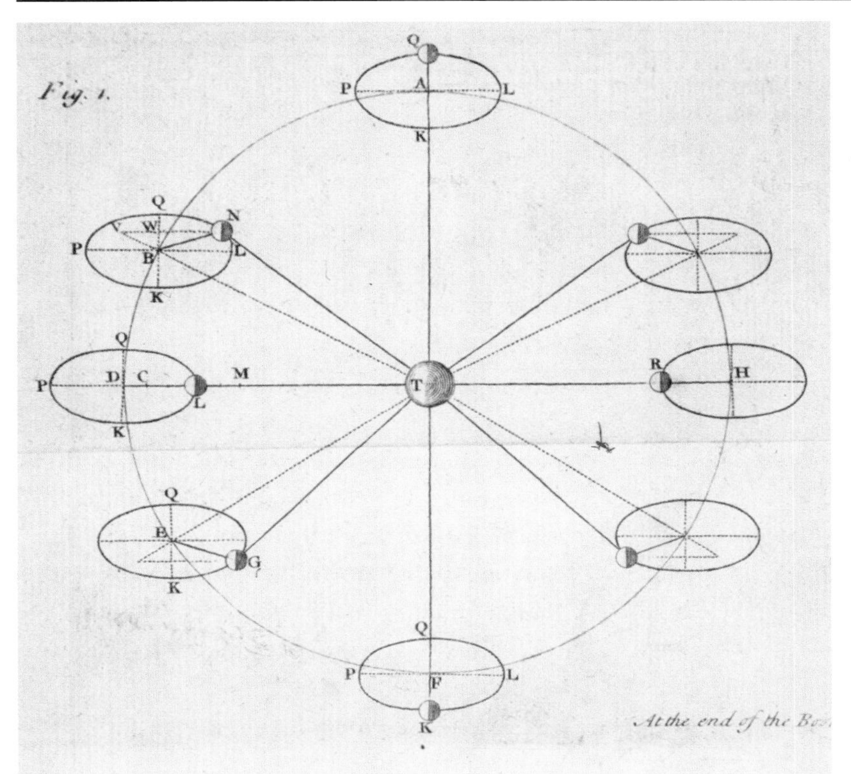

Newton's *Principia* generated an outburst of scholarly activity. The sketch (left) called *Moon's Motion According to Gravity*, by John Machin, appeared in 1729 in the first English translation of Newton's great work.

Adams, next-generation thinkers, are among its readers.)

The *Principia* is divided into three parts, or books. In book one, Newton lays a foundation for calculus and spells out the laws of motion.

In book two, he explains the motion of bodies in a resisting medium, like water. He also deals with sound and the movement of waves. He destroys Descartes's theory of swirling vortexes. Principles he sets forth in this book will become a foundation for the sciences of hydrodynamics and hydrostatics.

Both HYDRODYNAMICS and HYDROSTATICS are about water *(hydro-)* and other liquids. The difference between them is motion: Dynamic liquids are moving, and static ones are not. Hydrodynamics experts study the mechanics of water flowing through pipes or rivers, flooding over land, for example. Liquids at rest might not seem as exciting, but they keep hydrostatics experts busy pondering buoyant (floating) force, water pressure, and other important physical properties.

Book three is titled *The System of the World*. Later, the French-Italian mathematician Joseph-Louis Lagrange will write with some envy, "Newton was the greatest genius that ever existed, and the most fortunate, for we cannot find more than once a system of the world to establish." Newton's big idea in this book is that **nature has basic laws that are the same everywhere in the universe**. That revolutionary

When he was an old man, Newton wrote a letter to Edmond Halley and said of his gravity law, "I can affirm that I gathered it from Kepler's theorem about twenty years ago." Kepler had searched for patterns and symmetries in the world and, thanks to Tycho's precise charts, he'd found them in the movement of the planets. Newton had the genius to make Galileo and Kepler mathematical.

concept is the core of his philosophy.

To summarize Newton's achievements:

- He formulates the idea of universal gravitation.
- He invents the mathematics we call calculus.
- He describes motion with three basic laws.
- He makes important discoveries about the nature of light.
- He explains how to determine the speed of sound.
- He describes waves mathematically.
- He explains the motion of moons and planets.
- He tells how to find the masses of the Sun and planets.
- He understands the need for experimentation and proof in science.
- He ties the Earth to the whole universe.

Almost as soon as the *Principia* is published (1687), Newton becomes an international celebrity.

And it isn't just politicians and the scientifically minded who toast Newton. In London, Newton makes friends with the philosopher John Locke and the busy Admiralty official

The first home of the Royal Society, which published Newton's writing, was Gresham College (right). The original London landmark, a mansion, was destroyed in the eighteenth century.

Samuel Pepys (PEEPS).

Edmond Halley praises the great man in verse:

Come celebrate with me in song the name
Of Newton, to the Muses dear; for he
Unlocked the hidden treasuries of Truth. . . .
Nearer the gods no mortal may approach.

Newton is made warden of the Royal Mint (in charge of England's money), and he directs his anger at counterfeiters, sending a number of them to hang from the gallows.

BIG CENTURY

Put the seventeenth century (that's the 1600s) in your head as a dividing point between the ancient world and modern science. It was around 1600 when Tycho Brahe's observations, Galileo's experiments, and the insights of Kepler lit a fire. Newton's ideas—coming near the end of the seventeenth century—were the firecracker that exploded into a lasting scientific revolution.

England's coins are being debased. That means people are chipping their edges and selling the silver. They all need replacing. Newton gets 300 men and 50 horses (to turn the presses) and, in 3 years, mints twice as much currency as has been coined in the previous 30 years.

In 1689 and again in 1701, he is chosen to represent Cambridge in Parliament. When the solitary professor stands in that house, everyone falls silent. All want to hear what the great man has to say. Newton, who has never made a political speech, only asks that a window be closed; he can feel a draft. Two years later, after Robert Hooke dies, Isaac Newton is elected president of England's Royal Society. It is an elite but sleepy organization, composed of many of the nation's best-known thinkers. Newton turns it into a well-organized, purposeful institution.

During the Glorious Revolution (1688–1689), James II was replaced by William III and Mary II. But the big point is that Parliament became more powerful than the monarchs. It was "glorious" because no blood was shed. Newton, who was Protestant, attended Parliament on behalf of Cambridge. Had James still been on the throne, Newton's appointment would have gone against the king's wishes, for James had tried hard to fill Cambridge—and the army and the government—with Catholics.

In 1705, Queen Anne dusts off the royal sword, taps Newton on the shoulder, and dubs him a knight. It is the first English knighthood given to a scientist. He is now *Sir* Isaac Newton; he has not only tamed the world, he has made himself a star.

EDMOND HALLEY, MR. COMET

The Royal Observatory in Greenwich (GREN-itch), England, is shown now (photo below) and in about 1675 (large engraving). It's the reference point for the prime meridian—0° longitude on a globe. The imaginary geographic line is marked by a brass strip on the ground and a bright green laser shining into the sky.

The French had just built their grand observatory. England and France were rivals, so England had to have an observatory, too. In 1675, the Royal Society asked architect Christopher Wren to choose a site; he picked a hill in the Royal

CORRESPONDING ON COMETS

Before telephones and E-mail, people wrote long, thoughtful letters. Isaac Newton's correspondence has been collected in multivolume editions. You can find them at any good library. He wrote letters to John Locke (an important philosopher), to Samuel Pepys, who kept a gossipy diary, and to many of the scientific leaders of his time. Here are some lines from a letter he wrote to Edmond Halley on 17 October 1695:

> *Sr*
>
> *I had writ a letter to you last week but stopt it because I had inserted a passage I was uncertain of. Your calculations have satisfied me that the Orb* [orbit] *of the Comet of 1680/1 is Elliptical....* [Newton then goes into much detail about the comet, its orbit and times of transit. He points out some mistakes in Halley's calculations. He goes on to another comet.]
>
> *You have made ye Orb of the Comet of 1664 answer Observations much beyond my expectation tho with double pains in calculating all the Observations anew. I can never thank you sufficiently for this assistance & wish it in my way to serve you as much....*
>
> *I will send you to morrow by Will. Martin the Carrier the Box of brass rulers & beam compasses & 8s for Vlac's Trigonometry wth many thanks. Those edges of the brass Rulers wch look rough I ground true to one another wth sand....*
>
> *If it will not give you too much trouble to make an extract of your calculated places of the Moon you need send only those without your book....*
>
> *Your most humble Obliged Servant*
> *Is Newton*

Answering Newton's letter, on 21 October 1695, Halley talks of another comet, the one that is now known as Halley's comet. It had crossed England's sky in 1682.

> *Sr*
>
> *...I have almost finished the Comet of 1682 and the next you shall know whether that of 1607 were not the same, which I see more and more reason to suspect. I am now become so ready at the finding a Cometts orb by Calculation, that since you have not sent the rulers, as you wrote me, I think I can make a shift without them....*
>
> *I have sent you the book wherin I did most of my Lunar Computations, and would have sent an Extract therof, but am at present busy about the Society's Books, and withal I belive the whole work will be more satisfactory....*
>
> *Sr. I am With all imaginable respect*
> *Your most devoted friend and Servant*
> *Edm. Halley*

Note the imaginative spelling. Neither Newton nor Halley had a dictionary to look up spellings or word meanings. There was no solid dictionary of the English language until 1755, when Samuel Johnson published his famous one. (Actually there were a few early attempts at dictionaries, but they weren't comprehensive.) Before 1755, everyone just spelled words the way they sounded, which was an almost-anything-goes approach to language. (It also meant no spelling tests in school.)

Park of Greenwich (near London), and construction began. (You can visit the Royal Greenwich Observatory when you go to London.) In 1675, John Flamsteed (1646–1719) was named Astronomer Royal. His mission was to use modern telescopes to make Tycho Brahe's old sky catalog more precise.

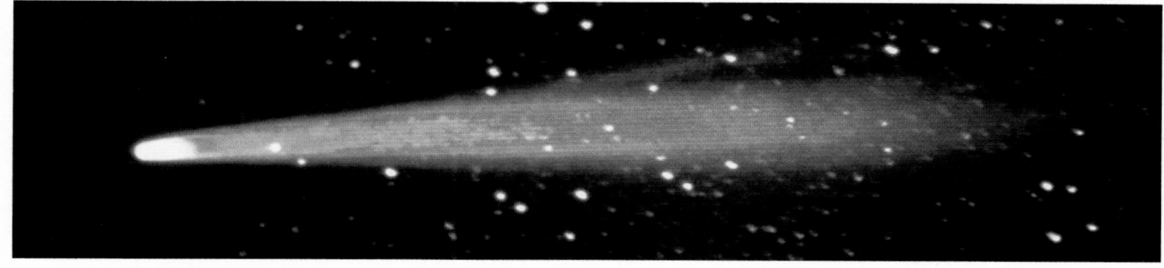

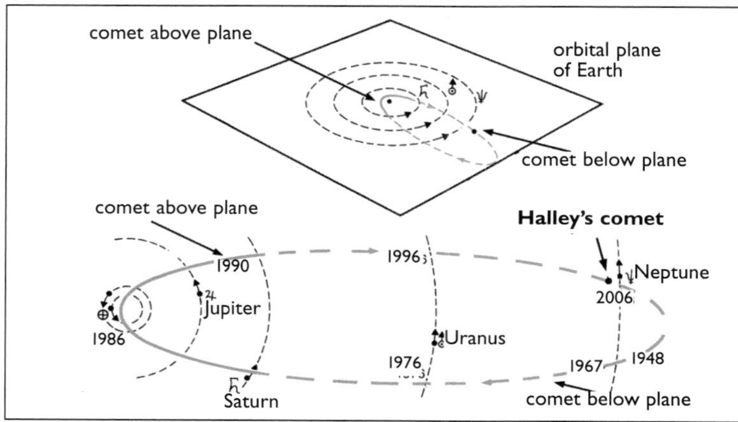

Halley's comet doesn't usually look this flashy, as shown above during its 1910 rendezvous with the Sun. (Stay tuned: Its next arrival is scheduled for 2061.) The spectacular tail doesn't even start growing until the nucleus—a core often described as a "dirty snowball"—passes just inside Jupiter's orbital path. There, the Sun's heat is strong enough to vaporize some of the icy materials, creating a bright streak of gas and dust particles. Solar wind keeps the tail perpetually pointing away from the Sun.

The path of Halley's comet is a long, eccentric ellipse (see pages 129–131). Some comets have open-ended trajectories—called parabolas or hyperbolas—and so never return. But Halley's comet graces our sky about every 76 years. Why "about"? The diagram holds a clue. The comet zips past the orbital paths of Saturn, then Jupiter—and vice versa on the way out. If one of those gas giants happens to be near the crossing point, it gives the comet a gravitational yank. That yank ever so slightly changes the comet's velocity (speed and direction) as it streaks toward the Sun.

So Flamsteed was delighted when a bright Oxford University student volunteered his time. "Edmond Halley, a talented young man of Oxford, was present at these observations and assisted carefully with many of them," he wrote to the Royal Society. Edmond Halley's father was a soap manufacturer, and soap, in an era of the plague, had made him rich. So he was able to give his talented son a good education. Edmond intended to be a poet, but at Oxford he became fascinated with astronomy. He found that European sky maps didn't include many stars visible south of the equator. European ships that ventured south had few celestial beacons to use as guides. Halley got his father to pay for a trip to the island of Saint Helena in the South Atlantic. There he set up an observatory and produced the first known map of the Southern skies, cataloging 341 stars. In a seafaring age, that was an important accomplishment; he was soon being called the "Southern Tycho."

Halley was just getting started. Comparing star positions of his time with those known from ancient Greece, Halley realized that **stars move; they are not fixed in the heavens.** For astronomers, it was a vital insight.

In 1682, Halley was at the French

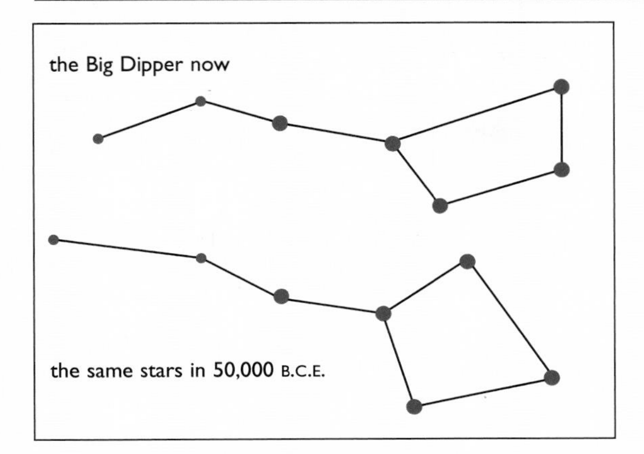

When Edmond Halley realized that the stars move, he wasn't talking about their nightly trek across the sky. By his time, astronomers knew that Earth's spin causes that apparent motion. Halley meant that the stars are actually traveling through space. They're so far away that the changes aren't noticeable on Earth for thousands of years, but here's how the stars of the Big Dipper have shifted in relation to each other (left).

observatory in Paris, where he watched a bright comet cross the sky. Later, he realized its trajectory matched that of comets seen in 1456, 1531, and 1607. Could they all be the same comet, following a path that brings it near Earth every 76 years? If so, it should return in 1758. Halley was fascinated. He studied every comet that came within his view and wondered Where do comets come from? Do they travel in a straight line? Or do they follow the elliptical path that Kepler described?

Halley came to believe that comets orbit in predictable ellipses. Isaac Newton helped him predict those elliptical orbits mathematically. Both knew that if Halley was right, and that if the comet of 1682 reappeared in 76 years, it would be clear that the bodies in the heavens follow rules, which would confirm Newton's theories. Would "Halley's comet" reappear as predicted? In 1758 lots of people gazed at the sky wanting to find out.

A German farmer, Johann Georg Palitzsch, was first to see the comet on Christmas night. He viewed the comet through a telescope he had made himself.

In France, the philosopher and writer Voltaire said astronomers didn't sleep that winter for fear of missing the comet's journey. It met their expectations.

In the British colonies, Ben Franklin was 52 and Thomas Jefferson was 15. They, like everyone else, watched to see if the comet would arrive on schedule. It did come, and it has come roughly every 76 years since then. Halley had proved his point.

By 1758, both Halley and Newton were dead, but they had become popular heroes (especially Newton); the comet's appearance just confirmed things. Comets had terrified people, but once they were understood, they were no longer feared. (Scientific knowledge usually does that for you. It's the unknown that is most frightening.)

Edmond Halley had put that thought in verse as part of an "Ode to Newton."

Matters that vexed the minds of
ancient seers,
And for our learned doctors often led
To loud and vain contention,
now are seen
In reason's light, the clouds of ignorance
Dispelled at last by science.

18

A Dane Lights the Way

God's first creature, which was Light.
—Francis Bacon (1561–1626), English philosopher, *The New Atlantis*

For light doth seize my brain
With frantic pain.
—William Blake (1757–1827), English poet, "Mad Song"

Today we know that light is more tangible than it first seemed. It exerts physical force, an effect that has inspired visions of spaceships with enormous sails pushed by sunlight. In Einstein's theory of relativity, light has weight of a sort, for it is affected by gravity. Nonetheless we do not yet have a unified understanding of light.
—Sidney Perkowitz (1939–), American physicist and author, *Empire of Light*

Can the speed of light be measured? Galileo tried. He stood on one hill and had an assistant stand on another. They exchanged lantern flashes. But light was just too fast, and their instruments couldn't catch it. Most people thought light could never be measured.

The French mathematician, René Descartes, said there was nothing to measure. He said light travels instantaneously; its speed is infinite.

Olaus "Ole" Christensen Roemer (who is two years younger than Isaac Newton) accepts Descartes's idea, until something happens that makes him wonder. He conjectures that light

Ole Roemer built a meridian telescope (right), which points only north or south and pivots only up or down. As stars pass into its field of view, the scope measures their altitudes, using its own fixed position for reference.

TELLING TIME

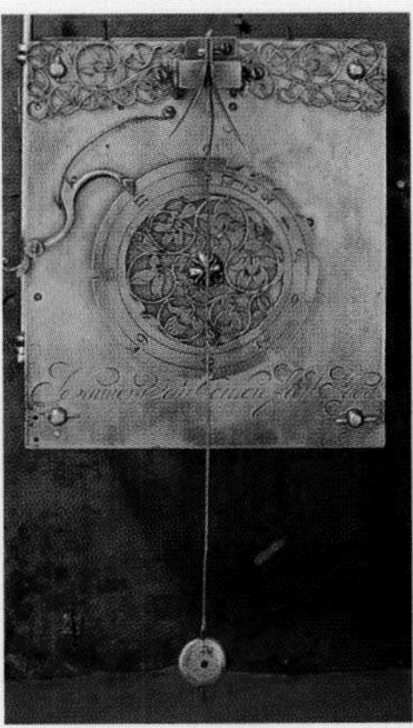

Christiaan Huygens designed this weight-driven pendulum clock.

In the late Middle Ages, mechanical clocks were developed that depended on a falling weight to move an hour hand. They weren't very accurate, but they were better than anything that had come before.

When Galileo discovered the regularity of a pendulum's swing, he figured out that it could be attached to gears that would turn the hands of a clock. Such a clock would have a new level of accuracy. He even designed one. But he never built it. Christiaan Huygens did.

Huygens's pendulum clock was more complicated than you might imagine. After you get a pendulum to swing in the proper arc, something is needed to keep it swinging. Huygens attached falling weights to prevent air resistance and friction from bringing the pendulum to a stop. In 1656, when he presented his clock to the Estates-General (the Dutch ruling body), the era of accurate timekeeping had begun. But the pendulum clock wasn't any good on a ship: it was thrown out of whack by heaving waves and changes of temperature.

In England, Robert Hooke collaborated with master clockmaker Thomas Tompion making a pocket watch with a balance spring to keep it running. They gave the watch to King Charles II, who was said to be "well pleased." Hooke said he could improve that watch, but he got hung up in patent negotiations and never followed through with this idea. (Hooke had a habit of half doing things.)

has a measurable velocity. Then he needs to prove it by finding that velocity. Ole Roemer realizes that he can't measure the speed of light on Earth—it travels too quickly—so, in 1676, he makes the heavens his laboratory.

Roemer (1644–1710) is a Dane. As a student at the University of Copenhagen he impresses a professor so much that he is given Tycho Brahe's manuscripts to edit for publication. A few years later, the French Academy of Sciences sends an astronomer to Denmark to pinpoint the exact longitude and latitude of Tycho's observatory (that's essential in using his observations). The French astronomer consults with Roemer and offers him a job at the Paris Observatory. It's a big opportunity. Paris is the place to be in the world of science, especially for a young man with ambition.

In France, he does scientific research, builds clocks and other high-tech measuring devices, and tutors the dauphin (the eldest son of the king).

Among other things (he is a workaholic), Roemer spends

DAUPHIN is a French word imported into English with a pronunciation twist. The English stress the first syllable and say the final *n* so that the word rhymes with "GO, fan!" The French stress the second syllable with a very nasal sound and pretty much leave out the "n" sound.

PRIME TIME FOR FRANCE

After Galileo, the center of European science shifted to France. Louis XIV, known as the Sun King, helped make it so. In 1667, in a splendid ceremony, he inaugurated the Paris Observatory (it's still in operation). The Sun King, who did nothing modestly, wanted to make sure the observatory was the world's best. He sent to Italy for Jean-Dominique Cassini, who became its first director.

The Paris Observatory is still in operation, but these days, a dome on its roof houses a large telescope.

Surveyors laid the walls along exact north–south and east–west lines of direction. Since they saw France as the center of the world, they made the north–south line the prime meridian (0° longitude), running it right through the observatory and on through Paris. (In 1506, with Portugal a dominant sea power, the prime meridian ran through the Portuguese Madeira Islands. In 1667, it went through Paris. In 1884, with England dominating science, the prime meridian was shifted to Greenwich, England, where it remains today.)

You can still find 135 brass medallions marking the historic Paris meridian in sites all over that city. When you're there, check the Louvre Museum and the Luxembourg Gardens as well as the Paris Observatory.

Those black stains on Io (below) are lava flows. The moon has a volcano named Loki that's more powerful than all the volcanoes on Earth, combined.

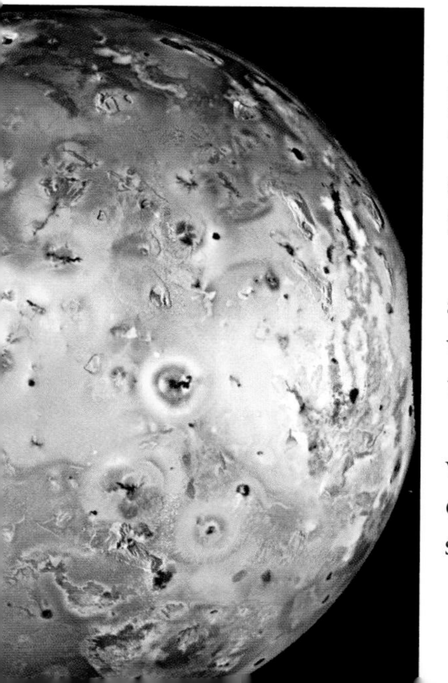

10 years observing Jupiter's moons. (They are often called Jovian moons; Jove is another name for Jupiter.) Roemer studies the moons with more care than anyone before him, and he keeps a record of what he sees.

He has a reason for that. Sailors at sea need to find a way to figure their longitude. Ships are hitting reefs and other hazards because they can't figure out their east–west location. There are no clocks that can keep time precisely on a rolling ship. Roemer (like most astronomers of his day) is looking for something in the sky that can serve as a sailor's clock.

The orbit of the Jovian moon Io (EYE-oh) has been accurately measured at the Paris Observatory. If sailors know when Io will enter Jupiter's shadow and disappear from view, they can use that eclipse as a clock.

Roemer believes he can predict it exactly. Eclipses happen with almost perfect regularity. And Descartes said the speed of light is instantaneous. So Io's eclipses should be easy to schedule. But to Roemer's astonishment, the timing between

LOST AT SEA

By the end of the fifteenth century, the Portuguese had drawn lines of latitude and longitude on their maps. That helped make the Age of Exploration possible. Sailors at sea, if they found the North Star, could figure out their latitude. But longitude is far more difficult. There was no known way to find the east–west position of a ship at sea. So lots of ships got lost, or worse, crashed into hazards.

In 1707, an English fleet was 100 miles off course and ran into rocks near the Isles of Scilly; 2,000 lives were lost. Everyone knew this great tragedy could have been prevented if the mariners had known their longitude.

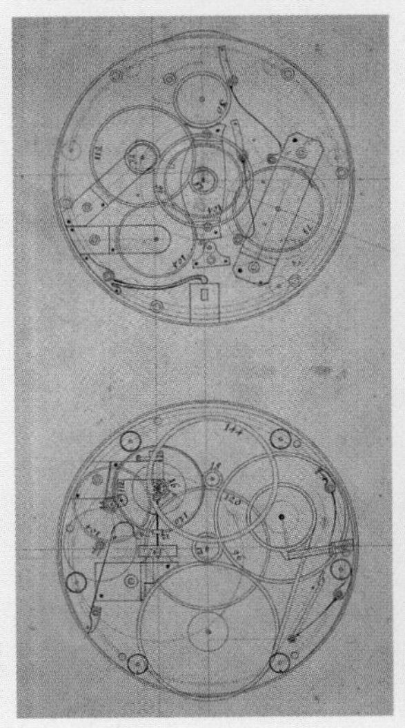

In 1714, England's Parliament offered an award of £20,000 (a huge sum in those days) to anyone who could solve the longitude problem. This was a seagoing era, and finding a way to pinpoint longitude was considered the most important scientific challenge of the time.

Astronomers searched the skies for an answer. An English clockmaker, John Harrison, looked elsewhere. He knew that sailors could figure their local time, wherever they were, by measuring the noon Sun. If they could compare that time with the time back in England, they could calculate their longitude. But how to know the time in England?

This was more difficult than you might think. Sailors needed to know the time in England within seconds. The best watches and clocks of the day lost or gained time in minutes—which could throw off longitude by many miles.

Harrison solved the problem, but not easily. For details of his amazing story, read *Sea Clocks: The Story of Longitude*, by Louise Borden, a young-adult book with engaging drawings and clear text. Dava Sobel's *Longitude*, especially the illustrated edition, is even better.

John Harrison saved many a sailor's life with his longitude timekeeper, also called H4. The sketch (far left) is by Harrison himself.

the onset of Io's eclipses keeps changing. Over a six-month span, he finds a difference of 22 minutes. (Today's measurements put the maximum time lag closer to 16.5 minutes—see page 197.) Roemer measures again and again; each measurement confirms his findings. Roemer thinks about that time difference, and suddenly he understands what is happening. He probably feels like Archimedes, the ancient Greek mathematician who said, "Eureka!" ("I have it!") at a breakthrough moment.

This is Roemer's big thought: If, as the great Descartes said,

SHEDDING SOME LIGHT ON HUYGENS

Christiaan Huygens had the bad luck to be a contemporary of Isaac Newton. Huygens was a brilliant scientist and a spectacular astronomer; in any other generation, he might have been *the* superstar. But it was tough competing with Isaac Newton.

Newton said that light exists in the form of tiny, bulletlike particles. "Multitudes of unimaginable small and swift corpuscles of various sizes, springing from shining bodies at great distances one after another," is the way he put it.

Huygens (and Newton's enemy Robert Hooke) said light isn't particles; it is an undulating wave, much like a sound wave or an ocean wave. You can't have a wave unless there is a medium or substance—something to wave.

Descartes had said the universe is filled with an invisible medium called ether. Huygens said light waves are produced when the ether vibrates.

Christiaan Huygens (1629–1695) pictured light as a wave. Newton saw it as "bulletlike particles." So, who was right?

Newton attacked Huygens's theory. Light is *not* a wave, he said.

This was an important issue. Is light made of particles, or is it waves? Hold on to this question; the surprising light story will continue into the twentieth century and beyond. As to the ether, is it there or isn't it? That too, will be a puzzle to solve.

light travels instantaneously, then Io's eclipses should always begin at the same time. But that isn't what happens. The eclipses come progressively earlier when Earth and Jupiter are approaching each other in their orbits, and progressively later when they are moving apart. Roemer figures out that light from Io is delayed when Jupiter is in a faraway location. That means that light is not instantaneous. It must have a measurable velocity.

But how can he measure it? He rethinks Galileo's experiment, making Earth and Jupiter two distant hills, and light from Io a lantern. Doing that, Roemer is able to come up with a figure that's slower than the modern measure, but not bad considering the equipment he has.

Roemer's breakthrough measurement of the speed of light comes 11 years *before* Newton publishes his *Principia*; but Newton doesn't make much of the news. Hardly anyone does, except for Christiaan Huygens (HY-genz). Huygens (1629–1695) is a Dutch scientist at a time when Holland is having a "golden age." He's a bright boy from a prosperous

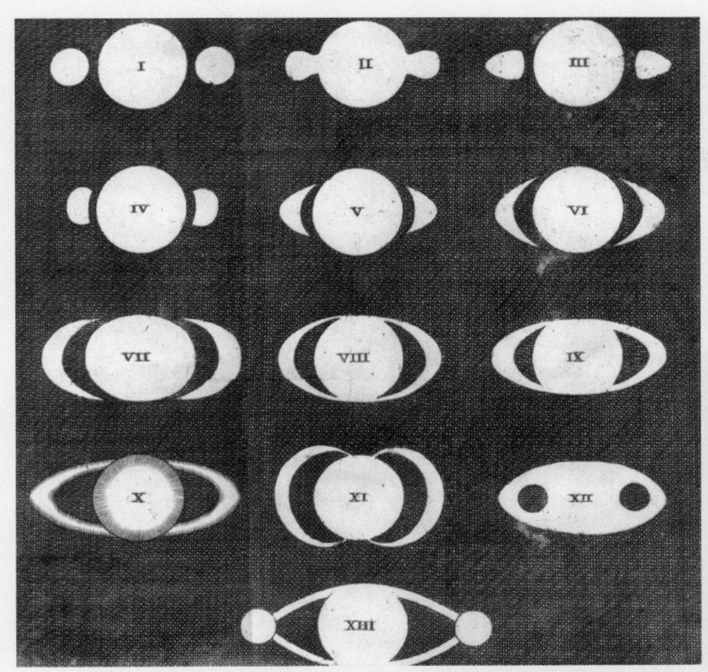

Through the first telescopes, astronomers gazed at Saturn in puzzlement. They saw orbs, ears, bullets, parentheses, loops, and other oddities—all incorrect (left). In 1659, Huygens figured out that a wide ring circles the planet. Spacecraft *Cassini* (artist's view, below), carrying a probe named *Huygens*, took this close-up in ultraviolet light (bottom).

SOLVING SATURN'S MYSTERIES

Even back in 1610, at the first telescopic glance, Galileo thought Saturn looked odd (drawing I, above). Other astronomers observed that the irregular shape changed over the years (drawings II to XIII). It even became a plain circle at times, like the other planets. No one could see—or imagine—what was really going on.

Then, Christiaan Huygens discovered a better way to grind lenses (with help from philosopher Benedict de Spinoza). He built a telescope whose lens had a 7-meter (23-foot) focal point, and, in 1659, became the first person to see Saturn's rings clearly. He then made a mental leap: He figured out that the shape changes because we view the planet from different angles. When Earth is above or below the plane of Saturn, the rings appear wide and full. On the same plane, the edge of the rings face Earth and are too thin to see.

Huygens also discovered Saturn's largest moon, Titan. In 2005, a probe bearing his name became the first Earth-made object to enter Titan's atmosphere.

Government Sponsored Science—a Good Idea?

A group of scientifically minded Englishmen started an informal club around 1645. Robert Boyle, Christopher Wren, John Wilkins, and Robert Hooke were founding members. Samuel Pepys, who kept a diary that would get published, later became one of its presidents. These thinkers met to talk about new discoveries and to share information. In 1662, the group received a charter from King Charles II and was named the Royal Society of London. Christiaan Huygens became one of the first foreign members of the society when he was admitted in 1663.

Later, Isaac Newton got the Royal Society to be more than a gentleman's club. Science, which had often been a lonely profession, was heading toward the modern idea of teamwork and coordinated research.

By the way, the motto of the Royal Society is the Latin expression (from the poet Horace) *Nullius in Verba*, which today is sometimes translated as "Don't take anyone's word for it."

Lord William Brouncker (left), the Royal Society's first president, and Francis Bacon (right) pay tribute to King Charles II (the bust on the column).

family; his dad calls him "my Archimedes." Through his father, he gets to meet Descartes and Rembrandt. In 1656, he invents the pendulum clock. In 1659, he discovers a "ring" around Saturn and becomes famous. Huygens confirms Roemer's light-speed measurement (and improves it a bit). Huygens understands that this is important information. Few others do.

Perhaps politics and personalities get in the way. In Paris, Roemer is an assistant to a famous astronomer, Jean-Dominique Cassini. Cassini is pompous and self-important and likes to pretend that he is French, although he was born Giovanni Domenico in Italy. Actually, Cassini makes several significant observations: One is that Saturn's ring is double and made of a myriad of small particles.

Cassini works out the scale of the Sun and planets. That turns out to be a shock. Almost everyone believes that the Earth is the largest heavenly body. Cassini makes it clear that

Swiftness is an Understatement

For most purposes we say that light travels at about 300,000 kilometers per second (about 186,000 miles per second) in a vacuum. (In water, in the atmosphere, or through a prism, it's a bit slower.)

Today, with clocks that tick off time by the vibrations of atoms—we can measure light's velocity with astonishing accuracy. We express it with a mathematical symbol, c, which stands for the Latin word *celeritas* ("swiftness") and equals 299,792.458 kilometers per second. (When using c in calculations, metric units are the standard.) That's fast enough to zip around Earth at the equator almost eight times *in one second!*

When you visit Copenhagen, be sure to climb the Round Tower. It was a cutting-edge science center with Ole Roemer as its director. He set up telescopes and other devices at the top of the tower. He also brought a new system of weights to Denmark, introduced the Gregorian calendar, invented a thermometer, and had enough time left over to be the mayor of Copenhagen, a Supreme Court judge, and a police and fire chief.

the Sun is much, much bigger, and so are Jupiter and Saturn. Jupiter is so large that all the other planets could fit inside it. Worth may not depend on size, but at the time this seems yet another blow to human self-esteem.

But Cassini isn't as great as he thinks he is; he is the last of the major astronomers to reject Copernicus: he still believes in an Earth-centered universe. And Cassini doesn't buy Roemer's astonishing concept. Or Huygens's verification. Cassini misses its importance. If light moved instantaneously, then when we look at the stars, we would only see the present moment. But that isn't what happens. Light from the stars is a tunnel into time. It can take a million or more years for light from some stars to reach us. We are seeing ancient history when we look at the sky.

Clocking Time (and Speed) with Io: Here's the Math

This brilliant graph records the orbits of the four Galilean moons as they swirl through time (the scale of days along the left) and space (the distance from the center, which represents Jupiter). Callisto, the yellow wave, is the farthest out and takes a couple weeks to complete an orbit. Io, the orange wave, zips around every couple of days.

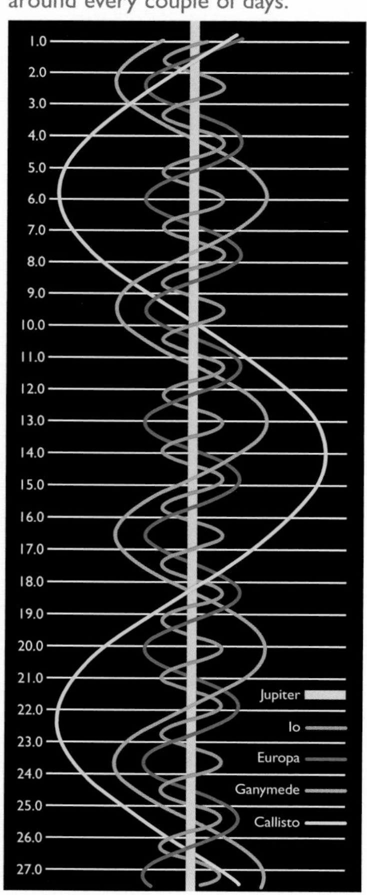

Of the four big Jovian moons, Io is the closest to Jupiter—and therefore the speediest. It zips around the giant planet in 42 hours, 28 minutes, and 21 seconds, like clockwork. Had Roemer's longitude scheme worked, seafarers could have fixed their location about every two days, each time Io entered Jupiter's shadow.

The problem was, most earlier observations had taken place when Jupiter was high and bright in the night sky. That only happens when Earth (A) is on the same side of the Sun (see diagram, opposite page). When Earth (B) is on the far side, Jupiter appears dim and low; it's hard to see. Roemer made the effort, which is why he found the lag in Io's eclipse before anyone else did.

That apparent delay happens when light from Jupiter and Io has to cross the diameter of Earth's orbit to reach us. (Io doesn't actually slow down; its light just takes longer to get here—like hearing thunder after lightning, even though both happen at the same time.)

Like all speeds, the speed of light is

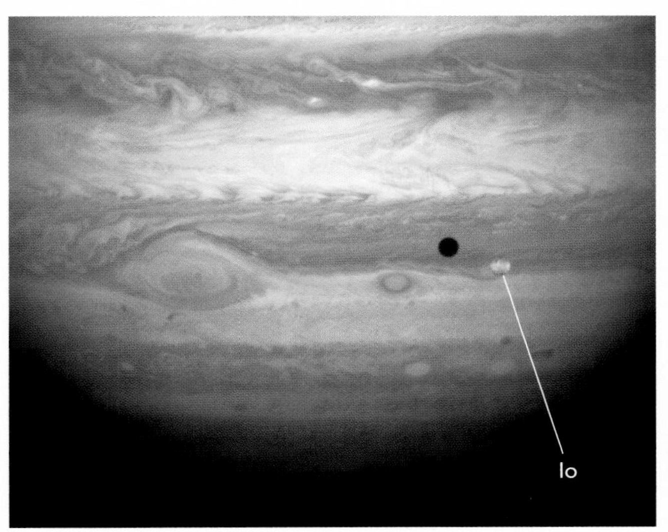

Even in this close-up photo taken by a visiting spacecraft, Io is tough to spot, but its dark shadow gives it away. Io is a little bigger than Earth's Moon, but Jupiter, the largest planet, dwarfs all of its traveling companions.

distance (the diameter of Earth's orbit) divided by time (how long it takes to cross the diameter). Roemer didn't have either number, exactly, though Cassini had come up with a close estimate for the diameter—292 million kilometers (181.5 million miles). As for time, Roemer used 1,320 seconds (22 minutes) instead of 996 seconds (16 minutes 36 seconds). (Remember: He was looking at a low, dim Jupiter with its tiny speck of a moon through a small telescope.) By dividing a slightly shorter distance by a longer time, his value for the speed of light came up short.

We have all the figures today: The average orbital diameter is just shy of 300 million kilometers (186 million miles). Round off 996 seconds to 1,000—an easy number to divide by. So what's the speed of light? Divide distance by time, and you'll get a ballpark figure, but the calculation isn't quite this easy. Jupiter and Io are moving targets, and Earth's orbit is elliptical, among other astronomical challenges. That makes Roemer's feat all the more impressive.

19 What's the Matter? (About Elements and Alchemy)

We stand on the elements, we eat the elements, we *are* the elements. Because our brains are made up of elements, even our opinions are, in a sense, properties of the elements.
—P. W. Atkins (1940–), British chemistry professor, *The Periodic Kingdom*

That is the essence of science: ask an impertinent question, and you are on the way to a pertinent answer.
—Jacob Bronowski (1908–1974), Polish-British mathematician and author, *The Ascent of Man*

The Egyptians had for centuries practiced embalming, dyeing, glassmaking, and metallurgy.... Perhaps it was only natural that people steeped in Greek philosophy would think of trying to make gold when they encountered the rich Egyptian tradition of practical chemistry.
—Richard Morris (1939–2003), American physicist and author, *The Last Sorcerers*

Aristotle thought everything on Earth was made from four elements—earth, air, fire, and water. Today we agree: There are basic substances out of which everything else is made—and we still call them elements. It just happens that the Greeks were wrong in those they chose. But you have to begin somewhere, and their idea got the search started.

We scientific-minded humans have now identified more than 100 elements. How did we find them?

To begin, we had to have a hypothesis that elements exist and that they are far more numerous than the Greek foursome. Then we could start looking for them.

For a long time, no one looked. As long as everyone believed in earth, air, fire, and water, there was no reason to

An ELEMENT is matter that is made up entirely of atoms with the same atomic number—that is, the same number of protons (positively charged particles) in the nucleus (core). Hydrogen has 1 proton, iron has 26, uranium has 92, to name a few.

Remember: A HYPOTHESIS is an idea, based on observations and data, that can be tested. A science theory is more than an idea; it's an explanation that fits known facts. When new facts come along (like the discovery of atoms), a theory can sometimes be proven wrong.

IT'S ALL GREEK TO YOU? (IF YOU'RE A LANGUAGE GEEK, THIS IS FOR YOU)

The noun *alchemy* comes from ancient Greek by way of Arabic. The Greek root word, using the Greek alphabet, is χημεια. That first Greek letter, χ, is usually called *chi* in English, but it doesn't sound like the beginning of *chair*; it's a thick, hissing *h*. So the word is pronounced hee-MEE-ah (adding the hiss). The Arabic addition is *al*, which just means "the." The Greeks had a related verb, χεω (HEH-oh). It has several meanings: to melt, to liquefy, to pour (as in a drink), and to mold. (This mold has nothing to do with fungus; it's about making objects by pouring molten metal or pressing clay into molds.) Do you see a connection between alchemy and molding things?

search further. That wrong four-element theory held sway for more than 2,000 years. It is one of the longest-lasting wrong theories of all time. There were explanations that made it seem logical, even to serious experimenters.

If you boil water in a metal pot, a residue—bits of earthy stuff—is left after all the water is gone. That seemed to prove that earth comes from water.

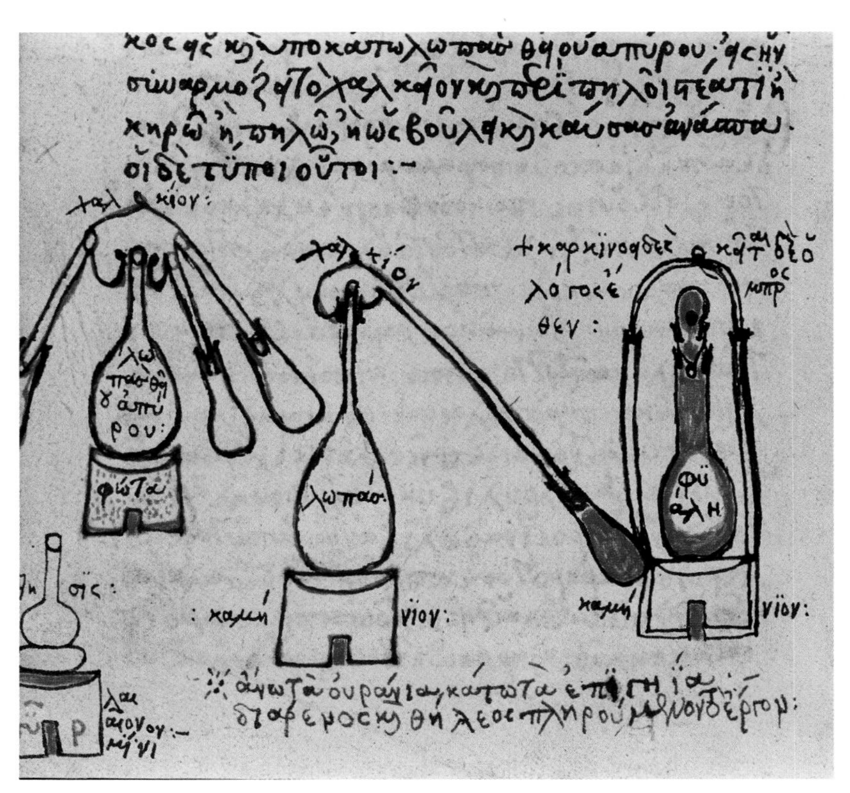

The alchemist's laboratory almost always included distilling equipment, as shown in this earliest known writing on the topic (left, a sixteenth-century copy of the original, which was lost). Distillation is about boiling liquids. The idea is to capture and condense the steam back into a liquid that's pure—free from minerals or other particles. It's also a way to concentrate a liquid by evaporating the water out of it. These instructions were written around the end of the third century by Zosimos of Panopolis and addressed to his sister, Theosebeia, a Hellenistic (Greek-speaking) Egyptian.

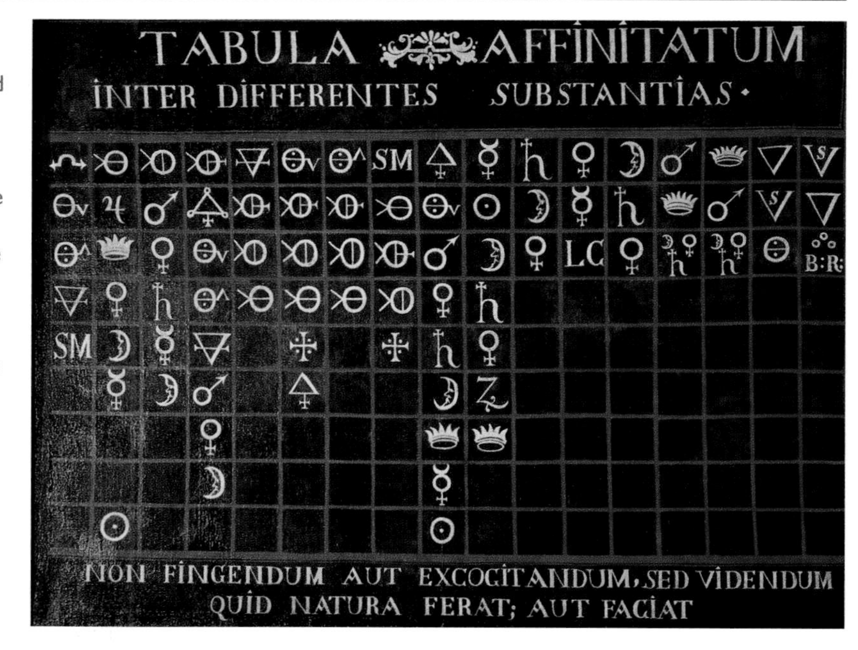

In searching for gold, most alchemists were hoping for a big-time payoff. They often used symbols, rather than words, to keep their findings secret. Here's a list of chemicals, titled "Table of Affinities Between the Different Substances," that once belonged to the archduke of Tuscany (in northern Italy). Note the triangular symbol in the fourth slot in the second row: it's phlogiston. The table is a work of an eighteenth-century Swedish chemist named Torbern Olof Bergman.

In the eyes of the ancients, the first chemists were the people who took ore from the earth and, by smelting it, made it into pure metals. That idea evolved into alchemy—attempting to turn base metals (lead, for example) into precious metals (like gold). It didn't work. But the alchemical experimenters were often serious thinkers, and their attempts taught them unexpected and valuable things.

Just about everyone thought that the process of burning released fire's element. Take a log, light a fire, and you get flames and smoke and then are left with a bit of ash. Where did the bulk of the log go? Trying to answer that question, the natural philosophers came to believe that there is an invisible—but real—substance in matter that emerges as flame in the course of burning. They called it *phlogiston* (floh-JIS-tuhn). It was said to be the fire element. Just about everyone believed in it; that's what was taught in schools.

For a long time, the phlogiston theory kept scientists heading in a wrong direction, but they didn't know that. Even those willing to question the idea didn't have good tools to work with. Equipment for precise measurement hadn't been developed, and measurement is essential for accurate research.

Those who were attempting to figure out the nature of matter had no idea of the composition of common substances. The three most abundant elements in the Earth's crust—oxygen, silicon, and aluminum—were unknown.

Chemistry wasn't a science. Those interested in the subject were alchemists. Alchemy was part hokum, part wishful thinking, and part serious science. It had roots in Greek Alexandria and in the Arab world. The greatest of the Arab alchemists was Jābir ibn Hayyān (ca. 721–ca. 815 C.E.), known to Europeans as Geber.

Geber kept trying to transmute (change) matter. He worked hard at it. Until his time, the strongest acid known was vinegar. By distilling vinegar, Geber got strong, pure samples of acetic acid. He discovered that it could cause chemical changes. Before that, only heat had been known as an agent of change. Because of Geber, experimenters had something else to work with.

Then a German alchemist, Andreas Libau (usually known as Libavius), writes a textbook (in 1597) called *Alchemia*. In it he gives clear directions for preparing strong acids, such as hydrochloric acid, sulfuric acid, and nitric acid. Those powerful agents can eat heavy metals. If you can destroy metals, maybe you can make them (or so the alchemists reason).

And why not? As long as there is a belief in four elements out of which comes everything else, it seems probable. If you

De A. Theuet, Liure II. 73

GEBER ALCHYMISTE ARABE.
Chap. 33.

For several centuries, a series of Arabic alchemists wrote under the name Jabir. The original one (above) lived in the eighth century in what is now Iraq. A fourteenth-century alchemist named Geber (the Latinized version) gained fame in Spain.

For more than a millennium, alchemists were often depicted working away in messy labs. An early illustration appeared with the writings of Boethius, a sixth-century Roman. A thousand years later German painter Hans Holbein, the Younger made a woodcut out of it. This engraving (left), based on Holbein's woodcut, was done still later, in the eighteenth century.

can just find the right combination of earthly things, you should be able to transform them. That's the idea that drives the alchemists.

ELIXIR comes from Arabic, which borrowed a Greek word meaning "a powder for drying wounds." Another word the alchemists sometimes used to describe a magical health-giving substance was *panacea* (pan-uh-SEE-uh)— from a Greek word meaning "all-healing." Today it means a cure-all for any problem you face.

They have two major quests: One, they want to find something to turn base metals, like iron and lead, into gold and silver. (That sought-after "something" is called the "philosophers' stone.") And two, they want to find the elixir of life—a substance that will heal all diseases and keep people young forever.

They are terrific goals. Wouldn't you like to be able to turn lead into gold? Until you try something, you don't know if it is possible or not. The alchemists try. They really try. But gold is an element. It can't be chemically made from anything else. The alchemists are doomed to fail, but they don't realize that.

Eternal life? When Spanish explorer Ponce de León discovers Florida (in March 1513) he is looking for a "fountain of youth." It is certainly tempting to seek everlasting youth. (Look at the advertising claims for some cosmetics, and you'll see we're still doing it.)

You might possess gold, but gold might also possess you. Gerard Dou's 1664 painting of a man weighing gold captures in a human face our fascination with the coveted metal.

Many alchemists are out-and-out phonies. They are especially good at promising kings and princes long life and riches. With gold lust in the air, lots of talented con men take advantage of wealthy believers. (For a whopper of a story, here's a name to research: Count Cagliostro of Italy.)

Some alchemists are serious workers; their experiments open windows that become useful when chemistry does become a science. Tycho and Sophie Brahe, along with Isaac Newton, are among those who work hard searching for the miraculous philosophers' stone. It is supposed to be a liquid or ground earth; some say it is a reddish powder.

> O cursed lust of gold! when for thy sake,
> The fool throws up his interest in both worlds;
> First starved in this, then damn'd in that to come.
> —Robert Blair (1699–1746), Scottish poet, *The Grave*

The alchemists thought the world was built on dualities: For every up, there's a down. Health means having a body in balance. This illustration, from a sixteenth-century manuscript, is called *Two Contraries*. The king and queen symbolize male and female; the suns and moons represent day and night. The dragon and the star seem to be unifying factors. The star is a symbol of the philosophers' stone.

Hardly anyone thinks it is really a stone. A few say they have seen it; others believe themselves close to finding it.

Bernard of Treves (born in Italy in 1406) spends a lifetime "and innumerable monies" looking for gold. Here are Bernard's words telling of his labors:

> *In dissolving and congealing common, ammoniacal, pineal, Saracen, and metallic salts, then more than a hundred times calcining them [in] blood, hair, urine, human dung and semen... and other infinite regimens of sophistications to which I stuck for twelve years having attained 38 years of age.... In my prayers I never forgot to beseech God that he would deign to assist my endeavors.*

Bernard keeps at it. When he is 48, he buys 2,000 hen's eggs, which he boils. Then he separates the whites from the yolks and putrefies them separately in horse's dung. No luck. No gold. When he hears that someone in Vienna has found the philosophers' stone by mixing silver, mercury, and olive oil, he sets out for Vienna. Meanwhile, rumors reach Vienna that Bernard knows the alchemical secret. He is welcomed with a big banquet. Bernard is an honorable man and confesses he has no miracle formula. But a Viennese alchemist tells him he has the magic recipe. If Bernard will put gold coins in a vessel and bury them in hot ashes, the gold will multiply. He does it. But when the vessel is dug up 20 days later, the gold coins have vanished. Bernard dies on the island of Rhodes at age 84, still trying to make gold.

PUTREFY means "to rot." Not surprisingly, a related Latin root, *putēre*, means "to stink." As for old smelly tennis shoes? *Putrid* is the word to describe them.

In North America, George Starkey, Harvard class of 1646, is dismayed by the teaching at that institution. He says the professors don't know what they are doing. Starkey rates the

WHY AREN'T ACIDS BASES (AND VICE VERSA)?

If a friend asks you what makes acids and bases different, you can feel smart by answering with a word—hydrogen. (That's the H in pH—the acid test.) You could even add, "It's the most abundant element in the universe!" But be prepared to delve deeper if your friend is the curious type. Start with another word—ion. It's an atom or molecule that has lost or gained an electron, giving it a positive or negative electric charge.

Ease into your explanation with a reminder: The chemical symbol for water is H_2O—two hydrogen atoms and one oxygen atom. Then get to it: If you add a substance to water, some H_2O molecules change. If the substance is an acid, it creates a lot of hydrogen ions—positively charged hydrogen (H^+). If the substance is a base, you end up instead with lots of negatively charged oxygen-hydrogen pairs (OH^-), or hydroxide ions. A neutral substance—neither acid nor base—has balanced ions. In fact, if you put an acid and a base together, they'll neutralize each other—that old vinegar and baking-soda trick.

It's amazing what a difference these little ions can make. This table lists a few ways to tell acids from bases based on their properties.

ACIDS	BASES (ALKALIES)
• Acids (like lemon juice) taste sour. (Beware: Many acids are lethal to taste—so don't do it.)	• Bases (like baking soda) taste salty; some others are bitter. (Beware: Many bases are also deadly to ingest.)
• Acids aren't usually slippery.	• Bases are often slippery (think soap).
• Strong acids are corrosive (they eat away metals).	• Strong bases are caustic (they can burn tissue).
• They turn universal-indicator paper red.	• They turn universal-indicator paper blue.
• Some acids react with bases to form salts.	• Some bases react with acids to form salts.
• The pH level (a measure of hydrogen and hydroxyl ions) is below 7 (neutral).	• The pH level is above 7.
• Acids produce protons (as in H^+) and take in electrons (negatively charged particles).	• Bases produce electrons (as in OH^-) and take in protons (positively charged particles).

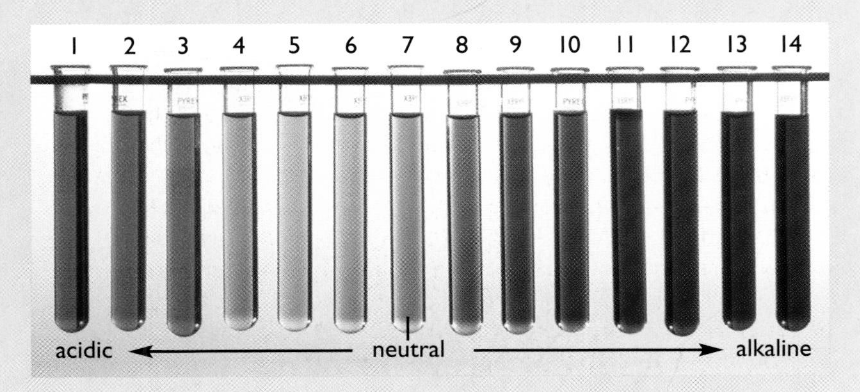

The pH scale measures the hydrogen ions (H^+) in a liquid. A liquid with a pH of 7 (center) is neutral, like water; it's neither acidic nor basic (alkaline).

natural philosophy curriculum at Harvard as "totally rotten." Why isn't the faculty pursuing alchemy more vigorously? Using the pen name Eirenaeus Philalethes ("a peaceful lover of truth"), Starkey writes a book on alchemy that purports to tell secrets "never yet so plainly discovered." He is determined to find the elusive philosophers' stone. Isaac Newton is one of those who reads his book.

Francis Bacon was disdainful of most "natural philosophy" (science). It was based on old ideas, he said. "All the philosophy of nature that is now received, is either the philosophy of the Grecians [Greeks] or that of the alchemists.... The one is gathered out of a few vulgar observations, and the other out of a few experiments of a furnace. The one never faileth to multiply words, and the other ever faileth to multiply gold." He argued that experiments (he called them "crucial instances") would separate the true from the false.

A half century later, in Germany, Johann Friedrich Böttger (1682–1719) comes along and boasts that he can make gold. King Augustus II—known as Augustus the Strong and ruler of Poland and several German states—believes Böttger. The greedy king throws the scared experimenter into a dungeon with orders to make gold or suffer the consequences. Böttger knows he can't make gold.

So, Is Alchemy for Real?

Thanks to experiments by chemist Marie Curie around 1900, we now know that nature transmutes (changes) one element into another through radioactive decay. Uranium, with 92 protons, sheds charged particles until its nucleus shrinks to 82 protons—which means it has become the element lead.

So, is alchemy—artificial transmutation—a ridiculous idea? It's still true that elements can't be changed by chemical reactions, but it's been done by nuclear reactions for nearly a century. So yes, scientists have turned lead into gold (along with other transmutations). But altering the nucleus of an atom is very expensive and technically challenging. Mining gold from the earth is still the cheaper and easier choice.

Watch out if you're tempted to show off. Johann Friedrich Böttger paid for a big false boast with his life: In this letter (right), he promised to give King Augustus II of Poland "arcanum," or the secret of gold making.

He escapes and flees the country. Augustus sends men after him. The king's guards capture Böttger; he is told to get busy experimenting. Augustus gives him a laboratory and funds his research. It pays off, but not as the king expects.

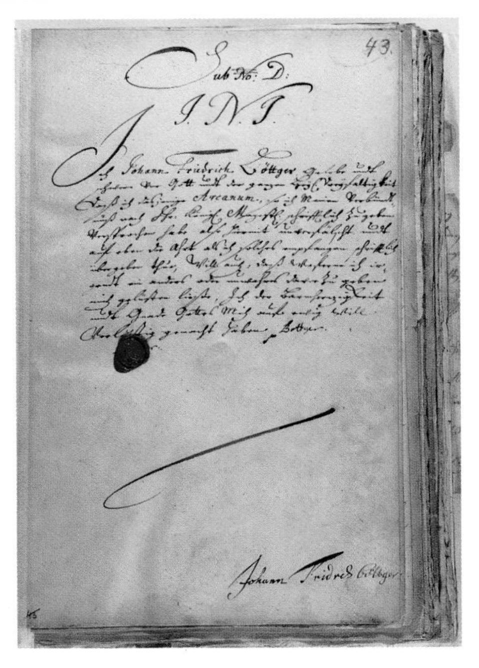

Böttger finds a way to make porcelain, not gold. Porcelain is beautiful, delicate, very expensive, and a Chinese secret. Wealthy Europeans are willing to spend huge sums on elegant pieces of porcelain. Thanks to Böttger (and some German artists and artisans), workers in Dresden are soon producing exquisite porcelain pieces; they bring unexpected riches to Augustus. Böttger has turned clay into a kind of gold; but he never makes the glistening metal, and the king never stops pressuring him. Böttger, a prisoner in his laboratory, desperately tries to please Augustus. He experiments with one bizarre chemical combination after another. He is 37, when he begins writhing with cramps. He can't stop coughing, and then he develops a raging fever.

The invention of [European porcelain] is owing to an Alchymist, or one that pretended to be such; who had persuaded a great many People he could make gold. The King of Poland believed it as well as others, and to make sure of his Person, caused him to be committed to the Castle of Konigstein three miles from Dresden. There, instead of making gold...he invented brittle Porcellane; by which in one sense he made Gold, because the great vend of that ware brings a great deal of Money into the country.

—Baron Carl Ludwig von Poellnitz (1692–1775), *Memoirs*

THE FOUR HUMORS AREN'T A JOKE

At about the same time that Böttger was turning clay into porcelain, a Dutch physician, Franz Deleboe (1614–1672), who was known by his Latin name Franciscus Sylvius, was studying saliva and the way the body digests food. He said that digestion is a chemical process (based on fermentation), not a mechanical one (based on grinding).

The Greek physician Hippocrates (ca. 460–370 B.C.E.) believed in four "humors" (blood, phlegm, yellow bile, and black bile) that, when balanced, were supposed to determine health. The four humors were accepted in the same way that Aristotle's four elements—earth, air, fire, and water—were accepted. Hardly anyone seems to have questioned them until Sylvius came along and threw out the concept of four humors. He said that good health depends on the balance of acids and bases in the body. It was a big step forward.

Snake venom is the prescribed treatment. Augustus does not attend his funeral.

Despite the lure of gold, alchemy is heading for an attic storeroom, along with witchcraft and magic. Those who make the transition to the scientific world are going to be known as chemists. Modern science is gearing up. But it doesn't happen overnight. More than a century after George Starkey, in 1771, a Harvard student will defend this thesis question, "Can real gold be made by the art of chemistry?" He will answer with a resounding "Yes."

In all of his travels through Asia, Italian explorer Marco Polo (ca. 1254–ca. 1324) found a city in China with a one-of-a-kind treasure: "In a city called Tinju, they make bowls of porcelain, large and small, of incomparable beauty. They are made nowhere else except in this city, and from here they are exported all over the world." This dragon vase (right) was made during the Yuan dynasty, around the time of Polo's travels.

AN ELEMENTAL TALE

Hennig Brandt was going for gold when he discovered an element that was completely unknown in his time. You can call Brandt an alchemist or an early chemist or a doctor or a military officer and be right on all counts.

Brandt was born in Hamburg, Germany, around 1630 (which makes him a contemporary of Newton and Leibniz). He was looking for gold in urine (yes, it's true) and managed to extract a waxy substance that glows in the dark. (This was about 1669.) He called it phosphorus ("light-bringer"), after the Greek name for the planet Venus. (The Greek stargazers knew that when Venus appears in the morning, the Sun will soon follow.)

Phosphorus has an eerie glow when it combines with air (at temperatures below 35°C or 95°F). It flames at room temperature. Brandt didn't understand that he had discovered an element, but he realized he had a recipe for something valuable. Europeans were soon fascinated with phosphorus, and the gentry were willing to pay big money for the stuff to

The element phosphorus (P) has two crystal forms—white and red. Brandt's urine experiment produced a waxy form (below right), which can burst into flame in the air and form a toxic gas. The red-brown powder (below left), which is not toxic and is used to make matches, was discovered in 1845 (well after Brandt and Boyle) by gently heating white phosphorus. Chemists now extract phosphorus (its pure form is actually colorless) from minerals—not urine. Phosphate—phosphorus combined with oxygen—is used in fertilizers, detergents, and metal coatings that prevent corrosion.

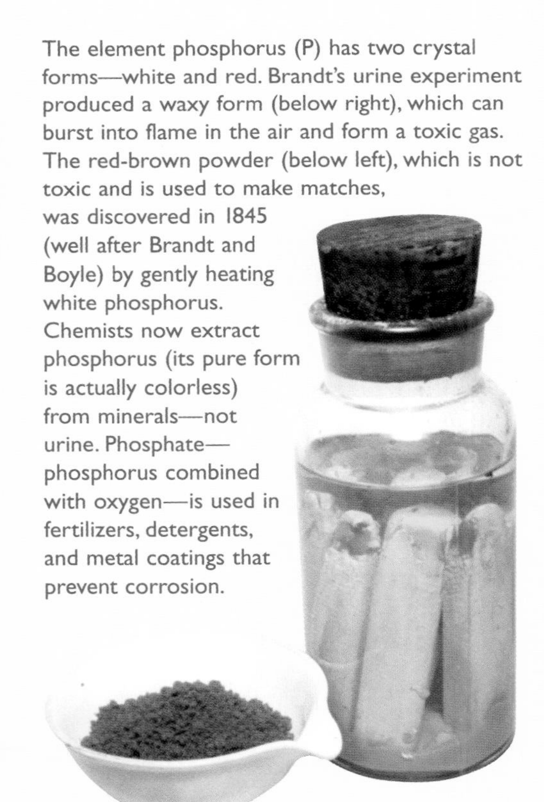

Joseph Wright of Derby, England, put an eyeful of symbolism in this 1771 painting. Its title is a mouthful: *The Alchymist, in Search of the Philosopher's Stone, Discovers Phosphorus, and prays for the successful Conclusion of his operation, as was the custom of the Ancient Chymical Astrologers.*

add atmosphere and glow to their evening parties.

But producing phosphorus wasn't fun. It took about 5,500 gallons of urine to make a pound of it. The urine had to "lie steeping in one or more tubs till it putrefy and breed worms, as it will do in 14 or 15 days." Brandt distilled the concentrated urine, boiled it until it was black, and finally he had a transparent waxy substance, which had to be kept under water to prevent it from burning up.

Chemists and physicians were soon prescribing phosphorus as a cure-all. In its pure form it is toxic—it can kill you. So they were probably killing some of their patients.

In England, Robert Boyle was fascinated with phosphorus, and he experimented with it again and again. Most alchemists kept their methods secret (they were trying to hit the gold jackpot). Boyle was different: he documented his careful research and published papers on his work, including one on phosphorus. By the end of the seventeenth century, a few other chemical experimenters were sharing their knowledge.

Of the substances we now call elements, nine were known to the ancients. They are gold, silver, copper, tin, iron, lead, mercury, carbon, and sulfur. Medieval alchemists added four others: antimony, bismuth, zinc, and arsenic. (No one understood that these are elements; they just knew they existed.) Arsenic was isolated by a German priest, Albertus Magnus, in 1250. Phosphorus is the second element we can credit to a specific discoverer.

20 Robert Boyle, Skeptic—or Airhead?

I am not ambitious to appear a man of letters: I could be very well content to be thought to have scarce looked upon any other book than that of nature.

—Robert Boyle (1627–1691), Irish-English chemist, *The Philosophical Works of the Honourable Robert Boyle*

Although he became the most respected scientist of his time, Boyle retained his retiring, modest nature and declined many honors....Throughout his life...he spread his almost indecently large income widely in charitable donations (he also left most of his property to charity when he died).

—John Gribbin (1946–), British science writer, *The Scientists*

Irish-born Robert Boyle is a prodigy (a young genius). In addition to being very smart, he is very rich and a good person too. Boyle is the fourteenth child of the earl of Cork, the wealthiest man in the British Isles. The earl, who has made his fortune through wit, skill, and marriage, has some strange ideas. He doesn't want his sons to be soft, so he sends them away as babies to live with country families. That's what happens to Robert, who never sees his mother again. She dies when he is barely three years old, about a year before he comes back home to the family castle in Ireland.

Robert Boyle was born in Lismore Castle in Waterford, Ireland. His father bought the estate from explorer Sir Walter Raleigh.

Boyle lives with his father for a few years and then, at age eight, is sent off again, this time to Eton, a fancy boarding school. (He has a gift for languages and is already reading Greek and Latin.)

Eton offers a solid education, but otherwise it's not a happy experience for Robert. Students are whipped with a cane (whether they've done anything wrong or not). Boyle develops a stutter and "forgets" his Latin.

A contemporary of Robert Boyle, John Aubrey, described him (in the fluid spelling of the time): "He is very tall (about six foot high) and streight, very temperate, and vertuose [virtuous], and frugall [stingy]: a Batcheler, keepes a Coach, sojournes with his Sister, the Lady Ranulagh. His greatest delight, is Chymistrey, he haz at his sister's a noble Laboratory, and severall servants (Prentices to him) to looke to it. He is charitable to ingeniose [ingenious] men that are in want."

He is now 12, and his father sends him traveling on the European continent with one of his brothers and a tutor who is fascinated by science. They are in Florence at the time Galileo dies (1642). Young Boyle is so impressed by the fuss made about the old stargazer that he reads all of Galileo's writings. Galileo seems to be telling him to study natural philosophy (science). He takes that advice.

Boyle is still traveling in Europe when Galileo's students Torricelli and Viviani do a famous experiment showing that air has pressure. They don't know, nor does anyone else, that air is made of several gases (including oxygen and nitrogen). No one yet has analyzed air or any gas. Boyle is fascinated with air. Later, he will experiment with it and realize that there is something in air that is essential for fire and for life itself.

When Boyle returned to England, the country was in political turmoil. Parliamentarians (who wanted more democracy) were pitted against Royalists (who believed in the divine right of kings). Most of Boyle's family supported the king, but his favorite sister, Katherine, was an avid Parliamentarian. Boyle tried to stay away from politics. But Oxford, where he settled, was a Royalist stronghold.

Meanwhile, a rebellion has broken out in Ireland and two of Robert's brothers have died in battle. The earl of Cork seems to be ruined financially. Young Boyle has a few uncertain years, but eventually things work out and he inherits enough income from his father's estates to make him a rich man. Boyle can do anything he wants with his life. What he chooses to do is research, study, and learn.

Some people call Robert Boyle the last alchemist; some call him the first chemist. They're both right. He has a foot in

Calcium shows its elemental color, emitting a characteristic red-orange color in a flame test. When heated, every metal element emits a different color (see pages 294–295).

VITRIOL, sulfuric acid (H_2SO_4), is a colorless, oily liquid with a glassy sheen (*vitrim* means "glass," and *vitreolus* means "glassy" in Latin). Green vitriol is sulfuric acid with iron. Both acids are caustic; they can eat away metal and, easily, your skin. (Never touch them.) A vitriolic person is caustic and corrosive toward other people.

each camp. Boyle spends years searching for the philosophers' stone and the elixir of life. He doesn't reject alchemy; what he does is try to weed out the "ignorant puffers" and support the "Chymical Philosophers" (those who, like himself, are attempting to be serious scientists).

Boyle, who calls himself a skeptic, isn't going to swallow things just because everyone else does. He begins to analyze matter using methods that will become standard in chemistry. One is a flame test. He finds that metals each burn with a different color flame. Lead is pale blue. Sodium is orange-yellow. Copper is bright green. And calcium is red-orange. Using fire, he begins identifying metallic elements. He is pioneering the scientific method in chemistry. And he writes about what he discovers.

His work is immediately useful. The European discovery of the New World and the search for gold and silver there have led to a need for assayers—people capable of testing, experimenting, and determining what is true gold and what is fool's gold (pyrite). Assaying is an in-demand skill, as mining is becoming a source of wealth in Europe. Assayers often deal with chemical reactions. They use potassium nitrate (they call it saltpeter), potassium aluminum sulfate (alum), and ferrous sulfate (known as green vitriol), without really understanding what they are. When they combine and distill those three chemicals, they have nitric acid (it is known as *aqua fortis*, "strong water"), and using it they can tell silver from gold. (Silver dissolves in nitric acid; gold does not.)

Boyle, who is considered an expert on all this, is refining his thinking. He begins to define elements as we do today—as basic substances that can't be separated into anything else. (He describes them as "irreducible.") He says

Pyrite is the most common among a host of metals mistaken for gold. Fool's gold is any mineral that some poor fool thinks is gold but isn't. Other imposters are chalcopyrite, pyrrhotite, marcasite, and biotite.

"Of course the elements are earth, water, fire and air. But what about chromium? Surely you can't ignore chromium."

elements are "certain primitive and simple, or perfectly unmingled bodies" and that elements can be combined to make compounds, and that compounds can be divided into elements. Can it be that the Greek foursome—earth, air, fire, and water—aren't elements? It's a big step to question those ancient "elements," but once you put something on paper and others read it—well, it gets them pondering.

In 1657, Boyle is living in Oxford when he asks his assistant, Robert Hooke (the man Newton will come to hate), to build an air pump. Hooke designs one that is much better than any previous model; then he pumps all the air out of a huge jar, creating a vacuum. Into that vacuum, Hooke drops a feather and a coin. To the amazement of most who hear of the experiment (except for Boyle and

Boyle's first air pump for creating a vacuum (airless space) is surrounded by detailed diagrams of all its parts (below). The Boylean vacuum, as it was called, is an upgrade of a pump made by German physicist Otto von Guericke. Robert Hooke had a hand in the design of Boyle's pump.

Suspecting that the common oil of vitriol [sulfuric acid] not to be altogether such a simple liquor as chymists presume it, I mingled it with an equal or double quantity of common oil of turpentine.... And having carefully (for the experiment is nice and somewhat dangerous) distilled the mixture in a small glass retort, I obtained [a substance which] discovered itself to be sulfur, not only by a very strong sulfurous smell, and by the color of the brimstone; but also by this, that being put upon a coal, it was immediately kindled and burned like common sulfur.
—Robert Boyle, *The Sceptical Chymist*

VOLUME is a measure of how much space a mass takes up. Isaac Newton (in his *Principia*) was the first to say that it's impossible to define volume without talking about **MASS** (he called it "quantity of matter"). That seems obvious—something has to be taking up the space you're measuring. But it took Newton years to come up with a formula for density.

DENSITY equals mass divided by volume:

$$d = m/v$$

Newton wrote, "If the density of air is doubled in a space that is also doubled, there is four times as much air.... The case is the same for...all bodies."

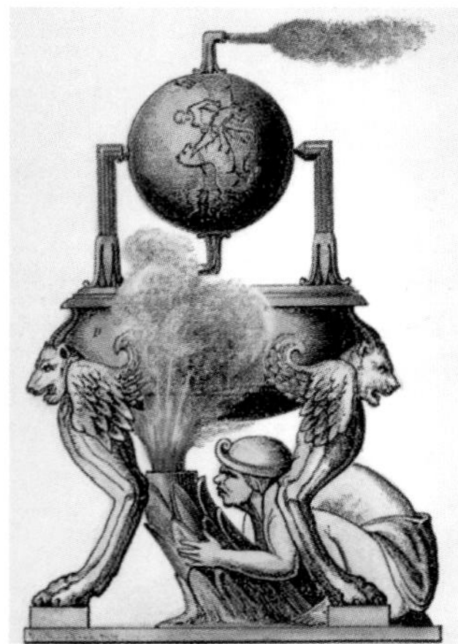

Around the first century C.E., long before Boyle, Hero of Alexandria realized that air is a substance. He observed that when trapped air is heated, its pressure increases. That led him to build this cross between a steam engine and a jet. Fire boils water in the tank; the steam escapes in high-pressure jets that spin the ball. It's a toy, not a tool. In a world powered by slaves, mechanical power was not a high priority.

Galileo's ghost), feather and coin fall together (at the same rate of acceleration).

Further experiments with the air pump make Hooke and Boyle realize that air can be compressed (squeezed into less space). In 1662, Robert Boyle does a famous experiment showing that if you compress air in a large container, it will be smaller in volume but not in mass, as long as the temperature stays the same. Then he comes up with what is known as Boyle's Law: **The volume of a gas at a constant temperature is inversely proportional to the pressure put on it.**

In other words, if you want to squeeze a volume of gas into half its space, you need to double the pressure put on it, and vice versa.

Boyle's Law, which is really quite simple, is a very important scientific milestone, although not everyone takes it seriously at the time. According to Samuel Pepys, the busy Londoner who keeps a diary, England's King Charles II "mightily laughed" when he heard that the scientists at the Royal Society are "spending time only in weighing of air, and doing nothing else since they sat."

But some thinkers understand the importance of Boyle's Law. (It is still the basis for much scientific research with gases, so it is worth memorizing.) Boyle is announcing, with this law, that science follows rules. He studies and experiments and carefully records what he discovers. He has listened to Francis Bacon, who wrote about the need for careful scientific investigation and record keeping. Both Boyle and Bacon will influence young Isaac Newton, who publishes his great *Principia* in 1687. (It is Boyle who helps Edmond Halley pay for its publication.)

Boyle understands that, in science, ideas are not enough: proofs are essential. He writes a landmark book, *The Sceptical Chymist* (which is fun to read because he writes so well). In the book, which is published in 1661, Boyle uses the dialogue form as Galileo did, and has three characters argue ideas. Galileo was gentle when he had Simplicio speak for the Aristotelians. Boyle isn't gentle at all. His Aristotelian is a pompous fool.

You can call Boyle a transitional figure. The very title of his book, *The Sceptical Chymist*, announces he is questioning medieval alchemy. In his book, Boyle separates chemistry from medicine and makes them distinct sciences.

Boyle's Law gets scientific thinkers asking, "What can air be made of, if you can change its size and shape without changing its weight?" Boyle says gases must be composed of tiny "corpuscles" and a lot of empty space, which is the reason a fixed amount of gas can be squeezed from a big container into a small one. With his corpuscle idea, he is going back to the Greek theory of atoms. With his law, he shows that air has elasticity; Boyle calls it "spring of the air."

Most people pay no attention to this. There doesn't seem to be any chance of seeing corpuscles/atoms. But in Switzerland, a young mathematician named Daniel Bernoulli reads Boyle's book and takes those tiny corpuscles seriously.

Boyle openly questioned the alchemists, who worked in secrecy, in his 1661 book *The Sceptical Chymist* (or *Skeptical Chemist*, as we would spell it today). With logic and scientific proof, he debunked the supernatural and mystical.

THE
SCEPTICAL CHYMIST:
OR
CHYMICO-PHYSICAL
Doubts & Paradoxes,
Touching the
SPAGYRIST'S PRINCIPLES
Commonly call'd
HYPOSTATICAL,
As they are wont to be Propos'd and
Defended by the Generality of
ALCHYMISTS.
Whereunto is præmis'd Part of another Discourſe
relating to the ſame Subject.

BY
The Honourable *ROBERT BOYLE*, Eſq;

LONDON,
Printed by *J. Cadwell* for *J. Crooke*, and are to be
Sold at the *Ship* in St. *Paul's* Church-Yard.
MDCLXI.

Today's atoms aren't the ultimate particles the Greeks had in mind. They can be cut. Inside an atom there is mostly empty space and extremely tiny subatomic particles.

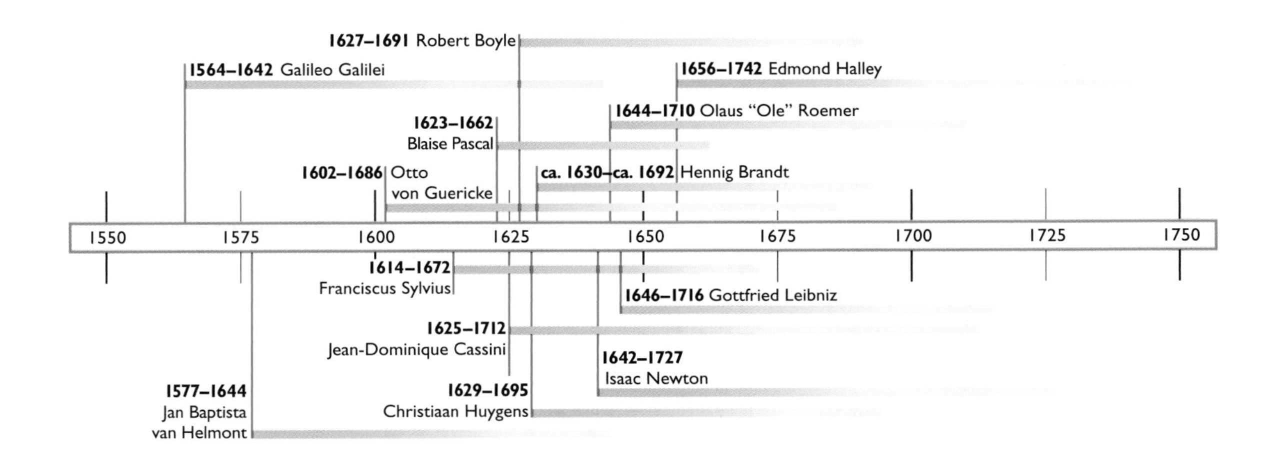

IS AIR SOMETHING—OR NOTHING?

This diagram of a Torricelli experiment shows how the level of mercury (F) remains constant whether the tube is vertical or angled.

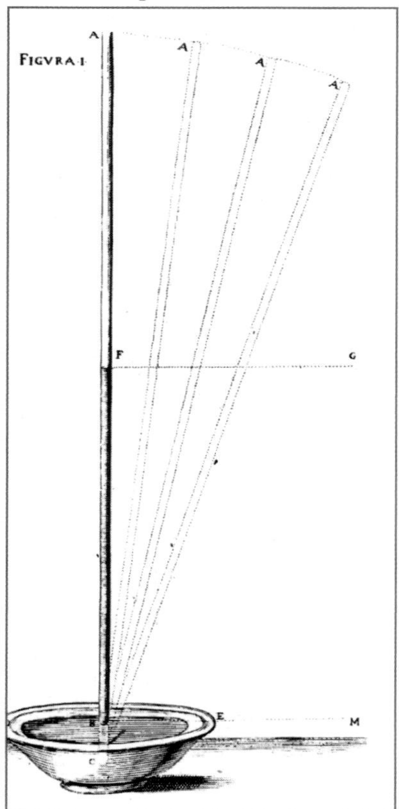

Air seems to just be there—invisible. But is it anything? The Greeks said it was one of four elements, so it had to be something. Aristotle said that air has no weight at all. Without weight, it can't be much. Without weight, it can't press on things or exert pressure. Hardly anyone bothered to think about air as a something rather than a nothing, until Galileo. He guessed that air may have mass and exert pressure. But Galileo wasn't sure of it.

Galileo encouraged two of his students, Evangelista Torricelli and Vincenzo Viviani, to do some experimenting with air. They filled a test tube with mercury. They chose mercury because it is heavy and they could get results with a small amount and a small tube. This was in about 1643.

Torricelli put his thumb on the open end of the tube. Then he turned it upside down into a bowl of mercury and slid his thumb away. Some mercury ran out, leaving a vacuum in the top of the tube. Why didn't all the mercury pour out of the tube? Viviani and Torricelli figured out that the mass of the air pressing down on the surface of the mercury in the dish reached a balance with the mass of the mercury in the tube. And that showed that air does have mass. (They were right; Aristotle was wrong.)

Torricelli, who was a good scientist, kept watching the height of the mercury in his tube. It varied slightly from day to day. He guessed (correctly) that the air in the atmosphere has a slightly different pressure at different times. He had invented the first barometer. A barometer measures atmospheric pressure, a clue to weather changes.

In France, a philosopher named Blaise Pascal heard of these experiments. (He was a friend of René Descartes.) Pascal guessed that the atmosphere is heavier at sea level than on top of a mountain. Pascal was in poor health, so he asked his brother-in-law and some friends to experiment for him. They got two tubes, two bowls, and about 7 kilograms (about 15 pounds) of mercury. They measured air pressure at the foot of a mountain. The column of mercury in the tube rose 67 centimeters (26.4 inches). Then they climbed 900 meters (about 3,000 feet) and repeated the experiment. The mercury only reached 59 centimeters (23.2 inches). They had proved the point: Air has mass. They had also shown that at sea level, the atmosphere has greater pressure than at higher elevations. Why? Because the atmospheric pressure at any point on the Earth's surface is due to the weight of the column of air above it—and at sea level there is more air pressing down than on a mountain peak.

(more about air on next page)

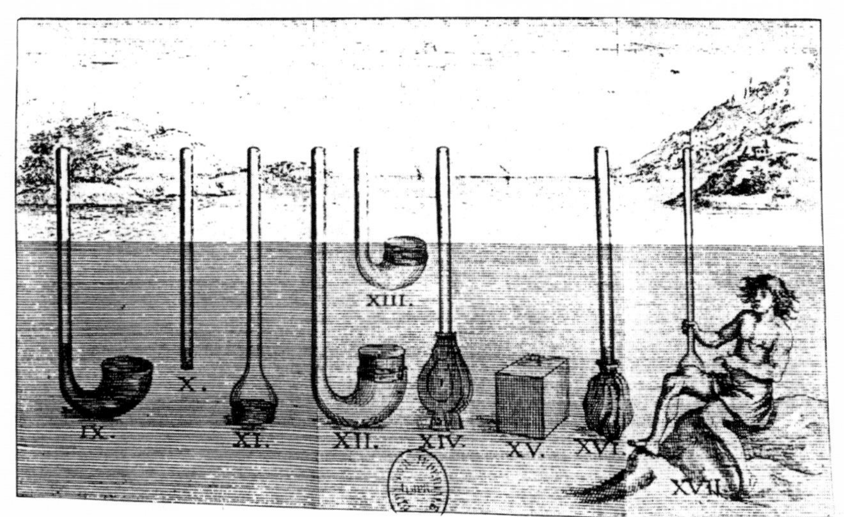

Blaise Pascal (1623–1662), a French mathematician, built on Robert Boyle's *Spring of the Air* book of experiments. His *Treatises on the Equilibrium of Liquids, and on the Weight of the Mass of the Air* features diagrams of 17 tests, some of which are shown in this detail.

Don't try this experiment. Mercury is lethal. All this experimenting with mercury made some scientists sick. A modern analysis of a lock of Newton's hair showed a high level of mercury. That helps explain his grouchy nature. He was poisoning himself!

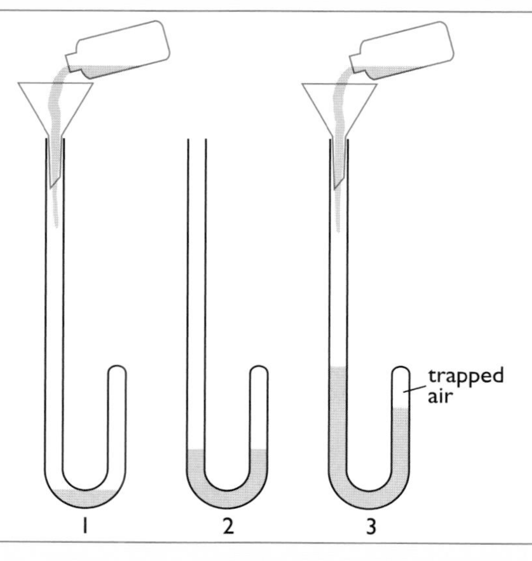

Boyle experimented using a J-shaped, 5-meter (16-foot) tube. He poured some mercury in the top (1). The mercury trapped air in the shorter part of the J (2). When Boyle added more mercury (3), the volume of the trapped air shrank in inverse proportion: Doubling the difference of height between mercury levels in the two arms cut the air volume in half; tripling it shrank the volume to ⅓.

diagram not to scale

All the King's Horses?

"Humpty Dumpty sat on a wall, Humpty Dumpty..." You know the rhyme. Where did it come from? Well, perhaps from a science experiment that drew big crowds all over Germany. Garbled tales of that experiment crossed the English Channel, and someone, trying to explain it to a child, is said to have turned it into a nursery rhyme. Here are some details on the experiment. You can decide if that explanation makes sense.

The experimenter was a German physicist with a talent for showmanship—Otto von Guericke (GAY-rik-uh). He was a colorful fellow who studied law and mathematics at the University of Leiden in the Netherlands, eventually became mayor of Magdeburg, Germany, and helped rebuild that city after the devastation of the Thirty Years' War.

Being mayor didn't keep him from thinking about scientific theories. He knew that Aristotle had said vacuums can't exist. That was expressed in a well-known phrase, "Nature abhors a vacuum." Guericke wanted to see for himself if it was true. He decided to build an air pump, empty the air out of a container, and thus make a vacuum. (It was 1650.) Aristotle had said if there were such a thing as a vacuum, sound couldn't travel through it. Guericke put a ringing bell in the container. The bell could no longer be heard. Guericke had proved (sorry, Aristotle) that nature does *not* abhor a vacuum (a vacuum can exist). But Aristotle had been right about sound and a vacuum.

Guericke's next big experiment with air was so dramatic he invited Emperor Ferdinand III to see it. (It was 1654.) He built a high platform with a fancy seat for the emperor. A crowd stood below. Guericke had cast two big copper hemispheres, shown in the illustration on the opposite page, top. They fit together making a tight sphere, especially after the mayor greased their rims. Guericke pulled the hemispheres apart and put them back together, to show how easily they came apart.

Using the air pump (it took a lot of muscle power), a blacksmith and an assistant sucked the air out of the big copper sphere. Then Guericke hitched 2 horses to opposite sides of the sphere and had them pull. They couldn't separate the 2 hemispheres. He added 2 more horses, and then 2 more. It was only when he had 16 horses—8 on a side—that they were able to pull the sphere apart.

Otto von Guericke's 1654 experiment with copper spheres and air pressure is depicted in this contemporary engraving. (The spheres didn't float in the sky. That's just the artist's way of showing a close-up of the copper hemispheres.)

Why was it so difficult? He had done it easily before pumping the air away. Could it be that air was pressing against the hemispheres holding them together? (The answer is yes.) When Guericke created a vacuum inside the sphere, all the air pressure was on the outside pushing in, with no pressure from inside to balance it. It took a team of horses to counter that air pressure. Eight years after Guericke's Humpty Dumpty experiment, Robert Boyle came up with his famous air pressure law.

If air exerts pressure, it has to be a substance, it has to be something. But what can air be? Finding an answer will lead scientists, unexpectedly, into the world of atoms. Boyle set up the problem in a scholarly manner; Guericke made it popular entertainment.

Boyle and Hooke improved Guericke's pump (see page 213) and conducted more airless experiments. They put mice and birds in a vacuum and watched them die; yet insects didn't die! They concluded that mammals and birds must breathe air from lungs. This 1768 painting dramatizes the moment of anticipation. The demonstrator, with his hand poised above an airless glass, seems to ask, "Will the little white bird inside die?" Candle-lit faces display a range of emotions: the fear and horror of two girls (right), the thoughtfulness of the man next to them, the awe of a boy (left), and a couple's self-absorbed indifference.

21 Daniel and the Old Lion Hunter

No fewer than 120 of the descendants of the mathematical Bernoullis have been traced genealogically, and of this considerable posterity the majority achieved distinction—sometimes amounting to eminence.... None were failures. The most significant thing about a majority of the mathematical members of this family...is that they did not deliberately choose mathematics as a profession but drifted into it in spite of themselves.

—E. T. Bell (1883–1960), British mathematician, *Men of Mathematics*

Of course these flying molecules must beat against whatever is placed among them, and the constant succession of these strokes is, according to our theory, the sole cause of what is called the pressure of air and other gases. This appears to have first been suspected by Daniel Bernoulli, but he had not the means which we now have of verifying the theory.

—James Clerk Maxwell (1831–1879), Scottish mathematician, "Molecules"

Daniel Bernoulli (ber-NOO-lee) wanted to be the Newton of the eighteenth century; he thought he could do it by studying fluids (by that he meant liquids and gases).

Daniel, who was born in 1700, had the genes to go for it. His father and an uncle were world-famous mathematicians. Another uncle, a cousin, an older brother, and some assorted relatives in this remarkable family were first-rate mathematicians. So it was no surprise that Daniel could deal with numbers at an amazingly early age. But his career wasn't as easy as you might think: his father, Johann, was not your normal, loving dad. He was jealous, nasty, and miserable—you'll see.

Who is the lion hunter in the chapter title? And who is the lion? There's a clue on page 179.

It is no accident that the words "ingenious" and "engineer" derive from the same root (*ingenium:* mental ability, cleverness, a naturally clever temperment).

—Lisa Jardine, English professor, *Ingenious Pursuits*

Almost all the Bernoullis were achievers; this family tree includes just the most prominent among them. Nicolaus (top), the patriarch, wasn't a mathematician or scientist. He was a spice merchant and druggist, like his father, Jacob the Elder. Nicolaus and his wife, Margaretha, had 10 children, including the feuding mathematician brothers, Jacob and Johann (the baby of that family). Johann and his wife, Dorothea, had three mathematician/scientist sons (including Daniel) and three grandsons who were mathematicians.

The Bernoulli story begins in Belgium with Jacob the Elder (Daniel's great-grandfather), who was a Huguenot, which means he was a Protestant follower of John Calvin—not a safe thing to be in seventeenth-century France or Belgium. Huguenots were branded, maimed, and sometimes killed by the majority Catholics. Many left for Switzerland or the Americas, which were havens for religious outcasts. Jacob fled to Basel, Switzerland in 1622, where he established a spice and drug business and prospered.

"Which Bernoulli do you wish to see—'Hydrodynamics' Bernoulli, 'Calculus' Bernoulli, 'Geodesic' Bernoulli, 'Large Numbers' Bernoulli, or 'Probability' Bernoulli?"

His son Nicolaus, also a prosperous druggist, has high hopes for his children. (He has 10 kids.) Jacob and Johann are expected to carry on the family traditions: Jacob is to study religion; Johann is to be a druggist. But both

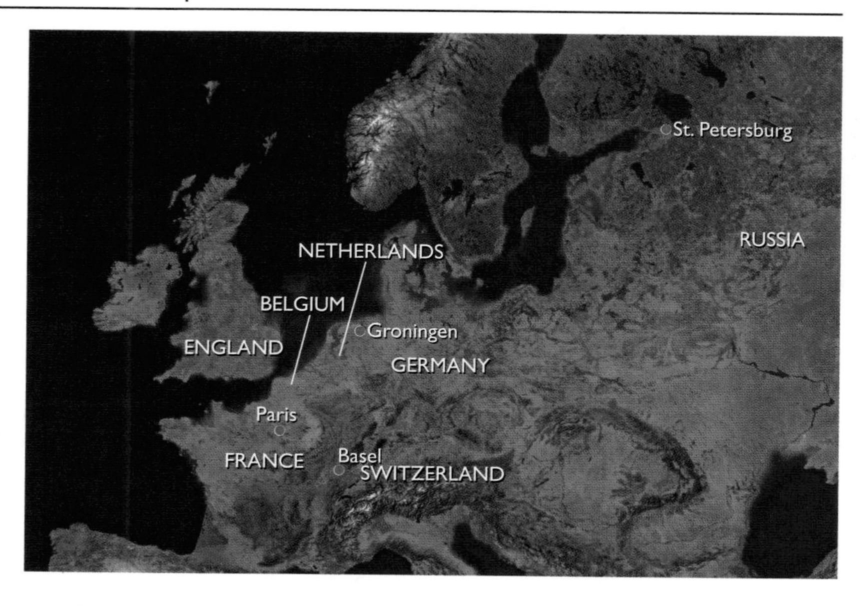

are fascinated with numbers and natural philosophy. Jacob, almost 13 years older than his brother, writes in his diary, "Against my father's will I study the stars."

Johann goes into business, but he doesn't have a talent for it, and soon even his father can see that. So Nicolaus agrees to let Johann study medicine; it has a connection to the pharmacy business. But like his brother Jacob, Johann is secretly studying numbers. "I've now turned to mathematics," he writes in his diary, "for which I feel a special joy."

When Daddy Nicolaus finds out what his sons are doing, he is furious. He says they will have to find jobs; he is not going to support their studies any longer. In 1676 Jacob leaves for Geneva, and 7 years later he becomes a professor of mathematics at the University of Basel. Brother Johann becomes chair of the mathematics department at Groningen University in the Netherlands in 1695.

Think of Gottfried Leibniz as a universal genius, and you won't be wrong. He wrote about math, philosophy, science, history, law, and politics, and has been called the Aristotle of the Seventeenth Century.

Meanwhile, the German mathematician Gottfried Wilhelm Leibniz (LYB-nits) publishes a paper on calculus (in 1684). As you know (I hope), Isaac Newton already worked out calculus in 1666—but he still hasn't published his results. Two great minds have independently found the same thing, which is not so unusual. But Newton feels his ideas have

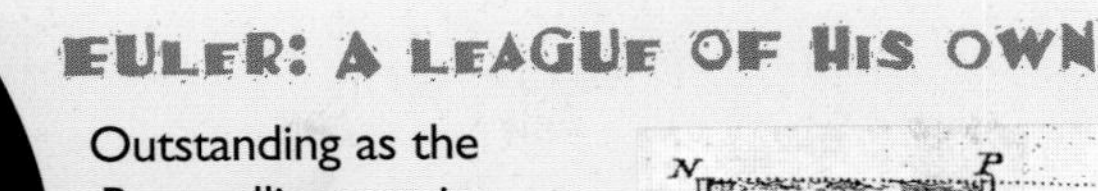

Outstanding as the Bernoullis were in mathematics, they aren't even in the same league as Leonhard Euler (say "oiler"), who was Johann's student and Daniel's friend. If this were a book on mathematics, there would be several chapters on Euler. But it isn't, so you'll have to find out about him on your own. As it happens, he was a kind, generous man (in contrast to grumpy Johann Bernoulli and miserable, secretive Isaac Newton). Euler loved to tell stories to his 13 children. William Dunham, in a book called *Journey Through Genius*, says, "The legacy of Leonhard Euler... is unsurpassed in the long history of mathematics.... Euler's collected works fill over 70 large volumes, a testament to the genius of this unassuming Swiss citizen who changed the face of mathematics so profoundly."

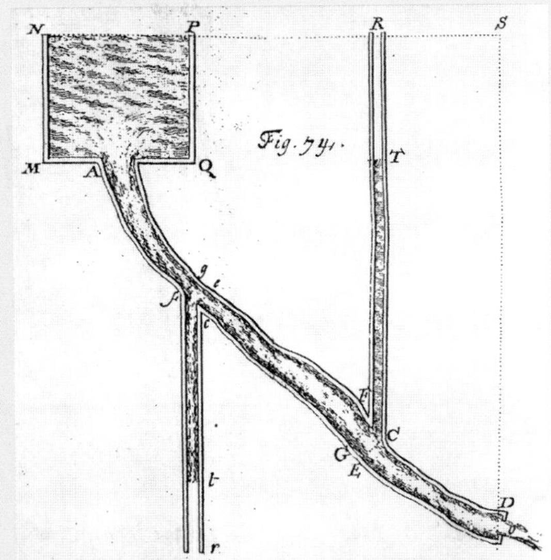

Daniel Bernoulli and Leonhard Euler found a way to measure the pressure of a fluid flowing in a pipe (A to D, above). The greater the pressure, the higher the liquid rose (T) in the glass tube (C). Doctors began measuring blood pressure by sticking a tube into a patient's artery.

been stolen. Some years later, Newton accuses Leibniz, saying that Leibniz had seen some of his (Newton's) work; he releases their correspondence. Leibniz retaliates by suggesting that it was Newton who actually used his (Leibniz's) methods.

Along come the two Bernoulli brothers. They read Leibniz's paper on calculus and at first have a hard time understanding it. (For more on calculus, see pages 160–163.)

The Bernoullis study Leibniz again and again until they figure out calculus for themselves; then they get involved in the Newton-versus-Leibniz argument—on Leibniz's side. To them (and some others), it seems to be British science against the science of the European continent. Johann takes a leading role in the controversy, writing with no modesty, "When in England war was declared against Monsieur Leibniz for the honor of the first invention of the new

TAKING A GAMBLE WITH MATH

Probability, a branch of mathematics, began with gambling. Pierre de Fermat (of the famous Last Theorem), Blaise Pascal, and the Bernoullis wanted to know the mathematical odds of winning at the card table. Probability didn't tell them for certain that they would or wouldn't draw an ace; it just told them how likely it was. A deck of 52 cards has 4 aces, so the odds of the first drawn card being an ace are 4 in 52 (or 1 in 13). If 20 cards have been played and not an ace among them, those odds improve to 4 in 32 (1 in 8).

Always keep in mind that probability is about the *likelihood* of outcomes, not the *certainty*. If there are only 4 cards left in the deck, and no aces have been played, you can predict with certainty that the next card will be an ace—but you're not using probability; you're using fact.

Probability is central to the physics that deals with the complex world inside atoms. We can't determine the action of an individual particle, but with a large number of atoms, predictions based on probability become very accurate.

Card games were widely popular in seventeenth-century France. Here, workers are making playing cards by hand in about 1680.

calculus of the infinitely small, I alone . . . kept at bay. . . the entire English army."

Fighting Isaac Newton doesn't seem to be enough for them: the brothers take to belittling each other's work. One prestigious mathematics journal prints their vitriol until finally the editor has had enough.

But the brothers also do serious mathematics. People of their time are fascinated with games of chance. When you roll of a pair of dice, can mathematics predict what numbers you'll roll? No. But looking for patterns in cards and in dice makes the Bernoullis realize that overall patterns can be described statistically. And that leads them to help develop a language of mathematics that deals with probabilities. (Probabilities will become essential in understanding the action of atoms.) The Bernoulli work in this field is 100 years ahead of the times.

Now, let's skip to Daniel's birth. He is the son of Johann, the younger brother. When he is five, his uncle Jacob dies. "This unexpected news bowled me over," says his daddy, Johann, who adds, "and then it entered my thoughts immediately. . . that I could succeed to my brother's position." So the family moves from Holland to Switzerland. Daniel grows up in Basel, where Johann is a university professor.

Johann begins acting just as his father, Nicolaus, had. Johann decides his son Daniel should become a merchant and enter the family pharmacy business. But Daniel wants to study mathematics. He fails as a pharmacist. (Does this sound familiar?)

Johann then insists that his son Daniel go to medical school, but he does allow him to study mathematics on the side, and he does answer his questions. And since Johann is one of the best mathematicians in the world, Daniel gets very good training.

One of the things that preoccupies the great professor Johann Bernoulli is a little-studied phenomenon called *vis viva* in Latin. It means "living force" and is what we call the energy of motion. No one understands it, but Daniel is fascinated. *Vis viva* is invisible but clearly powerful.

What do editors do? Here's an example: On the opposite page (line 5), I used the word *vitriol*. If you read the last chapter, you know what it means. But it wasn't the first word I chose to describe the Bernoullis' angry words. I called them "diatribes." This is what my editor wrote: "Joy, I find that, back then, a diatribe was just a discourse or study; it came to mean a bitter criticism in English in 1804. Rant would be an excellent choice because it's from Shakespeare, with a Dutch root." Good comment, but somehow *rant* didn't work for me. I tried *screed, harangue, tirade, censure, denunciation, attack,* and *invective*—but decided a word with a double meaning—like *vitriol*—was just what I wanted. That's what good editors do (like good teachers): they make you think.

That's the city of Basel, Switzerland (left), flanked by mountains and halved by the mighty river Rhine. Note how many (30!) churches appear in this 1740 engraving.

In a paint sprayer, high-speed air from a horizontal tube draws paint up a vertical tube. How? Try this: Snip a narrow straw *almost* all the way through and bend it at a 90° angle. Put the bottom end in water and blow—hard!—into the horizontal top half. Your breath zips across the open bend, reducing the air pressure and causing water to rise up the straw.

When Daniel finishes medical school—with top grades—he expects to get a professor's job in Basel. He gets no help from his father. Daniel finally takes a position in Russia as a mathematician at the Imperial Academy of Sciences and Arts in St. Petersburg. His experiments and writings there soon make him widely known; and he becomes a quiet fan of Isaac Newton. He doesn't realize it, but his father is fuming.

In 1734, both Daniel and his father, Johann, write papers for the Paris Academy of Sciences, which gives a big prize (much like today's Nobel Prize). That year, the top prize is split; it is awarded to the two Bernoullis—father and son. Daniel comes home to Basel. He thinks his father will be pleased. But Johann is furious. He says his son is trying to take over his position as Europe's top mathematician. Johann throws his son out of the house. Daniel never returns.

Now all of that is like gossip—interesting but not really important. What Daniel accomplishes becomes a landmark in science. Like so many achievements, it sounds simple, but no one else has figured it out.

Daniel Bernoulli comes up with a famous and very useful principle. It applies to moving fluids (liquids or gases) and the pressure put on them. (Don't confuse it with Boyle's Law, which is about gases that are compressed; Bernoulli's Principle relates to fluids that are flowing and incompressible.) Here it is: **When the speed of a moving fluid increases, pressure in the fluid decreases, and vice versa.**

If that theorem interests you, consider a career in engineering. You can think about designing airplanes or ships or even bridges if you understand Bernoulli's Principle. If you want to build a carburetor or an atomizer (a sprayer, like on a perfume bottle), where air and liquid droplets are the moving fluid, you'll use Bernoulli's Principle. In an aspirator (a suction pump), it is water (or another liquid) that does what Bernoulli said it should do.

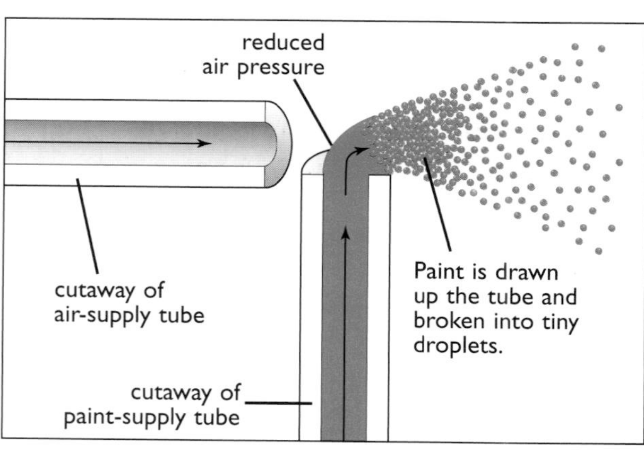

The principle, in simple language, is this: The faster a fluid (liquid or incompressible gas) is traveling, the lower its pressure. Here it is stated scientifically: **The pressure in a fluid varies inversely with its speed squared.**

That's what causes a house to lose its roof during a hurricane. The fast hurricane winds reduce the pressure on the top of the roof, and the air pressure *inside* the house pushes the roof off.

Engineers designing airplane wings know (thanks to Bernoulli) that the air flowing over the upper surface of an aircraft wing needs to travel at a faster rate than air flowing beneath the wing. Therefore, airplane wings are usually

Bernoulli's Principle doesn't describe the flow of fluids at or near the speed of sound. Supersonic fluids behave differently than those at slower velocities. Their density, temperature, and compressibility changes.

A Big, Imaginative Kinetic Idea—Before Science Was Ready for It

Daniel Bernoulli made a creative conjecture (guess or speculation). He said that a container of gas is mostly empty space and that the atoms making up the gas are always in rapid, chaotic motion bumping against each other and against the walls of their container. That turned out to be true. It is called the Kinetic Theory of Gases (*kinetic* is from a Greek word that means "motion"). But it would be more than 100 years before his conjecture could actually be proved.

Bernoulli's ideas didn't have much luck. It took a long time before they were accepted. Maybe it was because Bernoulli's writings were mostly in mathematical formulas. Maybe it was because he wasn't English, and for a while there was a rivalry between English science and that of the rest of the world. But I believe it was because his ideas came before the technology to prove them.

Kinetic Theory: Molecules in Motion

Bernoulli realized that when you heat a gas, its pressure increases. Heating the gas makes its atoms and molecules move faster, and they wham the sides of the container with more vigor.

The Kinetic Theory states that all atoms and molecules—in gases, liquids, and solids—are in constant, vigorous motion. According to Brockhampton's *Dictionary of Science*, "At standard temperature and pressure . . . a litre of air contains 30,000 million million million molecules, moving at an average speed of 450m/1,500 ft per second. Each molecule undergoes 5,000 million collisions every second."

Bernoulli was way ahead of his time in analyzing heat. Most scientists of his day thought of heat as an invisible substance; they had no idea it was a form of kinetic energy.

WHY DO PLANES STAY IN THE SKY?

What explains flight? Well, it happens that physicists argue about that. Some cite Bernoulli's Principle, and that does have something to do with it. Air travels faster over the top of the curved wing than under the flat bottom. The slower moving air below the wing exerts more pressure than the faster moving air on top, creating lift.

Airplane designers create curved wings to take advantage of this boost. Yet even a plane with flat wings or a plane that's upside down (with the curved side of the wings facing the ground) can still fly. The reason is Isaac Newton's third law: For every action there's an equal and opposite reaction. The wing is angled, and so its downward force on the air creates an equal and opposite upward force. This makes more sense if you never forget that air has mass. Think of an airplane riding on a massive cushion of air. If it dives, nose first, it slices through the air quickly, but then, as it flattens out, more of the airplane's surface pushes down on the air (there's more air resistance), and the plane bounces back up with more force.

The faster a plane is moving, the more force it exerts on the air (and vice versa). That's why, when an airplane slows down too much, it can stall out—stop and tumble from the sky.

Here's the simple reason airplanes can fly: Airplane wings push down on air, which pushes back. This plane just skimmed over some fog. Its downwash (downward motion of air) is recorded as a trough; the upwash is apparent in the rising, curled edges.

The angle of attack is the angle at which a wing slices through the air. Even a small 5° or 10° tilt adds enough air resistance (drag) under the wing to push it up. The greater the angle of attack, the greater both the drag and the lift. (To feel this for yourself, hold a cookie sheet edge-first into a strong wind and slowly angle it upward.) Airplanes need lots of lift to take off, so they usually do so at a high angle of attack and a high acceleration (to overcome the added drag).

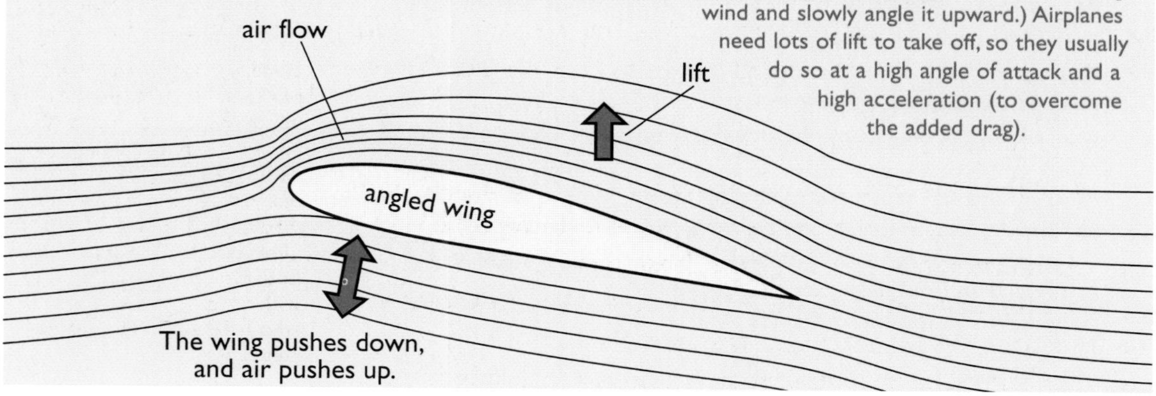

designed so that the upper surface of the wing is curved, which means air passing across it will move faster than the air flowing along the straight bottom. That makes the pressure on top of the wing lower than that below. And that helps the airplane to lift. (There are other factors involved with lift; see page 228.)

Bernoulli's Principle also led to a conservation law. Here it is: **The total energy in a fluid stays the same no matter what shape the fluid takes.** If liquid goes from a big bottle into a smaller container, the speed and pressure of its atoms will change but its total energy will not.

In physics, **CONSERVATION** basically means that nothing is lost or gained. More on that idea coming up.

Daniel Bernoulli goes even further than Robert Boyle in anticipating atoms. He says that gases are made up of a vast number of tiny particles, and it is the random, constant motion of those atoms hitting the walls of a container that explains pressure in a gas. (Later this will be called the Kinetic Theory of Matter.) It is a remarkable deduction, since no one in his time can be sure of atoms. The atomic idea, which goes back to the ancient Greeks, has mostly been discarded. But Bernoulli believes atoms exist. (Like Newton and Boyle, he thinks that atoms are hard, solid, impenetrable bits of matter—which they aren't.)

Believing in atoms and figuring out that they are in constant motion—well, that is an astonishing achievement. It will be several generations before there are tools to prove it true. Writing about atoms—getting the idea down on paper as Boyle and Bernoulli do (even if they called them "corpuscles")—means others will consider them.

This is happening at the same time that some people are saying science is a closed field. They think that Newton has said everything that needs to be said on the subject of science. They are wrong (understatement)! Energy is a wide-open field. So is the chemical world of elements. Atoms are going to be verified, and that will change the whole way we look at matter. Physics is just gearing up, and mathematics will provide its key tool.

Daniel Bernoulli seems to sense those possibilities (and probabilities). He isn't an Isaac Newton, but he is impressive. His thoughts and his principle will percolate in other heads and help bring changes in science and lift to technology.

OH, THOSE MATHEMATICIANS!

In Mathematicks he was greater
Than Tycho Brahe, or Erra Pater:
For he by Geometrick scale
Could take the size of Pots of Ale;
Resolve by Signes and Tangents straight,
If Bread or Butter wanted weight;
And wisely tell what hour o'th day
The Clock does strike, by Algebra.
—Samuel Butler (1612–1680), English poet, *Hudibras*, Part One

Hudibras is the comic hero of the poem, which makes fun of the serious Puritans. He is a buffoon, but the poem gives you an idea of the esteem in which mathematicians were held in his time. (This Samuel Butler is not to be confused with a more famous one who lived in the nineteenth century.)

22 Brains and Beauty Squared

Judge me for my own merits, or lack of them, but do not look upon me as a mere appendage to this great general or that great scholar, this star that shines at the court of France or that famed author. I am in my own right a whole person, responsible to myself alone for all that I am, all that I say, all that I do. It may be that there are metaphysicians and philosophers whose learning is greater than mine, although I have not met them. Yet, they are but frail humans, too, and have their faults; so, when I add the sum total of my graces, I confess I am inferior to no one.

—Émilie du Châtelet (1706–1749), French mathematician, letter to Frederick the Great of Prussia

Women in France were far more able to participate in intellectual activities than in England. . . . In Paris, influential women headed salons attended by men as well as women, and . . . were relatively free to engage in academic debates. Paradoxically, they were to lose this enlightened liberty after the [French] Revolution.

—Patricia Fara, British science historian, *An Entertainment for Angels: Electricity in the Enlightenment*

Common sense is not so common.

—Voltaire (1694–1778), French philosopher and writer, *Dictionnaire Philosophique* ("Philosophical Dictionary")

Think gorgeous (big brown eyes). Add brains (lots). Charm (much). Riches (a 30-room childhood home with 17 servants, for a start). Then include a desire to work hard and accomplish important things.

You're looking at Émilie du Châtelet (AY-mee-lee doo SHAH-t'lay). Émilie worries her father. Eighteenth-century French girls aren't supposed to be smart, or if they are, they aren't supposed to show it. "My youngest flaunts her mind, and frightens away the suitors. . . . We don't know what to do with her," says her dad. But he does understand

Some French women had *salons* (discussion groups) in their homes, but Émilie wanted to attend meetings of the premier science organization, the French Academy of Sciences, and women were excluded. They weren't allowed in the popular coffeehouses either. Gradot's was a Parisian coffeehouse where scientists, philosophers, and mathematicians met for regular sessions. Émilie was turned away. So she had a suit of men's clothes made, went to Gradot's, and was cheered by her scientist friends, who made a regular place for her at their table. As long as she wore men's clothing, the proprietors looked the other way. In this detail of a painting (left) by Henri Testelin, King Louis XIV meets with the all-male members of the academy at the height of his record-setting reign (1643–1715, longer than any European monarch). Émilie flourished under his successor, King Louis XV.

that he has an unusual daughter, and he sees that she has a good education.

Then, on her own (women aren't allowed in universities) and with tutors she finds, Émilie studies the mathematics of Newton, Leibniz, Descartes, and the Bernoullis. Only a few people, male or female, are able to do that. When she takes fencing lessons, she becomes good enough to scare off clumsy challengers. And as to games of cards—which are fascinating the rich of the time—Émilie easily memorizes every card played and figures out the mathematical

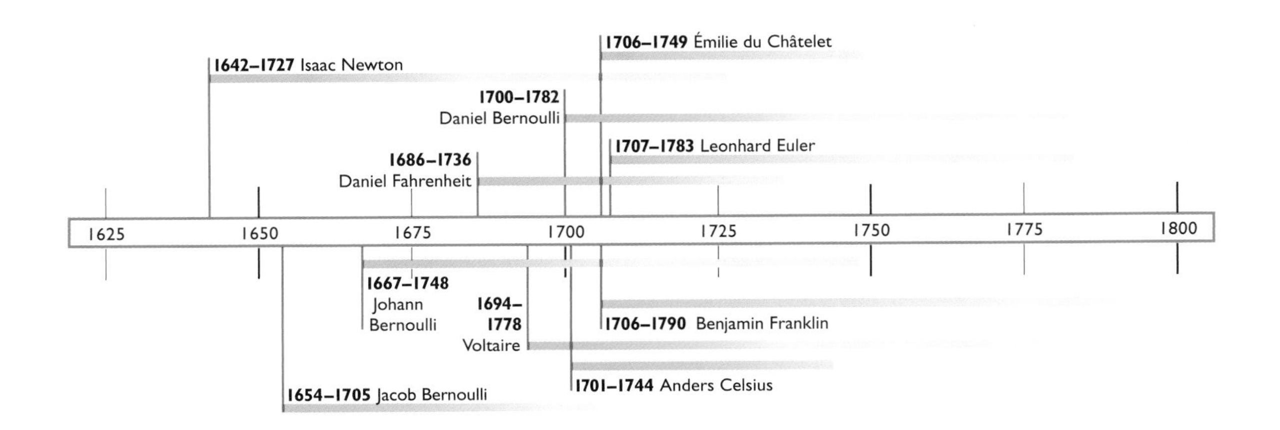

Blackjack is a card game called *vingt-et-un* ("21") in France.

probabilities of various moves. But gambling being chance, not science, she often loses at the blackjack table.

Still, she is in a league of her own. Can you imagine her at the French court of Versailles (ver-SY), where most beautiful young women just sigh and flutter their eyelashes?

"My daughter is mad," her father writes in frustration. "Last week she won more than two thousand *louis* [French money then] at the card tables, and after ordering new gowns . . . she spent the other half on new books. . . . She would not understand that no great lord will marry a woman who is seen reading every day."

Voltaire, describing life in Paris under Louis XV, makes it clear that magistrates of quite ordinary means lived in a style that few people could afford today. Their wives were covered with diamonds and dressed, as they themselves, in embroidered clothes worth a small fortune.

—Nancy Mitford (1904–1973), English novelist and biographer, *Voltaire in Love*

Émilie isn't eager to marry, but at 19 (a late age in her crowd), she is talked into marrying an aristocratic and handsome 34-year-old officer, Florent-Claude du Châtelet-Lomont. She becomes a marquise (mar-KEEZ; it's like a duchess) and has three children. But Florent-Claude spends most of his time soldiering; he is hardly ever home. They have little in common and agree to live separate lives.

Then she falls in love. It is with a Frenchman who is a renowned writer and philosopher. His name is François-Marie Arouet; his pen name is Voltaire. This is what he writes: "I was tired of the lazy, quarrelsome life led at

At this elegant French *soirée* (swah-RAY, an evening party), men and women play cards together. Notice the dog, the wigs, and the gestures in this French painting from the end of the seventeenth century.

Voltaire had the good taste to have his portrait painted by one of France's greatest artists, Maurice-Quentin de La Tour (1704–1788), a master of light and shadow. Voltaire is holding *La Henriade*, a book-length poem he wrote (of course). When it came to a battle of wits, he left his opponents in shreds. (No wonder they were eager to see him in jail.) Among people who saw as he saw, he was all charm and brilliance.

Paris. . . . In the year 1733 I met a young lady who happened to think nearly as I did." (When it comes to math and science, she thinks even better than he does.)

Voltaire may be the ultimate Enlightenment thinker: he combines a passion for literature and the arts with knowledge of science. He is a playwright, historian, poet, political essayist, and one of the best writers of satire of all time. He likes to poke fun at the decadent lifestyle and empty-headedness of the ruling classes. Satire is funny if you are a reader; it is not funny at all if you are being satirized.

Eighteenth-century France, like most nations then (and many now), didn't allow much freedom of expression. If you wrote or said the wrong thing, you could be thrown in jail, which happened to Voltaire more than once. One article he wrote got him sent to France's infamous political prison, the Bastille. He was locked in there for 11 months. But he did have his books and paper—and good food too. (This was France.) He managed to write an epic poem and revise a play.

Below is a clay model of the Bastille, the infamous prison where Voltaire and other political prisoners were locked up. Built from 1370 to 1382 as a fort, it became a symbol of absolute royal rule. On July 14, 1789 (a decade after Voltaire's death), a mob revolting against the king stormed the prison, which was soon torn down. July 14 is now Bastille Day, a national holiday in France.

But being in the Bastille didn't seem to teach Voltaire a lesson. A few years later, he got into a word duel with a rich aristocrat—who hired some thugs to beat him up. This was serious. A warrant was issued for Voltaire's arrest. After some plea bargaining and pleas from his friends, he agreed to leave the country. Voltaire went to England.

There he was a big celebrity and

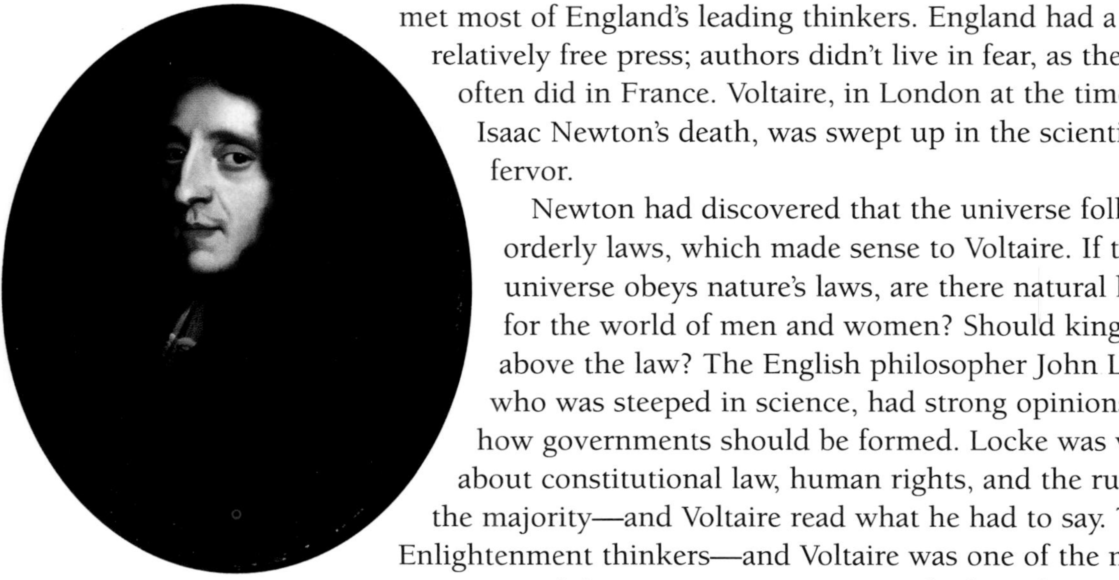

John Locke, who died in 1704, was a friend of Robert Boyle. Locke's ideas on liberty would influence the British colonists across the Atlantic.

met most of England's leading thinkers. England had a relatively free press; authors didn't live in fear, as they often did in France. Voltaire, in London at the time of Isaac Newton's death, was swept up in the scientific fervor.

Newton had discovered that the universe follows orderly laws, which made sense to Voltaire. If the universe obeys nature's laws, are there natural laws for the world of men and women? Should kings be above the law? The English philosopher John Locke, who was steeped in science, had strong opinions on how governments should be formed. Locke was writing about constitutional law, human rights, and the rule of the majority—and Voltaire read what he had to say. The Enlightenment thinkers—and Voltaire was one of the most important of them—were attempting to look at the world rationally and scientifically. They were obsessed with the idea of personal freedom.

But as much as he admired English ideas, Voltaire didn't much like the English weather or English food. He was homesick for France. Though he made fun of it, he missed French society. And despite its faults, the court of King Louis XV seemed to him the most powerful and exciting power center in the world. (Most Europeans agreed with that.)

Louis XIV added the glittering Hall of Mirrors to the Palace of Versailles, his lavish royal residence on the outskirts of Paris. Known as the Sun King, he took the throne at age four and believed he had the divine right to rule. He said, "*L'état c'est moi*," which means, "I am the state." Under his rule, the nobility paid no taxes, but peasants did. (Who got to use the hall?) King Louis XV, who reigned during the time of Émilie and Voltaire, inherited the splendors of Versailles.

So nearly three years after he fled from France, Voltaire is back in Paris. He's at an opera performance when he meets Émilie du Châtelet; it doesn't take long for him to fall in love. "Everything about her is noble, her countenance, her tastes, . . . her politeness," he says in a letter to a friend.

But Voltaire can't seem to stay out of trouble. He writes *Letters Concerning the English Nation*, which explains and praises English ideas on government and science. Members of the French parliament are not pleased; they see this as criticism of France. A friend warns Voltaire: the police are looking for him.

He heads for a castle that Émilie's husband (an agreeable fellow) owns at Cirey, in northeastern France, near the border with the independent state of Lorraine (which makes escape from France easy if necessary). Émilie soon follows. They decide to build a retreat at Cirey that will be a scientific research center as well as a place to enjoy life. Voltaire is rich (he is a brilliant investor) and he pays the bills. He writes to a friend that Émilie du Châtelet is, "changing staircases into chimneys and chimneys into staircases." She is also installing the latest laboratory equipment from London and building a scientific library that will include 21,000 books and be as good as any in France. Émilie and Voltaire add a small theater, rooms for Florent-Claude, rooms for her children, and lots of guest rooms.

When Voltaire first arrived at the Château de Cirey in 1734 (shortly after publishing his *Letters Concerning the English Nation*), it was so run-down, the wind gusted through cracks in the walls. He restored it and lived there with Émilie until her death in 1749. The castle is near Lorraine, which was then an independent province and is now in northeastern France.

A servant who worked at Cirey said, "Mme du Châtelet passed the greater part of the morning with her writings, and did not like to be disturbed. When she stopped work, however, she did not seem to be the same woman. The serious air gave place to gaiety and she gave herself up with the greatest enthusiasm to the delights of the society." This oil painting of her is by Carle van Loo.

Voltaire has a private wing. Émilie has her own lab and a reading room with wall paintings by Jean-Antoine Watteau (one of the best French artists of the day). Living in their fairy-tale castle (which you can visit today), they work hard during the day and play hard in the evenings. Émilie has a fine singing voice and a talent for drama; Voltaire writes plays. They put on dramatic performances in French; they talk science in English.

"We long employed all our attention and powers upon Leibniz and Newton; Mme. du Châtelet attached herself first to Leibniz, and explained one part of his system in a book exceedingly well written," Voltaire writes. Du Châtelet soon turns from Leibniz and focuses on Newton. She translates Isaac Newton's *Principia* into French (for a long time it is the standard translation), adding explanations so Newton can be more easily understood. She makes his ideas available on the European continent. Voltaire writes a popular work explaining Newton, and he keeps writing satire.

Those he makes fun of don't like it at all. (And some have influence.) To keep from getting arrested, Voltaire stays away from Paris. A few times, he has to flee to Holland. After one trip to Holland, he tells Émilie of some experiments he has seen there.

Meanwhile, she is writing mathematical papers, a textbook of science, a paper on the nature of fire, and others on philosophy and religion. But that isn't why she is included in this book. It is because of something she figures out after Voltaire describes the Dutch experiments. It is a link between nature and energy and mathematics.

The universe contains matter and energy. That's all there is. Newton analyzed the world of matter. Energy didn't get much attention until the nineteenth century.

The science of energy is just being born. Remember, the Bernoullis called energy *vis viva*. So does Émilie du Châtelet. Newton said all you have to do to figure out energy (E), is to multiply mass (m) by velocity (v). The formula $E = mv$ expresses his idea. So a 10-pound wagon going 5 miles an hour has, according to Newton, 50 units of energy. (In metric units, it is the same straightforward formula: just multiply any mass times any velocity.)

Du Châtelet realizes that when it comes to energy, Newton didn't have it right. (Mass times velocity is now used as the formula for momentum, not energy. See page 174.) She understands that there is something that is strange in nature that even the great Isaac Newton missed (although Gottfried Leibniz did not): **Nature seems to love to square things.**

At first, when Voltaire tells her of the experiments done in Holland, they don't make sense. The Dutch researcher has dropped heavy weights into soft clay. Everyone believes that a brass ball going twice as fast as another will sink two times deeper. That's what the formula $E = mv$ predicts. And a ball going three times as fast should sink three times as deep. But that isn't what happens in the Dutch experiment.

A brass ball going twice as fast as another one sinks *four* times as far into the clay. A ball going three times as fast sinks *nine* times farther. Nature is squaring itself. **Du Châtelet realizes that the formula for energy is actually $E = mv^2$.** (She repeats the experiment with ivory bullets, pendulums, and other objects—always with the same results.)

Émilie du Châtelet is a good writer and influential. She makes others aware of nature's tendency to square things. (In England, it will be a while before anyone questioning the great Newton will be taken seriously, especially a woman who isn't even English.)

Du Châtelet has come up with something important. Put it in your head. Her formula for energy is $E = mv^2$. Another formula for energy—one that will change the world—will be built on it.

The Dutch experimenter who shared his results with Voltaire and du Châtelet was Willem 'sGravesande. That's not a misprint; in Dutch, 's means "of." 'sGravesande was one of the new scientific thinkers who believed in experimenting. He wrote a book titled *Mathematical Elements of Natural Philosophy, Confirm'd by Experiments.*

To finish du Châtelet's story: At age 42, she became pregnant. Before she could give birth, an infection set in. Voltaire, her husband, and other friends and admirers were with Émilie when she died. Voltaire wrote to a friend, "I have lost . . . half of myself, a soul for which my soul seems to have been made." This eighteenth-century sketch (below) is a study for statues of soulmates Émilie and Voltaire.

23 It's a Gas! Take Its Temperature!

Heat and temperature...were considered to be one and the same thing until attempts to measure them precisely revealed that they were quite different....Heat is the total amount of molecular motion in something, while temperature is the average energy of that motion.
—K. C. Cole, American science writer, *Sympathetic Vibrations: Reflections on Physics as a Way of Life*

Once the Industrial Revolution got under way, it gave a huge boost to science, both by stimulating interest in topics such as heat and thermodynamics (of great practical and commercial importance in the steam age), and in providing new tools for scientists to use in their investigations of the world.
—John Gribbin (1946–), British astronomy professor, *The Scientists*

Just what is air? Boyle and Bernoulli thought about how it behaves, but what is it made of? Air is air, and that's it, say most people. There doesn't seem to be any reason to question the Greek idea that it's an element. But a few people do just that.

One of them is Joseph Black (1728–1799). Black comes from a family of Scottish descent, although his father was born in Ireland, and he was born and raised in Bordeaux, France (where his dad was a wine merchant).

Kidney stones are crystals, some as big as a golf ball, that form in the urinary tract. (Don't worry; they're pretty rare.) They are formed when minerals drop out of urine, like salt crystals from evaporating seawater. (Drinking lots of water can help prevent kidney stones.) A common kidney-stone mineral is the element calcium (Ca), which is present in your bones and teeth.

He becomes a real Scotsman when he goes off to the University of Glasgow (in Scotland). At that university he studies medicine and anatomy with a terrific teacher who is up-to-date on Boyle and his experiments with air. Although Black intends to be a doctor, he's fascinated with scientific research and does some original work that makes him realize that minerals in the landscape and minerals in the body are similar. (He's studying kidney stones.)

It's no secret why limestone has so much carbon in it. Carbon is the basic element of life (your body is 18 percent carbon by weight), and limestone is made of dead life-forms—the shells and skeletons of ancient sea creatures. Chalk is a soft, powdery form of limestone that is almost all calcium carbonate ($CaCO_3$). (Another limestone, called dolomite, contains magnesium.) The famous White Cliffs of Dover, England (left), are made of natural chalk. The sea creatures who created them shared the planet with the last of the dinosaurs—roughly 65 million years or so ago.

Black is a perfectionist; when he experiments he makes sure every component is precisely weighed and measured. He takes some chalk from the ground (mostly calcium carbonate, chemically known as $CaCO_3$) and superheats it, and finds that it gives off a gas that he calls "fixed air," with a leftover solid that is calcium oxide (lime). This is something new, his finding that gases can be formed from ordinary solids.

Black's fixed air is carbon dioxide, or CO_2 (he doesn't know that terminology). Black discovers that fixed air doesn't act like ordinary air. You can't burn substances in it, and you can't breathe much of it, either. He watches the leftover lime from his experiment slowly turn back into calcium carbonate—by nabbing some fixed air out of the air in the room, he deduces. He figures out that some calcium carbonate (especially chalk, a form of limestone) weathers away naturally, turns into fixed air, and becomes part of the air we breathe. All of these observations help him understand that air—which is supposed to be basic and elemental—is

If you're "in the limelight," you're on stage—the center of attention. The expression comes from pre–electric-light times, when theater stages were lit by the brilliant white flash of lime (calcium oxide) being flamed with a gas torch (oxygen and hydrogen). One of the comedians in this painting (below), by Jean-Antoine Watteau (1684–1721), is holding his own limelight.

WORDS THAT MATTER

Matter is divided into two categories—substances and mixtures.

A **substance** is made up of the same matter throughout. One kind of substance is an **element**. Elements can't be decomposed (broken apart) further. Gold is an element. Oxygen is an element. Another kind of substance is a **compound**. Compounds, such as water (H_2O), are made of elements bonded together. That bond can be broken chemically but not physically; water (H_2O) can be split into hydrogen and oxygen with an electric jolt.

A **mixture** contains two or more substances (either elements or compounds) whose atoms do not bond. In a sugar-and-water solution, the proportion of sugar to water can vary. (A water molecule, which is a compound, always has two hydrogen atoms and one oxygen atom.) Both the sugar and water keep their separate properties. (The hydrogen and oxygen in water don't.) You can separate a mixture of sugar and water physically—by evaporation, for example. In a mixture of oil and water, gravity does the separating: the lighter oil floats to the top.

These aren't tubby toys; they're models of water molecules. The "flying" ones are evaporating.

actually a mixture of gases. This is a totally new idea.

He's still a student when he publishes his findings in 1756. He not only gets his doctorate, he immediately becomes famous in the scientific community.

Black is soon a professor of medicine and a lecturer in chemistry at Glasgow, and later at the University of Edinburgh (also in Scotland). His lectures are so good, students come from far away to hear him. (One of his students is Benjamin Rush, who will become the first professor of chemistry in America.)

Black is also a practicing doctor whose patients include the famous Scottish economist Adam Smith, the Scottish philosopher David Hume, and the Scottish geologist James Hutton.

But his patients don't keep him from experimenting. Using one of Daniel Fahrenheit's thermometers, he discovers that there is a difference between heat and temperature (see box on page 242), and he makes that important difference clear.

Gases, which had always seemed mysterious, begin to be

What You Breathe

We now know that air is made up of a mixture of gases (not to mention water vapor, dust, plant spores, and bacteria). Here are its gassy ingredients.

Gas	Percent by Volume
Nitrogen (N_2)	78.08
Oxygen (O_2)	20.95
Argon (Ar)	0.93
Carbon Dioxide (CO_2)	0.033
Neon (Ne)	0.0018
Helium (He)	0.00052
Methane (CH_4)	0.0002
Krypton (Kr)	0.00011
Nitrous Oxide (N_2O)	0.00005
Hydrogen (H_2)	0.00005

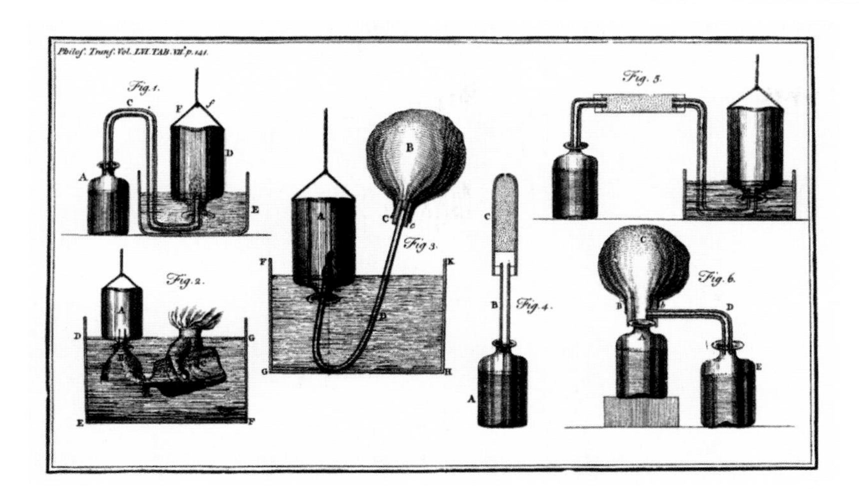

In 1766 Henry Cavendish published "Three Papers, containing Experiments on factitious Air." *Factitious* means "unnatural"—or in this case, laboratory fabricated. This figure (left) shows factitious ways that Cavendish trapped the gas created in chemical reactions.

taken seriously as states of matter, like solids and liquids. A few others join Black in seeing gases as chemicals that can be analyzed.

Englishman Henry Cavendish (1731–1810) is one of them. In 1766, he believes he has discovered the supposed fire element phlogiston. Some metals, acted on by an acid, release a gas that is very flammable. Cavendish calls it "fire air." There is no phlogiston, but he doesn't know that. What he's actually discovered is the gas hydrogen, a colorless, odorless element.

On Earth, hydrogen is usually found in combination with oxygen as water (H_2O). Its name refers to that water connection (*hydro-* comes from the Greek word for "water").

Hydrogen (H) is the most plentiful element in the universe, accounting for 93 percent of the universe's total atoms.

In 1772, a Swedish apothecary (druggist), Karl Scheele (1742–1786) discovers oxygen. Scheele finds that air contains oxygen, along with carbon dioxide and another newly discovered gas, nitrogen. (In England, Daniel Rutherford has also isolated nitrogen.) Clearly, air is a mixture, so it can't be an element.

Poor Scheele will discover eight elements (chlorine, fluorine, manganese, barium, molybdenum, tungsten, nitrogen, and oxygen) and get little credit for any of them. He finds a way to make phosphorus—without using urine—and Sweden becomes a leading maker of matches (but he never profits from his discovery). Scheele has an awful habit—he tastes everything he uses. No one is quite sure

Carl Wilhelm Scheele's
d. Königl. Schwed. Acad. d. Wissenschaft. Mitgliedes,
Chemische Abhandlung
von der
Luft und dem Feuer.
Nebst einem Vorbericht
von
Torbern Bergman,
Chem. und Pharm. Prof. und Ritter; verschied. Societ. Mitglied.

Upsala und Leipzig,
Verlegt von Magn. Swederus, Buchhändler;
zu finden bey S. L. Crusius.
1777.

Karl Scheele's discoveries were overlooked in his day, even though he published *Chemical Observations and Experiments on Air and Fire* (left) in 1777. Today, more than two centuries later, Scheele finally gets credit as the man who discovered oxygen. Original copies of his only book are selling in the neighborhood of $20,000.

what kills him, but at age 43, he is found dead at his workbench. His lab is full of toxic chemicals.

Not too long after Scheele, Joseph Priestley (1733–1804) also discovers oxygen. Since the publication of Scheele's results have been delayed by his publisher, Priestley doesn't know about them. Priestley does know that oxygen has a connection to fire because things burn brightly in this gas. He is convinced that the fire element phlogiston exists. Priestley also discovers that mice surrounded by this "new" gas are more

TALKING HOT AND COLD

The difference between heat and temperature is important, but subtle, so here's an attempt to clarify.

Everyone knows what **heat** is, but try to define it, and you'll find yourself going around in circles. It's **the internal energy possessed by a substance due to the motion** (known as kinetic energy) of its vibrating molecules or atoms. (See page 320 for why scientists prefer the term *thermal energy* instead of *heat*.)

Temperature is a measure of the *average* thermal energy (or heat) **of the particles.** The faster the molecules move, the higher the temperature.

This difference can be confusing, so here's a comparison: 2 gallons of water at a temperature of 10°C (50°F) have twice the heat of 1 gallon at 10°C (50°F). Both volumes of water have the same temperature but not the same amount of heat.

Here's another way to look at it: Temperature can be measured by a thermometer; it's a matter of degrees. Heat is a quantity thing. You can get the temperature of water in a container by measuring a small sample of it. To know its heat, you need to measure the whole quantity. Heat is a measure of amount. Temperature measures the intensity of a thing.

And here's one more *hot* concept: You boil water, and it gets to 100°C. But, if you continue to add heat, it doesn't get any hotter. Where does the heat energy go? It is used to change the liquid (water) to gas (steam). That's called *latent heat*, and it refers to the heat needed for a phase change without a change in temperature.

The boiling point of pure water is 100°C (212°F). But even if you turn up the burner, the boiling water won't get hotter.

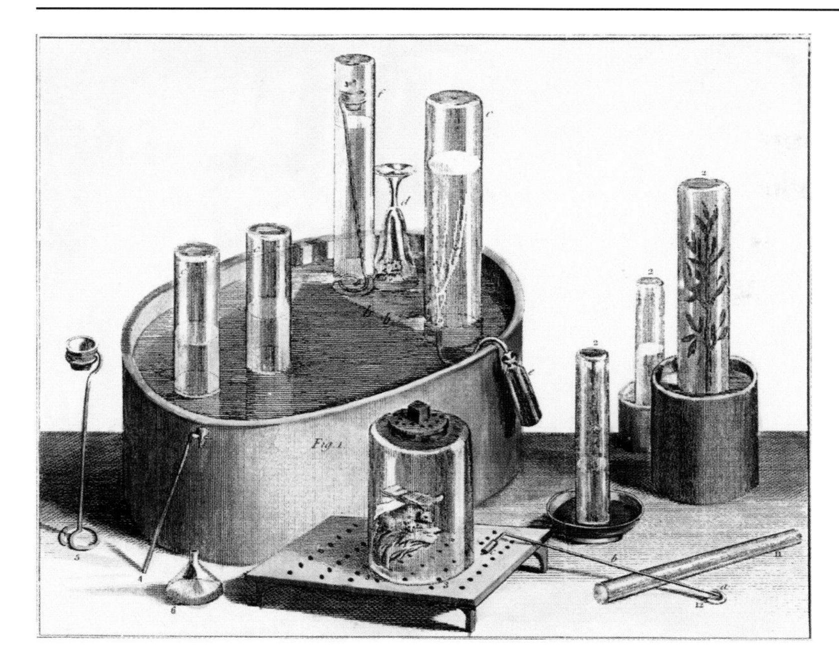

Here's a peek at Priestley's lab equipment from his 1774 book *Experiments and Observations on Different Kinds of Air*. The tub is a pneumatic trough. He used it to isolate gases, including oxygen (unaware of Scheele's discovery of the gas three years earlier). Priestley also experimented with the respiration of animals. Note the mice trapped in the glass in the foreground. Priestley is said to have always taken pains to keep his mice warm and comfortable.

frisky. He says he himself feels "light and easy" when he breathes it.

Joseph Priestley lives near a brewery, which gives off carbon-dioxide fumes. So he too experiments with CO_2. Priestley puts a lit candle in a container with normal air, encloses it, and watches as the candle burns out. The container is now filled with carbon dioxide. Then he puts a mint plant in a glass of water and places it in the closed container on a sunny windowsill. The plant flourishes and keeps flourishing. After several months, a mouse can live and breathe in the container and a fire will burn there again. What has happened? (We now know that plants, using sunlight, convert carbon dioxide into oxygen through photosynthesis.) Priestley has discovered a clue that plants and animals exist in a chemical balance, which keeps the Earth's atmosphere breathable. Besides oxygen, Priestley identifies another nine gases, including "alkaline air" (ammonia), "marine acid air" (hydrogen chloride), "phlogisticated nitrous air" (nitrous oxide—laughing gas), and "vitriolic air" (sulfur dioxide). (Those are his names in quotation marks.) But Priestley won't give up the phlogiston

In 1768 Joseph Priestley dissolved some carbon dioxide in water. When sugar and flavorings were added, the carbonated water turned into soda pop. Priestley sold his invention to a colleague named Mr. Schweppe. You can find Schweppes tonic water and ginger ale in some grocery stores today.

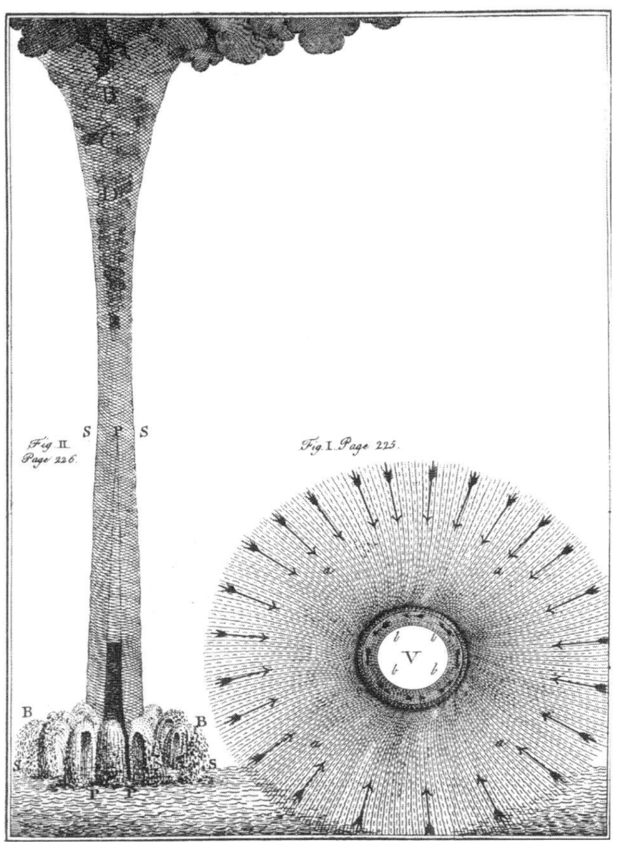

In 1743, Benjamin Franklin helped establish the American Philosophical Society in Philadelphia, the first scientific society in what is now the United States. The engraving above is from his paper *Experiments and Observations on Electricity* (1751). It's a diagram of a waterspout, which Franklin described as a whirlwind, except over water instead of land. He observed that, in both cases, a rising, twisting column of air forms when a light hot air mass collides with a cold heavy one. The air itself is invisible, but it churns up water at the bottom and forms clouds at the top as it cools. That wheel-shaped thing on the right is a cross-section of the air column.

idea; he keeps looking for the fire element that doesn't exist. Twenty-first-century writer John Gribbin will say, "As a chemist, Priestley was a great experimenter and a lousy theorist."

A big-hearted, nonconformist English clergyman, Priestley is a friend of the well-thought-of scientist Ben Franklin. Franklin is known as an "electrician," as are other experimenters investigating bolts from the heavens and static on Earth. In London, Priestley and Franklin share books and ideas. While this is going on, the British and the Americans are snarling at each other in Boston and Williamsburg. Then the fighting begins. Franklin says, "There never was a good war or a bad peace."

Priestley writes more than 150 books (one is *The History and Present*

Luck Happens to Those Who Are Ready

Do you think most scientists know what they're doing before they do it? Some do, but others, like Joseph Priestley, seem to combine serendipity (fate or luck) with curiosity. How did Priestley find oxygen? He heated mercuric oxide and then put a candle in the "air" he got. The candle flared like a torch. It was responding to the oxygen. Here are Priestley's words:

> *I cannot, at this instance of time, recollect what it was that I had in view in making this experiment; but I know that I had no expectation of the real issue of it. Having acquired a considerable degree of readiness in making experiments of this kind, a very slight... motive would be sufficient to induce me to do it. If, however, I had not happened... to have had a lighted a candle before me, I should probably never have made the trial; and the whole train of my future experiments relating to this kind of air might have been prevented.*

State of Electricity). He also writes books on religion and philosophy; they reflect his liberal beliefs, which enrage some people. In 1791, an English mob burns his house and laboratory during a three-day rampage that focuses on freethinkers and natural philosophers. The age of reason seems to be winding down. Priestley finds it hard to believe that the mindless mob has burned "manuscripts which have been the result of the laborious study of many years, and which I have never been able to recompose; and this has been done to one who never did, or imagined, you any harm." In 1794, he flees to the religiously tolerant new nation, the United States of America.

Joseph Priestley had some way-out beliefs. He supported the American colonists against King George III, he was against slavery, he hated religious bigotry, and he thought the French Revolution was a good thing. A mob burned down his house. He was a Unitarian minister, and the next Sunday, he preached, "Father, forgive them for they know not what they do." This 1791 cartoon (above), "Doctor Phlogiston, the Priestley Politician or the Political 'Priest'," shows Priestley trampling a pamphlet called "Bible explained Away" and holding flaming papers that say, "revolution toasts" and "political sermon."

Meanwhile, shy Henry Cavendish, who has no interest in politics, is busy experimenting with carbon dioxide. He finds that it is 1.5 times heavier than air. (The modern value is 1.52.)

Cavendish publishes a paper, "Experiments on Air" (in 1784), describing his discovery that water can be made by combining hydrogen and oxygen. But Cavendish never understands that his experiment shows that water is a compound of hydrogen and oxygen. Author Richard Morris will say, "He was hobbled by the phlogiston theory, the only theory that chemistry then had."

All this experimenting with gases and heat is disorganized—lots of people are doing lots of things. There's an important reason for the interest: It is fueled by a rising demand for power. Energy is still a baffling unknown *something*. Can it be harnessed? The public wants practical results from science. Will it have them?

TURNING ON THE HEAT

The engraving below shows several thermometers developed in the seventeenth and eighteenth centuries. The wacky, spiral-shaped one (second from left on the bottom) is from Florence and has 300 marked degrees.

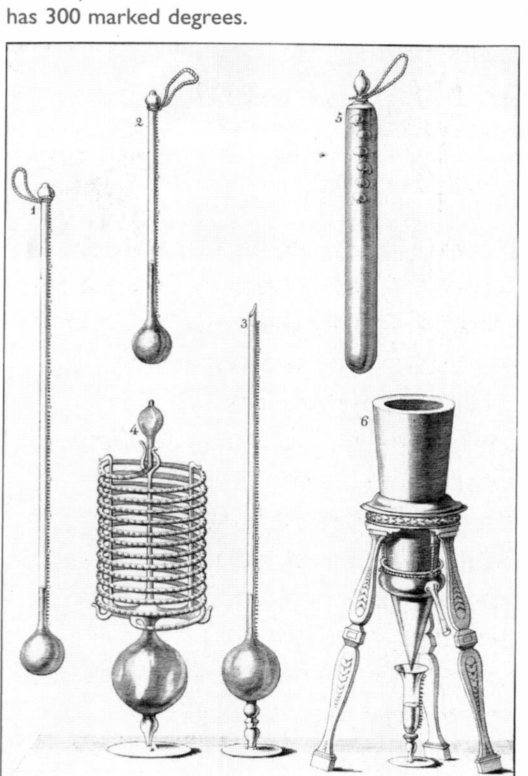

D**aniel Fahrenheit** (1686–1736) was on the run—running from the police—because he wanted to make a reliable thermometer. This is how it happened: When Fahrenheit was 15 and living in Danzig, Poland, both his mother and father died after eating toxic mushrooms. The Danzig city council had to figure out what to do with five orphans. The four youngest Fahrenheits were put in foster homes; the oldest brother, Daniel, who was quick and bright, was apprenticed to a merchant who took him to Amsterdam, Holland, to teach him his trade.

This was at the beginning of the eighteenth century, and thermometers were a hot item for merchants in busy, commercial Amsterdam. They weren't very accurate, though; no two were alike, and there was no uniform scale of measurement. Still, they seemed a marvel to young Fahrenheit. And when he learned that one thermometer showed that water always boils at the same temperature, he was amazed. Here are his words, written later:

Galileo's thermoscope was basically a thermometer without marked degrees. It measured relative temperatures—such as whether the temperature was rising or falling.

> *About ten years ago, I read in the* History of Science *issued by the Royal Academy of Paris, that the celebrated Amontons* [a French scientist] *using a thermometer of his own invention, had discovered that water boils at a fixed degree of heat. I was at once inflamed with a great desire to make for myself a thermometer of the same sort, so that I might with my eyes perceive this beautiful phenomenon of nature, and be convinced of the truth of the experiment.*

This was the age of reason, and there was a passion for knowledge and for precision. Surveyors and mapmakers were turning the Earth into a laboratory. Exact measurement was an obsession of the times. Scientific thinkers saw the need for reliable thermometers and for a standard temperature scale. Daniel, who didn't want to be a merchant, ran away from his apprenticeship in order to work on the problem. (Apprenticeships were a kind of legal bondage. An apprentice couldn't do what he wanted.) The Danzig city council sent the Dutch police after Fahrenheit. He took off, traveling to Denmark, Germany, and Sweden among other places, studying and learning wherever he went. When he turned 24, Daniel became an adult and was finally—and legally—free. He could concentrate on thermometers.

Galileo, who had lived a century earlier (1564–1642), is often called the inventor of the thermometer. But he built a thermoscope, a thermometer without a scale. It just indicated when the temperature went up or down. A contemporary of Galileo, Santorio Santorio, may have been the first to put a numbered scale on a thermometer (in about 1612). He used air as a measuring rod in his thermometer, but since he didn't

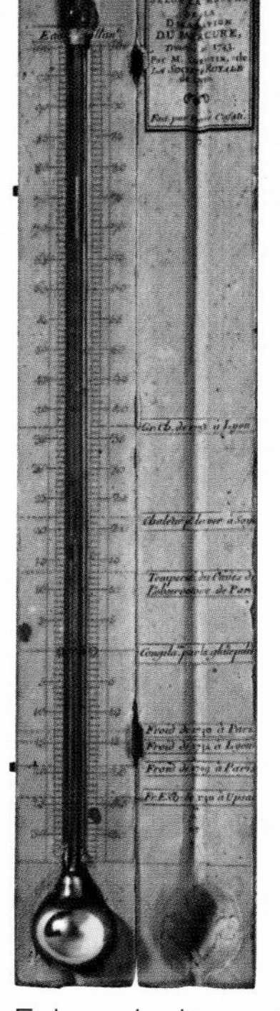

This French mercury thermometer (ca. 1790) is one of the first to use the Celsius scale. The degrees range from –35° (35 degrees below the freezing point of water) to 100° (the boiling point of water). The case has handwritten labels of memorable temperatures such as the sweltering 1738 summer in Lyon (37°C) and the frigid Uppsala winter of 1740 (–24°C).

understand the effects of air pressure, it gave a poor measure. In 1654, Ferdinand II, the Grand Duke of Tuscany, put alcohol in a glass tube and sealed the tube. That created a new model for thermometers, but, otherwise, the duke's device wasn't very accurate. Isaac Newton tried using linseed oil in a thermometer. Others used water or wine. All these substances expand when the temperature goes up. You can observe or measure the amount of expansion, which is why they work as thermometers. But none of the early models was very accurate. And because there was no uniform scale, one couldn't be compared to another.

Daniel was using an alcohol thermometer when he decided to figure out a scale that could be a standard for all thermometers. He mixed ice, water, and ammonium chloride (a salt compound) to get the lowest temperature he could manage and called it 0°. From there, he tried one scale of numbers and another and finally settled on 32° as the freezing point of water and 212° as its boiling point (a 180° difference, which is a convenient number for calculations). That became the Fahrenheit scale, which is still commonly used in the United States.

When he was 28, Fahrenheit put some mercury in a tube and sealed it tightly. Mercury—a silvery element sometimes called "quicksilver"—stays liquid at quite high and low temperatures and expands and contracts evenly as the temperature changes. Mercury thermometers had been used in the past, but Fahrenheit perfected them. With his marketing skills (perhaps learned as an apprentice), he got word of them out. Before long he was making thermometers that would set Europe's scientific standard. Émilie du Châtelet became one of his customers. So did Ole Roemer. (Then, as now, science knew no borders.)

But Daniel Fahrenheit's scale was unwieldy in mathematical calculations. In 1742, a Swedish astronomer, Anders Celsius came up with another temperature scale, which eventually set the freezing point of water at 0° and its boiling point at 100°, making calculation easy. At first, this scale was called centigrade (after Latin words meaning "a hundred steps"). In 1948, in an international agreement, it was renamed the Celsius scale. Except for the United States, most of the world

In Search of a Scale

At first, Fahrenheit chose small numbers for his temperature scale. He set the freezing temperature of a chemical mixture at 0°, the freezing point of water equal to 4°, and the human body temperature equal to 12°. There was too much of a leap in temperature, however, so Fahrenheit multiplied each degree by 8. That's how the freezing temperature of water became the now-familiar 32°F. Human body temperature became 96°F—lower than the actual figure of 98.6°F, but it was a nice, round number.

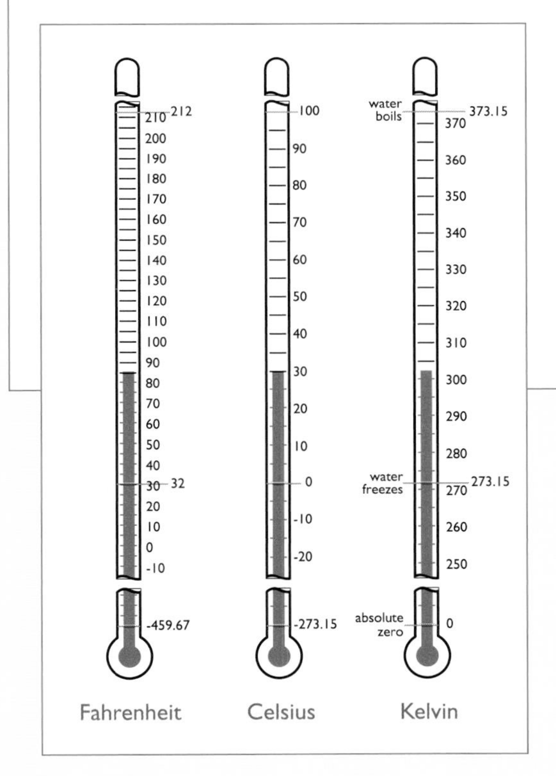

"Let's go over to Celsius's place. I hear it's only 36° over there."

officially adopted the Celsius scale that year.

Many scientists use yet another scale. In 1848, William Thomson (known as Lord Kelvin in Scotland and Baron Kelvin in England) took the Celsius scale and extended it to a base of absolute zero. (More on Kelvin in chapter 37.)

So Fahrenheit, Celsius, and Kelvin are the three scales to consider. As for thermometers, today's instruments often operate on principles different from Fahrenheit's mercury model. Some measure molecule speed. Others measure infrared radiation from warm bodies. A Yale University professor, Dr. Robert J. Schoelkopf, has invented a thermometer that can measure the noise of atoms bouncing off each other. The louder the noise, the higher the temperature.

24 Weighing the World

[Henry Cavendish's] first published work was issued in 1766, when he was thirty-five; it concerned chemical measurements. His last paper, released in 1809, one year before his death, concerned astronomical measurements. In between, he weighed and measured a lot of things very precisely. One of them was the world.

—Robert P. Crease (1953–), American philosopher and science historian, *The Prism and the Pendulum*

[Cavendish's] Theory of the Universe seems to have been, that it consisted solely of a multitude of objects which could be weighed, numbered, and measured; and the vocation to which he considered himself called was, to weigh, number, and measure as many of those objects as his allotted three-score years and ten would permit.

—George Wilson (1818–1859), Scottish biographer, *Life of the Hon. Henry Cavendish*

Henry Cavendish (1731–1810) has been called the greatest experimental scientist of the eighteenth century. But if you had lived in his time, you probably wouldn't have known him. Almost no one did. His first biographer, George Wilson, described him as "an intellectual head thinking, a pair of wonderfully acute eyes observing, and a pair of very skillful hands experimenting or recording."

To those who do know him, that's all he seems to be. Cavendish is shy almost to the point of illness. When he rides in a carriage, he scrunches himself in a corner so no one can see him. He takes his daily walks at night to hide in the shadows. He never marries, but he has a housekeeper. He leaves her written directions. Once, he accidentally meets her

Henry Cavendish was very rich. But he never cared about money. He spent only what he needed and passed most of his fortune on to family members. One descendant, William Cavendish, became chancellor of Cambridge University. In 1870, William endowed a laboratory in the Cavendish name. That now-famous lab was the site of great accomplishments in nuclear physics. It was also where DNA was discovered. James Clerk Maxwell was its first director (a lucky break for the lab; see page 364).

on the staircase and is so upset by that encounter, he has a back staircase installed so it won't happen again.

But he does belong to the Royal Society, as does his father, and they attend meetings and dinners. It is from recollections of its members that we know a bit about Henry. They tell us that he hardly ever spoke, and when he did, it was with a squeaky voice.

When he is 40, Henry inherits a great deal of money. It doesn't change his style of life, although he does build a platform in front of his house. From there, he can climb into a big tree to make astronomical observations. And he puts a large thermometer on his roof. It becomes a landmark. Otherwise, he lives a frugal life of exact routine. Every night he eats the same dinner—leg of mutton. He can't be bothered keeping up with fashions, so all his adult life, he has his tailor make him the same style suit. As he ages, the clothes he wears are more and more out-of-date; but he sticks with them, including a cocked three-cornered hat.

This, the only known portrait of Henry Cavendish, was painted secretly. Members of the Royal Society invited an artist to one of their dinners and sat him near the reclusive scientist. The artist studied Cavendish's face and form and later painted this picture without telling Henry what he was doing.

Some of the Cavendish clan: Henry's grandfather, William Cavendish, 2nd Duke of Devonshire, is in the red coat. Henry's great-uncle, James Cavendish, is standing. James's father-in-law, Elihu Yale (in the middle) gave some money to found a school in Connecticut. The figures on the right are a "Mr. Tunstal" and a page.

Reminder: In modern math terms, Newton's formula for the force (F) of gravitation is:

$$F = G\frac{m_1 m_2}{r^2}$$

m = mass
r = distance between the masses
G = the universal gravitational constant (a symbol that Newton didn't use, but the concept was in his formula)

According to Wilson,

> *He hung up his hat invariably on the same peg, when he went to the meetings of the Royal Society Club. His walking-stick was always placed in one of his boots, and always in the same one....His brain seems to have been but a calculating engine...his heart only an anatomical organ....Such was his life, a wonderful piece of intellectual clockwork....He was almost passionless.*

That may not be true. He does have a passion: he is in love with science, and that seems enough for him. But he doesn't publish all his work, and he certainly doesn't talk about his achievements, so it is a century after his death before many of his findings are uncovered.

Pardon this repetition, but some people (like me) need to be reminded of definitions.

MASS is the quantity of matter in a body as measured by its inertia. In a given place, equal masses experience equal gravitational pull (measured as weight).

WEIGHT is the force exerted on an object by gravity. On Earth, weight depends on the object's mass and the strength of Earth's gravitational pull—which decreases with distance from Earth's surface. (You weigh a tiny bit less on a mountaintop than at sea level.)

He is nearly 70 when he does his most famous experiment. He is determined to find the average density of the Earth. The size of the Earth was known, but not its weight (and therefore its density). How do you weigh the Earth?

According to Newton, if you know the force of attraction exerted by Earth on two objects—and you know the force of attraction between the two objects—you can use the ratio of those figures to find the density of the Earth.

Sound easy? The measure of Earth's gravitational attraction toward you is your weight. That is easy to find; a bathroom scale will measure it. The trick is measuring the mutual attraction between two objects that are on the Earth. You and the chair you are sitting on are attracting each other, but try measuring that attraction. Gravity is a very weak force. It seems strong only when we measure the gravitational attraction between huge objects, like a planet and the Sun. As to measuring the attraction between a cannonball and a cathedral? Newton seemed to think it would be impossible. "Whole mountains will not be sufficient to produce any sensible effect," Newton wrote of the possibility of measuring the gravitational force on Earth.

Newton could compute the gravitational attraction of one planet to another—that's a relative thing. But he couldn't figure the average density of Jupiter or Earth and be sure of his results. He estimated that Jupiter's density is one-fourth that of the Earth. He was very close—it's 24 percent rather than 25 percent.

So when Isaac wanted to know the density of the Earth, he had to settle for an informed guess. He knew the densities of

TWO SHOCKING EQUATIONS

Here's some sophisticated physics for those who want to know more:

Henry Cavendish, like most of the scientists of his time, was fascinated with electricity and did significant experimenting in that field. He discovered the electrostatic force. (When you rub your shoes on the carpet, touch a metal doorknob, and get a shock, that's the electrostatic force.) And he was first to figure out that electrical forces obey a rule similar to gravitational forces. But, typically for him, Cavendish didn't publish his results.

Both Charles-Augustin de Coulomb and Henry Cavendish built torsion balances to measure extremely weak forces of attraction. This small-scale model is based on Couloumb's design.

Charles-Augustin de Coulomb (1736–1806), a French military engineer who turned to science for its tranquillity, made the same discoveries—but he published. So he gets the credit.

Coulomb's formula is set up the same way as Newton's formula for universal gravitation. It, too, obeys the Inverse Square Law. What that means is that, like gravitation, the amount of electrostatic force depends on the inverse square distance between the two objects. If you double the distance, the force drops to one-fourth the strength. (How far from the doorknob can your finger be and still produce a shock?)

Here are the formulas, side by side, for the two kinds of force (F):

Electrostatic	**Gravitation**
$F = k \frac{q_1 q_2}{r^2}$	$F = G \frac{m_1 m_2}{r^2}$

You'll recall that r is the distance between the centers of the two objects. In place of mass (m), the two q's stand for the electric charges of the objects. Like G, k is a constant, but there's a gigantic difference. In metric units, k looks like this:

$$k = \frac{8{,}987{,}552{,}000 \text{ Nm}^2}{C^2}$$

Notice that there's no decimal—only commas in that number. Those zeroes mean that the value of k is extremely big. (G is extremely small.) Electrostatic force is stronger than gravitation, which explains why the socks from your dryer cling to you instead of falling to the floor. The capital C part of k is a unit of electric charge named the coulomb, after its discoverer. (The electrostatic force is called the coulomb force.)

There's another fascinating difference between these formulas. Gravitation is always a force of attraction, but an electrostatic force can attract or repel (push away). The reason is that q, an electric charge, can be positive or negative, but m (mass) can only be positive. A positive and negative charge (+ –) attract each other, but two objects with like charges (+ + or – –) repel each other.

Measurement Basics

Today's "system international"(abbreviated SI) begins with base units of mass, length, time, and so on. All other measurements are derived from these seven bases. Three of them include:

- the **kilogram** (about 2.2 pounds) for mass
- the **meter** (close to a yard) for length or distance
- the **second** for time

G (the universal gravitational constant) uses all three. It is expressed in newton meters squared per kilogram squared (Nm^2/kg^2). A newton is how much push or pull is needed to accelerate a mass of 1 kilogram (1 kg) 1 meter per second per second (1 m/s^2).

$G = 0.0000000000667\ Nm^2/kg^2$

That's 10 zeroes after the decimal point—gravity is extremely weak.

common substances—water, various rocks, metals—and, using them, he came up with an estimate for all of Earth. "It is likely that the total amount of matter in the Earth is about five or six times greater than it would be if the whole Earth consisted of water," he wrote. He was just about right, but not close enough for some of the scientists who followed him.

That's where the scientific problem stands in 1772 when the Royal Society appoints a "Committee of Attraction" to attempt to find the average density of the Earth. The committee's goal is to make the "universal gravitation of matter palpable." In other words, mountains, houses, dishes, and the Earth itself, all possess gravitational attraction. Is there any way those attractions can be measured? You need to start with their densities. Can Earth's density be determined? The Royal Society decides to try. It sends off an expedition to attempt to measure the mutual attraction of a Scottish mountain and a plumb line (a weighted cord). But the scientists, hoping to work with local farmers, bring a keg of whiskey. The locals drink up and burn the equipment—ending the experiment.

Giving up on the mountain approach, John Michell, a member of the Royal Society, decides to see if he can measure the gravitational attraction between two known objects separated by a known distance. If he can find that out and compare it with their attraction to Earth (which is the same as their weight), he can figure Earth's average density—and then, with simple multiplication, the mass of the Earth.

For most people this all seems to be about "weighing" the Earth—and since Earth can't be put on a scale, that sounds like quite a feat. (No one knows that the experiment will have a much bigger payoff.)

Michell designs a torsion balance to measure the attraction between two lead balls; then he dies. No one else wants to carry on with the experiment. (It seems hopelessly difficult.)

Henry Cavendish decides to go for it. Starting with Michell's concept (and giving Michell full credit), he selects two small metal balls and two large ones. He knows the force of their attraction to the Earth (their weight) and their distance apart. He needs to measure the force of their attraction to each other.

To make this measurement, Cavendish builds his own torsion balance (see illustration). Here are his words to describe it:

> *The apparatus is very simple; it consists of a wooden arm, 6 feet long, made so as to unite great strength with little weight. This arm is suspended in an horizontal position, by a slender wire 40 inches long, and to each extremity is hung a leaden ball, about 2 inches in diameter; and the whole is enclosed in a narrow wooden case, to defend it from the wind.*

Cavendish knows that any slight outside influence—even a human breath—will make the experiment meaningless. This is

A side view (top drawing) and an overhead view (bottom drawing) show the basic setup of Cavendish's torsion balance. A large dumbbell with two heavy lead spheres dangles from the top and can pivot freely. A tiny pair of 2-inch balls rest inside the frame. (In the overhead view, they are represented by the small circles on either side of the bar.) *Torsion* means "twist," and the attraction between the large and small masses causes the balls to move, which twists a string or wire.

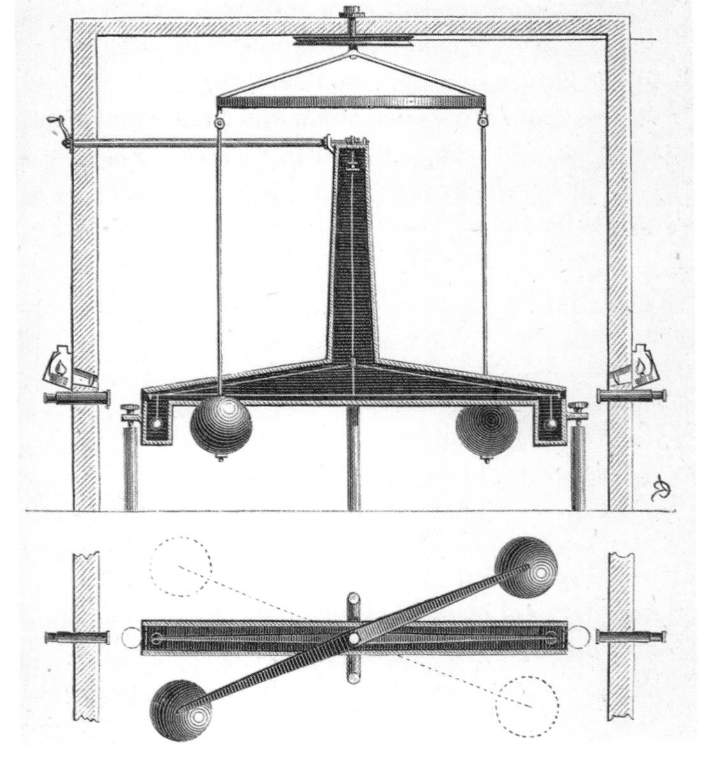

what he says: "I resolved to place the apparatus in a room which would remain constantly shut, and to observe the motion of the arm . . . from without by means of a telescope: and to suspend the leaden weights in such manner, that I could move them without entering the room."

He builds a pulley that can be used to slowly move the two heavy lead balls (weighing about 350 pounds each) and bring them near the very light 2-inch spheres. (We're skipping metric units here because Cavendish used English measures.) Will the gravitational attraction between the light and heavy balls swing the wooden arm enough to be measurable?

Cavendish puts tiny indicators on the arm; he fixes telescopes through the sealed room's walls. If there is any movement in the arm, he knows it will be slow and slight. He puts an eye to a telescope—and waits—often for hours at a time. And, yes, Cavendish is able to measure a very tiny oscillation (back-and-forth swing) in the wooden arm. Because he is a superb experimenter, and just to be sure, he does the experiment 29 times.

Knowing the mass of each ball and the distance between them, center to center, and the amount of movement of the arm, he calculates the gravitational attraction between the balls! "By means of the experiment," Cavendish says, "the density of the Earth comes out 5.48 times greater than that of water." His results are close to the modern figure (5.518) and the best by far until the twentieth century. (After experimenting with superb precision, he proves that he is human with a small error in his math; it is spotted by a later scientist.) With some pride, he writes that he has determined the result with "great exactness." He is right. He has weighed the Earth.

This is a small-scale model of Cavendish's torsion balance, which he designed in 1798. Compare it to the drawings on the previous page.

DOING A NUMBER ON DENSITY

To know the density (d) of anything, including Earth, you have to know both its mass (m) and volume (v):

$$d = \frac{m}{v}$$

The volume of the Earth is easy. Just plug Earth's radius (r) into this formula for the volume (v) of a sphere:

$$v = \frac{4\pi r^3}{3}$$

Earth isn't exactly a sphere (it bulges a little at the equator), but close enough.

For mass, the formula is just as easy—if you know the value of G, the gravitational constant for any object in the universe. Here's one way to compute the mass:

$$g = \frac{Gm}{r^2}$$

The little g, as you'll recall from Galileo's experiments in chapter 6, is the acceleration of a falling object due to Earth's gravity. The r is, again, the radius of the Earth. Knowing the value of G is critical to solving this equation for m (Earth's mass). Cavendish's experiment on gravitational attraction between heavy and light balls gave later scientists the tools to find this figure.

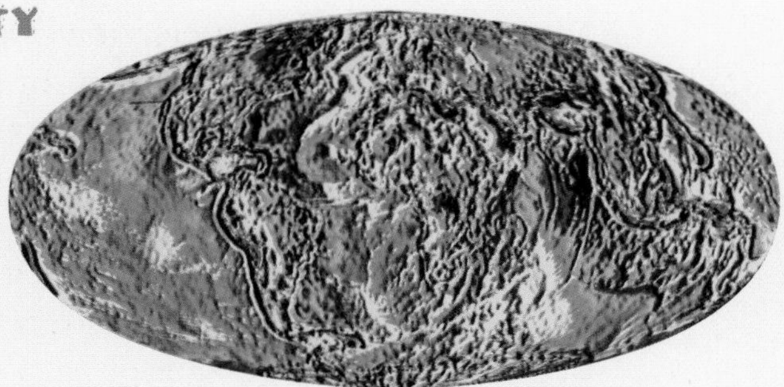

Imagine you're sailing on the Pacific Ocean. The deep seawater beneath your boat has a density a tad higher than 1 gram per cubic centimeter. That's pretty low when you consider many rocks are 2 to 5 times denser, and lead is a whopping 11 times more dense than water. Because of the ocean's low density, Earth's gravitational field (the pull of gravity on your body) is relatively weak there. If you stand instead on a massive slab of rock and metal, the gravitational pull is a little stronger (meaning you weigh a little more). The difference is supertiny but measurable in a unit called the milligal (named after Galileo). This computerized map (above) shows the varying gravitational fields (due to changing densities) of points on Earth. Dark blue is weak, and bright red is strong.

To find Earth's average density, divide mass by volume. Why "average"? One answer lies in this colorful world map (above). Earth is made of freshwater, seawater, rock, molten rock, heavy metals (like iron), featherlight gases, and all sorts of other matter. All of these materials have different densities, so there's no such thing as "the density of the Earth." There's only "average density"—the collective mass of all those things divided by their collective volume.

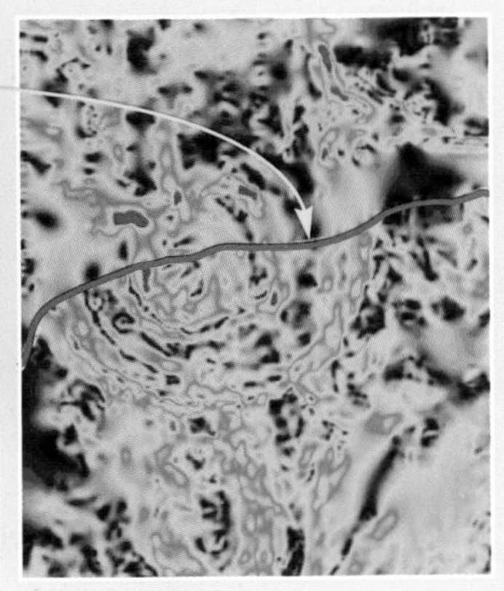

A gravity map (far right) revealed the distinct rings of an impact crater buried deep under the Yucatán Peninsula of Mexico—half under land and half under water. The crater's name is Chicxulub (CHEEK-shoo-loob), after a Mayan coastal village, but some people call it the Crater of Doom. Sixty-five million years ago, a giant asteroid blasted into the Earth here. The impact caused global changes in climate and atmosphere that ended the reign of the dinosaurs and many of their fellow life-forms.

But Cavendish and his contemporaries have missed something in Newton's formula for gravitation. It is right there but has been overlooked.

Newton's Law of Gravitation implies an unchanging number in the universe—a gravitational constant—today labeled G (see margin note below and the box on page 254). Ole Roemer had found a constant number for the speed of light; otherwise the idea of unchanging numbers doesn't get much attention until late in the nineteenth century. Mathematicians in Cavendish's day focus on ratios—not constants. As modern mathematics develops, Newton's formula is recast, and physicists soon realize that with Cavendish's experiment they can easily find G. Once they have the value of G, it's not difficult to calculate the force of gravitation.

Another reminder: In mathematics, a CONSTANT is a fixed quantity that does not change value in relation to numbers that do change (called variables). In physics, certain numbers—like the speed of light in a vacuum—are thought to be universal constants.

Scientists still measure G today, striving for ever-greater accuracy. In 2000, Jens Gundlach, a physicist at the University of Washington, used this instrument (right) to pin down G to within a .0014 percent uncertainty. The improved figure for G was used to recalculate the average mass of the Earth and the Sun.

Today, science students often replicate the Cavendish experiment (using laser beams and modern technology)—but they aren't looking for Earth's density; they're after G. The experiment has come to be ranked among the most important scientific achievements of all time.

That constant, G, defines the force that holds all matter in the universe together. "From this number, one can figure out the behavior of objects orbiting the earth, the motion of the planets in the solar system, and the motion of the galaxies from the time of the Big Bang onward," science historian Robert P. Crease writes.

Too bad Henry isn't around to learn what he's done. But given his feelings about not drawing attention to himself, maybe that's the way he would have wanted it.

Once you know the value of G, you can use the formula for gravitational force (along with Kepler's Laws of Planetary Motion) to calculate the mass of any planet that has a moon or the mass of any star (including the Sun) that has a planet—which means you can calculate the mass of the Earth.

WHO HAS MORE PULL—NORTH OR SOUTH?

It was 1763, and two of England's colonies in America—Maryland and Pennsylvania—had been fighting over their common border for years. Actually, it was the prominent Penn family versus the prominent Calverts, and they even took their quarrel to court in England. Two astronomers with surveying skills, Charles Mason and Jeremiah Dixon, were sent across the ocean to place the boundary according to lines of latitude. As it turned out, Mason and Dixon settled more than a dispute between rich landowners: Their boundary, known as the Mason-Dixon Line, would later divide free states from slave states, and North from South.

Politics, money, and borders were not things of concern to Henry Cavendish. He worried about something else. He believed that Mason and Dixon could not set an exact boundary. He knew that the massive Appalachian Mountains would exert a slight gravitational pull on their surveying instruments. That pull was not balanced by the less massive flat land and ocean water to the south and east. Thinking about that, with a mind that focused on precision, got Henry Cavendish thinking about the Earth's density.

The Mason-Dixon Line marks the border between Maryland, Pennsylvania (to the north), and Delaware (to the east, or right, of Maryland).

25 The Right Man for the Job

At this moment there emerged one of those men who can stand above the whole scene, look at the confused pieces of the jig-saw puzzle and see a way of turning them into a pattern....In 1772, when he was twenty-eight, he surveyed the whole history of the modern study of gases and said that what had hitherto been done was like the separate pieces of a great chain which required a monumental body of directed experiments to bring them into unity.

—Herbert Butterfield (1900–1979), British historian, *The Origins of Modern Science*

He really did two things. He drew the distinction between elements and compounds, so that people understood the way that the world was built much more clearly; and he found a way of attaching numbers to chemistry....As soon as you attach numbers to anything, you turn science into physical science and that gives you an enormous predictive power to investigate your ideas very precisely.

—P.W. Atkins (1940–), British chemist, as quoted in *On Giants' Shoulders* by Melvyn Bragg

It is the late 1700s, and on the American continent, a bunch of radicals—George Washington, Thomas Jefferson, and John Adams are some of their names—are getting fed up with British rule. They are imbued with scientific curiosity, as most thinking people are during the Enlightenment, but politics is taking much of their energy.

CHEMISTRY is sometimes defined as the study of the laws that govern the behavior of the elements. Here's a definition by English chemist John Read: "Chemistry is the branch of science which deals with the study of matter, or in other words with the character of the 'stuff' of which the material universe is composed." Today chemistry and physics often merge.

Still, they have time to follow the progress of a young French tax collector who is trying to devote as much time as he can to scientific experimentation. The Frenchman has a head for figures, and also for details. He designs his own superb scientific equipment and spends much of his personal wealth building it. He keeps a careful record of

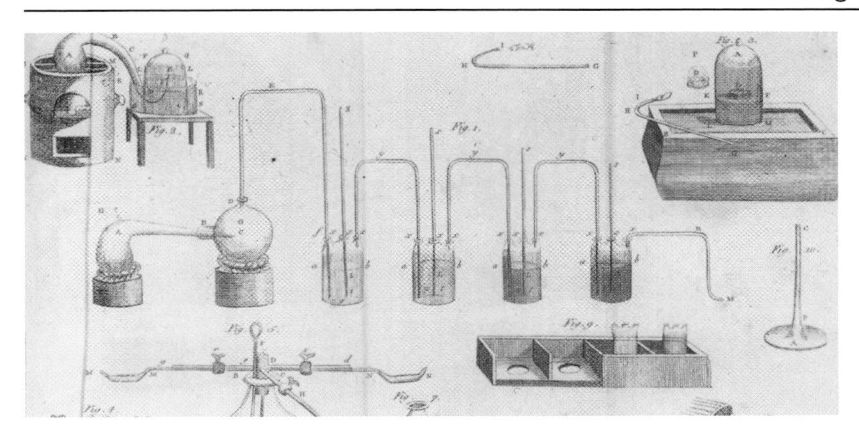

To distill water (or any liquid), you boil it, capture the vapor, and let it cool back into liquid form. Why? Because a lot of impurities—salt, minerals, dirt, and so on—are left behind. In this illustration of the process from *Elements of Chemistry*, note the pelican instrument in the upper left corner.

everything he does. This man is a real scientist, and although he studies the work of the best of the alchemists, he doesn't accept ideas he can't test and prove.

The alchemists combine ingredients to make their concoctions, but they rarely weigh things with precision. The Frenchman does. (He builds balance scales to do it.) Careful measurement is essential in science.

In this book, you will find me describing measurements as "precise" and "exact." Be aware: I don't really mean it. Today's scientists will tell you that—despite our most sophisticated equipment—there is no such thing as an exact measurement.

Does water turn into earth, as is widely believed? The Frenchman decides to test for himself. He weighs some distilled water and also weighs a glass vessel called a pelican because of its curvy, long-beaked shape: (see image at right, and above, top left). Then he pours the water into one flask of the pelican, seals the container, and boils the water so that it all condenses into a second flask. The sealed system never changes weight. But after 101 days, bits of residue have appeared in the flasks. He then weighs the flasks, the water, and the residue separately. The pelican has lost weight equal to the weight of the residue. The alchemists say that water is "transmuted" (changed) into earth. With his precise measurements, the Frenchman has shown that the boiling water broke away bits of the flasks, and the residue comes from the glass, not the water. Water does *not* turn into earth!

O NOTES

Oxygen, vital for life, is in almost every molecule in our bodies. It is the most abundant element in the Earth's crust, where it often combines with other elements to form oxides in rocks and minerals.

As a gas, oxygen makes up about 20.95 percent of Earth's atmosphere, where it is the diatomic molecule O_2. (*Diatomic* means each molecule is made up of two atoms of the same element.)

What's ozone?

It's O_3, a triatomic, or triple-atom molecule, that wasn't discovered until 1840. Ordinary oxygen is odorless and colorless, but ozone ranges from colorless to dark blue-violet, depending on density. It has a strong, sweet odor that you can sometimes smell after a thunderstorm: lightning creates ozone. Ozone can be a byproduct of automobile exhaust and is harmful to plants and lung tissue. Ozone in the stratosphere—the Earth's upper atmosphere—shields Earth from the Sun's harmful ultraviolet radiation.

At left is an engraving of scientist Antoine Lavoisier doing two things: combining hydrogen and oxygen into water and studying the respiration of a sleeping man. Several experimenters had exploded "inflammable air" and found that it condensed into water. Lavoisier named that air *hydrogène*, from the Greek for "water creator."

When he learns that British experimenters have separated water into hydrogen and oxygen, he does his own experiment and confirms their work. Now there is no question of it: Water can't be an element; it is composed of two gases.

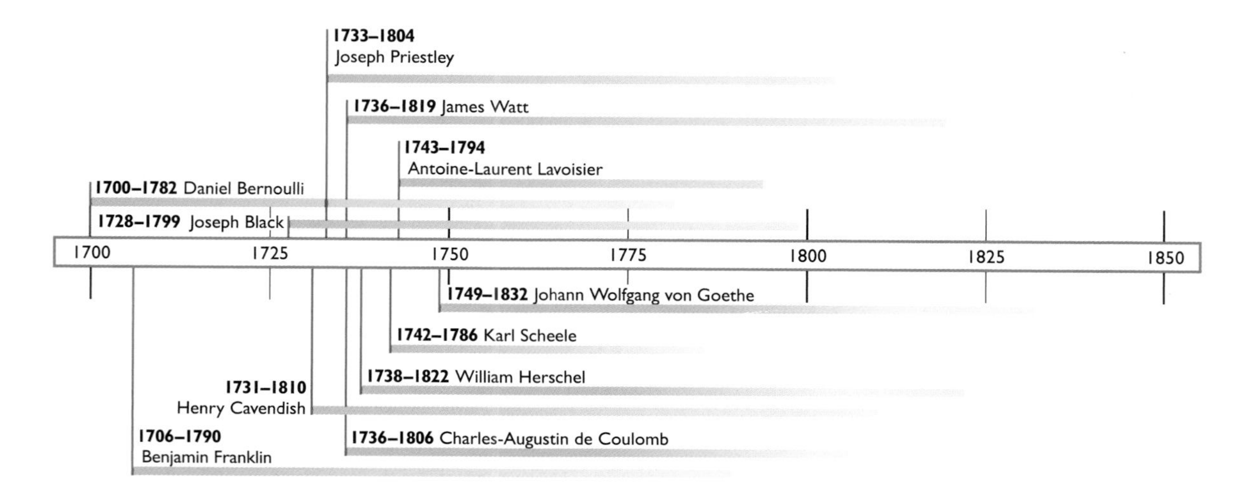

The Frenchman is sure that there *are* certain substances that can't be further divided: they are the elements. When elements combine, they create compounds (like water) which can't be easily broken apart.

He understands that air is neither an element nor a compound. He figures that out when he discovers that rusting objects do not lose weight, as everyone thinks; they gain weight. His measurements confirm this. But how can that be? He hypothesizes that the rusting object is attracting some kind of particle from the air. That turns out to be oxygen. If air can release the gas oxygen and still seem to be air, then air must be a mixture of gases and not a compound.

Some things never change. Mass is one of them. It may change shape, but not value. Its value remains constant. It is invariant. The idea that there is no loss of mass through combustion *in a closed system*—is called **the Law of Conservation of Mass**. It's an important law. Put it in your head. We'll get to it again.

RUSTING AND BURNING—YOU CAN BLAME (AND THANK) OXYGEN

Rust is a slow form of oxidation (oxygen bonding with a substance). A bike that rusts gains weight because oxygen combines with its steel (high in iron) to make iron oxide (Fe_2O_3). This happens through a series of chemical reactions that require both air and moisture.

When an apple bruises or a slice of apple turns brown, oxidation is happening. Damaged cells in the apple are breaking open and spilling their guts—including molecules of acids. There's oxygen in the air and in the apple (that's why apples float). It combines with the acids (with the help of an enzyme) and turns them brown.

When foods turn rancid (rotten), it is often because of oxidation. The idea behind vacuum-packed foods is to get rid of the oxygen.

Burning is a quick method of oxidation—and a major source of energy. When a car engine combusts (burns fuel), the fuel is undergoing an explosive process of oxidation—one that releases a lot of heat energy. In a more low-key oxidation process, your body burns the nutrients in food as fuel for energy. Oxygen in your cells combines with carbon chemicals like glucose (a sugar), releasing heat energy and carbon dioxide.

While oxygen bonds easily, it's not the only reactive substance. Today the term *oxidation* has broadened so that oxygen doesn't even have to be involved. Anytime one substance strips away the electrons of another, it's called oxidation. (More about electrons and other smaller-than-atom particles to come).

To the naked eye, a nail looks shiny. But during oxidation, seen under a high-power microscope (left), the surface is covered with a corrosive oxide layer known as rust.

Joseph Black's assistant, James Watt, used the new knowledge about gases (especially Boyle's Law) to design an improved steam engine. Steam engines turn the plentiful energy found in wood and coal into mechanical work. Those engines were first used to pump water out of coal mines. (Miners were dying in flooded mines.) Soon the engines were moving coal itself out of mines. But Watt's engine could do much more than just move coal: it could turn looms, power locomotives, and drive ships. When that was understood, the Industrial Revolution really steamed ahead. At right is a contemporary painting of one of Watt's steam engines.

He isn't the one who discovers oxygen—others beat him to that—but he is first to announce that oxygen is an element.

The Frenchman not only studies the experiments of Boyle, Black, Priestley, and Scheele—he does them all again, so he can be sure of the results. He says those experiments are like links in a giant chain that need to be welded together. He decides he is the person to hold the torch.

The Frenchman thought oxygen was essential for combustion. Later it was found that some other gases can also be used in the process.

When he burns a log of wood, he is left with ash and smoke. Most of the mass of the log seems to be gone. But when the Frenchman burns objects *in closed containers* and then weighs the residue—very carefully—the weight stays the same. He realizes that burning (combustion) doesn't change the amount of matter in the universe. It just changes its form. (This is a really important concept.)

Mass is never lost. It can change, but it can't be eliminated. Scientists call that a conservation law.

Mass is never lost? Thanks to Albert Einstein, we now know that mass and energy are interchangeable, according to his famous formula: $E=mc^2$ (E is energy; m is mass, and c is the speed of light). Right now, the Sun is changing tons of mass into the energy we call sunshine. But no one understood that in the nineteenth century.

And that still isn't all that this man uncovers. He finds that **combustion is a chemical reaction between two or more substances** (usually oxygen and something else) **that gives off heat and light.**

What about phlogiston? A belief in that theoretical substance is taught in schools. But his description of combustion leaves no need for phlogiston. He realizes: **There is no fire element; there is no phlogiston**. Discovering that is like taking chains off a captive: it sets

REVOLTING TIMES

Was revolutionary fever contagious? When the Americans rebelled against British rule in the 1770s, they were after political freedom. They would get what they wanted. Economic freedom was another issue. By 1790, the Industrial Revolution was accelerating in Great Britain. New textile machines had been invented there, and at about the same time, some technological geniuses, such as James Watt, figured out ways to use steam to power that machinery. This was a big deal. It would change the way goods are made and at the same time lower prices and spread prosperity. But without the secrets of the new technology, other nations, like the infant United States, had to buy goods from England.

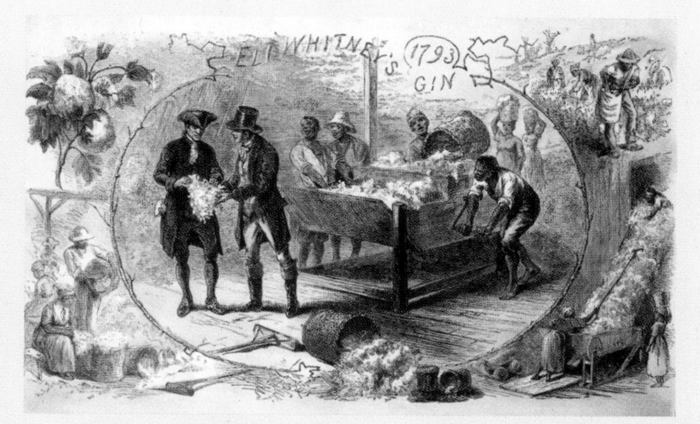

Eli Whitney was a Yankee who went south, saw the need for a way to separate cotton fibers from seeds, and invented a machine (above). Unexpectedly, it made slavery profitable. At right is his patent drawing.

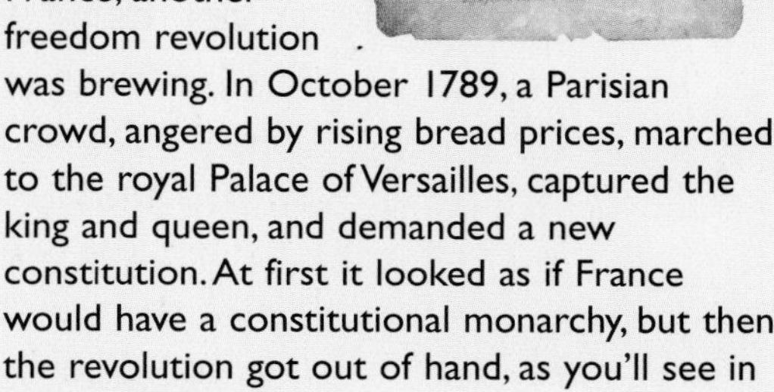

The British wanted to keep a monopoly on their new productivity; they tried to erect a wall of secrecy around their island. That's never easy. Merchants in the United States offered a big reward to anyone who could build spinning and weaving machines in the U.S. When cotton-spinner's apprentice Sam Slater arrived with plans for a spinning machine memorized in his head, the Industrial Revolution had made it across the ocean. Then in 1793, Eli Whitney invented the cotton gin (*gin* is short for "engine"), which furthered the revolution and changed the economy of the American South and the future of slavery.

Meanwhile in France, another freedom revolution was brewing. In October 1789, a Parisian crowd, angered by rising bread prices, marched to the royal Palace of Versailles, captured the king and queen, and demanded a new constitution. At first it looked as if France would have a constitutional monarchy, but then the revolution got out of hand, as you'll see in the next chapter.

science free from the burden of a wrong idea.

Who is this Frenchman? His name is Lavoisier (lah-vwah-zee-AY), Antoine-Laurent Lavoisier (1743–1794), and he is often called the Father of Chemistry. He is handsome, and his wife, who is part of the story, is beautiful. He intends to do for chemistry what Isaac Newton has done for physics—write a book that will bring together all the knowledge in the field. He will do what he intends and more. He will create a revolution in science.

When Lavoisier read his paper on phlogiston (saying that there is no such thing) to the Royal Academy, he was interrupted by jeers. The members had a hard time accepting a new idea.

26 A Man with a Powerful Head

We ought, in every instance, to submit our reasoning to the test of experiment, and never to search for truth but by the natural road of experiment and observation.
—Antoine-Laurent Lavoisier (1743–1794), French chemist, *Elements of Chemistry*

Hypotheses are the scaffolds which are erected in front of a building and removed when the building is completed. They are indispensable to the worker; but he must not mistake the scaffolding for the building.
—Johann Wolfgang von Goethe (1749–1832), German poet and scientist, *Maxims and Reflections*

The substances that fill our universe can be burned, squeezed, shredded, or hammered to bits, but they won't disappear. The different sorts floating around just combine or recombine. The total amount of mass, however, remains the same.... With all of Lavoisier's accurate weighing and chemical analysis, researchers were able to start tracing how that conservation happened.
—David Bodanis, (1956–), American science historian, $E=mc^2$

His parents expect Antoine-Laurent Lavoisier to be a lawyer, like his father and his grandfather before him. So he goes to Paris to study law. But this is the enlightened eighteenth century, and in most universities everyone seems to be talking about the latest scientific discoveries. Lavoisier hears a science professor lecture, and he is captivated. He goes back again and again to hear more lectures.

It is soon clear that his real interest lies in science, not law.

But scientists have no way to earn a living—at

We have just seen that all the oxides and acids from the animal and vegetable kingdoms are formed from a small number of simple elements.... We may justly admire the simplicity of the means employed by nature to multiply qualities and forms.
—Antoine-Laurent Lavoisier, *Elements of Chemistry*

In this 1789 French engraving, a hardworking peasant is carrying a nobleman and a priest on his back. The title: *Il faut espérer que le jeu finira bientôt.* It means "Let's hope the game will be over soon." The game? That's the political system that seemed to exploit the working class. It took the French Revolution (1789–1799) to change that game.

least not the kind of prosperous living Lavoisier has always enjoyed. He is determined to be a scientist and to live well too. So he makes what seems to him a smart investment. He buys a share in the company that collects taxes for the king. That brings him a handsome income, and it also allows him time to concentrate on science.

The king's taxes are especially onerous (OH-ner-uhs—it means "burdensome") to the peasants. Tax systems almost always seem to favor the rich (who have influence) and penalize the poor (who don't).

Lavoisier supports efforts to reform the French tax-collection system, but he doesn't get far. When crops fail and some peasants starve, Lavoisier loans them money without interest. He helps set up an old-age-pension system for farmers. He serves on education committees. He investigates hospitals for the poor and makes recommendations to improve them. But, no question about it, being a tax collector gives him wealth and privileges at the expense of others.

It gives him something else too—a wife.

The chairman of the French tax-collection firm has a daughter. She happens to be a beautiful and talented artist. When they marry, Marie-Anne Pierrette Paulze is 13 and Antoine-Laurent Lavoisier is more than twice her age. (Thirteen-year-old brides are not uncommon in these days.) The age difference doesn't matter: she has a mind that keeps pace with his.

The Lavoisier home becomes one of the liveliest and most elegant in Paris. Almost every night, guests are invited for

'Each day Lavoisier sacrificed some hours to the new affairs for which he was responsible. But science always claimed a large part of the day. He arose at six o'clock of the morning, and worked at science until eight and again in the evening from seven until ten. One whole day a week was devoted to experiments. It was, Lavoisier used to say, his day of happiness.

—Mme. Lavoisier (1758–1836), writing about her husband

This is a portrait of Antoine-Laurent and Marie Lavoisier painted by their friend Jacques-Louis David. Notice the portfolio of drawings on the chair. Those are Marie's drawings. If you study the work of most artists of this period, you will see stiff people dressed in classical gowns. David changed all that. He did the same kind of thing in art as Lavoisier did in science: he made it real. A director of the Metropolitan Museum of Art in New York City has said that this painting is the greatest of all neoclassical paintings. You can see it in that museum and decide if you agree.

dinner and conversation. Ben Franklin comes often, as does Jacques-Louis David (France's leading painter and Marie's teacher and friend). James Watt discusses technology (he is Joseph Black's assistant and a designer of steam engines). Gouverneur Morris (who wrote much of the American Constitution) adds charm as well as intellect. Felice Fontana (an Italian scientist renowned for his research on snake venom) is a frequent visitor. Thomas Jefferson, a special friend, brings his violin and astonishes everyone with the range of his interests. The physician Joseph Guillotin adds social concerns to the conversation. Guillotin proposes that a mechanical decapitation (head-chopping) machine be used in capital punishment. His idea seems more humane and technologically advanced than hanging people or whacking off their heads with an axe. The dinner guests talk of the latest technology as well as of science, art, music, politics, and fine foods.

During the day, the Lavoisiers work. When the king wants to improve gunpowder production, he asks Antoine for help, and the job gets done. Lavoisier studies street lighting in Paris. He puts shades over his windows, and for six weeks works in darkness studying all the fuels he can find. He decides that Paris's streets can be most efficiently lit with olive oil. (He wins a prize for that work.)

When he "sparks" hydrogen and oxygen with an electrostatic charge and produces water, he awes an audience. (Who would have thought that two gases could turn into a liquid?) Working with a geologist on land maps, he understands, before almost

HOW ELECTROLYSIS WORKS

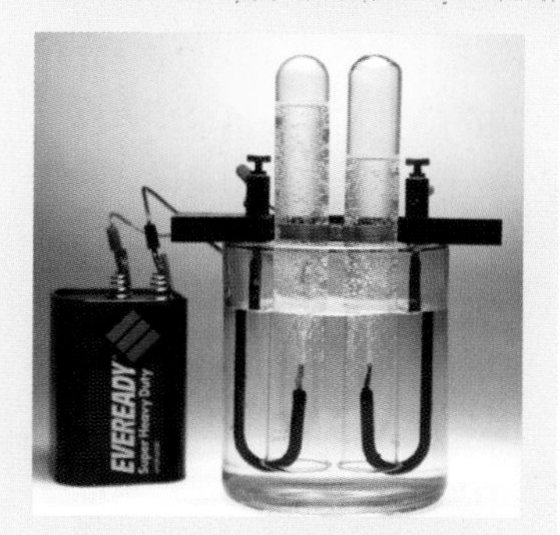

Henry Cavendish, Joseph Priestley, and Antoine-Laurent Lavoisier used an electrostatic device to "spark" oxygen and hydrogen and turn them into water. Only Lavoisier understood that water is a compound of the two gases; the others included phlogiston in their chemical equations. Today fuel cells turn hydrogen and oxygen into drinking water on spacecraft and power cars that run on hydrogen instead of gasoline.

Electrolysis is about sticking two rods called electrodes (the negative and positive poles of a battery) into a liquid—in this case, water. Electrons flow from the battery to the negative rod (the cathode), through the water, and back to the battery by way of the positive rod (the anode) in a circuit. At the anode, water molecules give up electrons (they become oxidized), producing oxygen. At the cathode, water molecules grab electrons and are reduced to hydrogen gas. (Reduction is the opposite of oxidation.) Water has two hydrogen atoms per oxygen atom, so electrolysis produces twice as much hydrogen gas (right tube).

Lavoisier's hydrogen burner

anyone else does, that Earth's layers tell a story of geological time and change. He designs a model farm using scientific farming methods: he weighs and measures the seeds he sows, the fertilizer he uses, and the crops that are yielded—at a time when most people are just guessing about those things. George Washington writes from Virginia, asking for news of Lavoisier's farming methods.

With his methodical mind, Lavoisier reorganizes the tax system and makes it more efficient. But when he decides to build a wall around Paris to make sure that no one gets in or out without paying a tax, he makes enemies. Not surprisingly, most Parisians hate that wall.

Joseph Guillotin (1738–1814) was a French physician and inventor of the guillotine, an execution machine designed to chop off heads quickly and cleanly.

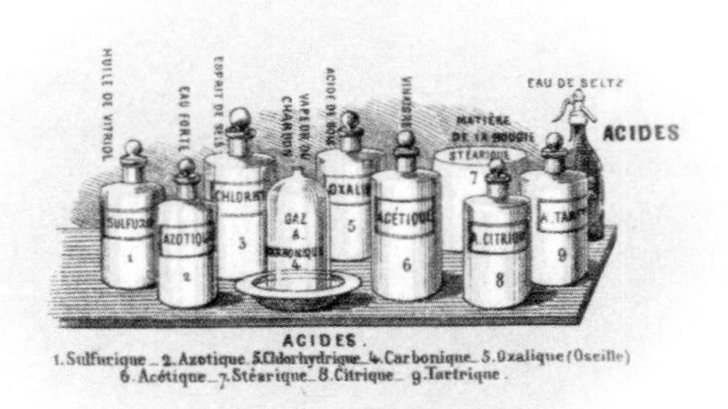

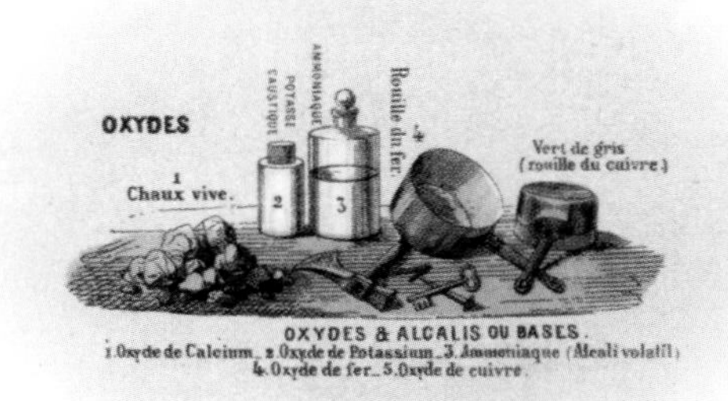

Above are examples of acids (left) and oxides (right) from a nineteenth-century French engraving. An oxide is a combination of oxygen and another element—calcium, potassium, iron, and copper in the detail on the right. Oxides of metals are normally bases (or alkalies) and react with an acid to produce a salt.

PURE SCIENCE is science for science's sake—researching to add to our knowledge of the world. APPLIED SCIENCE is using that knowledge toward a practical end—lighting streets efficiently, growing more crops, making better gunpowder, and so on. Lavoisier did both, but today, scientists tend to choose one or the other.

At age 25, he is accepted as a member of the prestigious French Academy of Sciences; the next youngest member is 50. That academy immediately puts him on committees to study gravity, bleaches, water supplies, prison conditions, and so on. He knows that clear writing will give his ideas power, so he writes his reports in elegant, lucid prose. Eventually he becomes head of the academy—which means serving on more committees and doing still more work. It doesn't seem to slow down his research into pure science.

He starts out where Boyle left off and demolishes the Greek concept of earth, air, fire and water—once and for all. Then he comes up with a clear definition of **an element: It's any substance that can't be broken down into simpler substances.** Next, he creates a useful classification system. He searches out, lists, and organizes all the known elements, grouping them in a table by their properties. He finds 33 elements. He isn't right about all of them. Some are compounds—like silica and magnesia—but this is groundbreaking work and not easy. His basic idea is right even if some of the details are wrong. He says light is an element (wrong) and also something he names "caloric" (wrong again).

Caloric? Lavoisier realizes that combustion (burning) is a chemical reaction that uses oxygen and gives off heat. But what is heat? He doesn't know. When he puts a hot coal next to a piece of ice, the coal's heat melts the ice. What's going on? Lavoisier says that an invisible fluid must move from the burning coal to the lump of ice. It's that invisible fluid (which he has theorized) that he calls "caloric."

George's Planet?

William Herschel (1738–1822) was a church organist who read a book about Newton that made him want to study the skies. But he couldn't afford a good telescope. So he studied optics, learned to grind lenses, and began building telescopes. Then his sister Caroline started helping him. Together they were soon making the best telescopes then in existence.

The Herschels decided to study the skies in a systematic fashion. He wrote papers on what they saw. On March 13, 1781, a Tuesday night, they trained a telescope at the sky and noted a "curious either nebulous star or perhaps a comet." It was actually a planet.

That's William Herschel's four-story telescope (left), but he didn't need it to spot Uranus. The seventh planet is even visible—a very dim dot—without a telescope. But no one else had given it a second look. Herschel realized the dot was a planet because it moved (very, very slowly—detectable after weeks of observation) against the backdrop of fixed stars. One orbit of Uranus around the Sun is, by coincidence, equal to Herschel's lifespan—84 years.

Uranus is twice as far from the Sun as Saturn, but the space probe *Voyager 2* was able to make a grand sweep of all four gas giants. After flying by Jupiter (1979) and Saturn (1981), it snapped the first close-up images of blue-green planet Uranus (above right) and its battered moon, Miranda, in 1986. *Voyager*'s last stop was the planet Neptune (1989), which was unknown in Herschel's time.

Five planets visible to the naked eye had been known since ancient times. Somewhat reluctantly, Earth had been added to that list. So there were six; Herschel's planet—which they soon identified by its orbit—made seven. Another planet! That was sensational news. Most people had thought that after Newton there was nothing left to discover.

Herschel, who was born in Germany but lived in Bath, England, wanted to call it *Georgium Sidus* ("George's Star"), after King George III. Some astronomers wanted to name it after Herschel. But tradition won out. The other planets are all named for characters in Greco-Roman mythology. So the newly discovered orb was called Uranus, after the mythological father of Saturn.

Herschel was made a member of the Royal Society, was knighted by the king, and became the most important astronomer of his day. Caroline (pictured above), who discovered eight comets, was awarded the Gold Medal of the Royal Astronomical Society. William's only child, his son John, also became a distinguished astronomer as well as a physicist and chemist of note.

He's really off-course when he calls caloric an element and includes it in his table. By now he has a great reputation, so most scientists go along with this theory. Actually the caloric idea does help scientists analyze heat's properties. When they think they are measuring "caloric," they are actually measuring heat. Hypotheses, even if wrong, are often an important step in the scientific process. They give you a place to start. Then you can prove or disprove them and go on from there.

Iron oxide was known in England as "astringent Mars saffron," and zinc oxide was "philosophic wool." But in France they had other names. When Joseph Priestley came to dinner at the Lavoisiers' house, he spoke of "red lead." No one knew what he meant. (He was talking of lead oxide.)

Clearly something had to be done. Chemistry needed a common language. Lavoisier and Guyton de Morveau established a system of chemical nomenclature. For instance, they gave a compound of a metal and a nonmetal the suffix *-ide* (like iron oxide). Their system is still used today.

If you don't come up with a hypothesis, or reasoned guess, you can't go anywhere and, mostly, Lavoisier is heading in the right direction. One of his most important ideas is that concept of **conservation of mass** (now expressed, thanks to Einstein, as conservation of mass-energy). He understands that while things may change shape or form, nothing is lost or gained in nature. (Think about that idea—it's a big one.)

Lavoisier serves on a commission that helps get the metric system started. He thinks up a naming system for chemicals. He brings order to a disorderly field. When he writes the book he had planned—bringing together everything known in chemistry—he and his wife celebrate by burning the books of the alchemists. (Book burners are usually nasty sorts, but the Lavoisiers are trying to make the point that alchemy belongs on the trash heap.)

TERRIFIC TEXTBOOK! (AN OXYMORON?)

Lavoisier's *Elements of Chemistry* laid the foundation for modern chemistry. Newton's *Principia* laid foundations in physics. Unlike Newton, Lavoisier wrote clearly. People could understand what he said. That gave his book special influence. Here's part of his preface:

> *Every branch of physical science must consist of three things; the series of facts which are the objects of the science, the ideas which represent these facts, and the words by which these ideas are expressed. Like three impressions of the same seal, the word ought to produce the idea, and the idea to be a picture of the fact. And, as ideas are preserved and communicated by means of words, it necessarily follows that we cannot improve the language of any science without at the same time improving the science itself.*

Lavoisier's book, *Traité Élémentaire de Chimie* (1789), or *Elements of Chemistry*, is sometimes compared to Newton's *Principia* (as Lavoisier intended). It lays the foundation for modern chemistry.

How does Lavoisier manage everything? He schedules his days

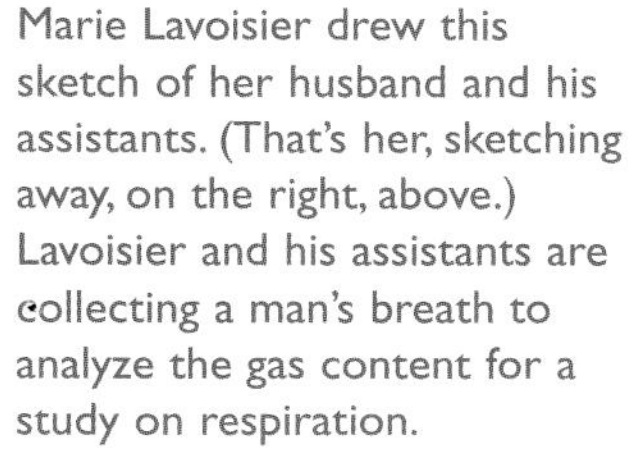

Marie Lavoisier drew this sketch of her husband and his assistants. (That's her, sketching away, on the right, above.) Lavoisier and his assistants are collecting a man's breath to analyze the gas content for a study on respiration.

carefully—and he works in tandem with his wife.

Antoine does precise experiments and records each of them carefully. Marie makes detailed drawings that are another clear record of their work. The alchemists have been haphazard about science; the Lavoisiers set ground rules for those who follow. They turn chemistry into an exact science.

Because she has a talent for languages—and Antoine doesn't—Marie learns English and then Latin. She reads and translates the work of scientists from other lands.

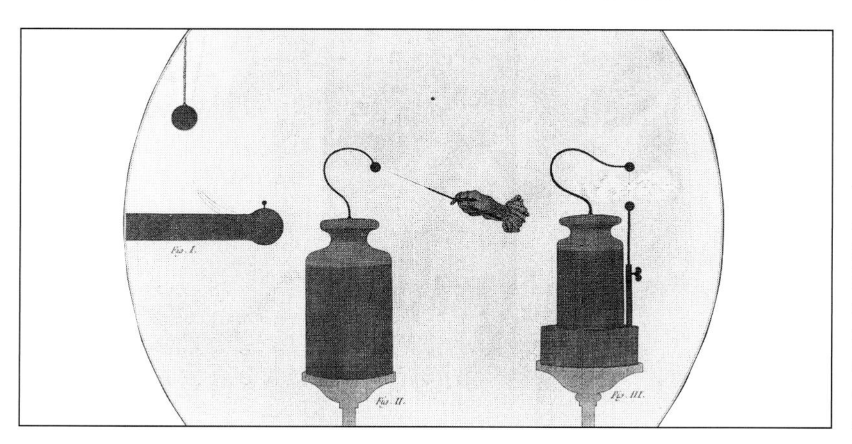

This illustration is from a book on electricity by Jean-Paul Marat (1743–1793), who believed he was snubbed by the scientific community. Shown here are Leyden jars, newfangled (at that time) devices that produced electric charges on demand.

Vive Marat (the sign on the left, above) means "Long live Marat." But he didn't have long to live. In 1793, a bloody year in the French Revolution, he was killed by Charlotte Corday as he sat in his bathtub. She hoped to keep his radical party, the Jacobins, from beheading more innocent citizens. Lavoisier was among the thousands of victims during that Reign of Terror.

With all their talent and energy—and the unselfishness of their civic work—the Lavoisiers have legions of admirers, but being part of the government tax system, they also have critics. And Antoine has an enemy with a personal grudge.

His name is Marat—Jean-Paul Marat.

When Marat tries to join the French Academy of Sciences, Lavoisier votes against him because he doesn't think he is a good scientist. Marat is actually a journalist, but he is enraged. This happens just after the American revolutionaries throw out the English king while crying, "No taxation without representation."

The French peasants hear that refrain. France explodes. It is the corruption of the king and his court. It is that tax thing. It is the times. France has a freedom revolution (in 1789).

Then it goes awry (as many revolutions do). Marat becomes a leader in a killing orgy called the Reign of Terror. People—all kinds of people, especially those connected to the king's tax-collection system—are hauled off and killed with the head-chopping machine designed by the kindly Dr. Joseph Guillotin. France loses its way.

Lavoisier loses his head. Science loses a brilliant mind. Marie loses a husband. "It took them only an instant to cut off that head, and a hundred years may not produce another like it," says mathematician Joseph-Louis Lagrange.

Later, the French people erect a statue in Paris in Lavoisier's honor. They are sorry. It is too late; the father of modern chemistry is gone. His ideas survive.

If you want to learn more about the French Revolution—and read one of the best books you will ever read—try Charles Dickens's *A Tale of Two Cities*.

FRANCE SINGS A METRIC TUNE

A French engraving from 1795 illustrates the use of the new metric system: (1) the liter, (2) the gram, (3) the meter, (4) degrees of an angle, (5) the franc coin, and (6) the stere, about a cubic meter of firewood. French francs replaced livres, or pounds, which are still used in England.

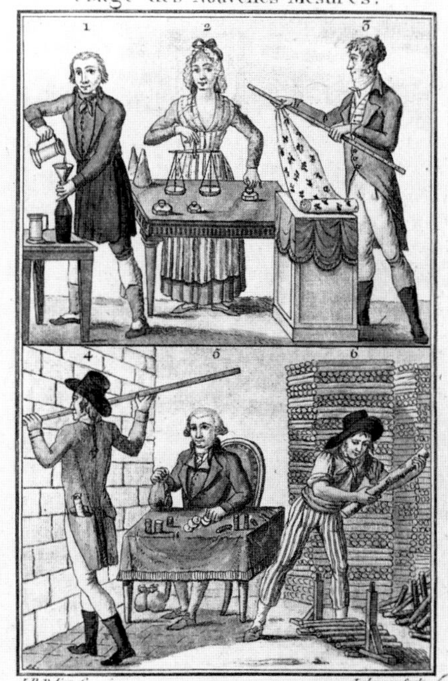

In June of 1792, two young Frenchmen—elegant Jean-Baptiste-Joseph Delambre and cautious Pierre-François-André Méchain—headed out of Paris. Both were skilled astronomers. Each rode in a carriage stocked with the latest scientific equipment. Each had a trained assistant. Delambre went north; Méchain headed south.

The Paris they left was in turmoil, but optimism and hope for new possibilities were in the air. The revolution promised equality, freedom, and justice for all humans. Delambre and Méchain believed in that revolutionary promise. So did their colleague Joseph-Jérôme Lefrançais de Lalande (a friend of Voltaire), who called himself, with no modesty, "the most famous astronomer in the universe." (His mistress, Louise-Elizabeth-Félicité du Piery, was the first woman to teach astronomy in Paris.) They had other supporters; many were well-known mathematicians or scientists. Lavoisier was one of them, as was Pierre-Simon Laplace, the leading mathematical physicist of the age. Most were part of the king's government (the *ancien régime*) and

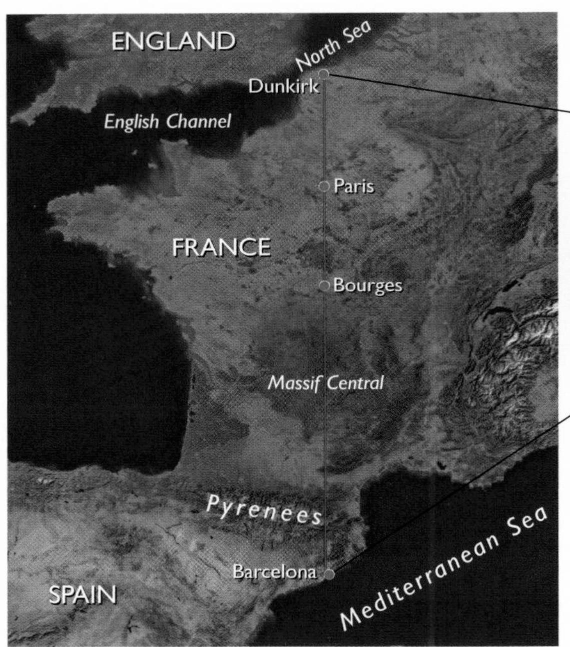

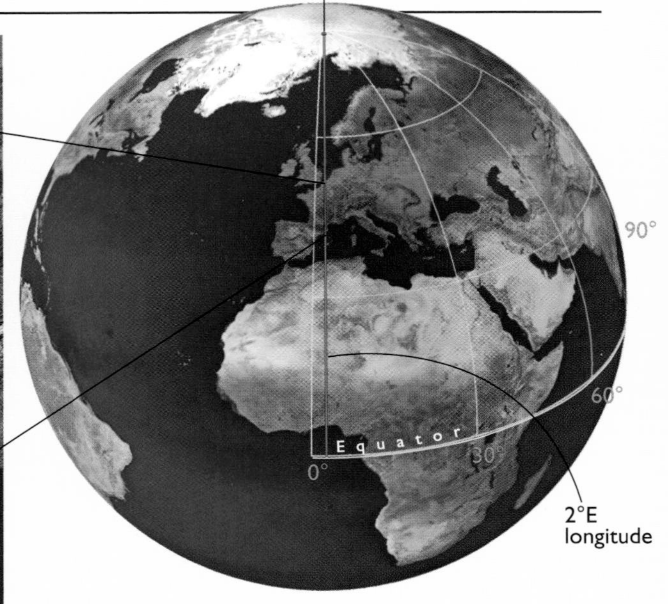

The globe above shows what the metric inventors pictured in their mathematical minds. The team decided 1meter would equal a 10-millionth of the distance from the North Pole to the equator (yellow line). No human had set foot on the North Pole—and wouldn't for two centuries. So the surveyors measured only a segment, walking along a longitude line (2°E) that follows Earth's curve through the heart of Europe (red line on map, left). (Try measuring a curve with a ruler, and you'll see part of the challenge.) They used the known circumference of the Earth to finish the calculation.

also ardent revolutionaries. They wanted to improve their world, and they thought they had the talents to do it.

Inspired by the revolution's promise of universal rights for all people, they planned to give the world's people a universal system of measure: "for all people, for all time." They would base that measure on a portion of the globe that was one 10-millionth (.0000001) of the distance from the North Pole to the equator. That measure was to be called a "meter." If they measured part of a meridian (a line of longitude), they thought they could calculate the length of the meter.

On a map of Europe (above), you can trace the meridian (now 2°E) that goes from Dunkirk at the northern tip of France straight south through Paris, through Bourges, through the Massif Central (mountains), then the Pyrenees (more mountains) and on to Barcelona in Spain. Delambre and Méchain spent seven years carefully measuring that segment of the meridian. It was an extraordinary endeavor. They climbed cathedral towers and volcano peaks. Because of the revolutionary turmoil, they barely escaped the guillotine. (Some of their supporters, like Lavoisier, were not so lucky.) Méchain died of malaria.

But, finally, they had their meter. At the world's first international scientific conference, held in Paris in 1799, the French scientists presented a meter bar of

Once they had the meter, the Parisian scientists need a unit of volume. They made a cube with sides .1 meter ($^1/_{10}$ meter, called a decimeter) long. They filled that cubic decimeter with water, weighed the water, and named that weight 1 kilogram. *Kilo-* means "thousand," so they subdivided the kilogram into 1,000 grams.

solid platinum to France's new ruler, Napoléon Bonaparte. Majestically he declared, "Conquests will come and go but this work will endure."

Napoléon didn't know it, but despite their care, there was an error in the measurement. The Earth isn't a sphere; it has an irregular shape. So 1 meter isn't exactly .0000001 of the distance from pole to equator. It didn't matter, their idea—to have a uniform worldwide system of measurement—made sense.

But change is often hard to accept. England, France's proud rival, wasn't ready to adopt metric units. France had helped the United States win its independence. Would that new nation show gratitude? The French sent a copper meterstick and a 1-kilogram weight across the Atlantic Ocean. (They are preserved in the National Institute of Standards and Technology in Washington, D.C.) Thomas Jefferson and George Washington tried, but they couldn't convince Congress to go along with the new measures.

Even in France, most people clung to known weights and measures. Napoléon began mocking the scientists. "It was not enough for them to make forty million people [France's population] happy, they wanted to sign up the whole universe," he sneered. It was the middle of the nineteenth century before France fully embraced the metric system. It wasn't until 1965 that England announced a transition to metric.

At the start of the twenty-first century, the metric system is in use worldwide—with the United States as the only major holdout. International business and science are making metric standards more and more necessary. "Few Americans realize that a silent revolution is finally underway in their nation, transforming their measures under the pressures of the new global economy," says Ken Alder in *The Measure of All Things*, a book about the making of the metric system. Today, more than 200 years after Delambre and Méchain set off on their historic odyssey, the scientific community—in the United States as elsewhere—measures metrically.

Metric on Your Mind

In this book, you'll find most measurements in metric—with inches and pounds in parenthesis where it is helpful or where English measures were used historically. Here are some quick mental comparisons:

- 1 meter is roughly 1 yard.
- 1 kilometer is a little more than ½ mile.
- 1 kilogram is just over 2 pounds.
- Volume is often measured in cubic centimeters (cm^3); 1,000 cubic centimeters equals 1 cubic decimeter (dm^3), which is 1 liter, or about 1 quart.

27 Dalton Takes Us Back to Greece—and Atoms

And the atoms move continuously for all time, some of them falling straight down, others swerving, and others recoiling from their collisions.... And these motions have no beginning, since the atoms and the void are the cause.
—Epicurus (341–270 B.C.E.), Greek philosopher, "Letter to Herodotus"

We might as well attempt to introduce a new planet into the solar system, or to annihilate one already in existence, as to create or destroy a particle of hydrogen.
—John Dalton (1766–1844), English chemist and physicist, *A New System of Chemical Philosophy*

Way back in the fifth century B.C.E. in Greece, Leucippus and his pupil Democritus said that if we want to comprehend the universe, we need to find its ultimate building block—the smallest particle possible—something that cannot be further divided. They called it an *atom*, after the Greek word *atomos*, for "uncuttable." Leucippus said only two things exist in the universe—atoms and empty space (the void). He said atoms are in perpetual motion.

A century later, the Chinese philosopher Hui Shih had a similar idea. He said there must be a smallest unit of nature.

Aristotle, of course, got into this act, too. He said there can't be any such thing as an atom. If you cut an element into small particles, said Aristotle, and divide each of those particles in half, and the result in half, and on and on indefinitely, you'll never finish. Aristotle said there is no ultimate particle. The ancient Greeks had no way to prove

To dissect the word ATOMOS, *a-* means "not" and *tomos* means "cuttable"—a popular term in surgery. Maybe you've had an appendectomy or a tonsillectomy—operations to cut out the appendix or tonsils. Computerized tomography is a three-dimensional X-ray technology that shows inside slices of the body—CT (CAT) scans.

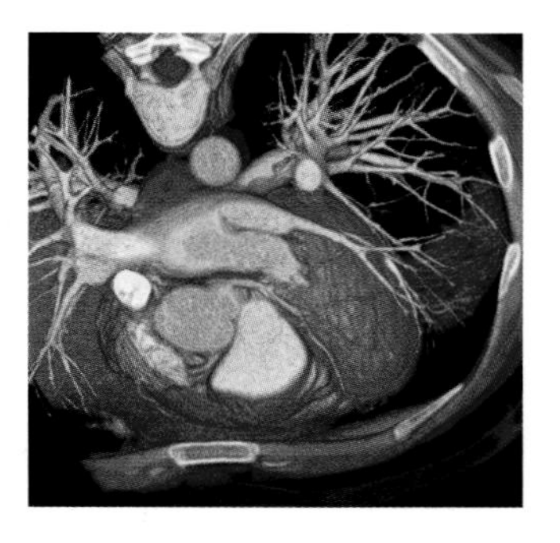

With a CT scan, you can "slice" the body at any angle to assemble a 3-D picture. At right, you're looking down from above at a normal heart.

A Poem Is Saved

Democritus (di-MAHK-ri-tuhs) was a Greek said to be the most learned man of his times. He was born about 460 B.C.E. and lived to a great age. Democritus wrote many books, but we have only a few fragments of them. He wrote on mathematics, music, ethics, and science. Mostly we know him through the writings of others. And we know he was very influential.

Democritus believed there is an ultimate invisible particle that is common to all matter. He said those particles are in constant motion. Because they are moving, they need empty space to move in. So there has to be a void (or vacuum). All of this was amazingly prescient (which means "having foresight").

Around 56 B.C.E., the Roman poet Lucretius wrote an important poem, *On the Nature of Things*, that was six volumes long and was based on the ideas of Democritus. It made a case for atoms.

Centuries and centuries passed, and every copy of *On the Nature of Things* was lost—*except for one*. But that was enough. When Johannes Gutenberg developed a movable-type printing press (1453), one of the first works printed, after the Bible, was Lucretius's poem. And that brought up the subject of atoms again.

Then things got complicated. Lucretius was an atheist—he didn't believe in God.

"Nature is free and uncontrolled . . . and runs the universe by herself without the aid of gods," said Lucretius.

Atomism (the belief in atoms) came to be called a form of atheism. In the church-dominated Middle Ages it was seen as heresy. So, in 1624, when three French scholars announced that they would give a lecture in Paris on the subject of atoms, church authorities canceled the lecture. They destroyed all the writings they could find on atoms. Jesuit priests (Jesuits are a scholarly Catholic order) were *prohibited* from teaching about atoms.

Democritus is meditating in this seventeenth-century engraving by Salvator Rosa.

Do you think believing that objects are made of particles that can't be divided has anything to do with believing in God? Lots of people did (and some still do).

either theory, so it seemed like an impossible argument.

As centuries passed, some thinkers believed in atoms, but most didn't. Thomas Harriot (1560–1621) did. Harriot, an

Thomas Harriot used a telescope to observe the moons of Jupiter the same year Galileo did. And he described the bending of red and green rays of light before Isaac Newton. He was an astronomer, mathematician, physicist, mapmaker, anthropologist, biologist, author, explorer, geographer, and historian who corresponded with Kepler and Galileo and was devoted to learning. You can call him a model Renaissance scholar.

Englishman who is hardly known today, was considered a great natural philosopher in his time. When Kepler asked him a question about optics, Harriot answered in a letter that described light as traveling through a vacuum, bouncing from atom to atom.

In that same letter he wrote: "I have now led you to the doors of nature's house, where its mysteries lie hidden. If you cannot enter, because the doors are too narrow, then abstract and contract yourself into an atom and you will enter easily. And when you later come out again, tell me what miraculous things you saw."

Now that's a terrific thought experiment: "contract yourself into an atom," observe, and report what you see! We're in the company of an inventive mind.

Harriot went to the New World with Sir Walter Raleigh and wrote a popular book about his adventures. He also wrote scholarly works on mathematics, astronomy, and physics. Sir Isaac Newton must have read Harriot. Like him (and unlike the followers of Aristotle), Newton believed in atoms. Newton even thought he could picture them. He wrote, "it seems probable to me, that God in the Beginning form'd Matter in solid, massy, hard, impenetrable, moveable Particles." (Atoms are not solid and impenetrable, but this was a start—and, he was right, they do move.)

Newton's enemy, the German philosopher-mathematician Gottfried Wilhelm Leibniz, thought the whole idea of atoms nonsense. "When I was a young man, I also gave in to the notion of a vacuum and atoms; but reason brought me into the right way," said Leibniz in a letter to Caroline, Princess of Wales, on May 12, 1716.

Two eighteenth-century French philosophers, Charles-Louis Montesquieu and Jean-Jacques Rousseau, looked to the natural sciences to create literature. For Enlightenment thinkers, all knowledge was interrelated. Science was of interest to everyone with an interest in ideas.

Robert Boyle (the Irishman who came up with the gas law) believed that gases must be made of tiny "corpuscles" with a lot of empty space between them. Could Boyle's corpuscles and Newton's particles be atoms? Except for Daniel Bernoulli, no one seemed to care.

Finally, near the end of the eighteenth century, an English

At a Quaker meeting, any member of the Society of Friends can speak, even if it puts some people to sleep (left). You can see this eighteenth-century painting at the Museum of Fine Arts in Boston.

Quaker named John Dalton comes along and decides to take atoms seriously. His timing is right.

Winds of change are blowing fresh air onto the European scene when Dalton is born. Newton's work has helped make science, politics, and philosophy hives of activity. So Dalton is lucky to arrive in the world when he does (in 1766).

John Dalton used this booklet of threads to test his own color blindness.

Otherwise, he doesn't start off with good fortune. His father is a poor weaver who works on a handloom and hardly earns enough to feed his family. Dalton is an awkward, color-blind boy with a weak voice. He is self-conscious and shy. But he is so bright that at age 12, he is teaching in a small Quaker school. How would you like a 12-year-old teacher? His students don't think much of the idea. They all drop out.

Dalton goes to a nearby village where he does some more studying and even teaches school again. At the same time, he keeps a journal that contains, along with other things, more than 200,000 meteorological notes. (He is fascinated with weather and makes his own instruments so he can record daily temperature, humidity, and wind and cloud conditions.)

METEOROLOGY, the study of weather, comes from Greek roots that describe anything lifted overhead—including "falling stars" or meteors. Today, the study of meteors, asteroids, and comets is part of astronomy.

Manchester, England, was a boomtown in Dalton's day. The cotton industry was moving from cottages (Dalton's dad kept his loom in their two-room home) into factories. This is an 1840 lithograph of a Manchester cotton mill.

His journal is published, and that gets him a job as a professor at New College in Manchester, England. New College was founded for Presbyterians, Quakers, and other "dissenters"—who aren't wanted at Oxford and Cambridge. Those two prestigious universities are open only to Church of England members. (Read some English history to understand why.)

But Dalton doesn't stay a professor for long; he wants to devote his time to experimentation and research, which he does by living modestly and by tutoring students. (His only recreation seems to be Thursday-night outdoor bowling.)

His studies in meteorology make him realize that evaporated water exists in the air as a separate gas. How can water stay in air? Dalton concludes that if air and water are each composed of discrete (separate) particles, then evaporation could be the mixing of water particles with air particles.

Thanks to Lavoisier, he knows that there are basic elements—like iron, oxygen, hydrogen, sulfur, and carbon—and that they *cannot* be broken down into simpler components. But what is it that makes one element different from another? Why is iron iron and not carbon? No one knows; that is Dalton's challenge.

A problem in meteorology has taken him to profound thoughts on the composition of matter.

Every atom in an element is alike and has the same weight as every other atom of that element, said Dalton, but atoms of different elements have different weights. (Today we think in terms of mass, not weight, and we know that isotopes of atoms have a slightly different mass.)

He carefully copies Newton's thoughts about particles from the *Principia* into his notebook. Gradually he expands on the idea. He starts by using Democritus's word *atoms* instead of *corpuscles*. He comes to believe that **all matter, not just gases, consists of those small particles.** Then Dalton takes a big leap of mind by hypothesizing that it is the weight of its atoms that makes one element (like iron) different from another (like carbon). From there it is a small step to figure out that chemical change (like hydrogen and oxygen turning into water) involves a partnering of atoms that we now call bonding.

In this painting, John Dalton is collecting marsh gas—colorless, odorless methane. The gas forms when bacteria decompose plant and animal matter, and carbon and hydrogen combine. (All life-forms contain carbon.) Dalton measured the ratio of the two elements (one to four, or CH_4). Today we're aware that methane contributes to global warming by preventing heat from radiating into space—called the greenhouse effect.

Considering this hypothesis, he writes: "Thus a train of investigation was laid for determining the number and weight of all chemical elementary particles which enter into any sort of combination one with another."

Now if he can find a connection between atomic weights and changes in weight when elements combine, he can prove his hypothesis. "It became an object to determine the relative sizes and weights, together with the relative *numbers* of atoms entering into such combinations," he writes.

Will his experiments tell him what atoms are made of? No. He knows he has no way of determining that, so he doesn't focus on it. He has hypothesized (brilliantly) that the difference between atoms is their weight. So how do you weigh an atom? Dalton knows that it can't be done. But he thinks he can find relative atomic weights. That's where he puts his attention.

No one has done that before. "An enquiry into the relative weights of the ultimate particles of bodies is a subject, as far as I know, entirely new," he writes, describing the path he has chosen. "I have lately been prosecuting this enquiry with remarkable success."

In the twentieth century, scientists discovered maverick atoms called isotopes, which have a slightly different mass from their brothers and sisters. (They all have the same number of protons—so they're the same element chemically—but they have a different number of neutrons.)

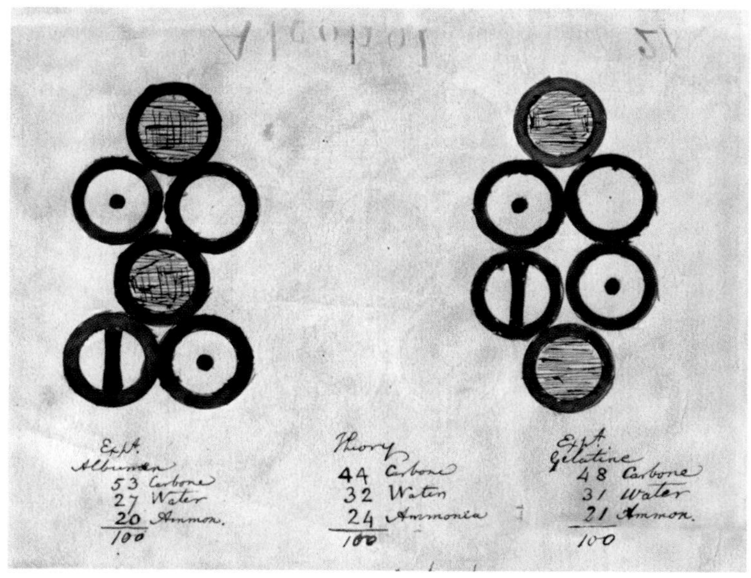

How do you picture something you can't see? Dalton drew atoms as circles and clumped them together in fixed ratios in order to envision chemical compounds. He gave each of these proteins (above) one "azote" atom (now called nitrogen), two carbons (gray circles), two hydrogens (dotted circles), and one oxygen (clear circle).

In every compound he analyzes, different elements always have the same ratio of weight. There is nothing random about it. (It's called the Law of Definite Proportions.) That gives him a breakthrough thought. Here's a way to picture what he does:

Imagine a crate with an equal number of red dinner plates and green coffee mugs. It falls off a forklift—*CRASH*. You now have a mess of broken pottery, which you can separate into a heap of green shards and another of red. You want to know the relative weight of a plate and of a mug, but you don't have either one. What do you do? You weigh each pile and compare their weights. That ratio between the pile of red bits and the pile of green bits is the same as the ratio of the weight of one plate to one mug.

Dalton knows if he weighs equal amounts of two elements, he can assume that each has a unique number of atoms. If he gets the ratio of their weights, he still won't know the exact weight of an atom of lead, but he will know how it compares

INFORMATION WORTH REPEATING

Atoms are the smallest form of an element (like oxygen or gold) having all the characteristics *of that element*.

We now know that atoms are filled with smaller particles. So how about Democritus's question? Is there a bottom-line, building-block particle?

Today, there are scientists who believe they have found the ultimate, uncuttable particle. They describe it as a vibrating string or membrane that is common to all matter and energy. But some scientists don't agree. They say Aristotle was right; you *can* always cut a particle in half. Keep reading to learn further developments in this quest.

DALTON'S DOTS

John Dalton drew symbols for atoms. He started with a circle for each atom and put markings inside to distinguish one from another. But his atomic symbols were close to impossible for typesetters to put in books.

Swedish chemist Jöns Jakob Berzelius (1779–1848) proposed that those symbols be replaced with alphabetic notation. He suggested that hydrogen become H and oxygen O. What about duplicate alphabetic initials? Berzelius recommended C for carbon, Co for cobalt, and Cu for copper (its Latin name is *cuprum*).

Berzelius said that when like atoms combine that should be shown with numbers that are superscripted (placed above): H^2O. Later that was changed to subscripted (placed below): H_2O.

Dalton hated Berzelius's notations and called them "horrifying." He said they would "perplex the adepts of science, discourage the learner, as well as to cloud the beauty and simplicity of the atomic theory." In other words, he liked his drawings and thought science should stick with them. He lost that contest.

ELEMENTS

	W.t		W.t
Hydrogen	1	Strontian	46
Azote	5	Barytes	68
Carbon	5.4	Iron	50
Oxygen	7	Zinc	56
Phosphorus	9	Copper	56
Sulphur	13	Lead	90
Magnesia	20	Silver	190
Lime	24	Gold	190
Soda	28	Platina	190
Potash	42	Mercury	167

John Dalton first published his table of elements in 1805. The numbers in the right column are the relative atomic weights, compared to hydrogen, which is set at 1. Azote (nitrogen) is listed as five times heavier, and oxygen as seven times heavier. Oddly, Dalton put carbon at 5.4—the only decimal value. He calculated the weights to the best of his ability, but he was mostly wrong. Nitrogen is a tad over 14, oxygen is just under 16, and carbon weighs in at a smidgen over 12 (see atomic masses on page 311). Cut Dalton some slack, though. Most of the 92 natural elements hadn't been discovered yet, and many compounds (like lime—calcium oxide) had been misidentified as elements.

to an atom of hydrogen (for example). He correctly hypothesizes that hydrogen is the lightest element, so he uses that as a standard; all the other elements become multiples of that lightest one. Once he figures out some relative atomic weights, he can draw conclusions.

Dalton prepares a table of atomic weights. (He is way off on some of them, but it is a start.) It gives scientists a way to begin to classify elements.

Dalton goes on to say that **atoms can neither be created nor destroyed** and that chemical reactions are just

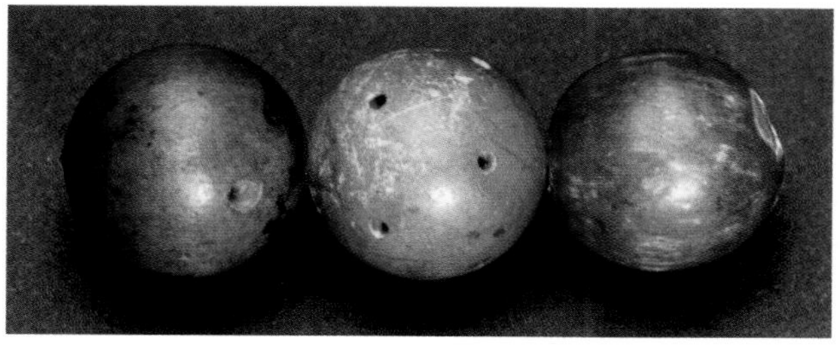

Dalton used these wooden balls as models of atoms. He said that matter is made up of solid, indivisible atoms. Atoms aren't solid and they *are* cuttable, as it turns out, but his book *A New System of Chemical Philosophy* (volume 1, 1808) had a huge influence on the advancement of physics and chemistry.

rearrangements of atoms. (He is expanding Lavoisier's conservation-of-mass idea.)

Dalton realizes that there is a difference between compounds and mixtures, and that is important information.

He doesn't understand molecules, which are two or more atoms bonded together. Knowledge of molecules, when it comes, will help explain a lot about nature's workings.

But the really big news, which will arrive a century after Dalton, is that atoms are not hard and impenetrable. They are not like wooden balls, as Dalton and Hui Shih and Newton thought. Still smaller particles (called neutrons, protons, and electrons) are inside atoms. Scientists will eventually count the protons inside each atom and use that number as a way to classify elements. Mostly that won't change the order of elements: Hydrogen still has the lightest mass and, with one proton, will remain number 1.

What size is an atom? An ordinary atom has a diameter of about 3-billionths of a meter, which means that a million of those atoms laid shoulder to shoulder would be about 3 millimeters long, or the length of this dash: –. To today's physicists, atoms seem quite big because a lot is packed inside them. They aren't the "ultimate" or "elementary particles" Dalton thought them to be.

Yes, he has a few details wrong, but Dalton gets the main part of the atomic story right. Before, atoms had been a philosophical and mathematical idea, but Dalton understands that they are real.

When he publishes his theories in 1808, people pay attention. The bookish scientist becomes a celebrity. Even the king asks to see Dalton. To be presented to the king requires wearing knee breeches, buckled shoes, and a sword. Quakers don't wear swords. And Dalton doesn't have fancy clothes.

Dalton's atomic theory allowed chemistry to become an exact science. The importance of making numerically precise measurements of chemical processes had been clear enough to Cavendish, Priestley, Lavoisier, and their contemporaries; but without an underlying theory of the elements, these numbers.... were like measurements of the depth of a river or the number of ants in a colony—they did not reveal anything about the fundamental constitution of the system.

—Philip Ball, columnist and science writer, *The Ingredients: A Guided Tour of the Elements*

To summarize some of Dalton's ideas:

- All matter consists of tiny particles.
- All the atoms in an element are alike and have the same mass. (Dalton didn't know about isotopes—the exceptions.)
- Atoms cannot be created, destroyed, or changed in a chemical process. (We now know an atom's nucleus can be split or fused with another nucleus.)
- Elements bond in simple whole-number ratios to form compounds.
- Compounds combine in varying proportions to form a mixture.
- Chemical reactions are the union and separation of atoms.

What is the shy, awkward scientist to do? He solves the problem by dressing in a university robe.

When he dies, 40,000 people file past his coffin. It's an over-the-top funeral, not exactly what a modest Quaker might wish. But the English are proud of their leadership in science. Many don't understand atoms, but they do understand that this man has helped explain their world.

28

A Molecule-and-Number Man

> If there were no real internal propensity to unite, even at a prodigiously rudimentary [basic] level—indeed in the molecule itself—it would be physically impossible for love to appear higher up.
>
> —Pierre Teilhard de Chardin (1881–1955), French philosopher, scientist, and Jesuit priest, *The Phenomenon of Man*

> Molecule: a group of two or more atoms bonded together. A molecule of an element consists of one or more like atoms; a molecule of a compound consists of two or more different atoms bonded together. Molecules vary in size and complexity from the hydrogen molecule (H_2) to the large macromolecules of proteins.
>
> —Brockhampton's *Dictionary of Science*

Amedeo Avogadro (ah-may-DAY-oh ah-vuh-GAH-droh) is a count from Italy's Piedmont (the northern foothills), but just think of him as the "molecule guy."

Avogadro (whose full name is Lorenzo Romano Amedeo Carlo Avogadro, count of Quaregna and Cerreto) is born in 1776 (an easy year to remember). He starts out as a lawyer but is so fascinated with scientific research that he gives up law to be a professor of physics (this is a familiar story).

Amedeo Avogadro earned his college degree at the age of 16 and a law degree at 20. While working as a lawyer, he took science lessons on the side and became hooked on chemistry and physics.

Avogadro figures out that most gases are made up of particles containing two or more atoms held in a tight embrace. He names those bonded particles "molecules" (from the Latin word for "small masses"). He is the first to make clear the distinction between atoms and molecules. (His ideas are later expanded to include liquids and solids.)

Avogadro is right, a few gas molecules have single atoms

(helium and argon), but most gas molecules consist of linked atoms. Hydrogen (an element whose symbol is H) is usually found in nature as two like atoms bonded together to make a molecule (H_2).

Carbon dioxide (CO_2) is a molecule made of unlike atoms—one carbon and two oxygen—wed to each other.

Understanding the difference between single atoms and combinations of atoms (molecules) sounds simple, but it is a big step. (Dalton never took that step.) The idea of atoms and molecules is the foundation of modern chemistry.

Dalton told us that **the smallest form of an element (with all the characteristics of that element) is an atom**. Avogadro tells us that **the smallest form of a compound is a molecule.** Break a molecule apart (water, H_2O, for instance) into atoms (two hydrogen and one oxygen), and you change its chemical properties: it's no longer what it was.

Avogadro comes up with a law of his own, known—naturally—as Avogadro's Law. Here it is: **Equal volumes of gas, at the same temperature and pressure, contain an equal number of particles** (they can be atoms or molecules). Think about that; it gives scientists a very useful measure to work with.

But the scientists who first hear Avogadro's hypothesis dismiss it; they must think he is crazy. They are aware that the particles that make up different gases have different sizes. They are right. Hydrogen molecules (H_2 with only two protons and two electrons) are small and light. Most other molecules are much heavier.

So how can equal volumes of gases at the same temperature and pressure contain an equal number of particles, if particles are different sizes and masses? You can't squeeze an equal number of oranges and olives into the same box. But Avogadro isn't concerned with squeezing. His idea only makes sense if the particles in a gas are far apart, with space between them (which is the case—even for liquids and solids). Working with gases, he has deduced that there are vast expanses of space between particles. (Imagine those olives or oranges in motion in a big room where they occasionally bump into each other or into a wall, but mostly they have space to roam.)

You might know frozen CARBON DIOXIDE (CO_2) as dry ice. Carbon dioxide in the atmosphere can cause acid rain or contribute to global warming through the "greenhouse effect" (trapping heat that otherwise radiates into space). In your soft drink, CO_2 adds bubbles through carbonation.

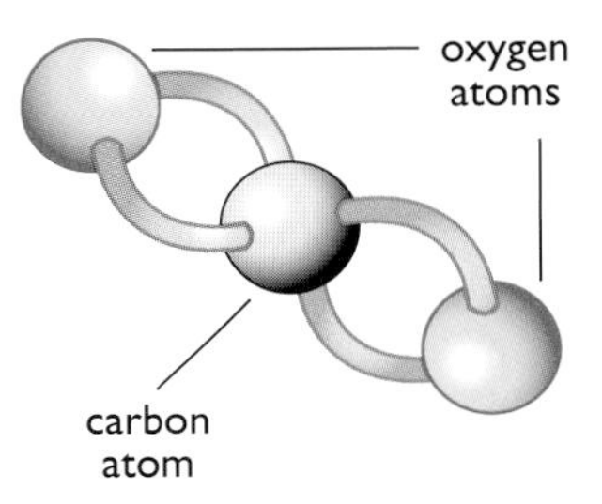

HAPPY MOLE DAY

October 23, from 6:02 A.M. to 6:02 P.M. is National Mole Day. No, it's not a day to go digging up moles; it's a day chemists celebrate (well, some chemists).

A mole is another name for Avogradro's number of atoms or molecules (6.02×10^{23}). The word was coined around 1900 by Wilhelm Ostwald, from the Latin for "mass, hump, or pile." (*Molecule* comes from the same root.)

National Mole Day enthusiasts suggest celebrating with "pi à la mole" and singing the "rock-'n'-mole" theme song "The Mole the Merrier!"

Given his vision, and his law (*equal volumes of gases at the same temperature and pressure must have an equal number of particles*), Avogadro is able to take a new look at the chemical world. Using his law, he finds the correct formula for water. No one has done that before. When water is broken apart into hydrogen and oxygen, and those gases are collected separately, the hydrogen takes up two times the space of the oxygen. According to Avogadro's Law, if the hydrogen occupies twice the volume, there have to be twice as many hydrogen molecules. And that's how Avogadro figures out that the formula for water is H_2O, not HO, as Dalton had believed. (Keep in mind, a molecule of water is always made of two hydrogen atoms and one oxygen atom. It is always a whole-number ratio.)

Avogadro's insight will lead to an unchanging number, a constant, that can be used to calculate the number and mass of atoms and molecules in a substance. (Today, it's called Avogadro's number. It's huge: 6.02×10^{23}.) But, for decades, no one gets his law—or realizes its importance. Avogadro has the bad luck to have his ideas mostly ignored while he is alive. Only later will he become famous. (A big part of the problem is that he lives in Turin, Italy, which is far from the center of science, in his day in England.)

The "Ancient Bridge" spans the Po River in Turin, Italy—just as Amedeo Avogadro must have seen it. This beautiful city is tucked in the northwest corner of Italy, near the French border. The painting at right is by Bernardo Bellotto (ca. 1721–1780).

COUNTING MOLES?

Is it possible to know the *number* of atoms or molecules in a substance? Amazingly it is. After Avogadro's time, Johann Joseph Loschmidt (1821–1895) worked it out in a series of experiments based on Avogadro's ideas and on a theory about moving atoms developed in the 1860s. Loschmidt, who was a high school teacher at the time, came up with a number known as Avogadro's number. Today, more conveniently, it's called a mole (in scientific notation it is *mol*). And it is the key to counting extremely small particles.

What is Avogadro's number? Take a deep breath. The current refined figure is 602,214,199,000,000,000,000,000. But scientists usually round it off to a more convenient 602,000,000,000,000,000,000,000, or 6.02×10^{23}, or a bit more than 602 billion trillion. (You can see why they decided to just call the whole thing a mole!)

Keep in mind, this is a number measure—it tells us the *number* of atoms or molecules or other particles in a given mass of a substance. It's essential to specify which elementary particle you are counting. You can't talk about 1 mole of oxygen; you need to say 1 mole of O atoms or of O_2 molecules.

In figuring this out, Loschmidt started with hydrogen gas. He knew it was composed of hydrogen molecules (H_2), formed from two hydrogen atoms. The hydrogen molecule has an atomic mass of 2. Loschmidt found that at standard temperature and pressure (known as STP), 22.4 liters (5.9 gallons) of hydrogen gas weighs 2 grams. And 2 grams of hydrogen gas contains Avogadro's number of hydrogen molecules.

Using Avogadro's number, Loschmidt could figure out the actual mass of one hydrogen molecule—it was 2 grams divided by 602 billion trillion! Since equal volumes of gases are made up of equal numbers of particles, 22.4 liters of *any* gas (at STP) contains 1 mole of particles. Considered another way: How many molecules are in 22.4 liters of hydrogen gas? Or in 22.4 liters of carbon dioxide (CO_2)? It's Avogadro's number—6.02×10^{23}.

Today Avogadro's number is used to measure a whole zoo of particles discovered in the twentieth century, like electrons and photons. And, if 1 mole is not enough for you (or maybe it's too much), you can turn to its relatives, creatures like the gigamole (Gmol), which is 10^9 moles, the nanomole (nmol), 10^{-9} moles, or the attomole (amol), 10^{-18} moles.

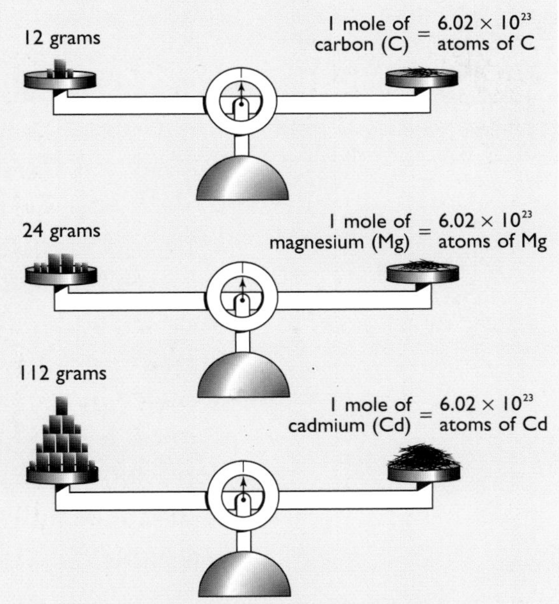

Remember this: A mole is a number. It's like saying, "a dozen"—only you mean many, many more than 12. By talking about a fixed number of items, you can compare the masses of things fairly. A dozen eggs weighs less than a dozen watermelons—and you can be sure because there are exactly the same number (12) of each. Likewise, 1 mole of carbon (C) weighs half as much as 1 mole of magnesium (Mg): 12 versus 24 grams (above). One mole of cadmium (Cd) is even heavier—112 grams. One *atom* of carbon also weighs half as much as 1 *atom* of magnesium. Atoms are so tiny, though, that measuring them requires a unit of mass much, much smaller than a gram—you need an amu, an atomic mass unit.

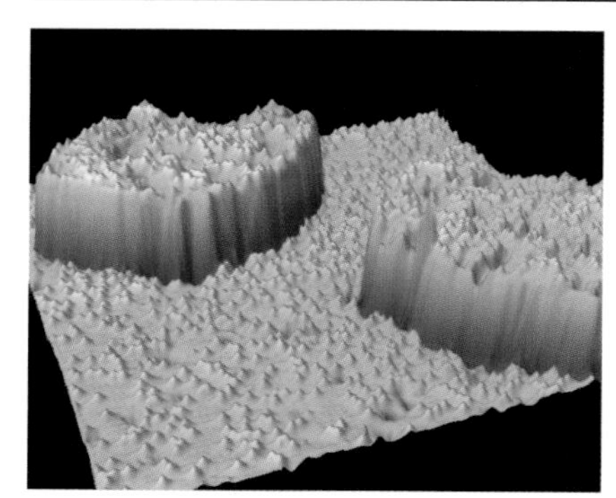

A scanning tunneling microscope (STM) revealed chromium atoms (the small yellow bumps, above) growing on an iron crystal. An STM has a needle almost too tiny to imagine, which can probe a surface on an atomic level. The tip of the needle creates a tunnel effect—making electrons travel narrow distances for a few atomic diameters. A computer turns the electrons' journeys into a three-dimensional image.

Dalton's atoms stay well known, but it takes about 50 years for science to pay attention to Avogadro's molecules. And if you don't understand molecules, you can't do much with atoms. So atomic research doesn't get anywhere, and as time passes, the atomic idea begins to be questioned. Atoms are called "a useful fiction." That's not hard to understand; atoms are beyond-belief small.

Just what size is an atom?

Imagine magnifying one drop of water until it is 24 kilometers (about 15 miles) wide; you will begin to see the atoms inside the water molecules (not clearly—that will take much greater magnification); that's why no optical (magnifying) microscope can see atoms. (Today, scanning tunneling microscopes "see" them electronically.)

Or picture an apple. Blow that apple up until it is the size of the Earth. Each of its atoms is now the size of a normal apple.

Here's another image: 250 million hydrogen atoms packed in a single-file line will stretch about 2.5 centimeters (1 inch) in length.

As for molecules, chemist Brian L. Silver writes, "Molecules tend to be very small entities. . . . If the whole population of Earth set out to count the molecules in a teaspoon of water, each person counting at the rate of one molecule per second, it would take over a million years."

Scientist Lewis Wolpert says, "There are many more molecules in a glass of water than there are glasses of water in the sea." How do we know things like that? It's because we have Avogadro's number to help with the calculating.

Imagine figuring out that atoms and molecules exist when they can't be seen. What Dalton and Avogadro do is astonishing. But, without proof, many scientists begin to reconsider Dalton's theory. They even make fun of it, just as scientists in ancient Greece ridiculed Democritus. No one will ever be able to see an atom, the skeptics say. They are absolutely sure of that. Would you believe in atoms and molecules if you had no concrete proof of their existence?

It wasn't necessary for Newton to understand the atom to figure out the laws of gravity and motion. But it *is* necessary to understand atoms in order to understand matter. It took a huge leap of mind to realize that matter is the same everywhere in the universe. Atoms in your toe are the same as the atoms of like elements that can be found out in space.

KEKULÉ'S SAUSAGES

Understanding that atoms and molecules exist was a big step up the ladder of knowledge. But why do atoms bond with each other and form molecules? Edward Frankland was one chemist trying to figure out the rules of engagement (see pages 294–297). Friedrich Kekulé was another. He started out studying architecture but then heard some chemistry lectures and decided to change his major. Perhaps it was because of his background in the arts that he had a graphic imagination, which helps in any field. One day he climbed on a bus and had a daydream:

> *During [a] stay in London, I lived for a considerable time in Clapham Road near the Common. But I often spent my evenings with my friend... at the other end of the great town. We spoke of many things, but chiefly of our beloved chemistry. One fine summer evening, I was returning by the last omnibus... through the deserted streets, at other times so full of life. I fell into a reverie. There before my eyes gamboled the atoms. I had often seen them moving before, each tiny being, but I had never succeeded in discerning the nature of their motion. This time I saw how frequently two smaller atoms united to form a pair; how a larger one embraced two smaller ones; how still larger ones... formed a chain dragging the smaller ones after them.... The cry of the conductor "Clapham Road" awoke me from my dream, but I spent a part of the night putting down on paper the sketches at least of these dream-forms. Thus began the structure theory.*

Kekulé's drawings, which were called "Kekulé's sausages" by other chemists, showed atoms bonding chemically. In an 1858 paper that accompanied his "sausages," Kekulé noted that carbon atoms tend to unite with other atoms in a ratio of one to four. (The valence—or bonding potential—of carbon is four.) It is as if carbon atoms have four slots for other atoms. That led him to picture chains of four-bonded carbon atoms that create huge molecules. He began writing out formulas. He was getting the same results as Frankland, but by a different route.

In 1865, Kekulé went still further. He said that carbon not only forms atomic chains, but that it also links into rings of atoms. He was right. Chemists were soon playing with carbon chains and links, which gave rise to organic (carbon-based) chemistry. Almost immediately it led chemists to create thousands and thousands of new carbon-based substances. Kekulé was ahead of his time with his mental pictures. Visualizing is a method many creative people use to "picture" complicated things.

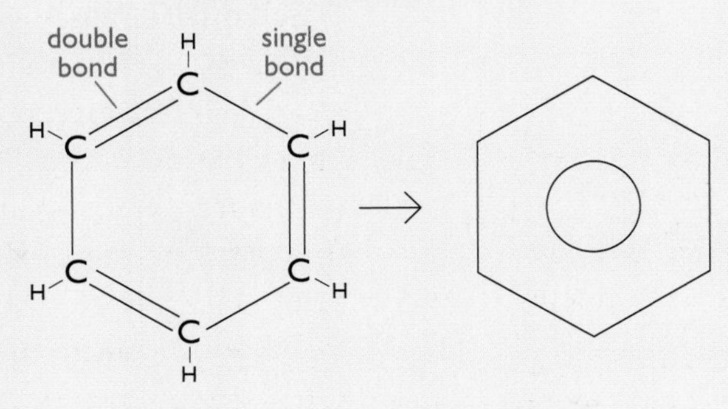

Benzene, a colorless and flammable liquid, was a Kekulé favorite. Its molecule has six carbon atoms and six hydrogen atoms (C_6H_6), and yet carbon bonds in a one-to-four ratio. Kekulé solved part of the puzzle by ditching his sausage links (below) and arranging the carbon atoms in a six-sided ring—a hexagon. He added alternating double and single bonds between the C's, giving each one four bonds. It was a start. Today's symbol for a benzene ring (far right) indicates that the bonds aren't double or single—they're somewhere in between.

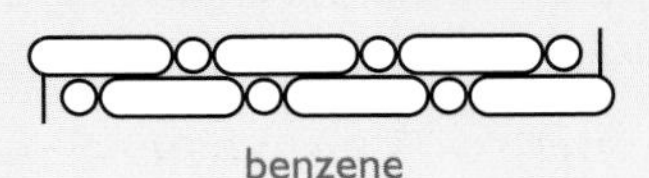

benzene

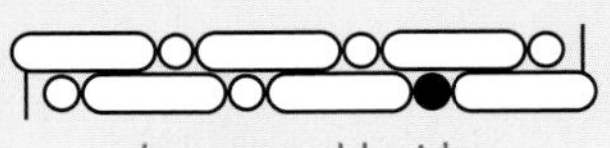

benzene chloride

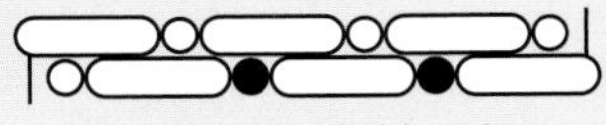

benzene dichloride

CHEMICAL BONDAGE

So what makes atoms stick together to form molecules? Are there any rules of engagement? Finding out was a scientific challenge.

In 1847, young Edward Frankland (1825–1899) was lucky enough to work for three months with Robert Bunsen, one of the world's outstanding chemists. Frankland, an Englishman, was a 22-year-old chemistry student studying for his Ph.D. at the University of Marburg in Germany. Now, three months isn't much time, but great teachers can change their students' lives. Bunsen seems to have been that kind of teacher.

Up until this time, chemists had worked mostly with nonmetallic elements like carbon, hydrogen, nitrogen, oxygen, sulfur, phosphorus, and so on. But metals make up more than three-fourths of the natural elements. It was a field crying for attention.

Bunsen, who was working with chemicals that cause metals to bond with other elements, believed in atoms and molecules. When Frankland arrived in

Flame tests help identify some metal elements by color. A Bunsen burner is igniting potassium, which gives off a distinct blue color.

Here's Sir Edward Frankland, Robert Bunsen's partner, in about 1880.

Germany, Bunsen was specifically looking at compounds that included zinc. Frankland said that when water was added to those zinc compounds, "a greenish-blue flame several feet long shot out of the tube, causing great excitement amongst those present and diffusing an abominable odour through the laboratory." Frankland noticed a pattern in those metal compounds. He realized that when elements combine (metallic or not), they always do so in whole-number ratios. Half an atom can't bond with three-quarters of another, for example. He figured out that the atoms of each element have a certain combining number, and that's it—no variation allowed. Or, as he put it, "The combining power of the attracting element . . . is always satisfied by the same number of atoms." Frankland called this property atomicity.

He used the word *bond* for the first time (that we know about) to describe the force of attraction that holds atoms together to make molecules. Every atom has a certain number of bonds that it can form, and only that number. That's the theory of valence, which Frankland proposed in 1852. (Today instead of valence, we usually think of bonding in terms of electronegativity, based on our modern knowledge of electrons.)

An element's valence is the number of chemical bonds that each of its atoms can make. Frankland worked that out by considering hydrogen and oxygen. They gave Frankland a starting point. When hydrogen and oxygen combine to make water, hydrogen forms one bond and oxygen two. So hydrogen was given a

(continued on page 297)

Scientists play with models because it's fun—and productive. This is a molecular modeling set from 1875. Each ball represents an atom, which can be bonded to other atoms with pegs to form molecules. The number of peg holes in a ball equals the number of molecular bonds (or valence) that an atom of a particular element can form.

Meddling with Metals

What is a metal? It's easy to name the popular ones: gold, silver, platinum, iron, copper, zinc, tin, lead, nickel. Here are some less famous ones: cadmium, chromium, manganese, sodium, potassium, lithium, calcium, strontium, barium, uranium. What do these and other metallic elements have in common? What makes a metal a metal?

These transition metals are clustered in the middle of the periodic table of elements (page 311) because they share properties. A key feature is that they easily form alloys (mixed metals). That's the transition metal copper in the center. Clockwise from left are: aluminum pellets, nickel-chrome ore, nickel bars, titanium bars, iron-nickel ore, niobium bars, chromium granules. That's gold (below) and droplets of mercury, (far right).

Nineteenth-century chemists identified metallic elements by observing their properties: Metals are lustrous (shiny), dense, and easy to shape. You can bend them, hammer them into thin sheets, or stretch them into long skinny wires. Most of them are gray or silver (copper and gold are exceptions). Metals conduct both electricity and thermal energy (heat) well. They mostly have high melting and boiling points (except for mercury, which is the only metal that is liquid at room temperature).

We now know that these properties aren't accidental or coincidental or incidental. There's one underlying cause—atomic structure. Electrons (negatively charged particles) orbit an atom's nucleus in layered shells. The atom of a metal has too few electrons in its outermost shell; the outer shell of a nonmetal atom, on the other hand, is packed with electrons.

Having lonely electrons in your outer shell—so distant from the nucleus with its positively charged protons—is not a stable situation. The outer electrons of a metallic atom easily jump ship; they often separate from the atom and become free agents or join other atoms. As you'll recall, electric current is made of electrons on the loose—which is why metals are so conductive. Footloose electrons also give metals their luster.

What happens when a metallic atom (iron) meets a nonmetallic atom (oxygen)? You get compounds such as oxides—iron oxide (rust), for example. Most metals bond easily with oxygen because their outer electrons are sparse and the outer electrons of oxygen, a nonmetal, are abundant.

Late-nineteenth-century chemists didn't know about the electron connection, but some of them began to suspect it. Keep reading to find out how and why.

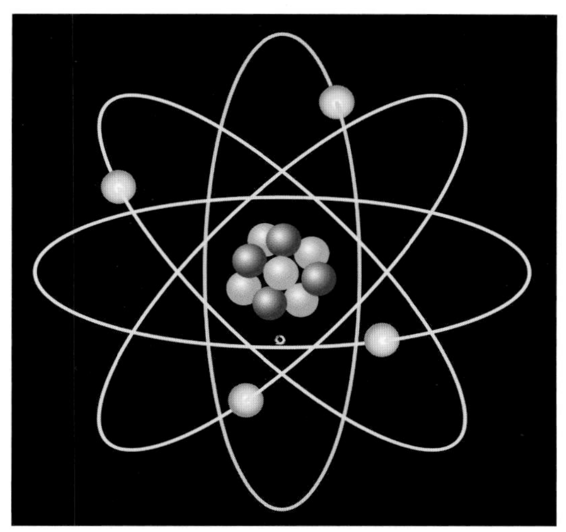

This symbolic atom is beryllium. How can you tell? Count the protons (in blue) and find that atomic number on the periodic table (page 311). The orange balls are neutrons, and the gray ones are electrons.

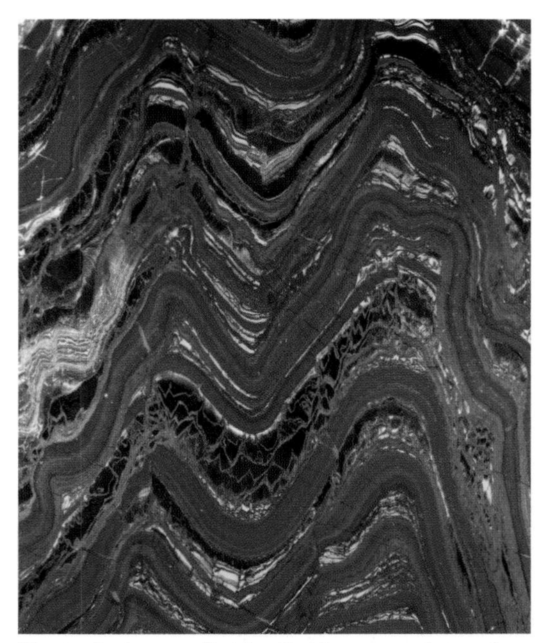

The rock above, called banded ironstone, has red layers of hematite, an iron oxide mineral.

(continued from page 295)

valence of 1 and oxygen a valence of 2. For other elements, valence is usually (but not always) the number of hydrogen atoms that will combine with that element's atoms. When nitrogen and hydrogen bond to form ammonia, nitrogen must bond with three hydrogen atoms (NH_3), so nitrogen has a valence of 3.

Frankland found that, to be stable, an element must use all the bonding options it has. When a hydrogen atom bonds with another hydrogen atom (forming H_2—a molecule of hydrogen), it has used up its bonding potential. Carbon has a potential (or valence) of 4, which gives it a lot of flexibility.

This was something that needed to be documented on paper so chemists could use it in their daily work. It was Frankland who came up with the idea of chemical notation, finding a written way to express the structure of molecules. In his notation, water was H-O-H. It would evolve into H_2O.

This system was useful for about 100 years, but no one knew why it worked. Then electrons were discovered and they turned out to be the glue in chemical bonds. For details, read the third book in The Story of Science series, *Einstein Adds a New Dimension*.

29 Putting Things in Order

The periodic table is arguably the most important concept in chemistry, both in principle and in practice.... It is a remarkable demonstration of the fact that the chemical elements are not a random clutter of entities but instead display trends and lie together in families.

—P. W. Atkins (1940–), British chemistry professor, *The Periodic Kingdom*

Mendeleyev was aware of the significance of his search. This could be the first step towards uncovering, in future centuries, the ultimate secret of matter, the pattern upon which life itself was based, and perhaps even the origins of the universe.

—Paul Strathern (1940–), British science writer, *Mendeleyev's Dream: The Quest for the Elements*

Scientists don't believe in magic—reason and proofs define science—and yet white-bearded Dmitri Ivanovich Mendeleyev (men-duh-LAY-ef), seemed to bring the magic of prophecy to science. He predicted that certain unknown elements would be found, he described their characteristics, and then he sat back and waited. Most people laughed. When the first of his predicted elements turned up, they explained it as coincidence. But when another was found, and another—well, the whole world soon paid attention.

Who is Dmitri Mendeleyev?

He begins life as the youngest son in a family of 14 children. His father is a high school principal. They live in Tobolsk, a small town in Siberia. His mother's father had been

In Tobolsk, the home town of Dmitri Mendeleyev, stands a gold-domed Orthodox Christian church.

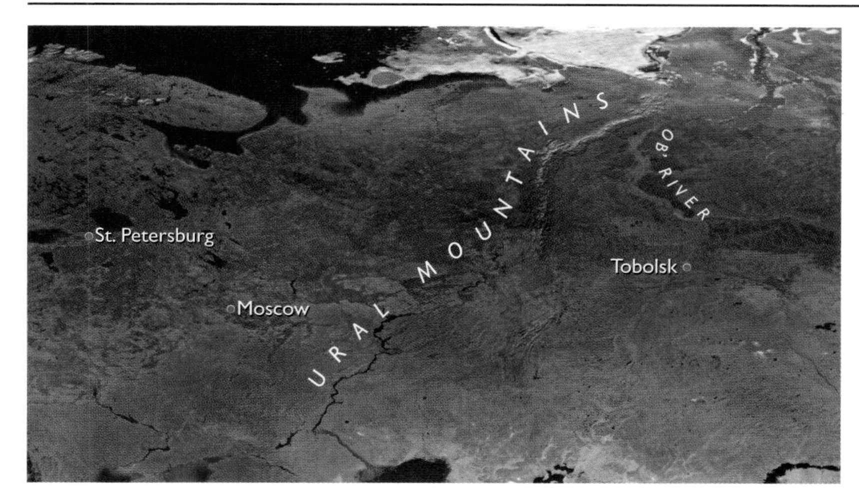

Seeking a better education, young Dmitri, his mother, and his sister walked and caught rides on horse carts to travel from Siberia to Moscow and, finally, to St. Petersburg. His mother and sister soon died, and he caught tuberculosis (a serious lung disease), but Dmitri threw himself into learning.

one of the Russian pioneers in that cold, isolated region. That grandfather published the first newspaper in Siberia, the *Irtysh*, and he established a glass factory in Tobolsk.

When Dmitri is very young, his father goes blind. Dmitri's mother—known for her beauty and strong-mindedness—reopens the glass factory, builds a church and school for its workers, and supports the family. When Dmitri is 13, his father dies.

Siberia is where the Russian leaders send their political prisoners—especially those who protest against the czar's rule. It seems like the end of the world to most who are forced to stay there. One of Dmitri's sisters marries an exile. That is lucky for Dmitri: His brother-in-law is a scientist who answers the boy's questions and teaches him about the world of science.

When a fire destroys the glass factory, Dmitri, his mother, and his youngest sister leave Siberia and head west to Moscow. Dmitri has done poorly in school in Tobolsk; his mother is determined that her son get a good education. They make the 2,000-kilometer (1,250-mile) journey by hitching rides on horse-drawn wagons.

In Moscow the schools won't recognize his Siberian qualifications. They turn him down. The trio moves on, this time to St. Petersburg, where his mom has a friend. In bureaucratic Russia, knowing someone helps. He is admitted to the Central Pedagogical Institute (a fine school). His mother tells him, "insist on work and not on words." And, "patiently seek divine and scientific truth." Ten weeks later she is dead. Her words will remain his lifelong inspiration.

But things don't get easy for him. His sister dies, and he is diagnosed with tuberculosis. No one expects him to live. Dmitri moves into a hospital and goes to classes from there. Actually that gives him some independence. The orphan becomes a kind of mascot at school and at the hospital. He's given freedom to study and experiment as he wishes. His mind catches fire, and he is soon doing original scientific research. In 1855, he graduates at the top of his class. That leads to opportunities. (And he seems to have recovered his health.) His friend Aleksandr Borodin, a first-rate chemist and a famous composer, encourages him to go to France for graduate work. There he has the good luck to study with a fine experimental chemist. That training will serve him well.

Robert Bunsen (center) poses with his friends in chemistry, Gustav Kirchhoff and Henry Roscoe. Besides the famous Bunsen burner, he invented the Bunsen ice calorimeter to measure the amount of thermal energy in a substance.

From France, he heads for Germany, where (more luck) he works with Robert Bunsen (a great teacher and the inventor of the Bunsen burner, still found in today's science labs). Bunsen is using a prism to split light into a color spectrum, as Newton did, but he takes it one step further. He heats elements with his burner, refracts the light they give off, and finds that each produces a unique spectrum—a color "fingerprint" that can be used to identify the elements. (This is the science of spectroscopy.) Dmitri Mendeleyev has put himself at a center of cutting-edge science.

But chemistry is in a confused state: a simple compound may be written 10 different ways by 10 different chemists. Coordination is close to impossible. When Mendeleyev studies John Dalton's atomic weights—numbers that

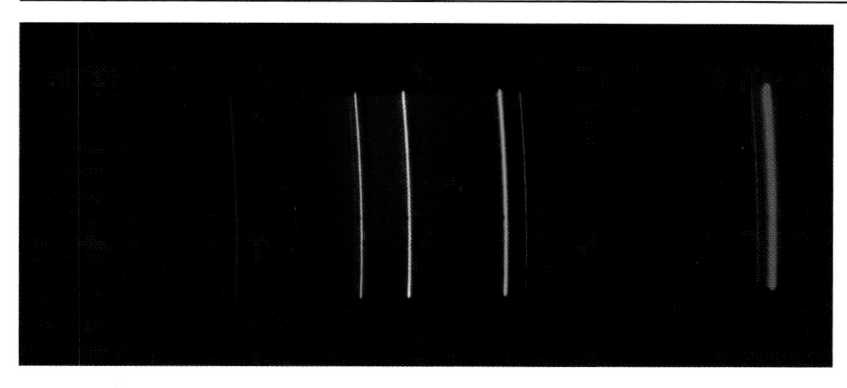

Only one element gives off this pattern of electromagnetic (EM) energy: cadmium (Cd), a soft and silvery transition metal similar to zinc, yet rarer. The pattern is a record of which wavelengths of radiation a cadmium atom absorbs and emits and which ones it doesn't. A spectrometer is an instrument that collects and measures the emitted radiation.

describe the mass of an atom in any given element—he realizes that hardly anyone agrees even on those weights. And Avogadro's ideas about atoms and molecules are mostly forgotten.

In 1860, the First International Chemical Congress is called to try to clear up the confusion. It is held in the little German town of Karlsruhe. The conference's goal is to standardize the atomic and molecular masses that are being used by chemists around the world. Friedrich Kekulé is the man who initiates the meeting.

Mendeleyev attends. He is awed by an Italian scientist who has discovered Avogadro's work and is bursting to tell his fellow scientists about it. (Avogadro had died four years earlier.) The Italian, Stanislao Cannizzaro, makes clear the important difference between atoms and molecules and explains how to use Avogadro's Law. He even passes out copies of a pamphlet he has written on the

"Bunsen, I must tell you how excellent your study of chemical spectroscopy is, as is your pioneer work in photochemistry—but what really impresses me is that cute little burner you've come up with."

In 1860, many leading chemists still thought the formula for water was HO (not H_2O)—just as John Dalton had six decades earlier. Hydrogen atoms naturally occur in bonded pairs (H_2). But the chemists figured that two like atoms, as with two like electric charges, could only repulse each other. Resistance to the idea of diatomic molecules was what begged the central question of the Karlsruhe conference: What is an atom and what is a molecule?

That's Mendeleyev (above) not long after an annual shearing.

subject. Cannizzaro is the talk of the Karlsruhe conference. After that, chemists begin to agree on formulas and on atomic weights. (It's the nineteenth century, and atomic masses are still called atomic weights.)

The next year, 1861, Mendeleyev is back in Russia as an instructor at St. Petersburg University. Tall and blue-eyed, he is an independent sort, radical, sometimes bad tempered, almost always outspoken. He is identified by his long beard and flowing hair; once a year (in the spring) a sheepherder cuts those locks with sheepshears. To a visiting Scotsman, Mendeleyev is "a peculiar foreigner, every hair of whose head acted in independence of every other."

To those who know his work, he is a genius. Mendeleyev's passion for his subject and his no-nonsense style inspires students; they come from around the world to study with him. One of those students, Prince Kropotkin (a future anarchist leader) later remembered this: "The hall was always crowded with something like two hundred students... but for the few of us who could [understand] it was a stimulant to the intellect and a lesson in scientific thinking which... left deep traces."

The American Civil War began in 1861. That was also the year Russia freed its serfs. The serfs, owned by Russian landlords, had been much like slaves. All at once, without a war, they were free.

In 1869, Mendeleyev is a 35-year-old chemistry professor with a problem that many teachers face. There isn't a good textbook to use with his students. So he decides to write one. But he doesn't know what order to use when dealing with the elements. He begins by writing about elements that are clearly linked—like the halogens (fluorine, chlorine, bromine, and iodine), which combine readily with sodium and potassium to make salts (the best known compound being sodium chloride, or table salt). Next, he describes the elements sodium and potassium. And then he hits a wall—he can't find a clear and logical choice for the next group of elements. He believes they must fall into groups, but how?

HALOGEN is from the Greek *halos*, meaning "salt."

Mendeleyev is struggling with a predicament that is haunting chemistry: Elements, which are basic to science, seem to have no coherent order. By 1869, 63 elements have been identified. They range from gold and silver (known to the ancients) to rubidium (found in 1861 by Robert Bunsen and Gustav Kirchhoff).

Rubidium (as in *ruby*, named for the deep-red line in its spectrum) is a rare metal that violently bursts into flames in water and burns up when exposed to air. Bizarre as this sounds, it can safely be stored in kerosene.

If there is a unifying pattern, it is elusive. Mendeleyev is sure that there is one. So are others. It's a mystery crying to be solved.

Clearly, atomic weight is important. Each of those 63 elements has atoms with a unique mass; it is the one property that is distinctive about every element. Mendeleyev knows that has to be a key in any blueprint. And there must be an explanation as to why some elements share properties (and some don't).

Dmitri searches for clues. He lists the elements by their atomic weight and he also lists them by other characteristics: specific gravity, specific heat, density, state of the element at room temperature (solid, liquid, gas), valence (ability to combine with other atoms), and so on. But one list doesn't seem to relate to the other.

Sodium, an alkali metal, is too chemically active to exist in its pure state near liquid or atmospheric water. Still, it's the sixth most abundant element in the Earth's crust and found in many useful chemical compounds. In its pure form, sodium is a silvery metal light enough to float on water, but it's so reactive that, like rubidium, it sizzles, bubbles, and burns when mixed with air or water (as at left).

SPECIFIC GRAVITY is the density of a substance compared to the density of water (1 gram per cubic centimeter). SPECIFIC HEAT is the heat needed to raise or lower 1 gram of a substance 1°C.

THE STORY OF CARBON

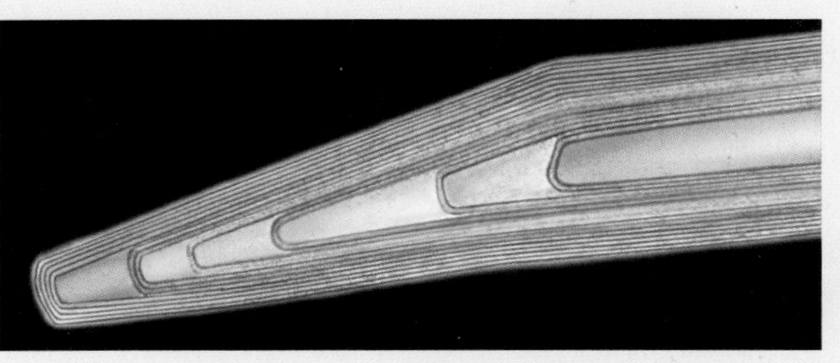
nanotube (electron micrograph)

Every element has a story. Carbon's tale is one to know: There would be no life without carbon. A whole branch of chemistry—organic chemistry—is about carbon compounds. Carbon atoms often make long chains or rings that form perches for other atoms. The backbone of a DNA double helix (bottom left photo) relies on the bonding ability of a five-carbon sugar called deoxyribose; deoxyribonucleicacid (DNA) is the genetic material in all living things.

Carbon atoms without partners (that is, pure carbon) can line up in different ways to form allotropes, like diamond and graphite. (When an element can exist in several natural forms, those forms are called allotropes.) Astonishingly, hard diamonds and soft graphite are both made entirely of carbon atoms.

One lab-created carbon allotrope is a molecule of 60 carbon atoms that looks a bit like a soccer ball (bottom right photo). It's nicknamed a "buckyball" after the inventive scientist Buckminster Fuller. Buckminsterfullerenes (that's the proper name) seem to have a future as efficient superconductors of electricity.

While experimenting with buckyballs, scientists found a fourth carbon allotrope called a nanotube (top photo). It consists of sheets of carbon atoms wrapped around each other to form a cylinder with a hollow core. In 2004, scientists announced a fifth allotrope of carbon, named nanofoam, that is a semiconductor and the first pure-carbon magnet.

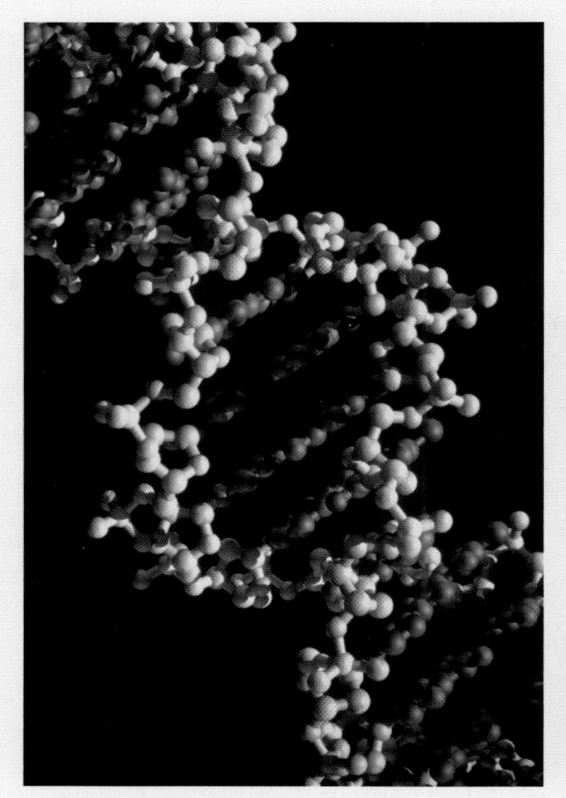
DNA double helix (model)

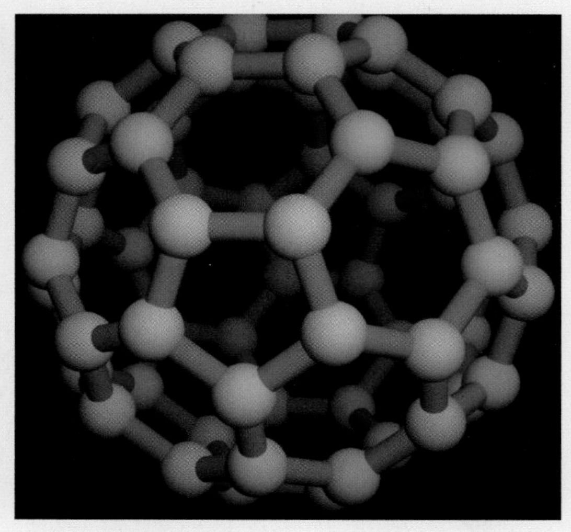
buckyball (model)

Mendeleyev sits at his desk under portraits of Newton, Descartes, and Galileo. For three days in February of 1869, he thinks about nothing else. He is expected to speak at a meeting of cheese makers, and a horse and carriage wait outside his door, but he is obsessed with this problem. He sees it as the most important question in the science of his day. "The edifice of science requires not only material, but also a plan," he says.

When Mendeleyev arranged the known 63 elements by atomic weight, they went from hydrogen (1), to lead (207). Today, hydrogen is still first, but lead is atomic number 82 out of 92 natural elements. By classifying chemicals, the periodic table not only brought order to chemistry, it allowed for predictions.

Lying on his desk is a letter from the cheese makers with an agenda for his visit. He drinks his morning tea and jots some notes on the back of the letter (the letter still exists with a ring from the teacup). He puts a few elements with similar properties on a horizontal line, including their atomic weights. Those weights vary widely. He sees no sense in them. On a second line, under the first, he lists a different group of elements and their atomic weights. Under that he puts a third group. Then his eye takes him vertically, straight down instead of across. Looked at this way, the weights, for no reason he can imagine, do seem to follow a pattern. Mendeleyev tells the driver of the carriage to leave. The cheese makers will be disappointed.

Mendeleyev, who often plays solitaire, now writes the atomic weights of all the elements on white cards—one element with its atomic weight on each card. He also puts properties of elements on the cards. Then he plays around with the cards, arranging them in rows and columns. He pins them to the laboratory wall and stares. He knows he is close, but he still can't quite get it.

Exhausted, he puts his bushy head on his hands and falls asleep at his desk. Then, in his words, "I saw in a dream a table where all the elements fell into place as required. Awakening, I immediately wrote it down on a piece of paper."

He draws a large rectangle with vertical columns and horizontal rows. In the columns, he lists the elements according to atomic weight—from light to heavy. In the rows, he lines up elements with similar chemical properties. There's a pattern! Reading from left to right across each row, the atomic weights increase by regular amounts. In a few places,

RUFFLING FEATHERS

Dmitri Mendeleyev was known as an expert on oil and chemicals. In 1876 he was invited to the United States, where he suggested that oil companies attempt to be more efficient rather than just keep looking for more sources of oil. Some experts are still making that suggestion.

A radical in his political beliefs, Mendeleyev allowed women into his lectures (although he wasn't so radical that he believed they are equal to men). He divorced and remarried, which was not approved of in czarist Russia.

Then, in 1890, he endorsed a petition to the Russian czar protesting his university's poor treatment of students. That was a mistake. The czar was a dictator; he didn't like protests. Mendeleyev was forced to resign and was never again allowed to hold a university job. No matter. The czar became a side note to history; Mendeleyev, a star.

where the pattern doesn't quite work, he has the imagination to place question marks. Mendeleyev sees his grid as a kind of jigsaw puzzle; he believes that missing elements will be found to fill the gaps. He calls what he draws a "periodic table"; he has taken chemistry's alphabet—the atoms of elements—and arranged them into words, sentences, and paragraphs.

To his critics, it all sounds like Russian mysticism. But when the elements gallium (now atomic number 31), scandium (21), and germanium (32) are discovered in 1875, 1879, and 1886 and fill spaces in the periodic table and have the properties Mendeleyev predicted, he becomes world famous.

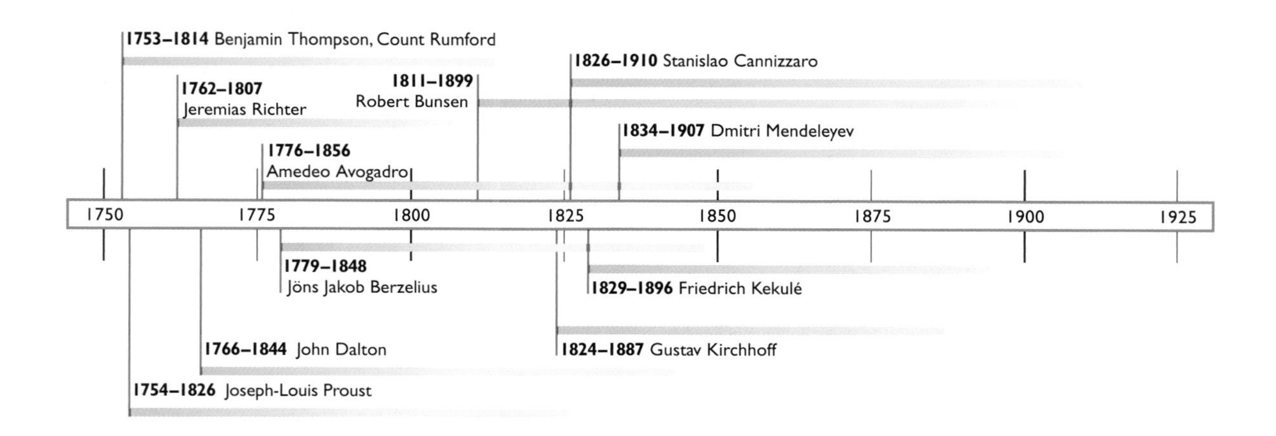

A Valiant Idea: Mendeleyev's Table

Critics asked Mendeleyev to explain a few atomic weights that didn't fit neatly into his table of elements. He didn't say, "I don't know" or "Maybe there's a mistake in my Periodic Law." He said, flatly, that the atomic weights were wrong. They had been miscalculated, he insisted.

Why was Mendeleyev so sure he was right and other brainy chemists were wrong? A key reason was valence. Like Edward Frankland, he had been a student of Robert Bunsen. He knew of Frankland's work establishing the idea of valence and accepted it fully. It was a key to his working out the periodic table. In nineteenth-century terminology, valence is the capacity of an atom to combine with other atoms. A hydrogen atom, with a valence of 1, can only combine with one other atom. Beryllium, with a valence of 2, can combine with two different atoms.

When Mendeleyev put the elements in order by atomic weight and then grouped them in rows by other properties, the columns fell neatly into place by valence. Hydrogen (H) is in the first column with lithium (Li = 7), which also has a valence of 1. Next to them is a column headed by beryllium (Be = 9), with a valence of 2. The valences continue in an astonishing pattern: 3 (boron), 4 (carbon), 3 (nitrogen), 2 (oxygen), and 1 (fluorine). This pattern (1, 2, 3, 4, 3, 2, 1) repeats throughout the elements listed (with a few exceptions).

That clear link between valence and weight, which couldn't possibly be due to chance, was what convinced Mendeleyev he had the order right. Except for hydrogen, which is in a class by itself, these elements are now listed in the second row of today's periodic table, from lithium (atomic number 3) to fluorine (atomic number 9). The last element in that row, neon, is one of the noble gases (all with a valence of 0), which weren't discovered until the end of the nineteenth century.

Today, we know that atoms combine—or not—because they lose, add, or share electrons (negatively charged particles). This discovery led scientists to replace the simple term *valence* (which has numbers 0 through 4) with *electronegativity* (which has a more complicated scale going up to 3.98).

но въ ней, мнѣ кажется, уже ясно выражается примѣнимость выставляемаго мною начала ко всей совокупности элементовъ, пай которыхъ извѣстенъ съ достовѣрностiю. На этотъ разъ я и желалъ преимущественно найдти общую систему элементовъ. Вотъ этотъ опытъ:

			Ti=50	Zr=90	?=180.
			V=51	Nb=94	Ta=182.
			Cr=52	Mo=96	W=186.
			Mn=55	Rh=104,4	Pt=197,4
			Fe=56	Ru=104,4	Ir=198.
			Ni=Co=59	Pl=106,6,	Os=199.
H=1			Cu=63,4	Ag=108	Hg=200.
	Be=9,4	Mg=24	Zn=65,2	Cd=112	
	B=11	Al=27,4	?=68	Ur=116	Au=197?
	C=12	Si=28	?=70	Sn=118	
	N=14	P=31	As=75	Sb=122	Bi=210
	O=16	S=32	Se=79,4	Te=128?	
	F=19	Cl=35,5	Br=80	I=127	
Li=7	Na=23	K=39	Rb=85,4	Cs=133	Tl=204
		Ca=40	Sr=87,6	Ba=137	Pb=207.
		?=45	Ce=92		
		?Er=56	La=94		
		?Yt=60	Di=95		
		?In=75,6	Th=118?		

а потому приходится въ разныхъ рядахъ имѣть различное измѣненie разностей, чего нѣтъ въ главныхъ числахъ предлагаемой таблицы. Или же придется предполагать при составленiи системы очень много недостающихъ членовъ. То и другое мало выгодно. Мнѣ кажется притомъ, наиболѣе естественнымъ составить кубическую систему (предлагаемая есть плоскостная), но и попытки для ея образованiя не повели къ надлежащимъ результатамъ. Слѣдующiя двѣ попытки могутъ показать то разнообразiе сопоставленiй, какое возможно при допущенiи основнаго начала, высказаннаго въ этой статьѣ.

Li	Na	K	Cu	Rb	Ag	Cs	—	Tl
7	23	39	63,4	85,4	108	133		204
Be	Mg	Ca	Zn	Sr	Cd	Ba	—	Pb
B	Al	—	—	—	Ur	—	—	Bi?
C	Si	Ti	—	Zr	Sn	—	—	—
N	P	V	As	Nb	Sb	—	Ta	—
O	S	—	Se	—	Te	—	W	—
F	Cl	—	Br	—	J	—	—	—
19	35,5	58	80	190	127	160	190	220.

Chemists refined this early periodic table by Mendeleyev (and so did he, in 1871), but it was well organized enough to foretell elements yet to be discovered. In the middle of the third column, with a confident "? = 68" and "? = 70," Mendeleyev predicted the discovery of gallium (atomic mass 69.723) and germanium (atomic mass 72.61 or 72.64). He also predicted they would share properties with the other elements in their rows.

Keep in mind: We've used *weight* here because that was the language of the time. Today's term is *mass*.

Two scientists (bottom of photo) are dwarfed by an illuminated periodic table of elements at the Lawrence Hall of Science in Berkeley, California. They're pointing at number 67, holmium, which is a silvery white metal.

By 1955, a modernized periodic table is hanging in almost every science classroom around the world. That year, a team of scientists at the University of California at Berkeley (including Albert Ghiorso and Glenn Seaborg), synthesize a new element with an atomic number of 101 and name it *mendelevium*.

In 2004, announcing the joint Russian-American discovery of two new elements, numbers 113 and 115, Dr. Joshua B. Patin says of the periodic table, "This is a working piece of art. We're not done yet. Nothing's been finished. What it could really mean down the road, nobody can tell. And that's the part that's exciting."

It sometimes takes a long time for the public to become aware of scientific findings. In the nineteenth century, most schoolchildren were still taught that the cosmos is made of those four "elements"—earth, air, fire, and water.

The Periodic Table: A Chemical Family Tree

In his book *Crucibles*, Bernard Jaffe writes, "The Russian peasant of his day never heard of the Periodic Law, but he remembered Dmitri Mendeleyev for another reason. One day, to photograph a solar eclipse, he shot into the air in a balloon, 'flew on a bubble and pierced the sky.'" Mendeleyev didn't know much about balloons, and he was alone in the rig, but he accomplished his mission and landed safely.

So far, we've discovered 92 elements in nature (see table). Smashing together the nuclei of atoms in a laboratory has brought forth others—from neptunium (93) to ununpentium (115) and counting.

Mendeleyev classified natural elements (both known and unknown) according to atomic weight and other properties. Hydrogen is the lightest—so it remains element 1. But variations of elements (called isotopes) have different atomic masses. So today, the periodic table is based on atomic number, the number of protons in the nucleus of the atom. Sodium, element 11, has 11 protons. *Always.* That's what makes it sodium.

It was a Dane, Niels Bohr, who guessed that atomic number determines an atom's (and therefore an element's) properties. His guess turned out to be on target!

These properties are like family traits, and the periodic table of elements is a kind of chemical family tree with branches in columns (called groups) and rows (called periods). As you look up and down each group, do you see any patterns in the

If the world is made up of a limited number of atoms and elements, how can there be so much diversity in matter?

numbers? Here's an interesting one: All the elements in group 1, headed by hydrogen (H), have odd atomic numbers. The next group, with beryllium at the helm, is entirely even. Then, group 3 is odd again. What's going on? Why does this matter?

In a normal atom, the number of protons matches the number of electrons in orbit around the nucleus. Hydrogen has one proton and one electron. All the elements in its group have one lone (and lonely) electron orbiting in the outer shell (inner shells can hold some electrons too). Single electrons strive to pair up, so the group 1 elements bond easily with other atoms. You'll recall that hydrogen is not H_1 but H_2—a pair of atoms sharing their lone electron so that they can both have two.

Even better than an even number, from the atom's point of view, is to have exactly eight electrons in the outer shell. The elements in the last group, 18, all have a complete set (except for helium, which only has two electrons). They're called noble (or inert) gases—and you can go ahead and think of them as snobs. They don't bond with other atoms (except in extreme conditions) because they don't need to. Helium won't even combine with itself.

"Odd seeks even and eight makes great"—such a simple idea. Yet it's at the core of what makes the periodic table work—and what makes elements behave the way they do.

UUP AND UUT

Standing in front of a cyclotron (a huge apparatus that accelerates particles at enormous speeds), a team of scientists—some American, some Russian—fired isotopes of the element calcium at the element americium. When they got lucky, a calcium nucleus (with 20 protons) fused with an americium nucleus (with 95) and created a new element bearing atomic number 115. It happened four times.

The element, never seen before, was named ununpentium (Uup). Seconds later, radioactive decay caused it to lose two protons and become element 113, named ununtrium (Uut). Both new elements, known as superheavies because of their enormous atomic mass, quickly decayed into known elements.

The scientists announced their new elements in 2004. Despite the short life, Uup and Uut are expected to help reveal secrets about the structure of the nucleus. According to *The New York Times*, "The discoveries fill a gap at the furthest edge of the periodic table and hint strongly at a weird landscape of undiscovered elements."

This table isn't finished yet. Elements higher than 92 are artificial—synthetic, or laboratory created—and scientists are still cooking them up. As we went to press (2005), numbers 117 and 118 had not yet been produced. Elements 112 through 116 were waiting for permanent names. (All the "Uu" names are placeholders.) The names darmstadtium (110), after its city of birth in Germany, and roentgenium (111), after the discoverer of X rays, had recently become official. More information on atoms and elements is coming in book three of this series, *Einstein Adds a New Dimension*.

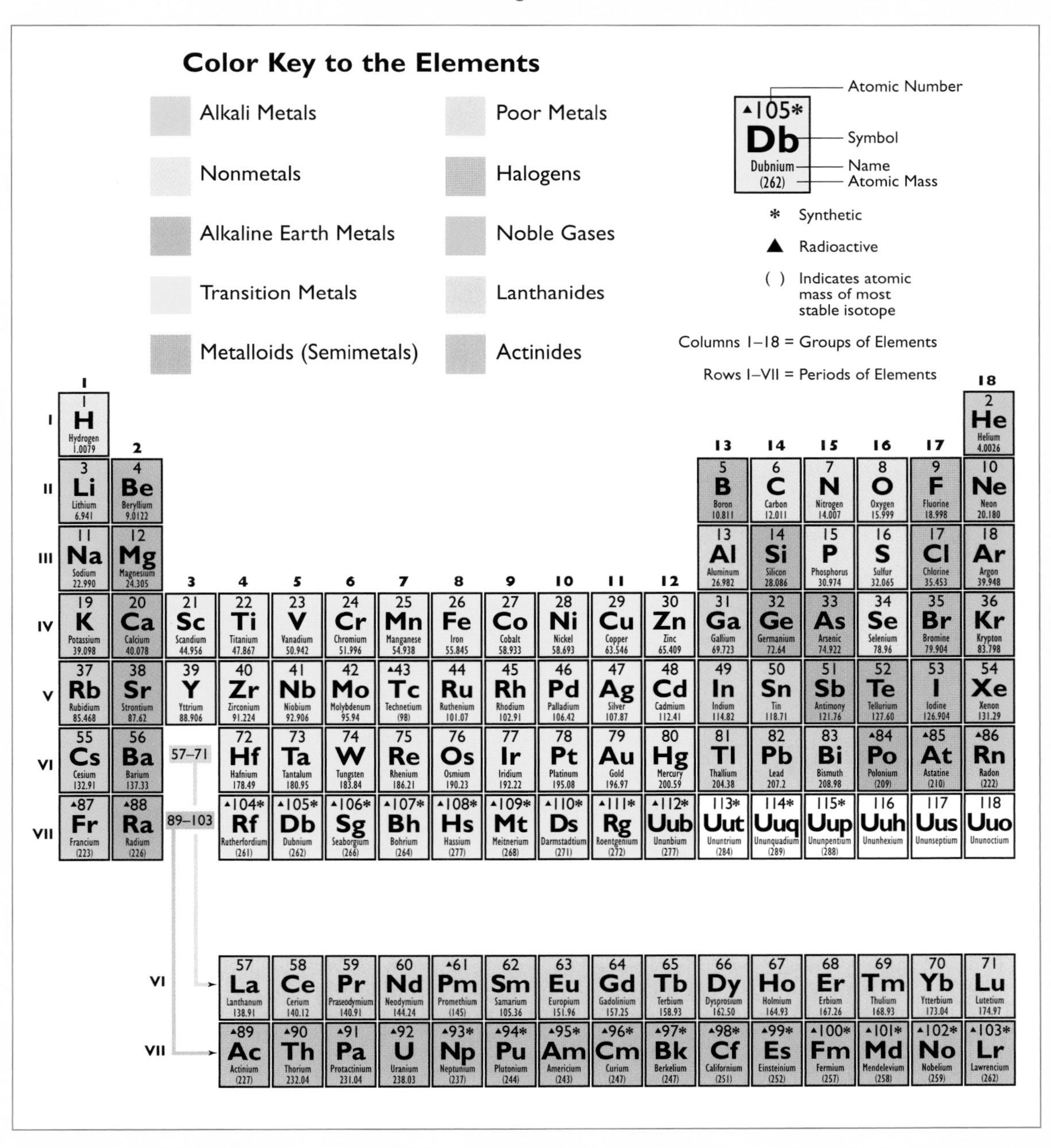

30 The Heated Story of an American Spy

We must look a long time before we can see.
—Henry David Thoreau (1817–1862), American naturalist and author, *Natural History of Massachusetts*

Being engaged, lately, in superintending the boring of cannon, in the workshops of the military arsenal at Munich, I was struck with the very considerable degree of Heat which a brass gun acquires, in a short time, in being bored; and with the still more intense Heat (much greater than that of boiling water as I found by experiment) of the metallic chips separated from it by the borer.
—Benjamin Thompson, Count Rumford (1753–1814), American scientist and inventor, "Heat Is a Form of Motion"

The...great barrier against the [idea that atoms exist]...was ignorance of the true nature of heat. At the beginning of the 19th century, opinion was divided. Some scientists thought that heat was a mechanical property...but others, perhaps a majority, subscribed to the notion that heat was a kind of vaporous fluid...that went by the name caloric.
—David Lindley (1956–), physicist and author, *Boltzmann's Atom*

Fill a teacup with hot tea. Put some ice cubes in a glass next to the hot tea. Let them stand. The tea will cool; the ice will melt. It's easy to explain what happens if you imagine heat as an invisible substance—"caloric"—that flows from hot to cold.

That caloric image is terrific, although it happens to be wrong. Heat isn't a substance. It is the vibration of molecules. It's a form of kinetic energy. Lavoisier's caloric idea, like the phlogiston idea, is all wrong.

An American makes that clear. He doesn't know about atoms and molecules, but he does figure out that caloric doesn't make sense. You won't find this Yank in most U.S.

Heat (what scientists now call "thermal energy") is a form of kinetic energy (meaning it's about motion). In a gas, atoms and molecules move wildly in all directions, and in a solid, they vibrate in place. In most other forms of kinetic energy (like a pitched baseball), molecules move in the same direction as the object that contains them. Today's scientists use *heat* as a verb, not a noun—as in "Heat up that pizza by adding some thermal energy."

history books, but visit Munich, Germany, and you can see him. He's a lean, dashing man with a military cloak on his shoulders, a knight's medals on his chest, a tasseled cane in one hand, and, in the other, plans for a great urban park. Cast in bronze, he stands with legendary self-confidence.

His name is Benjamin Thompson. He's a farm boy from Woburn, Massachusetts, who begins life in 1753 without a father. His mother doesn't have much money. What he has going for him is a good mind, driving ambition, a passion for learning—and, unfortunately, a sometimes unscrupulous nature. He will do anything he has to do to get ahead.

Count Rumford—an American scientist, inventor, and adventurer—looks dashing on a pedestal in Munich.

Of one thing Thompson is sure: he doesn't want to be a small-town farmer. So he manages to get working apprenticeships in Salem and Boston, the two most exciting cities in Massachusetts. There he grabs every opportunity he finds to read, study, go to lectures, learn languages, and do scientific experiments. At age 13, he is compiling notes on how to build firecrackers and rockets. One of his employers writes his mother that "he oftener found her son under the counter, with gimlets [hole borers], knife and saw, constructing some little machine, or looking over some book of science, [rather] than behind it . . . waiting on customers." He is fired; his next apprenticeship is to a doctor, who teaches him about medicine.

Here's the farm in Woburn, Massachusetts, where Benjamin Thompson was born. He didn't stay long. As a boy, he chose a more exciting life in the cities of Salem and Boston.

This famous engraving by Paul Revere, titled *The Bloody Massacre Perpetrated in King Street, Boston on March 5th 1770*, has been called America's first propaganda poster. The truth is that British soldiers shot some taunting colonists in Boston. This picture helped make American patriots angry enough to fight a war to be free from England's rule.

Thompson's interest in science is genuine and deep. He searches for opportunities to learn and finds a way to study with the Reverend Samuel Williams, who later will teach at Harvard. When he writes a scientific paper on the physics of electricity and builds an "Electrical Machine," the selectmen in the small town of Rumford, New Hampshire, ask him to be their schoolmaster. Three months later he marries the richest widow in New Hampshire. She is 33; he is 19. They set off for Boston in a fine carriage to buy him clothes suitable for a gentleman's life.

Thompson soon becomes a good friend of New Hampshire's young royal governor, who is also a science enthusiast. They go on a mountain-climbing expedition together and take careful notes of what they observe. But the American Revolution is brewing, and politics gets in the way of science. Thompson is a Loyalist (which means he is on the side of the British Redcoats). Quite a few Americans are Loyalists; many others are unsure of what is best for the colonies. But Thompson goes further than most other American Tories (another word for Loyalists): he is a spy for the British. To the patriots, those willing to fight Britain, that is treason. (One of his letters, written in invisible ink, still exists.) He could be hanged, but he manages to escape to England (leaving his wife and baby daughter behind, never to see them again).

In England he quickly makes powerful friends, including King George III. The king names him undersecretary of state for America's Northern Department. He is now a ranking British government official, but he still finds time to follow

THE SPY WAS A TRAITOR

Below is a page from Benjamin Thompson's secret message, written with invisible ink between the lines of an innocent-looking letter. It was sent to a staff member of General Gage's British army. Clearly, this was treason.

If you wanted to write in invisible ink at the time of the Revolutionary War, you'd probably use a mixture of ferrous sulfate and water. Ferrous sulfate ($FeSO_4$) is a compound of iron (Fe), sulfur (S), and oxygen (O). To read a hidden message, you either heated the paper over a flame (carefully) or treated it with sodium carbonate (Na_2CO_3). Both methods made the paper dark and brittle (as above), but also turned the ferrous sulfate brown, making the message visible. Both British and American spies used the technique.

...plan of operation they have formed—/ ...[D]unbar from Canada, & Ens. Hamilton of/ ...Reg [?] with their Servants are Prisoners in this/Town, But I have not permitted to see them tho' I/have made frequent applications for that purpose—/As to my own situation, it has been very disagreeable/since I left Boston, as upon my refusing to bear Arms a-/gainst the King I was more than ever suspected by the/People in this part of the Country—And it has been/with difficulty ... few friends that I have here/have more than once prevented my bein Asassinated./ I am extremely unhappy that my confinement to/ this Town (by this deluded People) should put it out .../power to do any thing for the good of the service, But .../soon to have an opportunity of giving convincing .../of my Loyalty to the King, & gratitude to all my benefac-/tors—In the mean time you will give me leave to as-/sure you in the most solemn manner possible, that neither/the threats nor promises of this wicked & Rebellious faction/shall ever induce me to do any thing contrary to my/professed Loyalty to his Majesty—But that on the contrary/I do with the greatest pleasure & alacrity dedicate my/Life & fortune to the service of my rightful sovereign/King George the third—/I am Sir with the greatest respect/ Your much obliged and/most obedient Servant/ (name cut out)/P.S. (name cut out) comes on purpose/to bring this & the Pistol you was kind/ enough to lend me—& I beg you would/be so good as to procure him a pass to/return

his scientific interests. His experiments—particularly with guns and explosives—are so impressive that he is elected a member of the Royal Society. (Yes, that's the society whose members included Boyle and Newton.)

King George III sends Thompson back to America—this

time as a lieutenant colonel to help Britain fight the patriots in the Revolutionary War. "To the inhabitants of Huntington, Long Island, Benjamin Thompson was the devil incarnate," his twentieth-century biographer, Sanborn C. Brown, will write.

> *He chose for the site of his fort two acres of land on which stood the First Presbyterian Church. He ordered the church razed* [torn down] *and its timbers used for constructing fortifications.... The gravestones were not used for construction, but for fireplaces, tables, and ovens.... Many of the inhabitants remembered being made sport of that winter by receiving from the soldiers loaves of bread with the reversed inscriptions from the tombstones of their deceased relatives.*

But no one is all good or all bad. One good thing about Thompson: he keeps busy. (Armies spend a lot of time sitting around.) While he is on Long Island, he designs a cork life preserver that lets a horse swim a river with a cannon on its back. And he invents a carriage that can transport a big cannon. The carriage can be taken apart, carried by three horses, and easily reassembled in a minute and a quarter.

When Thompson goes back to England, he makes himself out to be a great military leader—fair and very popular. And maybe he *is* with his men, but the Americans hate him. Whatever, King George III is pleased; the king dubs him a knight (which means he is now *Sir* Benjamin Thompson). But his ambitions still aren't satisfied, so he heads for the European continent, traveling with Edward Gibbon, who is writing a history of Rome (it will become famous). Sir Benjamin is supposed to do more spying for England, but he has other things in mind. In France, he gets the attention of Prince Maximilian. The prince, impressed, suggests that he visit Germany, and then he sends word ahead to his cousin, the duke of Bavaria.

The winds and waves are always on the side of the ablest navigators.... All that is human must retrograde if it does not advance.
—Edward Gibbon (1737–1794), *The Decline and Fall of the Roman Empire*

Benjamin Thompson made sure his portrait was painted by England's leading portrait artist, Thomas Gainsborough.

When Sir Benjamin arrives in Bavaria (a region in southeast Germany), he is in full military uniform riding astride a white horse. Before long, this knighted American seems to be running Bavaria: as a major general, chief of police, and all-round organizer. The man *is* talented. He puts soldiers to work draining the marshes near Munich, where he lays out a famous public park, the Englischer Garten. (It still exists.) He improves cattle breeds, encourages potato growing, reforms the army, reforms the civil service, founds a military academy, and tries to get Germans to eat more nutritiously. He rounds up all of Munich's homeless poor (2,600 people)—and puts them in a house of industry, where they are fed, clothed, housed, and trained. Some are given jobs producing goods (from which Munich profits); others find private jobs. "To make vicious and abandoned people happy, it has generally been supposed necessary, first to make them virtuous," he writes, explaining his actions. "But why not reverse this order? Why not make them first happy, and then virtuous?"

In his spare time, Sir Benjamin invents things: a double boiler, a drip coffeepot (see drawing at left), and a domestic stove—the first real kitchen range. (Since it both heats and cooks, it is more useful than the stove Ben Franklin invented.) He brings James Watt's steam engine into wide use in Germany. And he continues doing carefully documented and significant scientific experiments.

The Bavarians are awed. They make him a count of the Holy Roman Empire, which means he is now Count von Rumford. (He takes the name of the New Hampshire town where he had lived. Today that town is called Concord.)

Rumford sees an opportunity to make money and to improve Bavaria's artillery by manufacturing heavy cannons.

In 1799, Rumford used a very sensitive balance scale to show that there was no measurable change of weight when water was turned into ice or ice into water. Yet the heat exchange in that process was enough to turn a few ounces of gold from the freezing point to red hot. If caloric existed, it should have been measurable with his scale. "Heat," said Rumford, "... must be something so infinitely rare, even in its most condensed state, as to baffle all our attempts to discover its weight." Because he couldn't weigh it, he theorized that heat was not a substance but a "kind of motion."

(The French Revolution has made Europe's monarchs nervous; they are arming themselves.) One day he is watching a heavy, metal cannon barrel being drilled while it is in a tub of water. He can feel the heat produced in the process. He takes careful notes. In a report for the Royal Society in London he writes:

> *I perceived, by putting my hand into the water and touching the outside of the cylinder, that Heat was generated; and it was not long before the water... began to be sensibly warm... at 2 hours and 30 minutes it ACTUALLY BOILED. It would be difficult to describe the surprise and astonishment expressed in the countenances* [faces] *of the bystanders, on seeing so large a quantity of cold water heated, and actually made to boil, without any fire.*

Rumford notices that the metal can be drilled for hours (horses turn the drilling bore) and keep producing heat. If heat were a fluid, it would run out. But it never does. "Anything which any *insulated* body... can continue to furnish *without limitation*, cannot possibly be a material substance," he writes. He reasons that the motion of the bore rubbing against the cannon creates **friction**, which agitates the water and creates heat. It is the motion of the drilling process that produces heat, not an invisible liquid.

This delightful etching with watercolor, made in 1800, shows the benevolent inventor Count Rumford (Benjamin Thompson) warming himself in front of his stove. It's at the Royal Institution in London, which the count founded.

In 1798, Rumford writes a classic scientific paper called "An Experimental Enquiry Concerning the Source of the Heat which is Excited by Friction." In that paper he makes it clear: **Heat is related to motion.** There is no caloric. This is landmark science.

To finish Rumford's story: Eventually he leaves Bavaria and goes back to England. His invention—the Rumford stove—has made

FROM WHENCE CAME THE HEAT?

Rumford's famous experiment is good science—careful experimentation combined with solid thinking. Here's his description:

> *We have seen that a very considerable quantity of Heat may be excited in the Friction of two metallic surfaces, and given off in a constant stream or flux, in all directions, without interruption or intermission, and without any signs of diminution or exhaustion. From whence came the Heat . . . ?*
>
> *Was it furnished by the air? This could not have been the case; for, in three of the Experiments, the machinery being kept immersed in water, the access of the air of the atmosphere was completely prevented.*
>
> *Was it furnished by the water which surrounded the machinery? That this could not have been the case is evident: first, because this water was continually receiving Heat from the machinery, and could not, at the same time, be giving to, and receiving Heat from, the same body; and secondly, because there was not chemical decomposition of any part of this water.*
>
> *And, in reasoning on this subject, we must not forget to consider that most remarkable circumstance, that the source of the Heat generated by friction in these Experiments appeared evidently to be inexhaustible. . . .*
>
> *Anything which any insulated body, or system of bodies, can continue to furnish without limitation cannot possibly be a material substance: and it appears to me to be extremely difficult, if not quite impossible, to form any distinct idea of any thing capable of being excited and communicated in the manner the Heat was excited and communicated in these Experiments, except it be MOTION.*

That's Thompson/Rumford (above) in a Bavarian cannon foundry in 1798. He's pointing out that mechanical energy can be transformed into heat.

him wealthy. It burns wood or coal far more efficiently than any other stove and it doesn't fill a room with noxious smoke. Rumford is a hero in many households. He becomes interested in finding other ways to relate science to everyday life. In 1799, he founds the Royal Institution of Great Britain. (Don't confuse it with the elite Royal Society.) The Royal Institution, in London, is intended to show the general public how science can be used in their lives. In 1801,

What Rumford did was disprove a wrong theory—a major accomplishment—and an essential part of finding a right one.

Scientific Researches!...an Experimental Lecture on the Powers of Air!—this cartoon, also nicknamed *Fart Bellows*, pokes fun at a lecture at the Royal Institution demonstrating the effects of laughing gas (nitrous oxide). That's Count Rumford standing on the far right and Humphry Davy holding the bellows.

TURNING ON THE HEAT

A physicist friend of mine, John Hubisz, gets upset when I use *heat* as a noun, which is the way most dictionaries define it. "Heat is not a substance," he says. "It can't be a noun. We got rid of that idea when we got rid of caloric."

"I understand," I protest to him, "but we all talk about the heat in the oven, or the heat from the Sun. That noun *heat* describes our everyday use of the word."

Ah me, I don't get far with that argument. To a scientist, heat is action, and that makes it a verb, and only a verb. So be sure you are clear on its scientific meaning. And, in reading this book, if you see *heat* as a noun, you'll just have to understand it is a writer's way of using the breadth of the English language. If you're reading in a sunny chair, keep in mind, it is molecules in action that are toasting you—not an actual substance. If you're looking for a noun to use, my physicist friend (who likes to get in the last word) says, "Use *thermal energy*."

"I had this object . . . to determine whether the invisible rays which a warm body . . . gives out are not of the same character as those coming from the sun," said Rumford. (Are they?)

WAITING FOR ATOMS

Rumford could describe heat, but because he didn't understand atoms and molecules, he couldn't explain it. "I am very far from pretending to know how... that particular kind of motion in bodies which has been supposed to constitute heat is excited, continued, and propagated.... I shall not presume to trouble the [reader] with mere conjectures," Rumford wrote.

Heat is a form of energy, and no one in his time really knew what energy is, or that most of what goes on in the universe—from exploding stars to human growth to the action of machines—involves one kind of energy being changed into another. When that energy transfer occurs, heat is almost always a part of it.

Most people in Rumford's time ignored his theory and continued to believe that heat was a fluid called caloric. Antoine Lavoisier had come up with the idea of caloric. Lavoisier had died a martyr; he was a tragic hero. Rumford was arrogant and quarrelsome. Things like that can slow progress. Besides, science wasn't quite ready for this insight. Heat would have to wait until atoms were better understood. Then it would become a hot topic.

Rumford makes an inspired choice when he asks Humphry Davy to head the Royal Institution. Davy, a great lecturer and a good experimenter, attracts big audiences and turns the Royal Institution into a major popular scientific center. (A young man who attends Davy's lectures will be inspired to become a scientist. He will become world famous.)

Rumford also endows (it means "pays for") professorships in science: one at Harvard University, another at the American Academy of Arts and Sciences in Boston. (Could he be sorry about being a spy?) But he still has his arrogant ways; he quarrels with the directors of the Royal Institution and, in 1804, moves to Paris. In Paris he marries a very rich French widow: Her name is Marie Lavoisier.

In a letter to England's Lady Palmerston, Rumford writes, "I think I shall live to drive *caloric* off the stage as the late Lavoisier drove away *phlogiston*. What a singular destiny for the wife of two philosophers!!"

Rumford's achievements were many, his personality troublesome, his marriage to Marie Lavoisier unhappy. It didn't last long.

Atoms...were a complete unknown, and to explain something familiar yet enigmatic, such as heat, in terms of tiny, hard masses must have struck scientists of the early 19th century as too great a leap of imagination for them to follow.

—David Lindley, *Boltzmann's Atom*

31 A Shocking Science

The same generation that saw the development of steam power also saw the discovery of a means of transforming energy into ...a ready store...which could be delivered anywhere, in small amounts or large, at the push of a button. This form, of course, is electricity.

—Isaac Asimov (1920–1992), Russian-American science writer, *Asimov's Guide to Science*

No one would have thought up magnetism and electricity if their effects had never been detected; they are not a consequence of Newton's laws. So it's worth keeping an open mind regarding the existence of as-yet-undiscovered forces.

—Brian L. Silver (d. 1997), physical chemist and science historian, *The Ascent of Science*

You may remember Otto von Guericke. He's the German who built a vacuum pump, pumped air out of a sphere, attached a team of horses to each side, and had them pull. Guericke had a flair for drama and loved to do public demonstrations. (This was the mid-seventeenth century, and there was no TV to keep people indoors.)

In another popular experiment, Guericke put a ball of sulfur on a rod and turned it with a crank. Then he stroked the ball (with his hand or a piece of cloth) as it rotated. What happened was electrifying (literally). There were big sparks and a luminous glow. Guericke had built a friction machine that generated a static-electric charge—a greater charge than anyone had ever seen before (except from lightning). He liked to tie ribbons with threads to the apparatus; the threads would stick out straight. Guericke didn't know it, but he was showing that a charge moves on a conductor (the thread)—as current. (This experiment and those that follow are dangerous; don't try them.)

STATIC ELECTRICITY is an imbalance of positive and negative charges. Negative electrons separate from atoms, which then become positively charged. The electric shock you see, hear, and feel is those opposite charges getting back together. ("Opposites attract.") Some sources define static electricity as a buildup of electric charge. It's not. Before and after a shock, there are the same number of particles present; they just separate and then reunite.

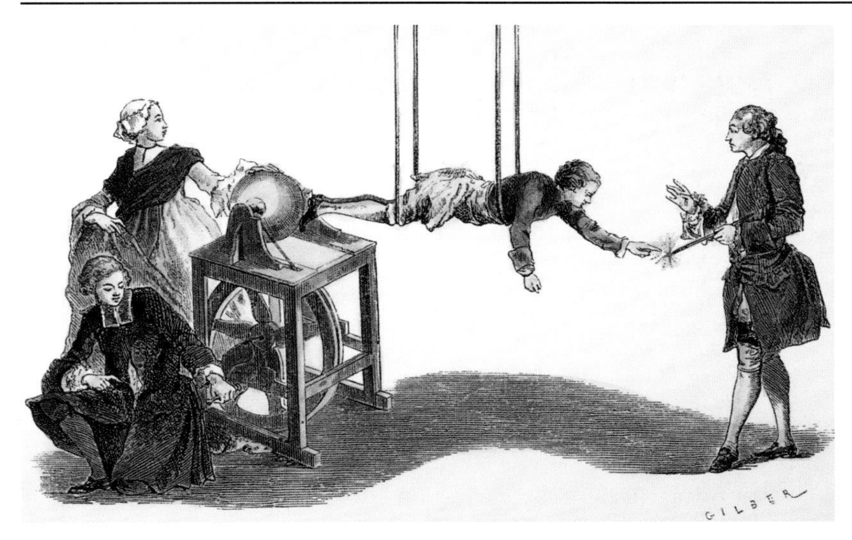

In eighteenth-century England, the wealthy and the curious bought electrical machines and endured shocks (as at left) for entertainment or for what they thought were health reasons. Women shot off miniature lightning flashes from whalebone corsets and gave their boyfriends shocking (and painful) kisses. Trees were given shots of electric current with the hope that they would bloom earlier. The Methodist preacher John Wesley carried his electrical machine with him as he made a circuit around England giving demonstrations. And the poet Percy Bysshe Shelley (1792–1822) kept an electrical machine on his desk. Sitting on an insulated stool, he charged himself until his long hair stood on end.

Some 70 years later, in the 1720s and 1730s, Stephen Gray does hair-raising experiments with static electricity. Demonstrating in front of big, awed English audiences, he shows that the human body conducts electricity. (He is lucky his assistants aren't killed!)

In one especially admired experiment, Gray brings a charged glass rod near a metal rod. A visible electric spark crackles as it jumps across the gap between the rod and the glass. Viewers find it astonishing. But does it have any meaning? Is there potential here for practical use?

Someone needs to see if this phenomenon can be explained. All of the experimenting so far has been with hard-to-contain static charges; a steady source of those charges is needed.

That's what leads to the invention of the Leyden (LY-dun) jar in the Netherlands in the mid-eighteenth century. It stores static-electric charges so they can be used at will.

Petrus van Musschenbroek, a professor at the University of Leiden (who meets Isaac Newton on a trip to England), makes one of these jars (in about 1745) and writes to a colleague describing the experience, "I would like to tell you about a new but terrible experiment, which I advise you

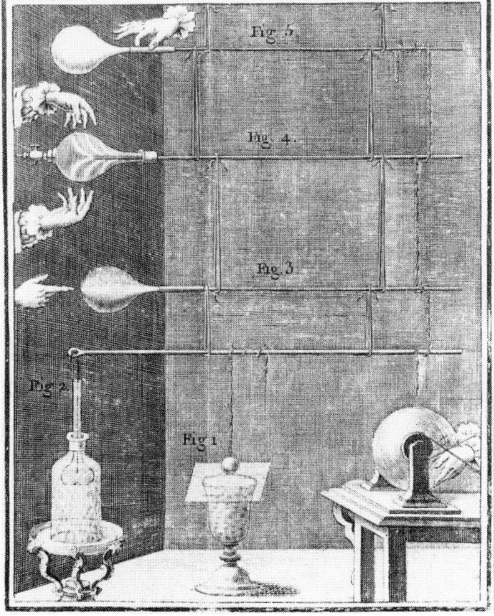

Feeling an electric shock is unpleasant, but in the early days of experimental science, curiosity won out over pain. Above is a friction wheel in the bottom right corner, among other "shocking" equipment.

THE POWER OF ATTRACTION

Amber—sticky, oily, tree resin that has hardened—is prized by jewelers for its beauty. It is actually a translucent fossil that often holds bugs and pollen. The ancient Greek scientist Thales noticed that amber has an interesting ability to attract feathers and fluff after it is rubbed with wool or fur. Other scientists wondered about that attractive power.

William Gilbert, who was physician to England's Queen Elizabeth I, took the Greek word for amber, which is *ēlektron*, and named

Amber (above) has attracted curiosity since ancient times. (That's an insect trapped inside.) In the painting (left), William Gilbert (1544–1603) demonstrates electricity to Queen Elizabeth I. Below is an eighteenth-century lodestone from Russia. Natural magnetic rocks were often carved and set into decorated brass frames.

that strange attractive effect "electricity." Gilbert found that some other materials, such as glass, also attract objects when rubbed.

It had long been known that lodestone, a form of the mineral magnetite (iron oxide), is a natural magnet. It attracts nails and anything else made of iron. When hung from a string, lodestone turns north, toward Earth's magnetic pole. That north-turning guided sailors who needed to know which way they were heading at sea. Ships' compasses were originally made of lodestone.

Magnetism was useful. Gilbert studied it for 18 years and in 1600 wrote a book about it. Why does a compass behave as it does? Gilbert concluded that the Earth is a giant magnet with northern and southern poles of attraction. But at that time, no one guessed that electricity and magnetism are related. Electrically charged amber doesn't attract nails, and magnetic lodestones don't attract feathers.

As to electricity, Gilbert didn't know any more about it than the ancient Greeks had. Too bad he couldn't stay around. By the mid-eighteenth century, electrical experiments were being performed all over Europe. Gilbert would have loved them.

There are plenty of ways to drum up static electricity—rubbing wool on glass, for example. A bigger challenge is to hold off the spark so that you can produce it on demand. That's what a Leyden jar does. In this early Dutch model (ca. 1745), a nail driven through the stopper carries a charge—say, a negative one (more electrons than protons)—to a metal lining inside. Meanwhile, a metal cup around the bottom of the jar holds the opposite charge (positive—more protons than electrons). The charges on either side of the glass attract and hold one another, building up, but the glass is an insulator (it keeps those charges separated). To create a spark, you use a conductor (like the two-pronged tool on the left of the jar) to bypass the glass. It connects the metal cup with the metal lining (by way of the nail and the stopper's knob), allowing the opposite charges to flow together and cancel out one another. Leyden jars are what today's electricians call a "capacitor"—an insulator sandwiched between two charged plates, one positive and one negative. (Your body is a conductor, too, but never use your hands to discharge a Leyden jar or a capacitor; the shock can be severe.)

never to try yourself, nor would I . . . do it again for all the kingdom of France."

Despite the warnings, experimenters around the world begin playing with static electricity. It is fun. Sometimes it is shocking. Too strong a charge, and the Leyden jar explodes. A few experimenters get carried away—to graveyards. But there is no going back. This marvel is too intriguing to ignore.

As the king of France watches, 180 soldiers are arranged so they hold hands in a circle. One of the soldiers is given a Leyden jar. He touches its knob and then he touches his neighbor—and one after another, all 180 soldiers are jolted. They leap into the air. The company has been shocked, and the king is amused.

A scientist/priest, Jean-Antoine Nollet, has arranged the experiment. It is 1746, and a lady admirer writes of Nollet to a friend, "Only the carriages of duchesses, peers, and beautiful women are lined up outside his door. Here, evidently, is sound philosophy which is going to make its name in Paris."

Sound philosophy? She means science when she says "philosophy." But is there a scientific principle in those jumps and shocks? No one knows. On a dry day, when you rub your feet on a carpet and then touch metal, you get a shock. That's static electricity. It's the form of electricity you get by rubbing glass or amber. A Leyden jar can hold and release the energy (all at once, not gradually).

In 1747, a committee of prominent English citizens decides to observe this new happening for themselves. A wire is stretched across the nearly quarter-mile-long Westminster Bridge, near the Palace of Westminster in London. On both sides of the bridge, the wire goes to the riverbank. On one

side, a man holds a Leyden jar, which is attached to the wire. He also holds an iron rod. He puts the rod in the water.

On the opposite bank, a man holds the other end of the wire. He too has an iron rod. That second man dips his iron rod into the river and—ZAP!—both men leap into the air. An electric charge from the Leyden jar has crossed the Thames River on a piece of wire. This is one of the first demonstrations showing that electricity needs a circuit in order to flow. Is there anything useful in this "electric" effect?

Benjamin Franklin thinks so. He believes there is more to electricity than parlor tricks. Franklin does his experiments carefully: he knows the dangers of electricity. He says that "electric fire is a common element." He means electric *charge* when he says "electric fire," and it isn't an element, but Franklin is a first-rate scientist. He notices that when a Leyden jar is discharged, it

In 1762, Mason Chamberlain painted Ben Franklin's portrait (above). A year later, Edward Fisher made engravings of it, which Ben Franklin sent from England to America to friends and others he wanted to impress. Those are electric bells in the upper left corner. To the right of Franklin is a scene contrasting the dangers of lightning with the protection of his lightning rod. In *The Autobiography of Benjamin Franklin* (1793), Franklin wrote: "What gave my book the more sudden and general celebrity, was the Success of one of its proposed experiments... for drawing lighting from the clouds. This engag'd the public attention every where." On the right is Franklin's famous kite experiment, depicted by the popular printers Currier & Ives.

gives off a spark of light and a crackle that make him think of lightning and thunder. Could a Leyden jar capture static electricity created by a thunderstorm? Franklin plans an experiment with a kite and a key to find out. French experimenters try it before he can and they prove that the electric charge in the stormy sky is the same as that from a friction machine. (Franklin, being a master of public relations, ends up getting all the glory.)

Think of LIGHTNING as a big static-electric shock across the sky. Rain separates charge in the clouds (often negative, with too many electrons) from charge on the wet surface of Earth (often positive, with too few electrons). Clouds and Earth form a giant Leyden jar or capacitor. When the electric field between them gets too large, the air breaks down, becomes a conductor and—boom! An upward stroke of lightning travels at almost half the speed of light. The downward stroke is slower.

Franklin's analysis of "electric fire" astonishes Europe. Ben's friend, the English chemist Joseph Priestley, compares America's philosopher to England's scientific idol, Isaac Newton. That the infant colonies can produce a scientist who thinks as profoundly as Ben Franklin does helps make the colonies worthy of respect. America's poets Philip Freneau and Hugh Henry Brackenridge join a chorus celebrating their hero:

. . . we boast
A Franklin, prince of all philosophy,
A genius piercing as the electric fire,
Bright as the lightning's flash, explain'd so well
By him, the rival of Britannia's sage.

That's overstating things. Franklin isn't Newton ("Britannia's sage"), but tying static electricity to lightning does link action in the heavens to that on Earth—which extends Newton's ideas.

Franklin isn't the only one doing serious experiments. A French chemist, Charles-François de Cisternay Du Fay, shows that two electrified glass rods repel each other. So do two electrified amber rods. But an electrified glass rod attracts an electrified amber rod. What does that mean?

Franklin suggests that when glass is rubbed, electricity flows into it, making it "positively charged." When amber is rubbed, electricity flows out of it, making it "negatively charged." He is hypothesizing, and he has it about right (the flow of electric charges is actually the reverse of what he thinks, but the basic idea is correct).

We still use Franklin's words—*positive charge* and *negative charge*—to describe electricity. Remember: Unlike charges (opposites) attract, and like charges repel each other.

All matter has electrical properties because all atoms contain electrons (negatively charged particles) and protons (positively charged particles). The atoms in metals give up their electrons easily, which is what makes them good conductors (carriers) of electric current; it's also what makes them shiny and metallic looking. The electric energy we use to power our homes zips along wires in waves through a slow-flowing sea of loose electrons—electrons in the wires that have escaped from their atoms.

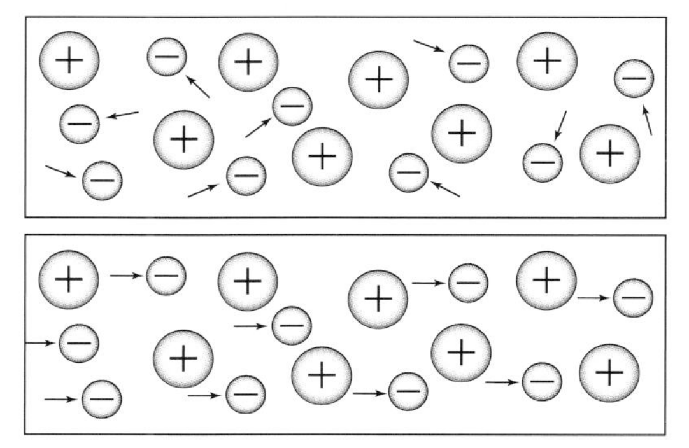

Another Frenchman, Jean Théophile Desaguliers (duh-sah-gyoo-lee-AY), discovers that electric current flows freely through some materials (like metals), which he calls "conductors." The current doesn't travel freely through other materials (like amber and glass), which he terms "insulators."

Franklin describes electricity as a "fluid." He thinks of it as invisible and flowing like water in a pipe. Franklin guesses that particles of electricity might exist in the flow. It is a good guess. Electrons—which are still unknown—are discrete (separate) particles of electricity with electric charges.

Benjamin Franklin was a big-time promoter. He built an array of models to show off his lightning rods. He connected a "thunder house" model to a Leyden jar (second from left). An electric charge from the jar sparked flammable gas that blew off the roof of the house. After he put a lightning rod on the house, there was no explosion.

Sensible Ben Franklin comes up with a practical idea. People are often killed when lightning hits homes or public buildings. No one knows what to do. It is common practice to ring church bells during thunderstorms to ward off evil spirits. But that doesn't work. Church-bell ringing is a killer profession (several hundred bell ringers have died in Germany because the metal bells, usually in the highest building in town, attract lightning).

Franklin notices that a Leyden jar is discharged more easily if a sharp needle is attached to it. More than that, with the needle in place, the jar can't be charged. Franklin uses that observation to design metal rods that can carry the electric charges in lightning safely to the ground. His lightning rods save lives and property, but at first, he is castigated for them in Europe as well as at home. When an earthquake strikes Boston, Franklin gets blamed. It is said that his rods steal lightning from the skies and drain it into the earth (which is what they do, but that isn't the cause of an earthquake).

Tall buildings, especially church steeples, were common targets of lightning strikes. Franklin's rods changed that. An 1820 French village (above) was so enamored with the long metal spikes that they put them on the church, on houses, and in the ground all around the town.

A well-known French priest attacks Franklin, saying it is impertinent to guard against the "thunders of heaven." Ben answers, "Surely the thunder of Heaven is no more supernatural than the rain, hail or sunshine of Heaven, against . . . which we guard by roofs & shades."

Franklin says that an electrical charge can neither be created nor destroyed. It just goes somewhere else. That is a conservation law, and he is right. Nothing is lost or gained in nature. The total charge in the universe remains the same.

Research in the field begins to heat up. In 1791, an Italian doctor, Luigi Galvani, finds that the muscles in a pile of frogs he is dissecting jump about when touched by two metal sticks with unlike charges. The same thing happens when he experiments with human corpses. (The English word *galvanize* comes from Mr. Galvani and his experiments.) He thinks electricity may have something to do with animal magnetism. He's wrong.

GALVANIZE means to stir up, or shock into action. It can also mean to shock with an electric current. Or it can mean to coat iron or steel with rust-resistant zinc.

Franklin became the modern incarnation of Prometheus, the Greek hero who aroused Zeus's fury by stealing fire from heaven. . . . Prometheus also represented freedom from political oppression. . . . [Franklin] the revolutionary politician became idolized in France for "snatching lightning from the sky and the scepter from the tyrant."

—Patricia Fara, *An Entertainment for Angels: Electricity in the Enlightenment*

Experimenters argued about why the leg muscles of dead frogs twitched when touched by two metal rods (right). Luigi Galvani figured out that the two metals rods formed a battery, and the resulting current caused contractions in the muscles.

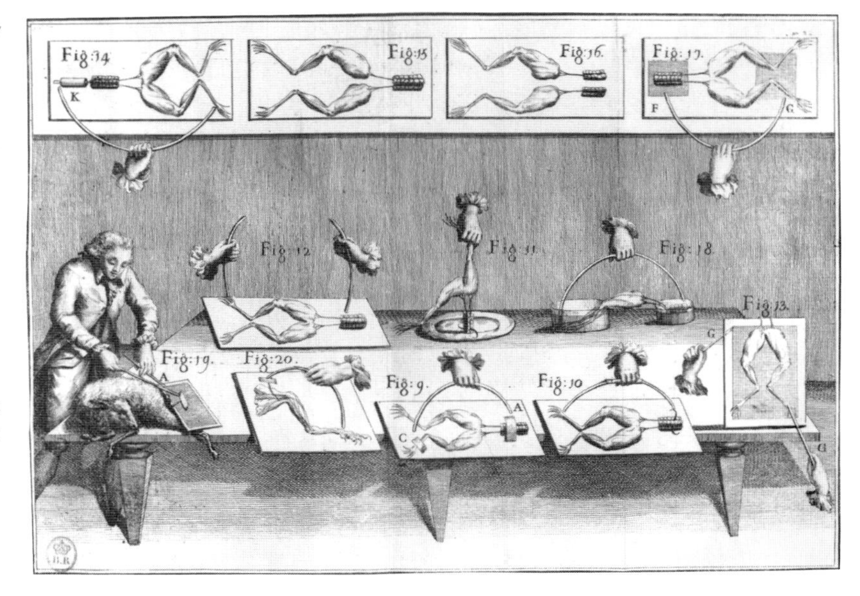

The idea of voltage came out of Volta's work. Think of a battery as a kind of pump for electrons: it gets them moving around a circuit. Voltage (V) is a measure of the strength of that push—called electric potential. Flashlight batteries have a potential of 1.5 volts; many car batteries have about 12 volts. An electric eel can build up about 500 volts. A high-tension transmission wire may have as many as 765,000 volts. At the other end of the scale, electrocardiographs (instruments for measuring heart-muscle activity) detect millivolts.

A friend of Galvani, Alessandro Volta (1745–1827), realizes that it isn't muscle tissue generating a charge; the source of the electric effect is Galvani's two metal rods. Volta experiments with metals soaked in simple solutions, which creates a chemical reaction that produces an electric charge.

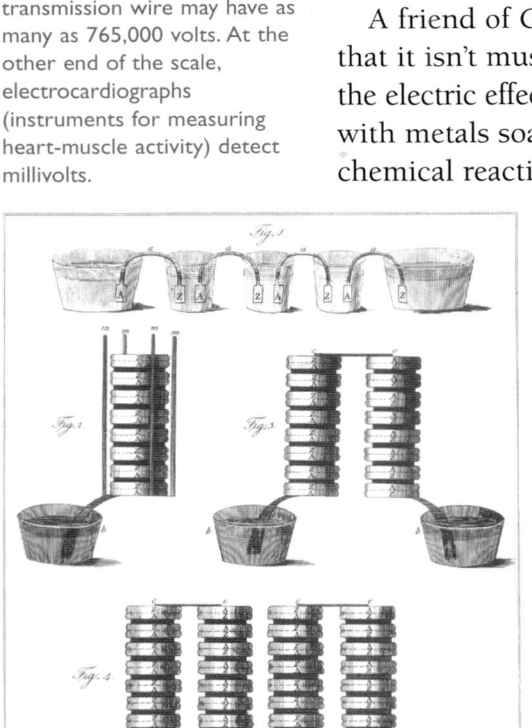

Alessandro Volta drew these voltaic piles, or batteries, with one, two, and four cells. The word *battery* means a set or series of something, and each "metal sandwich pile" is a set of alternating disks that can be connected in a series.

Volta sets out a row of cups and connects them with metal strips. Each cup holds alternating slices of copper and zinc soaking in salt water or a dilute (weak) acid. Those metal sandwiches yield a continuous current. He has made the first electric cell (two or more connected cells make a battery). Named a "voltaic pile" (see illustration at left), it gives scientists a reliable source of electric power. Instead of rushing out all at once, as the charge does from a Leyden jar, it flows steadily, like a river. Volta invents the phrase "electric current."

In 1800, two British

Electricity and Chemistry Get Hitched

Batteries have a disadvantage: The chemicals in them get used up. Still, the voltaic pile was the source of electricity for commercial use until well into the 1800s. The simple pile below (ca. 1800) is a stack of copper and zinc plates separated by sheets of cardboard soaked in a salt solution. To make electric current flow, the user connected the top and bottom of this crude battery with a wire.

Humphry Davy used a voltaic pile to spark his electrolysis experiments. In 1807, he built one with 250 metallic plates—the strongest battery of the day. Among other electrifying feats, he ran electric current through potassium carbonate (potash) to isolate the element potassium.

With his experiments, Davy united chemistry and electricity into the new science of electrochemistry. He hypothesized that electrical energy is more than simply related to chemicals: "May it not be identical with it, and an essential property of matter?" (It may.)

These tools for electrolysis are similar to Humphry Davy's equipment. They include a voltaic pile, on the left, along with an electrolysis cell (ca. 1810) and an electrolysis jar (ca. 1820).

scientists, William Nicholson and Anthony Carlisle, learn of Volta's experiments and build their own voltaic pile. When they place electrodes (positive and negative poles) from that electric battery in water, bubbles of hydrogen and oxygen form. They have created a chemical reaction using electric current, a process that will soon be common in chemical research. (It will be called "electrolysis.")

In 1807, Humphry Davy constructs the most powerful battery built so far. Davy, brought to London's Royal Institution by Count Rumford, uses electric current to break down compounds into elements. When he charges molten (liquid) potash, potassium emerges. Pure sodium comes from lye (sodium hydroxide). Using electrolysis, Davy isolates six new elements—potassium, sodium, calcium, strontium, barium, and magnesium. (Some of those elements had been

Sir Humphry Davy
Abominated gravy.
He lived in the odium
Of having discovered
sodium.
—Edmund Clerihew Bentley

Odium is a strong dislike. As to sodium, it was Davy who discovered in 1807 that it is an element. The connection between sodium and gravy? You can figure that one out.

Hans Christian Oersted is discovering that a magnetic needle moves when placed near a flowing current (above). This illustration is from an 1867 book, *The Marvels of Science.*

found before, but always in compounds. When Davy isolates them with electrolysis, he makes it clear that they can't be chemically broken down further and, therefore, are indeed elements.) Those electric jolts attract public attention to chemistry. Lots of people want to know more about the new sciences.

In Denmark, Professor Hans Christian Oersted (1777–1851), a friend of storyteller Hans Christian Andersen, conducts an experiment to prove to his students that there is no connection between magnetism and electricity. He holds a wire with electric current over a magnetic needle. He is sure the electricity will have no effect on the needle—but it does. The compass needle moves. Oersted understands at once that what is happening is the opposite of what he expected. He has demonstrated that electricity and magnetism *are* related.

When Oersted publishes his work (in 1820), he shakes the scientific world and sets off a surge of experimentation. It is quickly clear to thinkers and tinkerers that there is an exciting connection here. That unexpected link, between electricity and magnetism, will take science in a new direction. William Sturgeon, a soldier in the British Royal

THIS BOY WAS MOSTLY SELF-TAUGHT

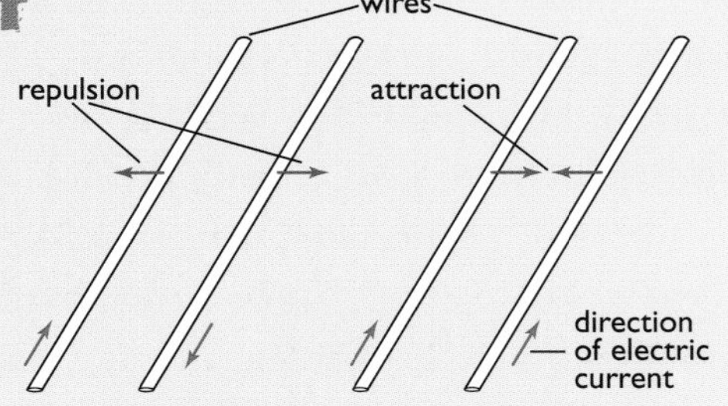

A French prodigy named André-Marie Ampère (1775–1836)—who mastered all known mathematics by the time he was 12—heard about Oersted's experiment and did his own research. Ampère showed that *two parallel wires carrying electric current attract each other if the currents flow in the same direction.* They repel each other if the currents flow in opposite directions. This is another link between electricity and magnetism. Ampère discovered that when current flows through a wire twisted into a helix (like a spring or corkscrew), the magnetic effect is strengthened with each turn of wire. Today, his name is a unit of electric current—ampere (or more familiarly, amp). When Ampère was 18 his father was guillotined along with 1,500 fellow citizens in Lyon, France.

Meanwhile, in the United States

As a boy in Albany, New York, Joseph Henry didn't do well in school, so he decided to write plays and be an actor. He grew up and did both. Then he read a book about science, and that changed his life's directions. He went to Albany Academy, did research on electricity, got a job at the College of New Jersey (it became Princeton University), and built an electric motor.

Soon after, an Englishman named Michael Faraday made a motor independently. Faraday published first, so he got most of the credit.

In 1835 Henry developed a switch, the electric relay, which allows current to go from one device to another. It made long-distance telegraphs possible. When he was 49, Joseph Henry was named the first director of the Smithsonian Institution. He was a model administrator and the first great American scientist after Ben Franklin. Tall, handsome, and healthy, Henry was still doing significant research when he was 80.

Joseph Henry claimed his giant Yale magnet (built for Yale while he was at Albany) could lift 2,063 pounds of weight, making it "probably, therefore, the most powerful magnet ever constructed." Henry became the first director of the Smithsonian Institution in 1846.

Artillery, learns of Oersted's experiment, winds an electrified wire around an iron horseshoe, and makes a magnet that lifts 4 kilograms (9 pounds)—20 times its own weight—as long as the current is running. Sturgeon has made the world's first "electro-magnet." (The word was hyphenated at this point in history. We're going to get rid of the hyphen soon.)

American physicist Joseph Henry builds another one. He wraps insulated wire around an iron core. That insulation means he can use many more turns of wire and not worry about one short-circuiting another. The more turns of wire, the stronger the magnet (as long as the current is running). Henry, being American, thinks big. In 1831 he uses a very long coil of wire and a huge piece of iron and creates a giant electro-magnet that lifts more than 1 ton of iron.

The science of electricity is charging ahead. And now, magnetism is moving with it. This is more than fun: it holds promise for the future.

In London, a bookbinder with an inventive mind ponders the connection between electric power and magnetism. This same young man is given tickets to hear Humphry Davy speak at the Royal Institution. Those lectures will change the direction of his life and, through him, the path of science.

A PENDULUM'S PROOF

Pierre-Simon Laplace was an assistant to Lavoisier, working in thermochemistry. Then he turned to astronomy. At a time of political turmoil, he got involved in politics and became a senator and was even made a marquis (it's a bit like an English lord). But it was his mathematics that brought him world renown.

ay back in the early seventeenth century, Cardinal Bellarmine, who took part in the prosecution of Giordano Bruno and was an important player in the trial of Galileo, wrote a letter to a colleague saying that if proof could be found that the Earth rotates, the church would change its view. It was a challenge. Galileo tried to find proof of that rotation; so did Descartes, Newton, Hooke, and others. Then one Frenchman, a mathematician, Pierre-Simon Laplace (1749–1827), wrote this in 1796:

> *Isn't it infinitely more simple to suppose that the globe we inhabit rotates about itself than to assume that a mass as considerable and as distant as the Sun moves with the extreme speed necessary for it to turn around the Earth in one day?*

But common sense is no proof, and Laplace knew it. He continued: "The rotation of the Earth must be established with all certitude [certainty].... A direct proof of this phenomenon should be of

That's Jean-Bernard-Léon Foucault in a daguerreotype, an early type of photographic image made on a light-sensitive, silver-coated metal plate.

Below is the first photograph of a sunspot. It was shot on April 2, 1846, by Foucault and his associate, Armand Hippolyte Louis Fizeau. Sunspots were discovered during that amazing year in astronomy, 1610, soon after early telescopes were first used for stargazing.

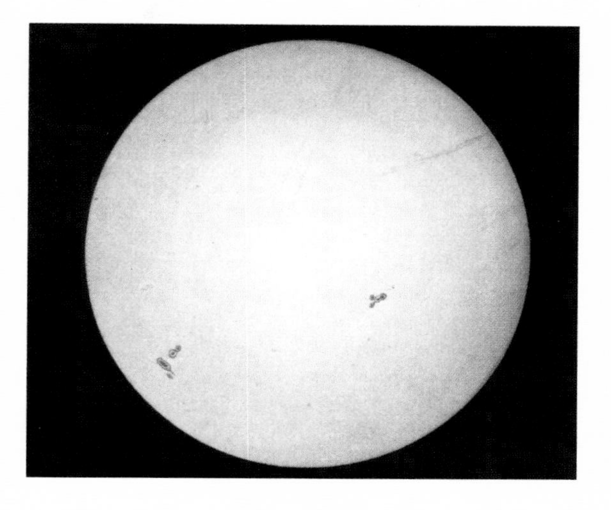

interest to geometers and physicists alike." That was an understatement.

By the mid-nineteenth century, the idea that the Earth rotates was generally accepted. That's what schools were teaching and what most educated people believed. Still, no one had actually found proof of it. Jean-Bernard-Léon Foucault (foo-KOH), a French journalist, thought he could do it.

Foucault (1819–1868) was self-educated when it came to physics. He wrote articles for a scientific journal, and he was a natural engineer and a genius at making things. He got involved with the new field of photography, and in 1845 Foucault and a friend took the first known photograph of the Sun. Besides that, he measured the velocity of light with a toothed wheel and a mirror that gave better results than any instrument used before.

Foucault's mother had wanted him to be a doctor, and he started out on that career path. It turned out to be a mistake: he got sick at the sight of blood. So he lived by his wits, which is what most writers and inventors do. Then in 1851, Foucault, who was quietly self-confident, became a celebrity. This is how he did it:

He hung a 67-meter (about 220-foot) pendulum from the ceiling of the Panthéon, a huge domed building then used as a church in Paris. (Emperor Napoléon III, an amateur scientist, helped arrange it.) Foucault's pendulum consisted of a steel wire with a heavy iron ball as a bob. The bob had a spike on it just long enough to make a mark in sand sprinkled on the church floor. The pendulum had been installed so it could swing freely without anything interfering with it. (That included the movement of the Panthéon itself, which would be rotating along with the Earth.)

Invited guests were watching as Foucault began his experiment. He pulled back the steel wire and tied it to the wall with a cord. Every effort had been made to see that the room was free of vibrations. The cord was lit with a match. (If it had been cut with a knife, vibrations might have spoiled the experiment.) The cord sizzled, the steel wire broke away, and the pendulum began to swing. Pendulums oscillate in a straight line, back and forth, and stick with that plane (unless nudged). Would this one behave differently?

The audience stood and stared, holding its collective breath. As they watched, the spike marking the sand went back and forth, and then it seemed to change direction. Foucault later described what happened in an article in the newspaper *Journal des Débats*:

> *After a double oscillation lasting sixteen seconds, we saw it return approximately 2.5 millimeters to the left of its starting point. As the same effect continued to take place with each new oscillation of the pendulum, this deviation increased continuously, in proportion to the passing of time.*

Foucault had put a large wooden disk on the floor of the Panthéon and marked it with degrees and quarter degrees. The sand was on top of that disk. The audience watched the bob etch long, narrow ellipses in the sand as it moved clockwise around the disk. The bob was eventually back where it started (having marked a 360° circle). It was doing exactly what Foucault

A Parisian crowd came to see Foucault's pendulum experiment. They knew they were seeing scientific history.

As a pendulum swings back and forth, the Earth rotates underneath it, tracing a pattern of rays like the one below. A computer is drawing a perfect pendulum pattern by simulating the motion in the absence of air resistance, which, in real life, eventually causes the pendulum to slow and stop.

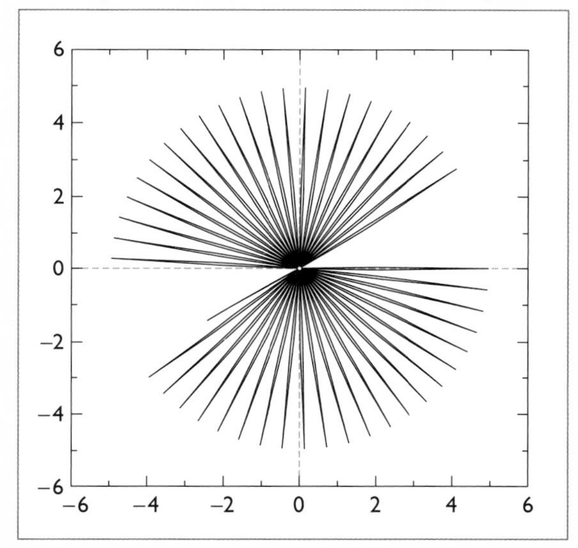

had predicted. The pendulum was not changing its orientation (even though it seemed as if it were). It was going straight back and forth, as pendulums do. **It was the Earth that was rotating under the pendulum.** The pendulum is "fixed in absolute space," said Foucault, "while . . . we and the planet rotate under it."

There's more to this story. Foucault came up with a mathematical formula to explain the action of the pendulum. That ruffled the feathers of Paris's mathematicians. Here was someone outside the academic community able to do something they hadn't been able to do. If you research Foucault, you'll read about jealous mathematicians and popular excitement worldwide. See if there is a Foucault's pendulum in a university or public building near you. Watching one is an awesome experience. (For a good book on the subject, read *Pendulum*, by Amir D. Aczel.)

What happens to a pendulum's swing when the Earth isn't steady? On February 28, 2001, an earthquake occurred under the Pacific Ocean, off the coast of Washington state. This pendulum (right) was hanging in a store in the city of Olympia. The Earth's sudden motion distorted the normal path of the swing to produce this unique pattern in sand (below). The "rose" pattern in the center was probably made during the aftershocks of the quake.

Michael Faraday Has a Field Day

It is the great beauty of our science...that advancement in it, whether in a degree great or small, instead of exhausting the subject of research, opens the door to further and more abundant knowledge, overflowing with beauty and utility, to those who will be...undertaking its experimental investigation.

—Michael Faraday (1791–1867), English chemist and physicist, *Experimental Researches in Electricity*

All the world's seemingly separate forces were slowly, majestically, being linked to create this masterpiece of the Victorian Age: the huge, unifying domain of Energy.... Faraday's vision of an unchanging total Energy was often felt to be...proof that the hand of God really had touched our world.

—David Bodanis (1956–), American science historian and author, *$E=mc^2$*

It is January 19, 1844, and the acclaimed Michael Faraday is giving a lecture at the Royal Institution (the place that Count Rumford founded). He asks his audience to do a thought experiment and imagine the Sun all by itself in space. Then he asks them to imagine Earth suddenly dropped into its present place. Will the Sun's gravity be there, ready to keep Earth from falling forever? Or will a special gravity force or "message" zoom out from the Sun to hold the Earth in place?

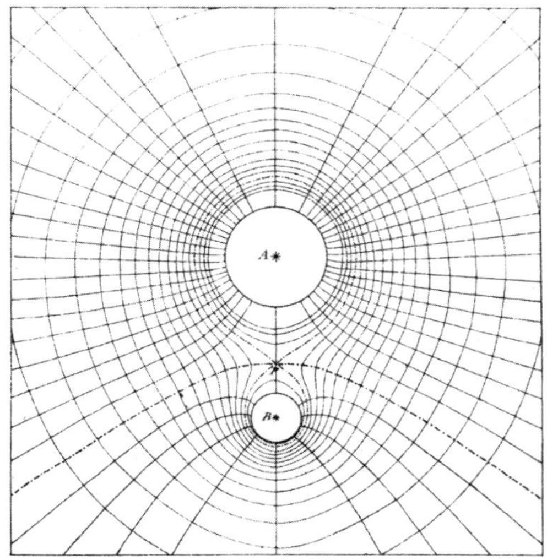

James Clerk Maxwell (coming in the next chapter) gave mathematical form (above) to Faraday's idea of a force field.

Faraday says that gravity's web—its *field* of force—will be there, *is* there, extending outward, and that any planet or asteroid placed in that field immediately disturbs and responds to it. There is no need for a special message from the Sun.

Faraday is on to something—something important. It's the

concept of fields. Everyone knows that the force of a magnet goes beyond the magnet itself. But how far does it go? Faraday says that tangible magnetic or electrical or gravitational fields extend right through to the end of space and can be tapped into and harnessed. That idea of fields is something to think about. Faraday is suggesting a new way of looking at the universe. Who is he? And from where does his inspiration come?

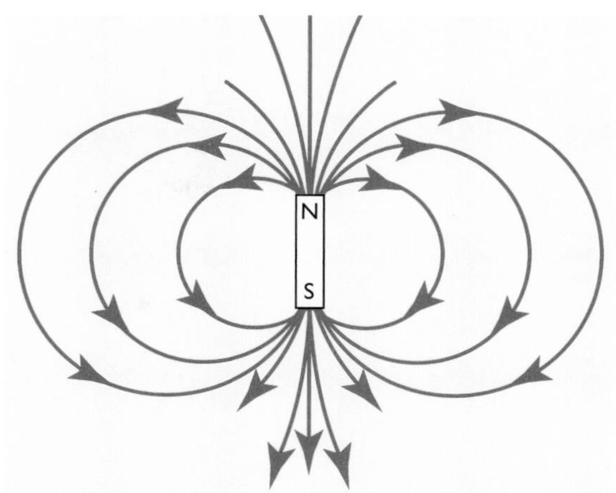

In the simplest terms, Earth can be thought of as a dipole (two-pole) magnet. Magnetic field lines radiate between Earth's north and south magnetic poles just as they do between the poles of a bar magnet. Charged particles become trapped, moving along the field lines in the pattern shown above.

Michael Faraday is an English blacksmith's son who, as a boy, doesn't have it easy. His father is often ill, and there is barely enough food to feed the children. But it's a big, loving, pious family. Michael learns to read and write and do basic arithmetic in what he calls a "common school." He will never get a high school or university degree, and, in often-snobby England, that will be held against him.

At age 13, he finds work as an errand boy for George Riebau, a London bookdealer with a shop just off Baker Street, not far from where the Faradays live. Riebau, a French immigrant, is kindhearted. When Faraday has free moments,

In the nineteenth century, the London cityscape was dominated by St. Paul's Cathedral and the Thames River (on the right). As a boy, Faraday carried a newspaper from home to home. Sharing papers was one way working-class families saved money.

Young Faraday saw Alessandro Volta's voltaic pile (right) in a chemistry book by Jane Marcet and decided to build one on his own. Faraday made his out of nickel and copper disks that he cut by hand.

he is able to read and learn on his own.

His main job is delivering newspapers. Paper and money are scarce in working-class London. Faraday takes a paper to a customer's house, leaves it for an hour, comes back, picks it up, and takes it to the next reader. And the next.

He impresses good Mr. Riebau and, at 14, is officially accepted as an apprentice in Riebau's bookbindery. He lives upstairs, over the shop, with two other apprentices. It's a seven-year commitment; they get no salary, but they do get room and board. "There were plenty of books there," he remembered later, "and I read them." Among the books he binds is a volume of the *Encyclopedia Britannica*. Young Faraday reads the whole 127-page entry on electricity. That subject becomes one of his life's interests.

He also reads with great care Jane Marcet's *Conversations on Chemistry*, a popular textbook. That leads him to cut disks out of nickel and copper and build a small voltaic pile (a battery), with which he is able to do some experimenting. And he manages to acquire a scholarly, four-volume chemistry text that includes the first printed account of John Dalton's theory of atoms.

This 1809 lithograph (below) shows the excellent library of the Royal Institution (RI). The RI was founded in 1799 as a place for working people to learn about science. Humphry Davy's dynamic lectures drew the elite. Anyone who wanted to be noticed in high society attended them. Soon most working people were elbowed out. The RI went into decline until Faraday's lectures revived it again.

Faraday has joined a group of young men who are fascinated with the new science; they do experiments and share ideas. Some will become lifelong friends. Faraday is interested in becoming a good writer; he searches out and finds help with grammar and style. He keeps journals of his scientific experiments and fills them with drawings as well as words. Then he turns those journals into bound books.

Mr. Riebau, who takes pride in his protégé's work, shows the bound volumes to a bookshop customer. The man is impressed and gives Faraday (who is now 21) tickets to hear Humphry Davy lecture at the Royal Institution. It is a thoughtful gift: tickets are hard to get because the lectures are enormously popular.

Sir Humphry Davy, who was a creative, important scientist, let his popularity go to his head. He was the only one who voted against Faraday's becoming a member of the Royal Society (twice). He had forgotten his roots. Davy was a poor boy when, at age 19, he was lent a copy of Lavoisier's *Elements of Chemistry*. It fascinated him. About the same time, his mother took a boarder into their house to help pay the bills. He was Gregory Watt, the son of James Watt, and he had studied chemistry at Glasgow University. They talked chemistry and did experiments, and without any formal education in science, Davy was soon on his way to scientific stardom. Later it was often said that his greatest discovery was Michael Faraday.

Davy is an eloquent speaker who does dramatic experiments in front of his audience. Faraday is awed. He has never before experienced this kind of excitement. He takes careful notes. Then he illustrates his notes, binds them in an elegant leather cover (remember, he works at a bindery), and sends them to Davy with a letter asking for a job. Later Faraday writes:

> *My desire to escape from trade, which I thought vicious and selfish, and to enter into the service of Science, which I imagined made its pursuers amiable and liberal, induced me at last to take the bold and simple step of writing to Sir H. Davy, expressing my wishes, and a hope that, if an opportunity came in his way, he would favor my views; at the same time I sent the notes I had taken of his lectures.*

Davy is busy with other things. The king has tapped him on the shoulder with a sword, making him a knight. Three days after becoming *Sir* Humphry, he marries a wealthy Scottish widow. (It is 1812.)

The next year, when Davy's laboratory assistant gets into a fight, Davy fires him and offers the job to Faraday. It will be menial work—mostly he will have to clean and repair

Sir Humphry Davy and Michael Faraday used this apparatus to focus sunlight and combust (burn) a diamond in Florence in 1814.

equipment—and the pay will be less than at the bindery, where he is now on salary. Still, Faraday grabs it.

Soon after, Davy and his wife decide to tour Europe to see volcanoes and other geological sites and meet with some of the continent's leading scientists. Faraday is asked to come along as a scientific and personal helper.

Unfortunately, Davy's wife is snooty. She won't eat with the lowly assistant and seems to go out of her way to annoy him. Faraday writes a letter from Rome to his friend Benjamin Abbott:

January 25th, 1815
Dear Benjamin,
I should have little to complain of were I traveling with Sir Humphry alone or were Lady Davy like him, but her temper makes it often times go wrong with me, with herself and with Sir Humphry.... She is haughty and proud to an excessive degree and delights in making her inferiors feel her power.

Faraday, who has never been more than 12 miles from London, is homesick and misses his friends and family. Nonetheless, this trip gives him an opportunity to learn. In Florence, Italy, he watches an expensive experiment. The Grand Duke of Tuscany has a burning glass—two huge lenses that focus the Sun's rays. They use it to burn a diamond. Faraday writes in his journal, "The diamond was all consumed.... From these experiments, according to Sir H Davy, it is probable that diamond is pure carbon." (Lavoisier had done a similar experiment in 1772. Whether you burn coal or graphite or a diamond, the carbon dioxide given off is identical. There's no ash; they're converted entirely to CO_2, which means they are each made of carbon.)

In Paris, the French show Davy and his aide a substance from seaweed. They think it is a compound. Davy gives it an electric jolt; it remains unchanged, and he declares it an element. He names it iodine.

IODINE, a dark purple element whose vapor is violet, comes from a Greek word meaning "violet colored."

Finally, after a year and a half, the trio heads home to

FARADAY'S LAWS OF ELECTROLYSIS

When Faraday went to Europe with Davy, he watched his mentor isolate metallic elements with electric current. Faraday (with his friend William Nicholl's help) was the one who named that process *electrolysis*, and he called any solution that conducts electricity an *electrolyte*.

In 1832, Faraday announced two laws of electrolysis. They're used to calculate how much mass of a substance you'll produce with a certain amount of electric charge. Here, they're reworded in modern language and simplified:

1. The mass of a substance liberated (chemically freed) by an electric charge is proportional to the quantity of electric charge used (the strength of the current multiplied by the time of the process).
2. The quantity of different elements liberated by a given quantity of electric current is in direct proportion to the atomic masses of the elements.

A German scientist, Hermann von Helmholtz (1821–1894), noticed an important idea that came out of Faraday's Laws of Electrolysis. He wrote, "If we accept the hypothesis that elementary substances are composed of atoms, we cannot well avoid concluding that electricity also is divided into elementary portions which behave like atoms of electricity." It was the first clue that there might be *particles* of electricity. (They were later found and called electrons; electric current is electrons moving in a wire.)

These identical glass cups with mahogany wood bases were handmade in the United States in the 1840s. In *Davis's Manual of Magnetism* (2nd ed., 1848), they're called "Decomposing Cells." That sounds yucky—as in rotting flesh—but in this case, *decomposing* describes the breaking down of water into hydrogen and oxygen through electrolysis; *cells* aren't the biological kind, but rather the kind run by batteries. The fancy-looking posts on either side of each cup are more than decorative; they're the anode-cathode pairs that attach to the battery.

London. Faraday goes to work at the Royal Institution as superintendent of apparatus. He also continues to work as Davy's assistant, learning science from England's most brilliant experimenter (although Davy spends less and less time at the RI).

Before long, Faraday is doing his own experiments. At first, hardly anyone pays attention to his results. Because he has no university degree, the scientific thinkers find it hard to take him seriously.

After a while, Faraday's genius can't be overlooked. His first fame comes as a chemist. He produces new compounds—extending knowledge of the way elements combine.

He analyzes steel alloys and lays the groundwork for the applied science of metallurgy. (The Industrial Revolution is in full swing, and metals are an essential part of the new technology.)

That's not actually Michael Faraday in the photo at right and on the opposite page. Both setups are reconstructions of his experiments on electromagnetism.

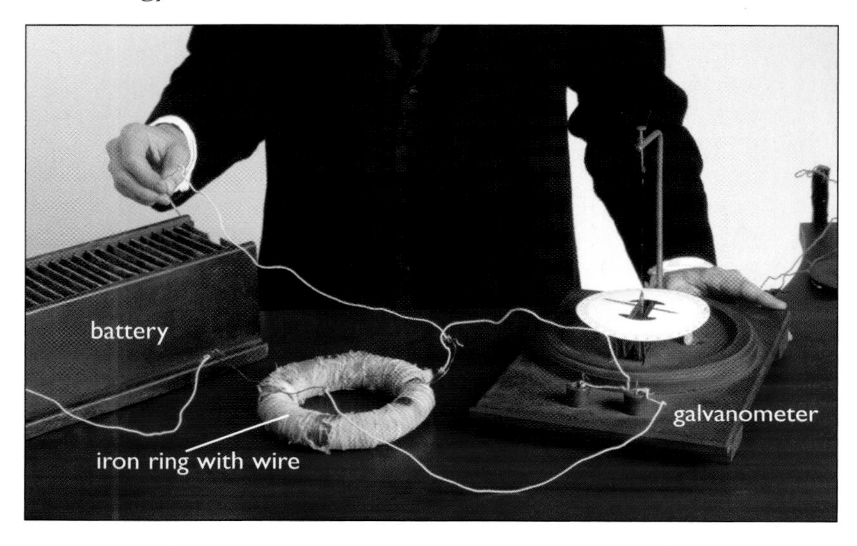

He liquefies some 20 gases under pressure. Refrigeration will come out of that work. So will compounds used in many chemical products (especially benzene, which he discovers).

But it is in the electrical sciences that Faraday lights the most fires. His experiments make it absolutely clear that "frictional" (static) electricity, animal electricity (as from an electric eel), and the electricity of a voltaic battery, are all the same. That is a big surprise in the nineteenth century.

When, in 1820, he learns of Hans Oersted's science-changing experiment in Denmark linking electricity and magnetism, he repeats the experiment. Oersted has made a magnet move in response to electric current. Can the opposite happen? Can a magnet create electric current?

Besides showing a connection between magnets and electricity, Oersted's discovery was surprising in another way: When he brought electric current near a compass needle (a magnet), the needle didn't point with the current or against it, but in a direction at right angles to it.

Oersted showed that moving current can produce a magnetic field. But can a moving magnetic field produce electric current? Faraday showed that it can.

Faraday decides to find out. He winds a strand of copper wire around an iron ring and connects the ends of that strand to a battery. He also attaches the iron ring to a device (a galvanometer) that measures electric current (photo above). Faraday hopes that current from the battery will magnetize the iron ring, which will then produce its own current, and that current will make the needle of the galvanometer swing.

And it does, briefly, as he adjusts the wires. And then it stops, not to start again until he begins to take the apparatus apart. He tries the experiment again and again and always gets the same result. A quick bit of current and then—nothing.

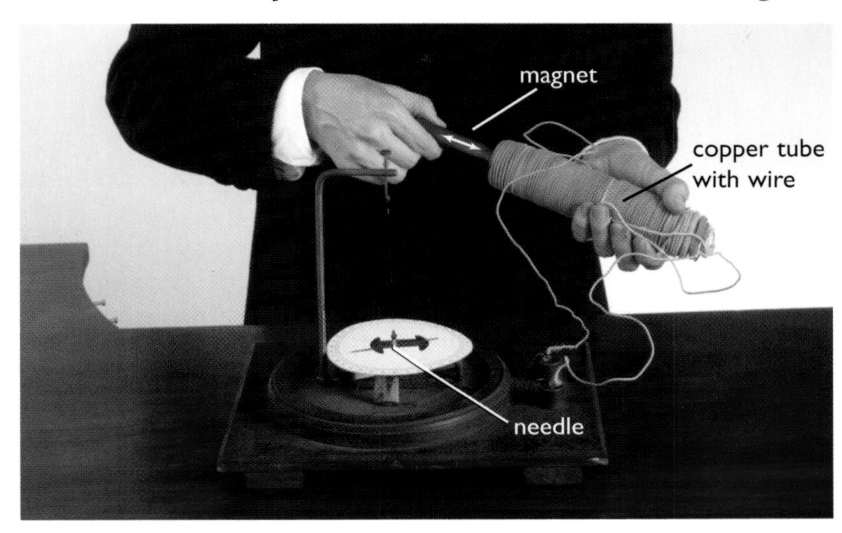

A magnet inserted into or removed from a coil of wire produces electric current—as long as the magnet and coil are moving toward or away from each other. This magnet and coil (below) belonged to Michael Faraday.

Faraday changes the experiment. This time he coils wire around a copper tube and pushes a magnet inside the tube (photo above). The needle moves—and stops. When he pulls the magnet out, it moves some more. He begins playing with the magnet. He moves it back and forth; the needle shows continuous current. Then he gets it. The secret is motion.

When the magnet and the electric coil are moved toward each other or away from each other, electric current is produced. When they are stationary, there is no electric current. He has found an efficient way to create electrical current! That is enough to insure his fame, but he isn't finished.

Nothing is too wonderful to be true, if it be consistent with the laws of nature, and in such things as these, experiment is the best test of such consistency.
—Michael Faraday, diary entry, 19 March 1849

Faraday is ahead of many of his peers in accepting Dalton's ideas and in assuming that matter is made up of small particles. (Davy

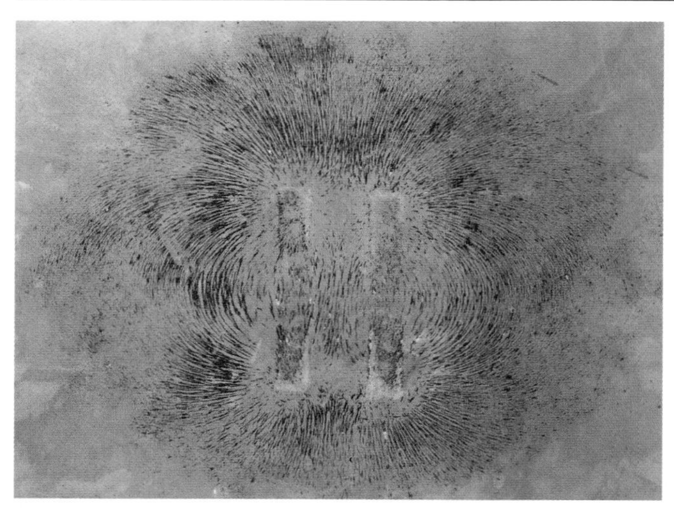

As far as we know, this sketch by Michael Faraday (above) shows the first pattern ever made by loose iron filings on a surface held over a magnet.

has rejected the idea of atoms.) Imagining atoms helps Faraday picture what is happening when electric current breaks apart compounds—like H_2O. And that leads him to come up with laws of electrolysis.

Have you ever sprinkled bits of iron on a piece of paper and put a magnet beneath the paper? When Faraday does that, he sees iron filings form themselves into a pattern—a field. He sees electric current do the same kind of thing. He throws out the old idea of electricity as a fluid. Electricity and magnetism travel in fields, fields of their own creation.

He soon comes to believe that electricity and magnetism are linked together as a force in the universe, like gravity. This is an astonishing insight, but he can't explain it mathematically and take it further.

Newton had been able to describe gravity mathematically, but he couldn't picture it. He said it was an "absurdity." Faraday has an amazing visual sense. He pictures gravitational fields that are something like spiders' webs. He says that those fields actually exist; they are real. This idea of fields is a new explanation of action in the universe. It takes time to be understood and accepted, but it will change the way people look at the universe.

Newton had said that gravity is action at a distance and that it is instantaneous. Faraday doesn't believe that. Traveling through fields, as he says the forces of gravity and electro-magnetism do, takes a bit of time, and that time must be measurable.

The electricity-magnetism connection is exciting, but someone has to prove it mathematically. That someone is coming. He will make Faraday's theories mathematical and therefore broadly useful.

I must confess I am jealous of the term "atom," for though it is very easy to talk of atoms, it is very difficult to form a clear idea of their nature, especially when compound bodies are under consideration.

—Michael Faraday, as quoted in *A History of Chemistry*, by J. R. Partington

All this breakthrough thinking should be enough for one person, but there is still more in Faraday's head. He has new ideas about light (also known as "radiation"). In a lecture in 1846, Faraday says, "The view which I am so bold as to put forth considers therefore, radiation as a high species of vibration in the lines of force."

In other words, light is radiation, electro-magnetic radiation, and it travels as vibrations (disturbances) in fields.

As soon as scientists understand what Faraday is saying—and when they develop the mathematics to explain it—there will be electric lights, radio, television, the telephone, and more. Theories that lead to future discoveries make you important in history books, but concrete results are more likely to please your contemporaries. Michael Faraday's brain is full of practical ideas that soon lead to new inventions.

Check the intensity in Michael Faraday's eyes. It's clear he was a man determined to accomplish things.

A Visual Guy

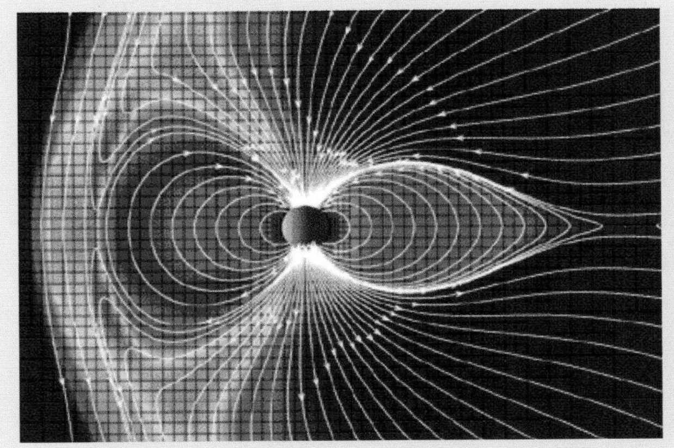

Warped space? You bet. Here's a computer model of "space weather," a scientific term that refers to surges of matter and electromagnetic energy blown off the Sun. Those surges periodically travel across interplanetary space—warping it.

Since Michael Faraday had limited math skills, he was forced to approach science visually, which gave him some advantages. He could do thought experiments, picture fields, and see them extending throughout the universe. (Scientists are still picturing science, as the computer model at left shows.) When Faraday saw electricity and magnetism connected as a universal force, most experts laughed. When he said they were linked to gravity, the experts really thought he had gone too far. Seventy years later, Albert Einstein showed that light rays are bent by the pull of the Sun's gravity.

The giant wheels in this 1887 woodcut are connected to generators (called dynamos), which sent electrical power a distance away. Bright and imaginative people, like Albert Einstein's father and uncle, were attracted to this astounding new technology.

He knows that getting current from batteries is expensive. He wants to find a low-cost source of electricity. Faraday designs a machine with a copper wheel that turns so its rim cuts through magnetic lines of force—and that makes current flow (it's that motion thing). Being a big thinker, he makes his wheel large, but turning a huge wheel isn't easy with muscle or horsepower. So he tries steam and falling water and blowing wind, and finds that each of them can get a wheel turning, although water seems easiest to harness. As soon as he has his big wheel turning efficiently, he has made a giant electric generator—a dynamo. It is a source of cheap electric current.

A GENERATOR (a DYNAMO, for example) turns mechanical energy into electrical energy. An ELECTRIC MOTOR changes electrical energy into mechanical energy.

There's an often-told story that when England's prime minister, Robert Peel, comes to the Royal Institution to observe Faraday's dynamo, he asks, "What is the use of it?" Faraday replies, "I know not, but I wager that one day your government will tax it."

The story may be apocryphal (which means "questionable," or "maybe fictitious"), but the point is right. Dynamos are soon lighting cities. Does electricity create taxable wealth? More than even the prime minister can imagine.

Besides dynamos, Faraday develops the electric motor. (Joseph Henry, in the United States, may have done this first,

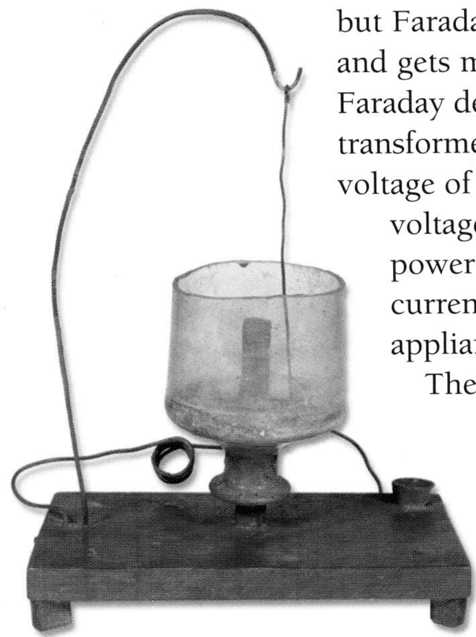

but Faraday is at the center of science and gets most of the acclaim.) Then Faraday designs the first electrical transformer. A transformer changes the voltage of a current—reducing high-voltage current generated by a power company to low-voltage current that won't fry your electrical appliances.

The dynamo, the transformer, and the electric motor become the foundation of an electrical revolution. Another self-educated physicist and inventor, Thomas Edison (1847–1931) working in the United States a generation later, will invent the incandescent lightbulb and plan the first electricity distribution system (with dynamos, insulated underground cables, meters for measuring consumption, outlets, and switches). For the first

At left is an early motor from the 1830s, called a Faraday motor. It doesn't look like the motor under the hood of a car? Well, this was a start. A motor converts electrical energy into mechanical energy (moving parts), which is what this machine does. A magnet stands rigidly in a cup of mercury with a wire conducting current to it from below; the circuit is completed by a wire dipped into the mercury while hanging loosely from a hook above. When current flows, the end of the hanging wire in the mercury circles round and round the magnet. Amazing! At first, most people just thought it was an amusing gadget.

ABOUT WORDS AND A WORDSMITH

Cambridge professor William Whewell (1794–1866) was a classics scholar who loved language. Michael Faraday went to him when he was looking for words to describe electrical processes. Whewell (right) came up with *anode* and *cathode* for the positive and negative poles of a battery. (They're from Greek words meaning "way up" and "way down"—the paths that the electric currents were thought to take from the poles.) Together they were termed *electrodes*.

Whewell supplied geologist Charles Lyell with the words *Eocene* and *Miocene* for geological time periods. And he coined the word *consilience*, which describes the coming together of many fields—like literature, art, and science—into one endeavor. It's a dandy word that twenty-first-century scientist E. O. Wilson has resurrected.

Whewell was Master of Trinity College at Cambridge and much loved by his students, some of whom became "natural philosophers." Maybe it was their influence that led him to come up with his most influential word, *scientist*. First published in 1840, it took decades for it to really catch on.

HOT STUFF! YOU'RE COMBUSTIBLE!

From Faraday's famous lecture on a candle, given in 1849:

> *Now I must take you to a very interesting part of our subject—to the relation between the combustion of a candle and that living kind of combustion which goes on within us. In every one of us there is a living process of combustion going on very similar to that of the candle, and I must try to make that plain to you. For it is not merely true in a poetical sense—the relation of the life of man to a taper; and if you will follow, I think I can make this clear. . . . You will be astonished when I tell you what this curious plan of carbon amounts to. A candle will burn some four, five, six or seven hours. What, then, must be the daily amount of carbon going up into the air in the way of carbonic acid! What a quantity of carbon must go from each of us in respiration! What a wonderful change of carbon must take place under these circumstances of combustion or respiration! A man in twenty-four hours converts as much as seven ounces of carbon into carbonic acid; a milch cow will convert seventy ounces, and a horse seventy-nine ounces; solely by the act of respiration. That is, the horse in twenty-four hours burns seventy-nine ounces of charcoal, or carbon, in his organs of respiration to supply his natural warmth in that time.*

An 1856 engraving from the *Illustrated London News* shows Faraday delivering one of his legendary lectures: the Christmas lecture on December 27, 1855. The audience included royalty—Prince Albert with his young sons Prince Edward (a king-to-be) and Prince Alfred. The Royal Institution's tradition of Christmas lectures with renowned scientists continues today.

time in history, human beings are able to adjust the length of their day.

Twelve years after Davy hired him, Michael Faraday towers over the field of electricity. He is named director of the laboratory at the Royal Institution. The lowly assistant has far surpassed his mentor.

But Faraday never forgets how inspired he was by the Davy lectures he heard when he was young. He studies public speaking (then called "elocution") and begins a 30-year series of Christmas lectures for children. An adult audience hears him at regular Friday-evening talks. Charles Dickens comes to some of the lectures, so does Prince Albert (Queen Victoria's husband). Charles Darwin attends all he can. One week, Faraday puts water in iron containers that are taken to an icy basement. When the water freezes, it expands and bursts the iron containers. The next week an astonished audience sees the split containers—and the power of moving atoms.

Faraday, now famous, refuses most awards, including the presidency of the Royal Society and knighthood. When a shop owner addresses him as "Sir," he says, "You do me too much honor in your address, for I am plain Michael Faraday and shall so remain."

Perhaps it is his natural modesty or perhaps it is his religion that makes him turn away honors. He belongs to a small Protestant sect, the Sandemanians, who believe in sharing their earnings with the poor; they don't pay much attention to worldly goods. Faraday is a church elder. Once, when he misses Sunday worship, he is reprimanded by the other elders for not having a good enough excuse. He had been dining with Queen Victoria.

In Faraday's time, people wrote letters, lots of them, even to those who lived nearby. (There were no telephones!) Faraday and his friend Benjamin Abbott corresponded for 50 years. Once Faraday explained with this riddle why he was late answering a letter:

Dear Abbott,
What is the longest, and the shortest thing in the world: the swiftest and the most slow: the most divisible and the most extended: the least valued and the most regretted: without which nothing can be done: which devours all that is small and gives life and spirits to everything that is great?

Answer: Time

IS EVERYTHING CONNECTED?

Faraday's curiosity was inexhaustible. He wasn't content with just the practical; he considered the whole universe. And that led him to an extraordinary vision—perhaps the most important thought to come out of nineteenth-century science: **The whole universe is tied together through matter and energy, and all of that matter and energy was there when the universe was created** (although not in the same form it is today).

Here it is in Faraday's words: "The various forms under which the forces of matter are made manifest have one common origin; . . . are so directly related and mutually dependent . . . that they are convertible . . . one into another."

In the twentieth century, Albert Einstein came up with a famous formula that simplified and unified things still further. He linked matter to energy—which allowed one to be turned into the other (that's what led to the atomic bomb).

Scientists are still working to find Faraday's "one common origin," something that will unite the four major forces that we now believe power the universe (see red note on page 365). Today that quest is known as the search for a unified field theory (and we seem to be almost there).

Physicist Stephen Hawking (holding the same job title Newton had at Cambridge University, Lucasian professor) writes, "The discovery of a complete unified theory . . . may not aid the survival of our species. . . . But ever since the dawn of civilization, people have not been content to see events as unconnected and inexplicable. They have craved an understanding of the underlying order in the world."

TURNING ON THE LIGHT

The Bible says of the Sun and the Moon, "Let them be for lights in the firmament of the heaven to give light upon the Earth." But what is light? On that subject the ancients were . . . well, mostly in the dark.

Isaac Newton believed that light is made up of tiny bulletlike particles. That particle idea explains why light bounces off

In this 1690 illustration by Christiaan Huygens, picture yourself as the tiny observer (B) on the left. You're looking at those three buildings, but building A appears taller (point D) than it actually is because the air is refracting (bending) the light.

Sir Thomas Young was a medical doctor who studied the way the lens of the eye changes shape as it focuses light from near and distant objects. From there it was just a step to analyze light itself. Young combined the wave theory of Christiaan Huygens with the color theories of Isaac Newton to explain the interference pattern (see next page) produced by slits, gratings, and the rainbow.

a hard opaque surface (like a mirror). It also explains why it bends (is refracted) in water, a medium that slows particles.

Still, that light-as-bullets theory doesn't explain everything. Why do some colors refract more than others? Particle theory doesn't make sense of that.

Newton's contemporary, the Dutch astronomer Christiaan Huygens, had a different theory. (Remember, Huygens was the man who built the first pendulum clock.) He said that light is made up of tiny waves with different colors having different wavelengths. That explains why some colors refract differently than others. But it can't explain why light bounces straight off a mirror and back to your eyes. So the scientific thinkers argued about this. Some said light is particles, and some said it is waves.

Thomas Young (1773–1829), an English doctor, got caught up in the problem. He was a Quaker (a member of the Society of Friends) at a time when Quakers were barely tolerated in England. Young actually wasn't very religious, but his Quaker background formed his personality. He had an independent mind.

Young knew that sound travels as waves, so he set out to prove that light does the same thing. He realized that Isaac Newton didn't think so. Newton had written: "Sounds are propagated just as readily through crooked pipes as through straight ones. But light is never known to follow crooked passages." Thomas Young thought the great Newton might be wrong. And, being a Quaker, he was used to questioning authorities.

In about 1801, Young shone a narrow beam of light through two holes and looked at the screen behind them. He expected to see two points of light and an overlap between—which is what would happen if light were particles. But what he saw were bands of light and dark. It looked exactly like the waves and depressions that form when you toss a stone into a pond. That experiment became famous. Young seemed to have proved that light is waves.

Seventeen years later, a Frenchman, Augustin-Jean Fresnel (1788–1827), found

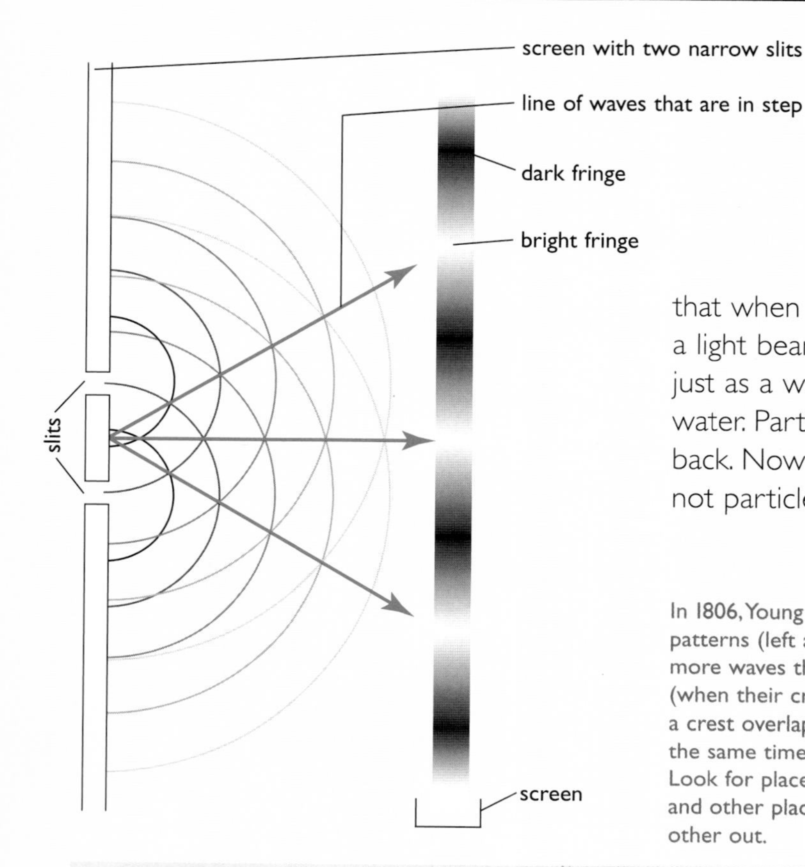

that when a small object gets in the way of a light beam, the beam will go around it—just as a wave goes around a stick in the water. Particles don't do that; they bounce back. Now it seemed sure: Light is waves, not particles.

In 1806, Young pioneered the study of interference patterns (left and below). Interference is about two or more waves that overlap, either enhancing each other (when their crests join) or disrupting each other (when a crest overlaps a valley). Observe: Drop two pebbles at the same time near each other on a calm water surface. Look for places where crests join to make a supercrest and other places where crests and valleys cancel each other out.

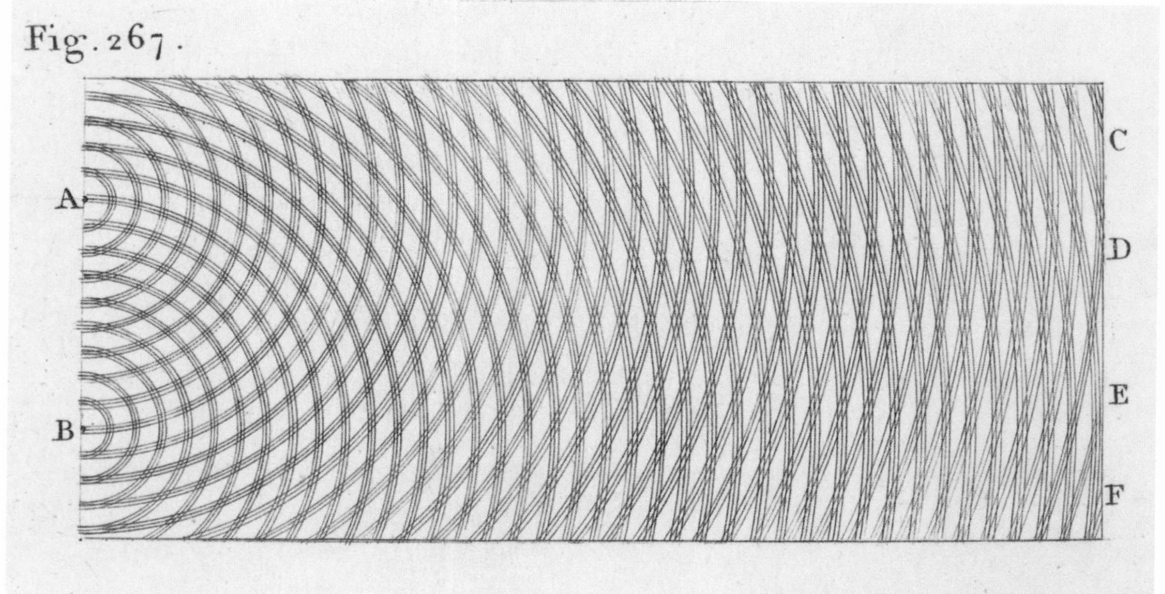

SUNBEAMS MOVED BY INVISIBLE STUFF? REALLY?

Waves travel not as matter, but as vibrations (scientists call it a disturbance) through a material. Waves need to have something to disturb. We hear thunder because its sound disturbs Earth's atmosphere. But sound can't go through the vacuum of outer space. If Mars were to explode, you would see it, but you wouldn't hear it. If light is waves, what medium do those sunbeams travel through?

"No problem," said the scientific community; "the Greeks answered that one. They said that there is an invisible substance called *aether* (more recently spelled *ether*) filling space, and light waves move through it." Aether seemed to answer a lot of questions. In the nineteenth century, most scientists talked of the "luminiferous ether" to explain light. They were convinced that gravity as well as light travels through the ether. Were they right? And—is light composed of particles, or is it energy waves? In the twentieth century, scientists would find answers to those questions. To say those answers were a surprise—well, that's an understatement.

The English painter, J. M. W. Turner (1775–1851) loved to watch what happens to light when it hits water waves—and then paint his vision (above). Turner, who found mystery and passion in wind and water, had his first art exhibition when he was 15. Then he went off to Paris and Italy, saw some Renaissance paintings, and grew as an artist. He handled paint his own way, with great, sweeping strokes that leave his viewers free to imagine more. You can see this luminous painting, *Yacht Approaching the Coast*, at the Tate Gallery in London.

33

Maxwell's Charges

Lives of great men all remind us
We can make our lives sublime,
And, departing, leave behind us
Footprints on the sands of time.

—Henry Wadsworth Longfellow (1807–1822), American poet, "A Psalm of Life"

Faraday and Maxwell showed us that particles were not enough, and we must consider, also, that there are continuous fields pervading space, with a reality as great as that of the particles themselves. Those fields were combined into a single all-pervasive entity, referred to as the electromagnetic field.

—Roger Penrose (1931–), English mathematical physicist, foreword to *Einstein's Miraculous Year*

James Clerk Maxwell, who is born in Scotland in 1831, is not at all like grumpy, withdrawn Newton or too-sure-of-himself Galileo. Quiet and likable, Maxwell is brought up on a country estate, Glenlair, among the lakes and hills of Galloway (in southwest Scotland). His childhood is filled with books, horses, fishing, swimming, and games. It is idyllic, until he is eight, when his mother dies of cancer. "I felt her loss for many years," he writes later. Still, he adores his father, who loves him back.

He is a boy who is full of questions and likes to putter and take things apart. "What's the go o' that? What does it doo?" he asks his dad in their Scottish way of talking.

When James is 12, he gets a toy that some call a "devil-on-two-sticks." It's a popular Chinese toy first brought to Europe by mariners around 1790. In China, its name is *Fong Lai*, which means "thunder wind," perhaps because, when it spins fast, it makes a humming noise like a strong wind. In

James's mother wrote proudly to her sister-in-law about her happy, inquisitive three-year-old son:

> *He has great work with doors, locks, keys, etc., and "Show me how it doos" is never out of his mouth. He also investigates the hidden course of streams and bell-wires, the way water gets from the pond through the wall... and down a drain.... As to the bells... he drags papa all over to show him the holes where the wires go through.*

He used to pick up frogs—"clean dirt" was his generic description of anything that crawled in which he was interested—stroke them, hear them croak, pop them in his mouth and let them jump out again to terrify his parents. His frog imitations were to become famous.

—Martin Goldman, *The Demon in the Aether: The Story of James Clerk Maxwell*

James's cousin Jemima Wedderburn was a gifted artist. As a child, she painted pictures of the family almost every day. She was 10 when she made this lovely sketch of infant James with a toy owl and one of his notorious frogs.

Europe, it was first called *diavola*, from Greek words meaning "throw" and "through." But *diavola* sounds a lot like the Spanish word for "devil," *diabolo*, and since this toy is a devil to tame, that became its name.

Diabolo tests a person's juggling skills. To handle it takes concentration, balance, and agility. The "devil" becomes the rage, especially in France, where there are tournaments and team play, although most players just compete against themselves, trying to master tougher tricks.

The object is to toss and catch a wooden piece—a "devil"—and keep it rolling back and forth on a string attached to two sticks. Most European "devils" are made of two wooden cones attached at the points. That hourglass shape helps with the catching. But young Maxwell seems to have a Chinese model; it takes more skill to handle. His "devil" is two disks with a dowel between, making it look a

In this nineteenth-century engraving, young and old enjoy playing with diabolos. For Maxwell (according to Hans Christian von Baeyer), "It was a matter of of playing with the problem, tossing it around, looking at it from different points of view, poking at it and putting it through its paces until it became docile in his hands—until he became its master. He turned Devils of that sort all his life."

While still in school in Edinburgh (pictured on the right), Maxwell was encouraged by a college professor to visit his laboratory and try his own experiments. Later, Maxwell wrote:

> *If a child has any latent capacity for the study of nature, a visit to a real man of science at work in his laboratory may be a turning point in his life. He may not understand a word of what the man of science says to explain his operations; but he sees the operations themselves, and the pains and patience which are bestowed on them; and when they fail, he sees how the man of science instead of getting angry, searches for the cause of failure among the conditions of the operation.*

As a student, Maxwell wrote letters home to his father in mirror writing (just for fun) and signed them Jas. Alex McMerkwell (an anagram of James Clerk Maxwell). He was descended from two Scottish families, the Maxwells of Middlebie and the Clerks of Penicuik. During his lifetime, he was known as Clerk Maxwell. Today we usually just call him Maxwell.

lot like a yo-yo. With practice, he gets the piece to spin, to hum, to climb into the air, and to do other tricks. Sometimes he can flip the "devil" and jump over the string before he catches the piece again. The "devil" seems to defy gravity—but clearly it follows some kind of rule. Controlling the "devil" means experimenting, practicing, and thinking. James becomes good at those things.

But when he goes away to school in Scotland's capital, Edinburgh, the city boys make fun of him. They nickname their curly haired, shy classmate "Dafty." Perhaps it is because of the loose, strange-looking smock and square-toed shoes his practical father has designed for him. He goes to his aunts' house after his first day of school with his clothes in tatters. The two sensible aunts see that he dresses like the other boys. "They never understood me, but I understood them," he says later of his classmates.

He shows those boys that he is anything but daft. He becomes an outstanding student, a fine gymnast, and an expert horseman.

When he is 14, he invents a way of drawing an ellipse by using two pins, a loop of thread, and a pencil (see page 129). As you know (if you've been reading this book), an ellipse is not just any old oval—it has a precise curve. The members of the Royal Society of Edinburgh are so impressed they publish a paper on his method.

In 1860, 29-year-old James Clerk Maxwell became a professor at King's College in London. He was able to go to lectures and discussions at the Royal Society and the Royal Institution. He especially enjoyed walks with his hero, Michael Faraday. Faraday, in his 70s, was suffering from failing memory (perhaps due to exposure to dangerous chemicals). The two modest, gracious men shared a passion for science, and Maxwell understood Faraday's genius in a way that few others did. Too bad we can't know what they said to each other. A watercolor painting from 1852 (left) shows Faraday experimenting in his lab.

When Maxwell grows up, he becomes a professor at King's College in London and first head of the Cavendish Laboratory at Cambridge. But for relaxation, he plays devil-on-two-sticks all his life; it seems to help him solve problems. "At practical Mechanics I have been turning Devils of sorts," he writes to a friend.

For someone who likes to play with ideas and objects, electricity and magnetism are irresistible. He pays attention to Faraday's experiments and is aware that electricity and magnetism are different faces of the same force. He learns that moving magnets create electric current, and that moving electrical charges create magnetic fields. The whole idea of electro-magnetic fields is something new. Maxwell takes that idea seriously and goes with it. (Most people think it's curious, but no more than that.)

Faraday is the century's outstanding experimenter. His experiments have led him to come up with brilliant hypotheses, but someone needs to prove his thoughts mathematically. Without mathematics, it's hard for anyone

Maxwell believed in the ether, a kind of invisible cosmic broth through which light and other electromagnetic waves travel. Faraday didn't agree (he would turn out to be right). Here are Faraday's words:

> *The view which I am so bold as to put forth considers radiation as a high species of vibration in the lines of force which are known to connect particles, and also masses of matter together. It endeavours to dismiss the ether but not the vibrations.*

In the decades to come, the search for nonexistent "ether" will take a lot of scientific time and energy.

Maxwell is holding a top invented by his teacher J. D. Forbes. It has two sets of tinted papers. When the top is spun, the colors appear to blend together. Maxwell came up with an equation to explain why. His research into color showed that all the colors can be made from three primary colors of light (red, green, and blue). In 1861, he used this three-color process to produce the first color photograph.

to apply or verify scientific knowledge.

James Clerk Maxwell is a skilled mathematician. He provides the mathematics that links electricity and magnetism, and he is the one who removes the hyphen and coins the word *electromagnetism* (EM). He comes along at just the right time and is to Faraday as Johannes Kepler was to Tycho Brahe. (It's tricky ranking scientists, but Faraday is a much greater scientist than Tycho, and Maxwell is at least on a par with Kepler.)

Maxwell's Cambridge University thesis (which he must prepare in order to graduate) is about Faraday's lines of force. That gets him started. He concentrates on electromagnetism for 10 years and turns it into a science. He shows, with mathematical formulas, that electricity and magnetism do not exist in isolation from each other. Where one is, the other will be found. He gives scientists four formulas, called "Maxwell's equations," that link the two phenomena.

"Every problem involving electricity and magnetism (at the level of classical physics) can be solved by using Maxwell's equations," astrophysicist John Gribbin will write more than 100 years later. "Maxwell's work was, indeed, the greatest step forward in physics since Newton's work."

RING MYSTERY

If Saturn's ring system were the diameter of a football field, it would be as thin as paper. Are these amazingly slim bands solid, liquid, or gaseous? Maxwell worked on that problem for a decade. He wrote to a friend, "I am very busy with Saturn on the top of my regular work. He is all remodeled and recast, but I have more to do to him yet." (That's Maxwell's Saturn model on the far right.) In 1859, by plotting the probable motions of molecules, he predicted that the rings could only be made of lots of very small particles. Any other form, his equations showed, and they'd break apart. His prediction was verified by direct observation in 1895. In the 1980s, the two *Voyager* spacecraft took our first close-up views (above left). The images show details of the rings' separate bands and that most of its particles are smaller than soccer balls.

Newton's mathematical formulas made gravity understandable and usable; Maxwell's equations do the same for electromagnetism.

There is something else. When Maxwell realizes that electromagnetic waves travel at the same speed as light, he immediately infers that light is electromagnetic. Then he confirms it mathematically. This is a huge step. Kepler wondered about light, and Newton did too. But they didn't know that it is electromagnetic.

That isn't all. Maxwell realizes that light and other forms of electromagnetism travel through fields, much as Faraday said they do, but Maxwell adds an astonishing detail. Here are Maxwell's words and his italics: "We can scarcely avoid the inference that *light consists in the transverse undulations of the same medium which is the cause of electric and magnetic phenomena.*" Take a deep breath and read that line a few times.

What he is saying is that **light is electromagnetic and it travels as undulations (wavelike movements)**.

Light is undulations. In simple words: An electromagnetic wave is made up of waves that are at right angles to each other (see diagram). One is the electric field; the other is the magnetic field.

In Oersted's experiment, the compass needle hadn't pointed with the current or against it, but in a direction at right angles to it. A magnetic field creates an electric field that creates a magnetic field and on and on—always at right angles to the previous field. (In contrast, sound waves move in a straight path, usually described as longitudinal.)

In a letter to his cousin, Maxwell describes the idea of waves traveling across fields as "great guns." That's an understatement. But there is another question to be asked.

A HIGH-LEVEL PHYSICS CONCEPT

Maxwell's electromagnetic equations show that as an electric field changes from moment to moment, its magnetic field changes from point to point in space (or vice versa). In other words, if one of the fields changes with time, the other field changes in space in a definite way. One can't act without the other reacting.

Maxwell's mathematical link between time and space would provide a foundation for the physicists of the twentieth century. Albert Einstein said, "One scientific epoch ended and another began with James Clerk Maxwell."

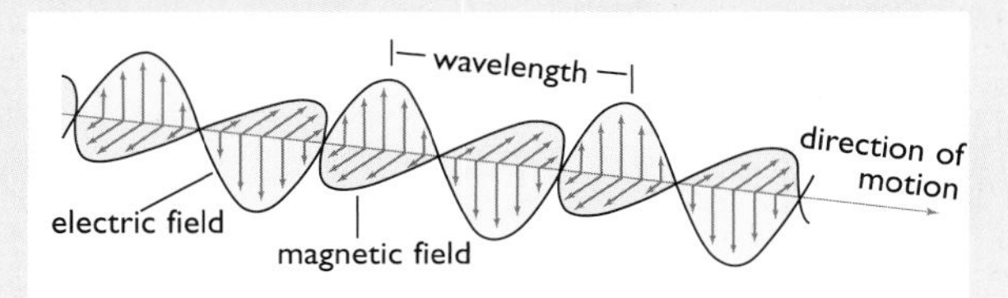

In an electromagnetic (EM) wave, electric and magnetic fields point at right angles to one another. (EM waves are light, infrared, radio, and so on.)

In his inaugural lecture, at age 25, as a professor at Marischal College in Aberdeen, Scotland, Maxwell made it clear that he planned to have his students experiment and learn for themselves: "I have no reason to believe that the human intellect is able to weave a system of physics out of its own resources without experimental labour." As it turned out, he had some problems with his fellow Scots. "No jokes of any kind are understood here. I have not made one for two months, and if I feel one coming I shall bite my tongue." But he did share jokes with the daughter of the college principal, Katherine Mary Dewar, and married her in 1858.

TURNING ON THE LIGHT CONNECTION

James Clerk Maxwell was named the first professor of experimental physics at Cambridge University in 1871. He was much loved, but he didn't have the gift of simple language, so most of his lectures went right over the heads of his students. Like Isaac Newton, he often spoke to an audience of only three or four. (Actually, Newton sometimes talked to the walls.)

Remember, Maxwell wrote, "We can scarcely avoid the inference that *light consists in the transverse undulations of the same medium which is the cause of electric and magnetic phenomena.*" (That was in *A Treatise on Electricity and Magnetism*, published in 1873.)

He was announcing the most important scientific idea since Newton. He was also demonstrating his problem with language. Those "transverse undulations" are just waves. In that astonishing (and obtuse) sentence, he was linking light to electromagnetism. How did he make that connection?

Maxwell realized that electromagnetic fields radiate from the Sun, or from any source, at an unchanging speed. But what is that speed? He realized he could calculate it using the ratio of units of magnetic fields to units of electric fields. This involved higher math, which was his specialty. The important point is that when he worked the math, he got a figure of about 300,000 kilometers (or 186,300 miles) per second. He knew that was just about the same as the velocity of light (because of measurements by others). That seemed too much of a coincidence not to have meaning. That's what made him realize that light *is* electromagnetism.

Scientists understood the importance of that idea. But Maxwell never reached a general audience with this enormously important idea: The Sun's rays, and a host of invisible rays, are electromagnetic. We can track them and, with equations that Maxwell gave us, we can use those rays. That has changed the way we humans live.

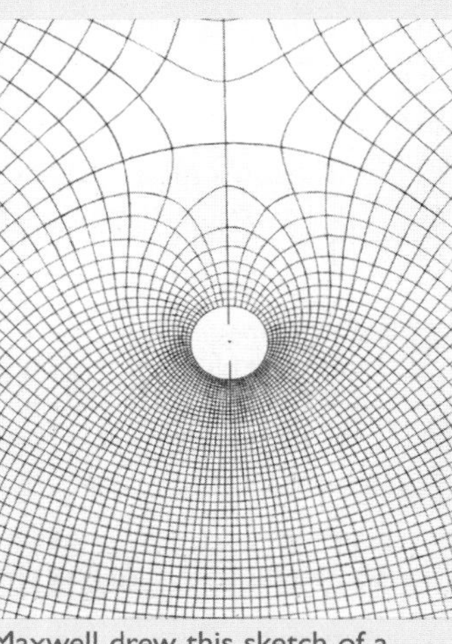

Maxwell drew this sketch of a uniform magnetic field disturbed by an electric current.

Why can EM waves travel through the vacuum of space, but sound waves can't? Sound energy exerts pressure—it squeezes (compresses) molecules, making them vibrate. With nothing to squeeze, sound waves can't form in space. EM waves vibrate electric and magnetic fields instead of matter. In fact, matter slows them down, which is why I keep adding "in a vacuum" when I talk about the speed of light. In physics terms, sound waves are mechanical, and EM waves are not.

How do those undulations of light travel through space? Don't waves need something to travel through? A medium? Something to wave?

As it happens, EM rays make their own fields and then travel through them—that's enough. EM radiation can travel in a vacuum. And space is a near-vacuum. Faraday seems to have understood that. No other major thinker in the nineteenth century figures it out. Not even Maxwell. Most people believe an invisible medium fills space, and that light undulates through it. It's that stuff called "the ether."

When it comes to the electromagnetic road, Maxwell is on target. He realizes that colors are waves of electromagnetism of differing lengths (red is long waves; blue is short waves). He understands that there must be a spectrum of electromagnetic rays, and that light and color are the part of the spectrum that we humans can see. The other waves, those we can't see, are the same type of waves, except they come in different lengths. Maxwell realizes that the longest and shortest waves are invisible to our eyes. He predicts that those other rays will be found—he even suggests where to look. (It is a bit like Mendeleyev predicting undiscovered elements.)

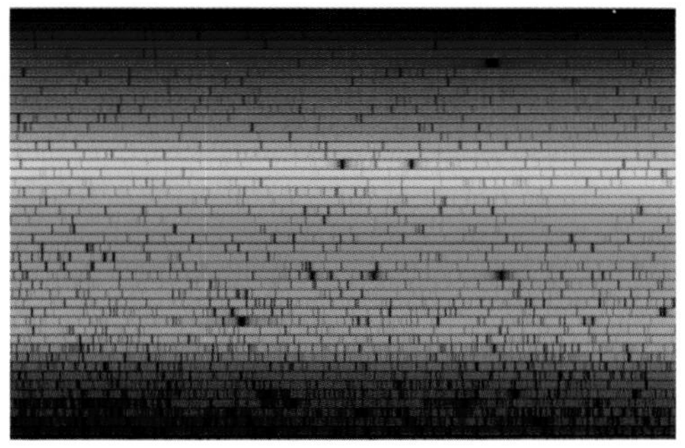

This spectrum of visible light from the Sun shows the shortest wavelength we can see in the bottom left corner to the longest one in the top right corner. Ultraviolet waves are even shorter than the violet ones at the bottom. Infrared waves are longer than the red waves at the top.

Scientists go looking. Radio waves are discovered in 1888 by Heinrich Hertz, and then microwaves and X rays and infrared rays and gamma rays and TV signals and more. All those *invisible* rays travel through space at the same speed. Remember the Danish astronomer, Ole Roemer, who measured the speed of light in 1676 (when Newton was alive) and got a reasonable estimate? At the end of the 1840s, two French physicists, Armand Fizeau and Léon Foucault ("Mr. Pendulum"), send a beam of light through a gap in a rotating wheel to a mirror and back through another gap in the wheel, and measure the journey. They get a number for the speed of light that is within 5 percent of the figure we now use.

So Maxwell knows that light travels at about 300,000 kilometers (about 186,300 miles) per second (in a vacuum). Maxwell's equations confirm that all electromagnetic rays—

FOLLOW THE HEAT

If you want to understand the physical world, follow the heat. Maxwell did. He understood that the dance of molecules creates thermal energy (heat). Maxwell figured out (in a brilliant bit of imagination) that molecules don't all move at the same velocity. A thermometer measures an average of the kinetic energies of the molecules in a given space.

Maxwell had started thinking seriously about the minuscule world of atoms and molecules after studying the mathematical ideas of Daniel Bernoulli. (His investigation of Saturn's rings also played a role.) See more on Maxwell's molecular ideas in the next chapter.

A LUCKY LAB

In the nineteenth century, in good part because of the Industrial Revolution, scientists began to concern themselves with experimentation, invention, and "real-world" problems. Gone were the days of Galileo's ball-and-ramp experiments. Experimental apparatuses were becoming more and more sophisticated. Lone individuals were less likely to set up their own laboratories; the new equipment required special training and teamwork. It also required more money. (This was happening at a time when the business community began to see the value of science and would help finance research.)

Students are experimenting in the Cavendish laboratory in the 1930s.

Universities began to change, too. After Oxford established an undergraduate course in experimental physics, Cambridge grudgingly had to offer one. This was annoying to some conservative professors who liked things as they were. But when William Cavendish (a descendant of Henry Cavendish) made a generous donation to get things going, there was no stopping the new science.

Cambridge asked some scientific stars to head the new Cavendish Laboratory. They were turned down. Then Cambridge went begging to Maxwell. He declined at first, writing from Glenlair, Scotland: "Though I feel much interest in the proposed Chair of Experimental Physics, I had no intention of applying for it when I got your letter, and I have none now, unless I come to see that I can do some good by it."

Cambridge was lucky. Maxwell, who had started experimental curricula at two universities, changed his mind and decided he *could* "do some good by it." He started Cambridge's program from the ground up. He designed and built much of the experimental equipment. Collaborating with the architect, Maxwell made plans for a comfortable building with state-of-the art labs. It would house Cambridge's physics department for almost 100 years.

When the electron was discovered at the Cavendish Lab in 1897, Cambridge was the site of a jolt in physics like the one Newton had delivered when he sat in his lonely rooms at Cambridge's Trinity College and wrote the *Principia*. As to the electronic revolution, it was the discovery of the electron in that lab and that town that got it started. That should have been enough for any place, but Cambridge and the Cavendish Lab would initiate yet another world-changing scientific revolution when DNA was discovered there in 1953.

short and long—travel at that same speed. Always. It is a constant. This is a big thought. Remember it; it will become significant. But in Maxwell's time no one pays much attention

Decades after her earliest sketch of James (page 357), cousin Jemima painted this watercolor of the family at a seashore gathering. James Clerk Maxwell is the second seated man on the left.

to it. Even Maxwell doesn't seem to realize its importance.

Maxwell isn't around when radio waves are discovered; he would have been excited but not surprised. He dies of cancer when he is 48, the same age as his mother when she died of the same disease. It is 1879, the year Albert Einstein is born.

To sum up: Maxwell's mathematics, based on Faraday's ideas, lays the foundation for an electromagnetic revolution, and that leads to an electrically powered world beyond imagining in the nineteenth century. Today, James Clerk Maxwell is ranked with the greatest scientists of all time.

Gravitation keeps the planets in orbit around the Sun and acts between all particles that have mass. The electromagnetic force also works on both a universal scale and within the atom. It keeps solids from falling apart and acts between all particles with electric charge. Two other forces (the strong and weak nuclear forces) will be found in the twentieth century; both operate inside the nucleus of an atom. The four forces seem to explain most of nature's workings, but there is hope that a superforce will be found that unites all four forces.

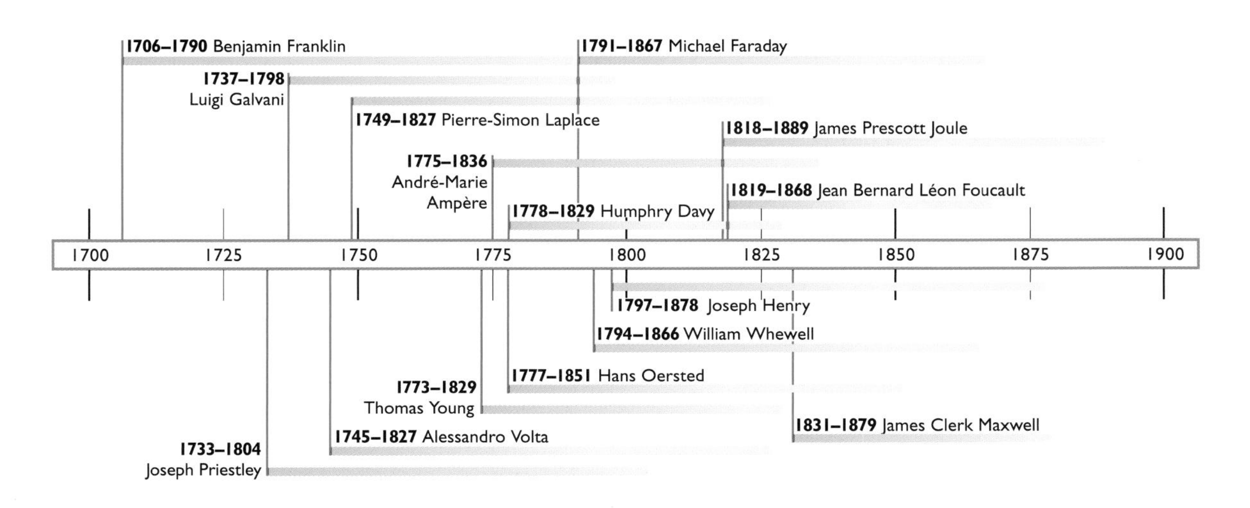

AGAIN AND AGAIN AND AGAIN—THAT'S FREQUENCY

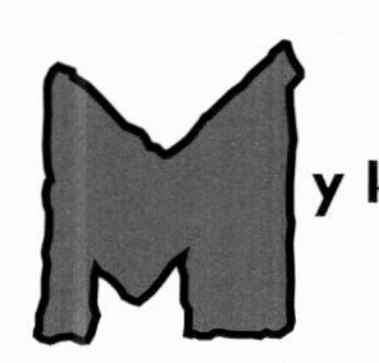

y kids tell me I repeat some things with frequency. "Be polite"; "Stand up straight"—you know, the mommy messages. So I can relate to the scientific meaning of the word *frequency*. It is just nature repeating itself at regular intervals. Here's a science dictionary definition of *frequency*: *In physics, the number of periodic oscillations, vibrations, or waves occurring per unit of time.* The repetitions can be vibrations caused by sound pressure (sound energy squeezing molecules), disturbances of an electromagnetic field (causing EM waves), or even plain old water waves.

The scientific measure of frequency—those regularly repeating vibrations or waves—is known as the hertz (Hz). One hertz is equal to one cycle per second. A cycle is the repetition of a wavelength—the distance from peak to peak in a water wave, for example. The hertz is named after Heinrich Rudolf Hertz (1857–1894), the man who discovered radio waves. He confirmed what James Clerk Maxwell had predicted—that there are more

(continued on page 368)

German physicist Heinrich Hertz proved with a series of experiments that the electromagnetic waves predicted by Maxwell exist. When he demonstrated that they could be detected at a distance, he had discovered radio waves.

SOUNDING OFF AND ON

Sound is vibration. The frequency of a sound wave is a measure of vibrations per second. For a sound to be audible, the vibrations must be between about 20 and 20,000 Hz. Higher frequencies (faster vibrations) create higher pitches; some animals hear frequencies too high or too low for us to hear—called ultrasonic or infrasonic. (Would you call frequencies that we can't hear "sounds"?)

Want to hear (and feel) sound frequencies in action? Press a piano key, and a little hammer strikes a string, causing it to vibrate. (If you put your other hand on top of the piano, you can feel the vibrations.) The length and tightness of the string determine the number of vibrations per second—namely, its frequency. The table (right) lists frequencies for the middle keys of a piano—both black and white.

Notice anything refreshing about the A keys? The second A (440) has twice the frequency of the first A (220)—which is why they're both called A. All octave pairs (A-A, B-B, and so on) have frequencies in simple ratios—2:1, 4:1 (two octaves apart), 8:1 (three octaves apart), and so on. Because their vibrating strings are very much in sync with each other, these notes sound harmonious when played together.

In fact, the same is true for any notes whose frequencies reduce to low whole numbers. The first A (220) and the E (330) vibrate in a nice, round 2:3 ratio, called a "fifth." (The interval has five whole notes; each sharp—♯—has a half-step interval.) If you have access to a piano (or a calculator), you can easily find other pleasing pairs. Just listen (and look) for notes with frequencies that reduce to single-digit whole numbers (or thereabouts).

Music fans: The second A is the note to which the others are tuned; the C, at about 262 Hz, is middle C. Some of the sharps are better known as flats—B♭ instead of A♯ for instance. Language fans: *Octave* means "eight"; there are 12 notes total, but eight whole notes, per octave. Math fans: The difference in the frequency between each successive pair of notes increases by the 12th root of 2—or roughly 1.059463094359295....The numbers in the table are rounded off.

Piano Key	Frequency in Hertz (vibrations per second)
A	220
A♯	233
B	247
C	262
C♯	277
D	294
D♯	311
E	330
F	349
F♯	370
G	392
G♯	415
A	440

The number of vibrations per second determines the pitch—high (on the right) or low (on the left). But sound also has amplitude—or loudness. Strings that vibrate widely, like the one on the right, pass more energy to the surrounding air, and so sound louder. It's easy to see and hear the difference by plucking a string lightly and then hard.

(continued from page 366)

electromagnetic waves out there than just the light we see, and they all behave the same way—except for the length and frequency of their waves.

Warning: Don't confuse frequency with speed. The speed of light and of all electromagnetic waves (in a vacuum) doesn't vary. But the speed of the repetitions (the frequency) does. It varies because the length of the waves vary. Frequency and wavelength are what make radio waves different from microwaves.

If you find this note repetitive, keep in mind that repeating important information with some frequency, well, makes it stick in your head. Here's one more thing to remember: The frequency of a sound wave determines its pitch; the frequency of a light wave tells you its color.

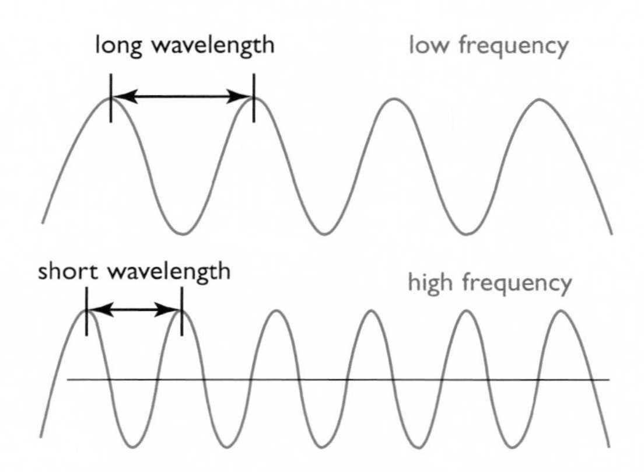

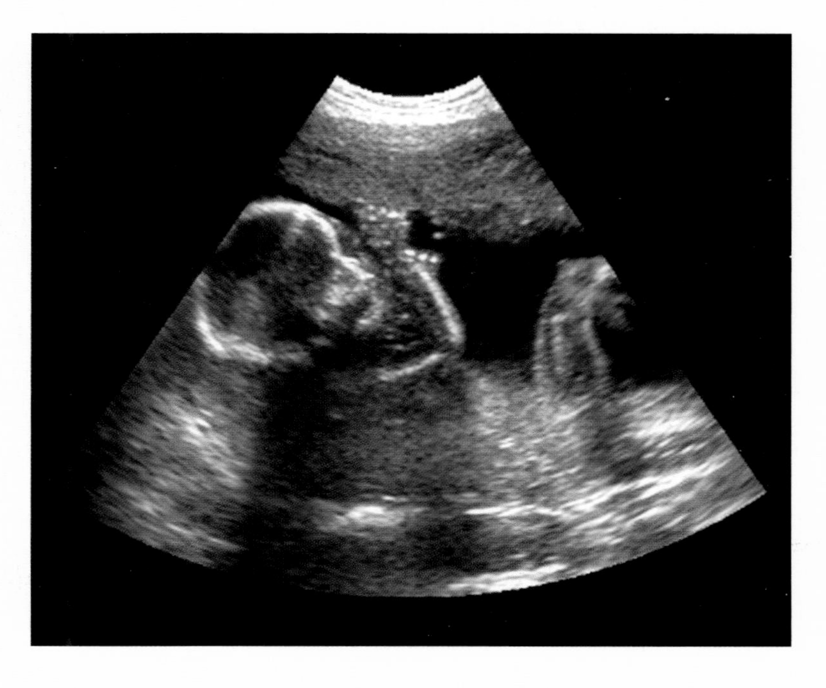

Maybe you've read that bats use sonar—sound waves—to navigate. The longer it takes their high-pitched sounds to bounce off an object and return, the farther away the object is. Ultrasound machines work the same way. (*Ultrasound* means frequencies too high for humans to hear—above 20,000 hertz.) This sonogram (sound image) at right reveals the head, arms, and legs of a fetus.

THE EM DANCE, OR AN OSCILLATING STORY TOLD WITH FREQUENCY

The electric current in the walls of your home (if you live in America) has a frequency of exactly 60 cycles per second (60 Hz). When waves travel with a frequency of 500,000 cycles per second, you've got radio waves (considered slow frequencies). Get into really high frequencies, and you have radar, then ultraviolet (UV) light, then X rays, then gamma rays. To repeat: The shorter the wavelength, the higher the frequency. Gamma rays (on the left side of the chart) have the shortest wavelengths—as short as the distance between two atoms. Radio waves (on the right side) can be hundreds of miles long, but are typically the length of a football field.

The electromagnetic (EM) picture gets really impressive when you consider that these oscillating waves of magnetism and electricity zip along at about 300,000 kilometers (about 186,300 miles) per second and account for all electromagnetic radiation—light, microwaves, radio waves, TV signals, X rays, and gamma rays.

At the end of the nineteenth century, this spectrum of frequencies was just beginning to be understood. In our technological world, it is important stuff, so pay attention to the EM radiation all around you. You can't see the waves, but you can see their effects—like a microwaved hamburger (lower frequencies than visible light) or a sunburn from too many UV rays (higher frequencies than light).

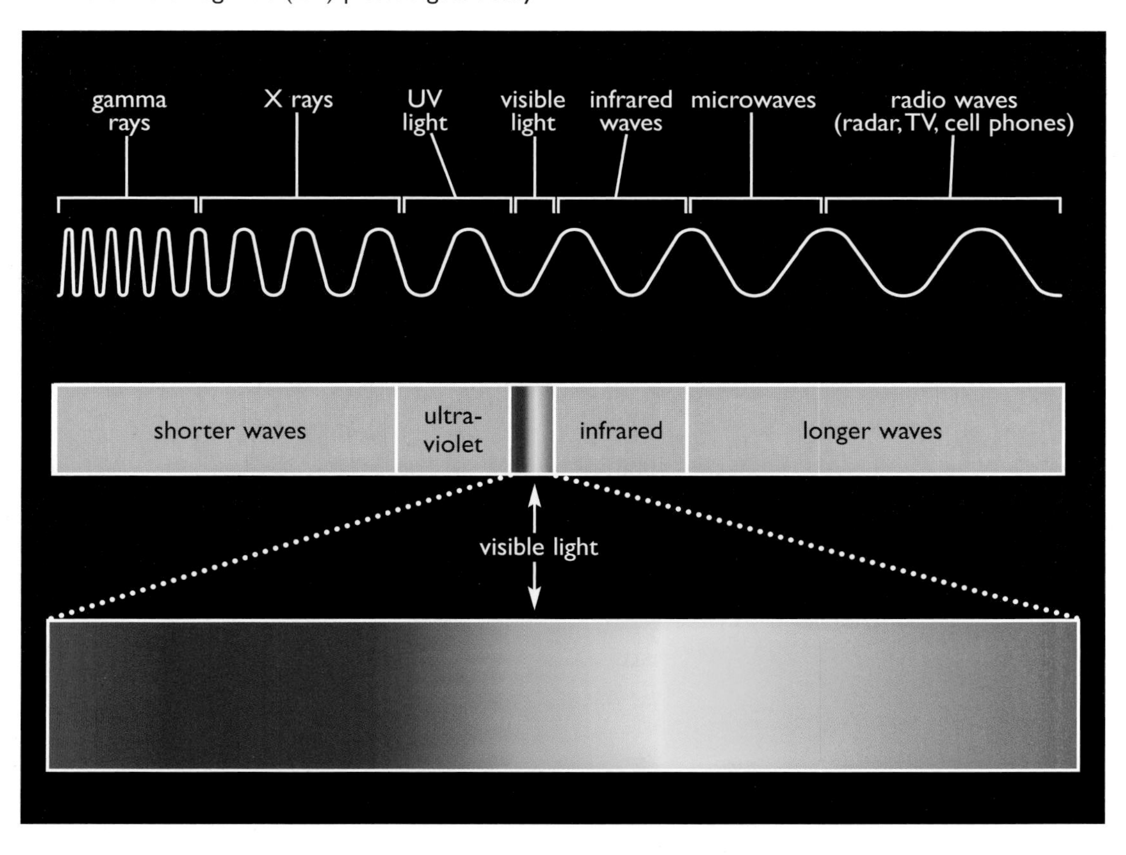

34 Bulldog Boltzmann

The scientist asks not what are the currently most important questions, but "which are at present solvable?" or sometimes merely "in which can we make some small but genuine advance?" . . . All the more splendid is the success when, groping in the thicket of special questions, we suddenly find a small opening that allows a hitherto undreamt of outlook on the whole.

—Ludwig Boltzmann (1844–1906), Austrian physicist, *Theoretical Physics and Philosophical Problems*

Boltzmann was a wizard of a mathematician and a physicist of international renown. The magnitude of his output of scientific work was positively unnerving. He would publish two, three, sometimes four technical books a year. . . . He was what we would now call a hot property, and was continually receiving job offers. He bounced from university to university throughout Germany and Austria.

—George Greenstein (1940–), American astrophysicist, , *Portraits of Discovery: Profiles in Scientific Genius*

Ludwig Eduard Boltzmann has a round face, big ears, a bushy red beard, and a head thick with curly brown hair. Only his glasses are small. They have wire rims, and he peers through them with nearsighted eyes. A student sketches him on a bicycle—a portly professor with coattails in the air. There's charm in the sketch, as there is in the man.

Here's Professor Boltzmann lecturing. Has the artist captured the man for you with his rumpled clothes and intense focus?

That's a 1860 watercolor of the metropolis Vienna (above). Boltzmann was a graduate student teaching mathematics in Vienna when he accepted a position as a physics professor at the University at Graz (GRAHTZ). Johannes Kepler had taught there. Graz, Austria's second largest town, is in the mountains and is bucolic (pastoral). Boltzmann decided to live a country life, so he bought a farmhouse, went to the town market, bought a cow, and led it home himself. He had already consulted a professor of zoology to learn how to milk it.

Boltzmann is a physicist and, in the late nineteenth century, one of the stars of his profession. He can be intimidating—he is intense and very learned—but he is also kindhearted. He can't bear to give his students low grades, and he usually doesn't. They adore him. One of them, Lise Meitner (who will become a famous physicist), describes him as "in a way a 'pure soul,' full of goodness of heart, idealism, and reverence for the wonder of the natural order of things." For Ludwig Boltzmann, physics is a battle for ultimate truth.

Boltzmann's story begins in Vienna, where he is born in 1844. Vienna is a world metropolis, a cultural magnet like Paris or Rome. But the family soon moves to the small Austrian city of Wels, and later to Linz, some 200 kilometers (about 125 miles) west of Vienna. Ludwig's mother is from a merchant family; his father is a tax officer. You might describe them as moderately affluent.

For a while, Ludwig's mother hires tutors to school her three children at home. One is the composer Anton Bruckner

$$\nabla \times E = -\frac{1}{c}\frac{dB}{dt}$$

$$\nabla \times B = -\frac{\mu}{c}(4^{\pi}i + \frac{dD}{dt})$$

$$\nabla \cdot D = -4\pi\rho$$

$$\nabla \cdot B = 0$$

More than 100 years after James Clerk Maxwell died, a professor writing in *Physics First* magazine (in 2004) asked his readers to name the greatest mathematical equations of all time. Maxwell's four equations (above) and Leonhard Euler's most famous equation ($e^{i\pi} + 1 = 0$) tied for first place, beating out Albert Einstein's famous equation ($E = mc^2$), the Pythagorean Theorem ($a^2 + b^2 = c^2$), Isaac Newton's second law ($F = ma$), Ludwig Boltzmann's equation ($S = k \ln W$) and arithmetic's basic ($1 + 1 = 2$).

(then a cathedral organist in Linz), but she fires Bruckner when he throws his wet raincoat on a bed. Nonetheless, Ludwig becomes an accomplished pianist.

But Boltzmann's childhood is not trouble-free. When he is 15, his father dies, probably of tuberculosis. A year later, his 14-year-old brother dies of the same disease. Perhaps those losses help explain his determination to do well and to make his mother proud. He is first in his class at school in Linz.

When it is time for him to go to college, in 1863, the small family—Ludwig, his mother, and his sister—moves back to Vienna. Ludwig enrolls at the University of Vienna as expected. The school has an outstanding faculty, and the new ideas in physics—Maxwell's electromagnetism and Dalton's atoms—are studied and debated. Ludwig is soon doing original scientific research.

He is still a student when a professor hands him some of Maxwell's papers and, as he will write later, "an English grammar in addition, since at that time I did not understand one word of English." Boltzmann doesn't need to translate the mathematical equations—they are in a universal language—but he struggles with the English until he understands that too. Maxwell impresses him above all other scientists of his day. "As with a magic stroke, everything that earlier seemed intractable falls into place. . . . Whoever doesn't feel it should put the book away; Maxwell is no composer of program music who has to add an explanation to the notes."

Not many nineteenth-century scientists understand the importance of Maxwell's equations, but to Boltzmann, they sing a sweet song.

Maxwell's theory of gases describes molecules in action. Some scientists believe those particles are just a metaphor—a convenient fiction that helps explain things mathematically. It's hard for anyone to imagine that atoms and molecules really do exist—or, if they do, that humans can ever prove it.

Boltzmann is convinced that Maxwell's equations define a real world of atoms and molecules. He studies gases and their behaviors in detail and learns that the precise amount that a gas can be compressed can be explained mathematically, but only if you accept the idea that gases are made of tiny entities (atoms and molecules). Starting there, he realizes that if those atoms and molecules exist in mostly empty space, it explains the elasticity of a quantity of gas. And if they bounce around and collide with each other and the walls of their container, that helps explains the pressure gases can exert. (Moving molecules create the kind of pressure that keeps a balloon inflated.) Bernoulli understood that back in the eighteenth century; Boltzmann pays attention to their ideas.

While Boltzmann can't see atoms (except in his mind), and he certainly can't measure an atom, he can calculate their behavior in gases. So he approaches the atomic world with mathematics and statistics. Out of that method he develops the Kinetic Theory of Gases, which becomes an important part of the science of thermodynamics.

Boltzmann is fascinated by thermodynamics. The steam engine, a technological marvel that has changed the nineteenth-century world, has gotten scientists thinking

(continued on page 377)

A computer model shows the paths of three colliding molecules. The number of strands equals the number of atoms in the molecule. Follow one from arrival, to collision, to exit.

ELASTICITY is easy to picture in a rubber band: stretch it, let go, and it snaps back in shape. It's very elastic. Elasticity is the ability of a material to maintain its shape after a force has acted on it and then been removed. How can a gas, which has no shape, be elastic? The force is the collision of molecules, and, even as these molecules keep banging into each other, the total kinetic energy of the gas stays the same. In an inelastic collision, some kinetic energy changes into heat energy—like a baseball stopping when caught in a glove.

KINETIC, as you'll recall, comes from the Greek word for "motion."

THERMODYNAMICS (from the Greek words for "heat power") is the study of thermal energy (heat) in action, especially its conversion to and from other forms of energy.

For more information on the early history of thermodynamics, find out about Frenchman Nicolas Léonard Sadi Carnot (shown at left at the age of 10). He lived a short life (1796–1832) but published an important paper on thermal energy. Some twenty years after it was published, James Joule used Sadi Carnot's ideas to do groundbreaking work on heat and other forms of energy.

HOW FAST IS A GAS?

Gas molecules that are zipping around and colliding with each other and the sides of their container aren't easy to measure. And it's not just because they're invisible to the eye. The molecules are traveling at different velocities, and those velocities change at each collision. (Reminder: Velocity is speed *and* direction.) Boltzmann couldn't possibly clock the speed of one gas molecule, let alone all of them. So he didn't try.

Instead, he assumed that, like all matter, those tiny particles follow the universal laws of physics. According to Newton's second law of motion, they go in a straight line at a constant velocity until a force (another moving molecule) hits them. The equation for an object's momentum is mass times velocity (mv). In chapter 22, Émilie du Châtelet promoted the idea of a squared relationship between mass and velocity—mv^2. Remember the balls dropped into soft clay by a Dutch scientist? Those that went *twice* as fast went *four* times deeper into the clay. In any case, the key point is that energy is about mass and velocity.

In this 1850 drawing, a balloon is filling with hot gases and smoke. The balloon will rise because the heated gases and smoke—with their high-speed, hard-colliding molecules— are less dense than the colder air around it.

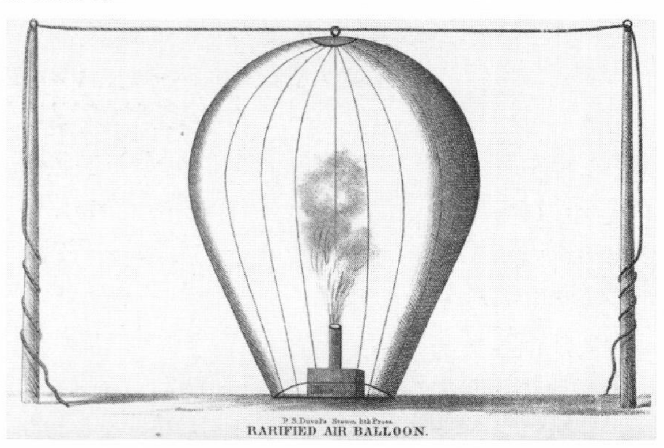

Most things, including mathematical formulas, take time to evolve. The whole idea of energy did so dramatically in Boltzmann's time. Scientists discovered that, even as individual gas molecules change velocity at every collision, the *total* kinetic energy doesn't change. Kinetic energy is the energy of motion. Heat is a form of kinetic energy—molecules and atoms in motion. So this next statement makes a lot of sense: The total kinetic energy is equal to the gas's temperature. If you raise the temperature, the molecules move faster: they have more kinetic energy. And vice versa: If the molecules speed up, the temperature of the gas goes up.

So, Newton's laws, Émilie's squaring, and the thermodynamics of gases led to this simple formula for kinetic energy (E_k):

$$E_k = \frac{1}{2}mv^2$$

Look familiar? The ½ is another one of those constants—numbers that don't vary. (Energy formulas all seem to have one.) Adding that constant allowed scientists to calculate the value of kinetic energy for the first time. In simple words, the formula says this:

Double the *mass* (m), and you double the kinetic energy.

Double the *velocity* (v^2), and you *quadruple* the kinetic energy.

What does this have to do with recklessly speeding gas molecules? If you know their atomic masses and the total kinetic energy of the gas (which is just the temperature), then you can compute the average speed of all the molecules—without even seeing them.

How fast are gases? The table (below) lists the average speeds of a few popular ones, all at a temperature of 25°C (77°F). Oxygen and hydrogen are usually found in nature as diatomic (two-atom) molecules (H_2 and O_2). The average speed of a molecule of hydrogen (H_2) at room temperature is about 6,400 kilometers (about 4,000 miles) per hour! Oxygen (O_2) molecules, which are 16 times more massive than H_2, move at one-fourth that speed—1,600 kilometers (about 1,000 miles) per hour.

Gas	**Atomic Mass**	**Average Speed** (kilometers per hour/ miles per hour*)
hydrogen (H)	1	9,000/5,600
hydrogen (H_2)	2	6,400/4,000
helium (He)	4	4,500/2,800
oxygen (O)	16	2,260/1,400
oxygen (O_2)	32	1,600/1,000

*Figures for average speed are rounded off.

(more about gas speed on next page)

TAKING CHANCES WITH THE SPEED OF GAS MOLECULES

Imagine oxygen (or any pure gas) trapped in a bottle at room temperature. The O_2 molecules are zinging around in all directions, at different speeds. The curve on this graph (below)—called the Maxwell-Boltzmann distribution—plots how fast those various molecules are probably moving. By *probably*, I don't mean "maybe." I mean it in a statistical sense. That's a lot surer and more precise than "maybe," because statistics are based on mathematical formulas—like the one for kinetic energy.

The bottom line of the graph is speed, which increases from left to right. The left-hand side of the graph is the number of molecules. Not many molecules are traveling at very slow speeds (beginning of the curve). A very low portion are traveling at super-high speed (the right side of the curve, farthest from 0). The highest number of molecules are likely zipping around at the speed marked P, at the peak of the curve. To restate: The peak is the *probable* speed of the greatest number of the molecules. And, relatively speaking, it's on the slow side (more to the left of the graph than the right).

The second point, marked A, is the *average* speed of all the molecules—which is a little faster. That's the figure used on the table (page 375)—1,600 kilometers (1,000 miles) per hour for O_2 molecules at room temperature. It's like adding all the speeds and dividing by the number of molecules—only tougher, since there are an infinite number of speeds (points on the curve). If you read the feature on page 163, you'll recognize that as a calculus operation.

There's one more thing. When you're talking about speed, and all the masses are the same, you might as well be talking about kinetic energy—the energy of motion. Here's that formula again:

$$E_k = \frac{1}{2}mv^2$$

If you substitute the atomic mass of oxygen, 32, for m (mass), and divide by 2, the kinetic energy equals 16 times the speed squared. In other words, take m out of the equation, and there's a direct relationship between the speed (squared) and kinetic energy. (Note: Technically, kinetic energy deals with speed—how fast an object is moving. Momentum, the "oomph" of an object, deals with velocity—both speed and direction.)

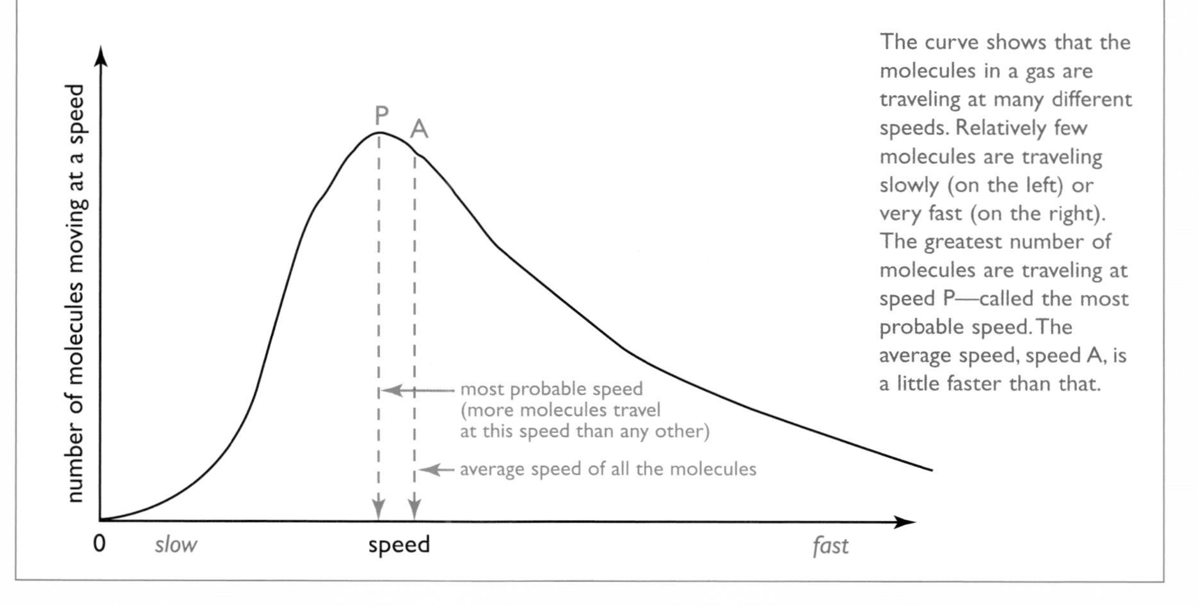

The curve shows that the molecules in a gas are traveling at many different speeds. Relatively few molecules are traveling slowly (on the left) or very fast (on the right). The greatest number of molecules are traveling at speed P—called the most probable speed. The average speed, speed A, is a little faster than that.

seriously about heat. They know steam has power: it can move an engine. But they can't agree on what heat is, scientifically. It is a big question that needs answering. Is heat a property of matter, or is it matter itself? Is it an attribute or an entity? The debate is heated.

"[I shall] live a sufficiently long time to have the satisfaction of seeing caloric interred with phlogiston in the same tomb," said Rumford. But he didn't make it.

Rumford had it right (mostly). He showed that heat isn't an element, but that it is "excited" by motion. Despite that, his evidence isn't proof for some scientists. They stick with "caloric," arguing that heat is a form of matter and a liquid element. It makes a difference that Rumford isn't much liked in England; scientists there tend to ignore his work. Boltzmann, an Austrian, doesn't.

But if heat comes about through motion, *what* is moving? Rumford didn't answer that question, because he didn't know the answer.

Boltzmann figures out that **heat is moving atoms and molecules**. Extending the work of James Clerk Maxwell, he comes up with a formula to measure the speeds of molecules in a gas (it's called the Maxwell-Boltzmann distribution). He has it right, but his ideas get a mixed response. Most scientists of Boltzmann's day, especially his peers in Austria, just will not believe in atoms.

Have you ever had an idea that seems perfectly clear and true and yet no one else seems to get it? Talk about frustration! That's what Boltzmann faces. Dalton (who died in 1844, the year Boltzmann was born) had come up with an explanation of atoms that made sense. But history is full of cycles and, by the late-nineteenth century, the spirit of open inquiry that marked the Enlightenment is slipping away. Getting new ideas accepted is no longer easy.

So Boltzmann becomes a battler. He fights for that atomic idea. Ernst Mach (pronounced "mock"), another well-known Austrian physicist, says atoms are a convenient fiction. "Have

The first image of an atom was made at the University of Heidelberg in Germany and published in 1980. This more recent picture (below) shows a zigzag of cesium atoms (colored red in this computer image) on a sea of gallium arsenide (used to make computer chips; colored green here). Cesium is a silvery metal, but its name comes from the Latin for "sky blue." Blue is the color of its flame test, which is how Robert Bunsen and his partner discovered the element.

HORRORS, SAID MACH; THIS ISN'T SCIENCE

For more than two centuries, scientists had worked hard to be, well, scientific. They'd learned to measure exactly, to categorize, to investigate directly. By the mid-nineteenth century, a few scientists realized that some things can't be exactly measured, or even seen. What should they do? Ignore those things? Yes, said traditionalists like Ernst Mach; science is based on precise measurement.

What about atoms? No one expected to ever see them. To understand them required an imaginative leap of mind. Was it scientific to do that? Ernst Mach was one of many scientists who couldn't abide the idea. Mach also rejected ideas like absolute time and absolute space because they can't be proved for all time and all space. That thinking prepared the way for Albert Einstein, who would also reject those absolutes (but didn't reject atoms).

Ernst Mach was another professor at the University of Graz. You know him because of the Mach number—Mach 1, Mach 2, etc.—the ratio of the speed of an object to the local speed of sound. (The speed of sound varies.) The Mach gauge above measures the speed of a jet. Mach was an interesting fellow who, like Boltzmann, went from university to university because he was in such demand. He was a professor of math and philosophy. He believed that nothing in science is acceptable unless it can be proved empirically (by experience or observation).

you ever seen one?" he taunts when Boltzmann lectures. If something can't be observed directly, says Mach, then scientists shouldn't treat it seriously. That does make sense. England's esteemed Lord Kelvin (William Thomson) also rejects the idea of atoms and molecules, and Kelvin is a first-rate scientist with international clout. (So is Mach.)

It would be easy to explain heat if atoms could be seen with a microscope, but that's impossible. And heat needs explaining. It is not only mysterious, it is important. Even the most primitive organisms depend on heat to survive. But if you're looking for an invisible fluid that doesn't exist, you can't get far. The arguments get angry. "Caloric" has a tenacious ghost. Finally, most scientists who are working on thermodynamics just give up on the whole question of what heat is. Instead, they measure thermal energy mathematically and find the work that can be obtained from a given amount of it. In other words, they measured what heat *does*, not what it *is*. Looked at that way, it doesn't matter why heat works the way it does.

It does matter to Ludwig Boltzmann. He understands that heat is moving atoms and molecules. And he has it right.

But both approaches to understanding heat (through studying what it is and what it does) turn out to be useful. The mainstream nineteenth-century work on thermodynamics will lead to two important scientific laws. (Keep reading, we're about to get to them.)

As to atoms, the concrete proof that they exist is coming. Unfortunately, it will happen after Ludwig Boltzmann dies of a depression-related suicide.

Ludwig Boltzmann was a stout, Santa Claus–looking Austrian who was once compared to a bulldog, probably because bulldogs are stubborn and tough. Actually, he was more like a puppy—sometimes gruff on the outside but sweet-natured inside. It was only when faced with atom doubters that he got growly. He believed in atoms. He really believed.

Boltzmann laid a foundation for Albert Einstein, with his studies of something called black-body radiation. A black body is a perfect (but imaginary) container that absorbs all the heat radiation that strikes it. When it gives off radiation—well, that's where mathematical problems appear that neither Boltzmann nor any of his contemporaries could solve. The explanation of black-body radiation had to wait for quantum theory (about subatomic particles) to appear at the beginning of the twentieth century.

INSIDE THE ATOM

In this book we've been traveling with scientific explorers—climbing the ladder of knowledge one rung at a time. That those pioneers "discovered" atoms when there was no way to see them is one of the all-time astonishing feats of the human mind. Physicist Gerald Feinberg, writing in 1978, said, "Even the smallest bacterium that can be seen with an ordinary microscope contains 10^{19} atoms, while a person contains about 10^{28}. This means that there are more atoms in each of us than there are stars in the whole Universe."

Those squiggly green lines (below) are tracks made by subatomic particles as they zip through superheated hydrogen. The particles make the liquid hydrogen boil—or bubble up—which is why the apparatus for recording their paths is called a bubble chamber.

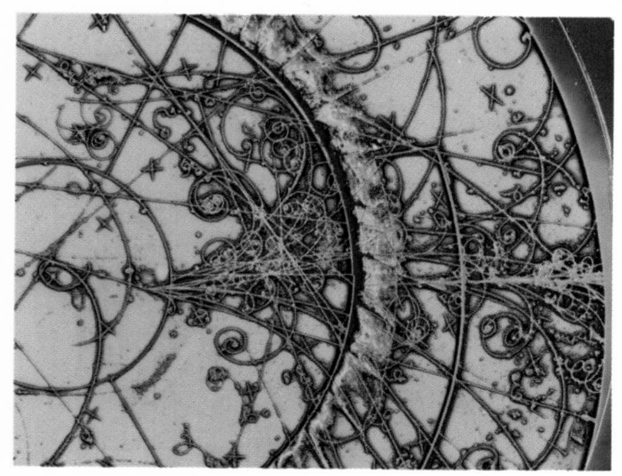

How we found out about atoms is a true adventure story. You and I can now know things that would have seemed impossible to the most brilliant thinkers of the past. Here is a summary of today's basic information on atoms, to help you make sense of the chapters to come.

The Essence of Atomic Theory

All things are made of atoms.

Atoms are complex entities, each with a structure and parts. Atoms seem small to us, but in the microscopic world,

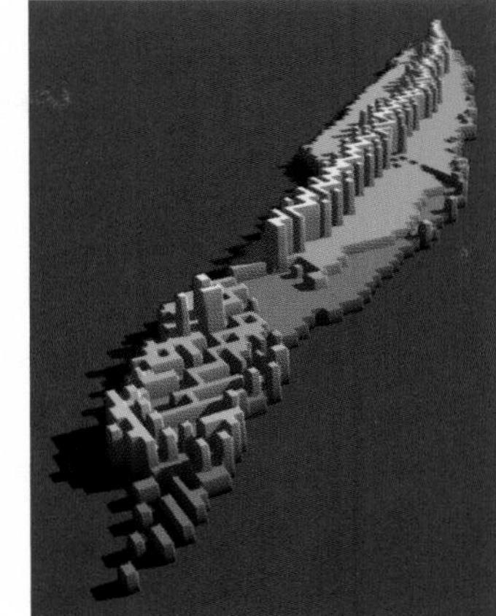

This is a picture graph of the properties of the nuclei of atoms. The gray ones are natural and stable and range from hydrogen (at the top) to uranium (in the middle). The orangish dots on the bottom are unstable; they have extra neutrons and are radioactive.

they are giants. There are smaller-than-atoms (subatomic) particles. Most of the mass of an atom is in the nucleus, which seems huge on an atomic level. It is composed of very tiny particles called neutrons and protons. Orbiting the nucleus are much, much smaller electrons. The neutron has no electric charge. Protons have a positive charge; electrons have a negative charge. In a stable atom, they are in balance—one proton for each electron. An *ion* is an atom that has gained or lost one or more electrons and so is electrically charged.

Each element is made up of atoms with a unique number of protons. Hydrogen atoms have one proton; uranium atoms have 92 protons. The number of protons is the element's atomic number. An element can have *isotopes*—atoms with a different number of neutrons. An isotope is the same element as its brother and sister atoms (because it has the same number of protons), but extra neutrons can make it heavier and sometimes radioactive. Radioactive atoms spontaneously emit bursts of radiation (electromagnetic energy or particles), a process called radioactive decay. Radioactive decay results in an atom transmuting (changing) into another type of atom—sometimes almost instantly, sometimes after waiting for millions of years.

Atoms never stop moving. Another way to say that is: Atoms are in perpetual motion.

Here is Richard Feynman (1918–1988) speaking at the California Institute of Technology in 1961 and putting our knowledge of atoms in perspective. (Some say that Feynman is the greatest physicist America has produced so far.)

> *If, in some cataclysm, all of scientific knowledge were to be destroyed, and only one sentence passed on to the next generations of creatures, what statement would contain the most information in the fewest words? I believe it is the atomic hypothesis that all things are made of atoms—little particles that move around in perpetual motion, attracting each other when they are a little distance apart, but repelling upon being squeezed into one another.*

How Scientists Got to Know About Those Tiny Entities Called Atoms

Studying gases gave them clues. The behavior of a gas only makes sense if you

You've seen these computer models of water molecules before (pages 123 and 240). That's water vapor (a gas) on the immediate right, then on page 383 liquid water, and, finally, an ice crystal. Notice all the open space in the crystal, especially compared to liquid water. That's why ice is less dense than water.

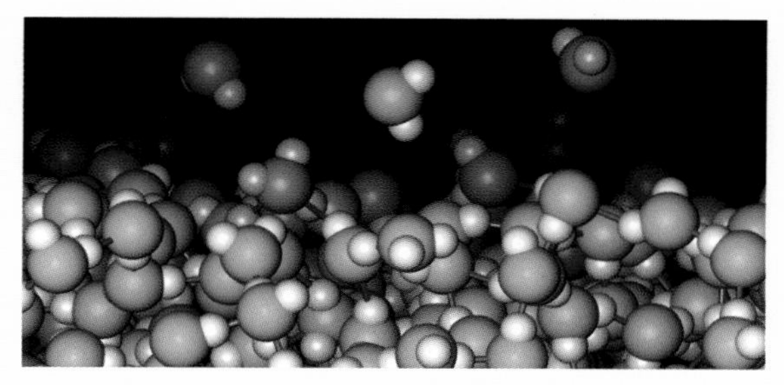

do a thought experiment and imagine that it is composed of tiny particles. Then a gas can be measured and analyzed. Its behavior can be predicted mathematically. One of the pioneers in atomic research was that Italian count, Amedeo Avogadro, who came up with the idea of *molecules*. (His story is in chapter 28.)

By studying the behavior of gases intently and imagining small particles, Avogadro figured out that equal volumes of gases at the same temperature and pressure have equal numbers of molecules (or atoms). He realized that if you knew the number of atoms for a certain volume, you could compute the masses of the atoms in other volumes. Later experimenters calculated the number of atoms in 1 gram of hydrogen or 8 grams of oxygen or 12 grams of carbon (6.02×10^{23} atoms, or 602,000,000,000,000,000,000,000 atoms) and called it Avogadro's number. (Today, the textbook definition of Avogadro's number is the number of atoms in 12 grams of carbon.) What that gigantic number tells us is that each atom is very, very small. Today, Avogadro's number, also called a *mole*, is a foundation of atomic theory.

Atoms Move, and That Explains Heat

Remember, when we say all atoms are in perpetual motion, we mean *all* atoms—in a drop of water, in the cup that holds the water, in the metal pipes that carry the water.

The energy in the dancing motion of atoms is known scientifically as thermal energy and informally as heat. If thermal energy moves from one object to another (from a burner to the water in a teakettle), the dance quickens (in the kettle). To say it another way: Speed up the atoms, and you increase the heat.

Water, H_2O, as an Example to Further Explain Atomic Theory

One atom of oxygen and two atoms of hydrogen, bound together, make up a molecule of water. The oxygen and hydrogen atoms act as if they are glued to each other, but the truth is, they have found a balance of electric charge. If you try to squeeze the atoms closer together, they will repel each other.

If you could watch the molecules in hot water, you would see them flying this way and that. The hotter the water gets, the more they rush away from each other. Some zoom right out of the water container as a gas—steam. In a gas, some molecules move as fast as a supersonic plane.

What happens when you cool down water? Molecules in a solid—like ice—don't move around because they're locked in place. Instead, they vibrate.

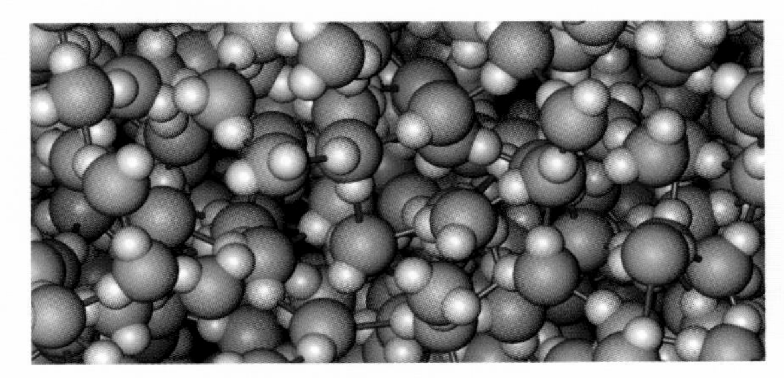

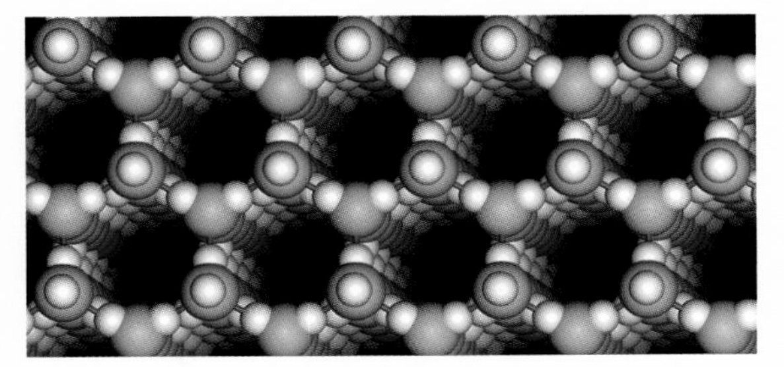

Pressure and Density

Heat water in a closed container, and the movement of those tiny molecules, dancing wildly, will create *pressure*. Pressure can blow the whistle on a teakettle or the lid off a container.

Turn up the heat, and the molecules go wild. They hit harder and move faster, and they hit more often. **Pressure increases as temperature increases.**

Suppose your container has a lid that slides up and down (a piston). Push down on the piston and compress the steam. Those bouncing molecules have less room—so they hit each other with more frequency.

When we compress gas slowly, its pressure increases.

When we slowly expand the space a gas occupies, we decrease the gas's pressure.

Solids, Liquids, and Gases

To repeat: Atoms are in constant motion—this way and that. In a liquid, they seem to move at random at a moderate speed. In a gas such as steam, the speed of their dance becomes frantic.

When water becomes a solid (ice), something interesting happens. **Each molecule has a specific place in a solid.** The molecules no longer move around at random. They make a pattern that is hexagonal (six-sided) and symmetrical. It is called a *crystalline array*. But even though there is a symmetrical organization of molecules in a solid, don't think the jiggling of the atoms stops. They vibrate in place within that crystal pattern they've formed.

Near a temperature of absolute zero (which is unattainable), there is hardly any vibration, but **there is never zero vibration.** Those atoms don't give up! If we provide some heat, the jiggling increases, the ice melts, and the molecules zip around freely again.

Why Is This Important to You?

To begin, you're a mass of atoms. (Or are you a mess of atoms?) As to the stars and planets, they are just collections of atoms—the same kinds of atoms you're made of. You can't understand the world you inhabit without knowing about atoms.

You have now learned the basics of atomic theory. Keep in mind, those who brought us knowledge of atoms changed human history. Those pioneers were exploring with their eyes closed. They couldn't see atoms, let alone see inside atoms. Read on, and you'll learn more about how they managed it.

Wake Up! This Is About Work, Which Takes Energy

All science is the search for unity in hidden likenesses.... Science is nothing else than the search to discover unity in the wild variety of nature.... Poetry, painting, the arts are the same search.
—Jacob Bronowski (1908–1974), Polish-British scientist, *Science and Human Values*

What Joule's experiments showed above all was that when one form of energy was converted into another; no new energy was created and no old energy destroyed.... Whenever a certain amount of energy seems to disappear in one place, an equivalent amount must appear in another.
—Isaac Asimov (1920–1992), American science writer, *The Intelligent Man's Guide to Science*

This is a tough chapter, so get prepared. We're going to turn up the heat, do some work, and define energy. And that isn't easy. Energy is not like matter. You can touch, taste, hold, and smell matter. It occupies space. It is a substance. Energy is less tangible to our senses (but not less important). It's the mover of things. Energy and matter are all there is that's scientifically knowable in the universe.

Did you catch the title of Asimov's book, above? It's a good book, but in 1960, when it was published, many men didn't notice the very intelligent *women* who make up half of the population.

But there wasn't even a scientific word for energy until 1807, when an English scientist, Thomas Young, took a similar Greek word that means "work" and defined energy as we do today. It's "the capacity to do work."

What's work? Watch out: Scientists use words differently than they are used in everyday conversation. To a scientist, **work is force times distance**. If you spend a day pushing on

WORK (w) equals force (F) times distance (d):
$$w = Fd$$
ENERGY (E) is the "capacity" to do work. Capacity can be measured, it can be quantified (given a number), and its relationship to mass and volume (and so on) can be calculated.

a huge boulder but never budge it even a bit, you may think you have worked hard, but scientifically you have not accomplished work. There needs to be both effort and movement.

To repeat: *Work* is a word with a scientific meaning: force times distance. That definition makes work exact and able to be calculated mathematically. If you double the force applied to an object, you will double the work. If you double the distance over which the force is applied, you will double the work.

A unit of work—the product of force times distance—is called a **joule**—(sounds like "jewel") and is abbreviated J. It's named after an English scientist, James Prescott Joule, who was born in 1818 to a wealthy beer brewer. Papa Joule was able to give his son a private tutor (the family had six live-in servants). The tutor's name was John Dalton! Dalton was a dedicated, patient teacher, and Joule an eager student. Joule learned up-to-date science. The family business taught Joule things, too. Brewery technology was cutting edge in the nineteenth century; when you brew beer it's important to have exact temperatures. As a scientist, Joule became famous for his careful measurements.

James Joule was never a professor; he remained a brewer all his life. In the democratic world of science, that didn't matter. A modest, religious man, toward the end of his life he protested the increasing use of scientific discoveries to make war.

In 1599, ENERGY meant "force of expression" in general. It's easy to picture Galileo, in the prime of his life, putting energy into his explanations of gravity and acceleration, but he didn't put energy (in a literal sense, anyway) into an equation. Thomas Young's 1807 definition of *energy* made it measurable. Scientists could now write out equations that included energy.

In nineteenth-century breweries, steam engines powered the machines that mixed the beer ingredients. Keeping the mixture at a constant temperature was tricky because, as yeast turns sugar into alcohol, heat energy is released.

FOLLOW THE MONEY

In the sixteenth century, wealth came mostly from trade. So when the Scientific Revolution got revved up, it focused on navigation and an understanding of the stars. That's why Galileo's telescope was so important.

In the eighteenth and nineteenth centuries, wealth shifted. Now manufactured goods brought riches. Machines began to replace muscles. Science changed too. It began to focus on mechanical energy. Thermodynamics and electromagnetism became "hot" fields. The knowledge from these studies helped with the building of machines.

James Joule was born into this whirlwind of change. The animal and vegetable kingdoms had always supplied humans with all their energy needs (think basics—food, transportation, warmth). Beginning at the end of the eighteenth century, coal began to be tapped, and then oil, electricity, and, later, the atom. By the mid-nineteenth century, you didn't need a horse to transport you; trains could do it. Life expectancy in the pre-industrial age was about 30 years. But by 1850, most men and women in industrialized nations were blowing out lots of birthday candles. As to personal income, many ordinary people, who a century earlier might have been hungry peasants, had full cupboards, comfortable homes, and even leisure time.

Today in this information age, we're producing intelligent machines. They're doing the work of the brain. And they've created new wealth. Where do you think science will head next?

Power looms and express trains! Imagine a machine that could do quickly and well what a lone worker did slowly and laboriously on a hand loom. As to travel? Those trains could climb mountains and out-race the fastest horse and buggy. Now ordinary people could travel and wear fine fabrics.

He was fascinated with thermal energy (heat) and spent 35 years analyzing and measuring it. Joule measured the heat generated from electric current, waterwheels, compressed air, muscle power, and more. He discovered that a given amount of work always produces a given amount of thermal energy. His precise measurements became standard and are used today to measure work and heat.

One J is equal to the amount of energy needed to lift 1 pound of water 9 inches off the ground, or to accelerate it to a speed of 7 feet per second (walking speed), or to raise its temperature 1/1,000° Fahrenheit. In metric terms, a joule of work is done when an object weighing 1 newton (N) is lifted straight up a distance of 1 meter (that's equivalent to lifting an apple over your head).

The measure of how fast we do work—we call it power—is the joule per second, also known as a watt, named in honor of James Watt (you know him, the man who designed the steam engine). One watt (W) of power equals 1 joule of work done in 1 second.

Scientists describe two kinds of energy: Anything in motion—like the wind or a falling rock—has kinetic energy. The formula for **kinetic** energy is $E_k = \frac{1}{2}mv^2$. The m stands for mass, and the v stands for speed, not velocity, because the direction of the object doesn't matter here. Yes, that formula was in the previous chapter, but it's important and worth repeating. (Besides, it's not difficult. Take a deep breath if you have a math phobia, and think of this as a simplified expression of a natural phenomenon.)

Still air, a stationary rock, water in a placid lake, and you sitting in a chair—all have **potential** energy. Potential energy is energy-in-waiting. It has the potential to be converted into kinetic energy. Start moving the air, the rock, the water, and you—and that potential energy turns into kinetic energy. The formula for gravitational potential energy is $E_p = mgh$ (g is acceleration due to gravity; h is height).

A ball thrown straight up into the air has

Power is the amount of work done over a period of time. If you need to do a lot of work—fast—then get a high-powered machine to do it. Another way to think of power is the amount of energy that's changed or transferred, divided by the time.

A European tree frog resting on a twig has chemical potential energy stored in its muscles. As soon as the frog leaps up, its chemical potential energy goes down, and gravitational potential energy and kinetic energy enter the equation.

Picture yourself pulling back an arrow on a bowstring, as young Franklin Delano Roosevelt is doing here (right). If you calculate the force you use to pull the bowstring and multiply it by the distance you pull, you can figure out the amount of work you've converted to potential energy—work done by the muscles in your arm. Now release the string, and, as the arrow flies, that potential energy becomes kinetic energy.

kinetic energy. As it rises, gravity slows it down, and the ball loses kinetic energy, until, at the top of its flight, all it has is potential energy. Then, as the ball descends, it gradually gives up its potential, which turns into kinetic energy.

You have the potential energy to clean your room. Will you use the electric energy of your brain, nerves, and muscles and convert that potential into kinetic energy and do the job? (Of course you will . . . won't you?) Your energy

PREVENTING HEADACHES, AND OTHER PRACTICAL REASONS FOR BEING SCIENTIFICALLY SAVVY

If one day you happen to walk under a crane and look up, would you rather see a piano or a paper clip dangling over your head, about to be dropped? That's a no-brainer—more mass equals more potential pain.

Now, suppose you're standing next to a six-story building and you look up. It's just not your day: someone is holding an old shoe out the window—again, right above your head. Would you rather that this strange, careless person live on the first floor or the sixth floor? Another no-brainer. The greater the height, the louder the possible "Ouch!"

So there you have it: Mass and height are what gravitational potential energy is all about. As long as we keep that energy as potential (a dangling piano), you don't have to worry. But if it should become kinetic (a falling piano), start running. If you want to know *how much* mass and *how much* height is cause for concern, feel free to use this equation (which you'll also find on page 387):

$$E_p = mgh$$

The E_p stands for the amount of potential energy (or potential pain). Picture that piano, and you won't forget mass (m) and height (h). (The height, by the way, is relative to any surface that you decide to call "zero"—the ground, a tabletop, your head, etc.) The g is acceleration due to gravity—a constant (unchanging number) for all objects on Earth, which is why you need to worry only about mass and height.

If you think of height as a distance and remember that we're talking about force (gravity) acting over that distance, potential energy begins to sound more like work—force times distance. The formulas are closely related.

Like its alter ego kinetic energy, potential energy is measured in joules. That old shoe has a potential energy of about 15 joules at the first-floor window or about 100 joules dangling from the sixth floor. And the piano? We're talking five figures, maybe 60,000 joules hanging from the sixth floor.

—LJH

comes from the hamburger you ate for lunch, its energy came from a steer, whose energy came from grass it chewed, and the grass absorbed its energy from sunlight. One form of energy can turn into another.

Energy is often classified in terms of its nature. We speak of

- **mechanical energy**—a big category—*mechanics* being the scientific word for motion
- **gravitational energy**
- **electrical energy**
- **chemical energy** (especially from coal and oil)
- **thermal energy** (the energy of heat, which is the kinetic energy of atoms and molecules)
- **nuclear energy** (from the atom's nucleus; it fuels power plants and the Sun)

Still other forms of energy are **radiant** (electromagnetic waves from the Sun) and **mass**. How can mass have energy? Hold that question. Physicist Albert Einstein will answer it in the twentieth century. (Here's a hint: In their ultimate particle essence, matter and energy are two forms of the same thing.)

The thing to remember here is that, scientifically, energy is anything that can be converted into work. Even work itself—which can be turned into thermal energy and then back to work—is a form of energy. And work always involves a transfer of energy. So to sum up: Energy is a measure of work; it can be kinetic or potential; and it has many sources, from coal and oil to the Sun's rays to human effort.

All change is a miracle to contemplate; but it is a miracle which is taking place every instant.
—Henry David Thoreau (1817–1862), American essayist and poet, *Walden*

The thermal energy needed to make a geyser explode comes from the molten rock deep inside the Earth. Above is a geyser in the Whakarewarewa Valley in New Zealand.

Energy is one or the other—potential or kinetic—because something is either moving or it's not. Kinetic energy is about *motion*, and potential energy is about *position*. Potential energy relates to the position of an object in a field where a force is acting on it. A rock sitting on the edge of a cliff has *gravitational* potential energy because the force of gravity is acting on it. *Elastic* potential energy describes stretched rubber bands, wound-up springs, and pulled-back bowstrings about to let arrows fly. *Electric* potential energy is voltage—as in the potential of a battery to get electric current moving.

INFORMATION-AGE THINKING IN THE INDUSTRIAL ERA

The world was changing. Everything seemed to be moving, and moving fast, in ways that had never happened before. The Industrial Revolution was doing it, and whether you were a surveyor, a banker, an insurer, a sailor, or an astronomer, early in the nineteenth century you were suddenly confronted with mountains of statistics, graphs, and tables.

Among other changes, Europe's population had jumped from about 160 million in 1750 to about 280 million in 1850 (without counting Russians). And the percentage growth in productivity was even greater than the population growth. No one was prepared for this—especially the accountants. Before the industrial era, they had dealt with small numbers; now there was a need for sophisticated and extensive calculations.

In this 1845 engraving of a factory in Germany, men and women are nowhere in sight. Only children work the machines.

The only "computers" were humans. Large companies were hiring teenagers as adding machines. (It kept down expenses; child labor was cheap.) The teenagers made mistakes, sometimes costly mistakes. But don't blame the kids; even skilled

"The whole of arithmetic now appeared within the grasp of mechanism," Charles Babbage wrote in his 1864 autobiography, *Passages from the Life of a Philosopher*. He had just solved a key part of his Analytical Engine invention (left). After 20 or so plans, "each succeeding improvement advancing me a step or two," his calculating machine could finally carry tens from one column to another.

mathematicians make mistakes.

In 1821, two friends, astronomer John Herschel (1792–1871) and mathematician Charles Babbage (1791–1871) were examining some mathematical tables. The young men had been undergraduates together at Cambridge University and were now working on a project for the Royal Astronomical Society (which they had helped found). As they checked the mathematical tables, they grew increasingly agitated: they could see the tables were filled with errors.

They knew this wasn't an isolated problem. In the first edition of the Royal Society's *Nautical Ephemeris for Finding Latitude and Longitude at Sea*, there were more than 1,000 mistakes. Sailors depended on those astronomical tables. England's economy was tied to its dominance at sea. The errors were more than annoying: they could cause shipwrecks and engineering disasters.

As for the tables compiled for Herschel and Babbage, unless they were corrected, the tables would be useless. Babbage said, "I wish to God these calculations had been executed by steam!" Steam was powering the engines that were changing and obsessing the times.

Charles Babbage decided to build a machine that would calculate math tables automatically and print out results (eliminating the possibility of human error). His machine, with a massive cast-iron frame, would be unlike anything built before. It would require hundreds, in some cases thousands, of near-identical parts—at a time when mass production was unheard of and machines were still a craftsman's pride.

Babbage, now revered as the granddaddy of computer gurus, was an independent thinker, a gentleman philosopher, a mesmerizing storyteller, and the delight of socialite hostesses (he kept their parties lively). He invented skeleton keys and the speedometer, helped England devise the first modern postal system, and tried (but failed) to come up with a foolproof system for winning at horse races. (He was attempting to please the mathematically brilliant Ada Byron, Countess of Lovelace, daughter of the poet Lord Byron, who had lost a fortune betting.)

When Charles Babbage was a student at Cambridge University, he urged his teachers to strive for higher standards in math teaching. Then, to help them, he translated some textbooks used in Germany.

The son of a wealthy banker, Babbage was Lucasian Professor of Mathematics at Cambridge, a title that Isaac Newton had held. But he spent most of his fortune on his calculating machines, and that's where he put his heart and his energy. Babbage was a perfectionist, and before he had finished what he named a "Difference Engine," he had an idea for an "Analytical Engine" that would be even better. It was a computing machine fed with punch cards, and Babbage said it could eat "its own tail." By that he meant the machine could feed the results of a calculation in one column into other columns. In other words, it could create its own instructions.

Babbage was on to something, but his machines were costly, and the government funding he fought for was never enough. It didn't help that, like many innovators, he kept changing and reengineering his ideas. He never got the practical side of the machine quite right; his vision was ahead of the mechanics of the time, and he died thinking himself a failure. A century later, when the times and technology were right (after an electronic revolution), engineers turned to Charles Babbage's machines for some of the basic ideas that guide today's computers.

MR. BOOLE, NO FOOL, DID IT BY TWOS

George Boole (1815–1864) was a shoemaker's son and so poor he couldn't afford college. But Boole's mind was too good to go unnoticed. He taught himself math and became a professor of mathematics at a university in Cork (a city in Ireland). In 1854 he published a book called *An Investigation of the Laws of Thought*. Philosophers, since the time of Aristotle, had come up with systems of logical thought, but they always did it with words. Boole took the ideas of logic and turned them into a mathematical system with an arithmetic and an algebra similar to the familiar arithmetic and algebra of numbers. (It is called symbolic logic.)

In Boole's algebra (known as Boolean), the only digits are 0 or 1. That's called a base-2 or binary system. Boole gave each false statement the value 0 and each true statement the value 1. When you give concepts number values, you're on your way to doing arithmetic with them. It can get quite sophisticated. Binary has turned out to be the best system for programming computers.

Charles Babbage never used binary arithmetic or Boolean algebra in his computers. He stuck with the familiar decimal, or base-10, system. The mechanical switches he had to use were slow, and base 10 made them slower still. When the transistor was invented in 1947 (almost a century later), electronic switches became possible; those switches turn on and off so fast they seem instantaneous. A binary system, with only two possibilities (0 or 1) is perfect for that switching process. So, when twentieth-century computer engineers searched for a mathematical base for their computer programs, they found it in the work of George Boole.

Today, computers store information as a series of 0's and 1's, known as *bits* of data. (A byte is a series of 8 bits.) In binary, 1 is 1, 2 is 10, 3 is 11, 4 is 100, 5 is 101, 6 is 110, and 7 is 111. For most of us, the problem with using base 2 is that it takes a lot of digits to express most numbers; 729 is 1011011001. That is no problem for a computer.

In modern computers the electronic switches are either on or off. Technicians describe that switching as "up or down" or "true or false" or "zero or one." They all refer to base 2 and Boolean logic.

Think of the following sentence: I am happy *and* I am hungry. We seem to have added together two sentences with the word *and*. If you combine two true sentences in this way, you get another true sentence. What happens if you combine a true and a false sentence? What happens if you replace the word *and* with the word *or*? Boole's algebra answers even the most complicated questions of this nature. Words like *and* and *or* are called logical operators.

Inside modern computers, microscopic switches—based on Boole's logical operators—are known as "gates." (No, they are not named after computer billionaire Bill Gates.)

To picture a computer gate, think of a light switch. Suppose that when the switch is up, the light is on. What happens to a light if two switches are connected to it? Here's a simple table that summarizes possibilities (there are actually a few more ups and downs that can be charted):

First Switch	Second Switch	Light
Down	Down	*Off*
Down	Up	*On*
Up	Down	*On*
Up	Up	*Off*

Tables like the one above (but often more complicated) are an important part of the process. Electricians use such tables for wiring circuits. For computer engineers, tables describe the gates that make the brains of a computer work. Logicians call them "truth tables." What do they have to do with truth? Take the following sentences. They can't all be true, but if you make your own chart, you can figure out the true-false possibilities:

Sentence 1: It will rain today.
Sentence 2: I will get a letter in the mail today.
Sentence 3: Either sentence 1 or sentence 2 is true, but not both.

Lewis Carroll (1832–1898), who wrote *Alice in Wonderland*, was a younger contemporary of George Boole, and a mathematics professor at Oxford. Carroll also wrote a book on symbolic logic for young readers. It's full of fun, as you might expect from the man who created the White Rabbit and Tweedledee and Tweedledum. Those memorable chaps (below) are talking to Alice about our subject: "I know what you're thinking about," said Tweedledum: "but it isn't so, nohow." "Contrariwise," continued Tweedledee, "if it was so, it might be; and if it were so, it would be; but as it isn't, it ain't. That's logic."

Using gates, computers deal with possibilities. Boolean logic makes computers appear as if they can think. Every time you use a search engine, like Google, you are using Boole's logical operators.

Boole had no idea that his thinking would help unleash a multibillion-dollar industry and a new era in history—the information age. He died long before the modern computer could be invented.

Poor George Boole. One night, he hiked 2 miles to deliver a lecture. It was raining those proverbial cats and dogs. He got soaked and came down with pneumonia. Boole's wife, following advice of the time, believed that to get rid of an illness you exposed the patient to its cause. That may sound strange, but it is what makes a vaccine work. Boole's wife didn't get the idea behind vaccines. She kept drenching her husband with water as he lay in bed. No surprise, he didn't make it out of that bed. He was 49 when he died.

A Number-One Law, Thermodynamically Speaking

[Energy] once in existence cannot be annihilated; it can only change its form.

—Julius Robert von Mayer (1814–1878), German physician and physicist, "Remarks on the Forces of Inorganic Nature"

This creative association of the two seemingly incongruous concepts of heat and motion is so rewarding…that it ranks with the supreme achievements of physics.

—Hans Christian von Baeyer (1938–), American physicist, *Rainbows, Snowflakes, and Quarks*

We have now arrived at one of science's most important laws. It sounds simple, but sometimes the most obvious things sit under your nose and you don't notice them. This law is considered one of the great achievements of nineteenth-century science. Put it in your head.

It is called **the first law of thermodynamics**, or the Law of Conservation of Energy. It says **Energy can be changed from one form to another, but it can't be used up**. The nineteenth-century physicists who were studying heat discovered this concept—and it has turned out to be a fundamental law in physics. Stated again, it is *energy can neither be created nor destroyed—it can only be changed.* Or, said still another way, *The total amount of energy in the universe is always the same.*

Remember: ENERGY is the capacity for doing work.

Albert Einstein said the first law of thermodynamics "is the only physical theory of a general nature of which I am convinced that it will never be overthrown."

In essence, the First Law of Thermodynamics says heat is just another form of energy, and energy as a whole is conserved.

—Emilio Gino Segrè (1938–), Italian-American physicist and Nobel Prize winner, *A Matter of Degrees*

But, hold on; this law isn't quite as easy and clear-cut as it seems. When one form of energy turns into another, some energy always escapes. For example, electrical energy lights your lamp, but some of that energy radiates away as heat (you can feel it near the lamp). When you boil water, changing it into steam, some of the energy goes out the teakettle's spout as moving molecules of vapor. It seems to be lost, but it isn't. That "lost" energy stays in the universe in some other form; that's what the first law of thermodynamics tells us.

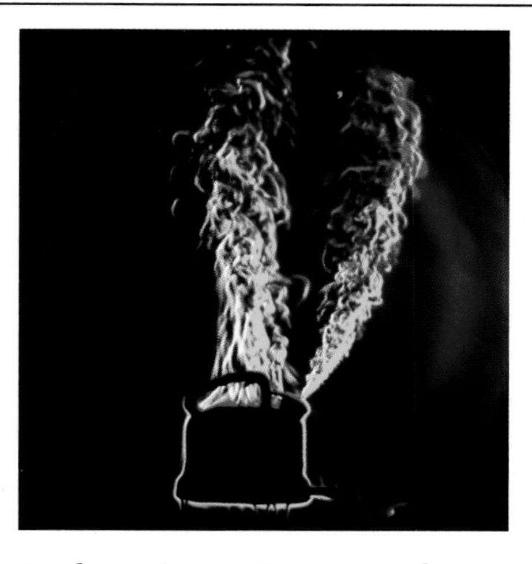

As hot steam rises from a teakettle, it changes the density of the surrounding air. A technique called Schlieren (SHLEER-en) photography recorded the tiny differences in density to create this image.

Now, thermodynamics does have a problem. How would you like a T-shirt that says *The total amount of energy in the universe is always the same*? Yawn. That's the problem. The subject doesn't excite—didn't in Boltzmann's day, doesn't now. But it should. That idea is right up there with Newton's laws of motion and Maxwell's equations. It's something every scientifically literate person needs to have in his or her mental backpack (or on a T-shirt). So memorize the first law. It's part of your intellectual heritage. Now, to know how we managed to figure out about it, read on.

LAWS FOR CONSERVATIVES—AND CONSERVATIONISTS TOO

Science's conservation laws go a long way toward helping us make sense of the universe. Here are two others.

From Lavoisier, the **Law of Conservation of Mass** (1789): In any closed system (the universe is a closed system), the total amount of mass stays the same no matter what kind of chemical or physical changes occur.

From John Wallis, the **Law of Conservation of Momentum** (1668): The total momentum in a closed system always stays the same. Momentum? It's mass times velocity, as you'll recall (see page 174). Newton discovered that the rate of change of momentum is equal to the force applied. Hit a golf ball, and the momentum of your body and of the club is transferred to the ball.

Julius Robert von Mayer, a doctor, was the first to say, "Energy can be changed from one form to another, but it can't be used up." Too bad no one listened.

In 1842, Julius Robert von Mayer (1814–1878) becomes convinced that heat and motion are two forms of the same thing (kinetic energy). He realizes that one can be changed into the other, and that when that happens, the universe doesn't lose any energy. In other words, he figures out the first law of thermodynamics. But he doesn't do it with conventional physics experiments. Mayer is a medical doctor, and he gets his ideas by examining people. So when he announces his findings, the academic physics community just doesn't pay attention.

Mayer is probably in the wrong profession. He is more interested in pure science than in practicing medicine. As a young man just out of medical school in Germany, he takes a job as a doctor on a Dutch ship heading for Indonesia. Before he leaves Germany, he watches a horse turn a machine that is stirring a vat of paper pulp. As the horse works, the temperature of the pulp rises. With long periods at sea to think, Mayer considers that increase of temperature.

Mayer knows that Count Rumford measured the temperature of a turning tool—a bore drilling the barrel of a cannon—and found that the temperature of the bore rose as it was turned. That wasn't all. Remember, the cannon was immersed in water and, after two and a half hours of drilling, the heat became so intense that the water actually boiled! Rumford's big mental leap was to connect motion and heat and to understand that one can change to the other, on a simple level, by friction.

As you'll recall, when you measure heat, you are totaling all the kinetic energy of all the vibrating atoms and molecules in a given body. Quantities of heat are usually measured in joules or calories. Temperature is not a measure of the total heat in a body; it's a measure of the average kinetic energy of its particles. Temperature is usually measured in degrees Celsius, Fahrenheit, or Kelvin. The distinction between heat and temperature is subtle, but important. Here's an example that I hope will help make it clear: A pail of water at room temperature has *half the heat* of two pails at room temperature.

Since Mayer is aware of Rumford's experiments, they must be on his mind as he watches the paper pulp being stirred. So he can't be surprised by the movement of energy from the churning paddle to the pulp. He considers something else: It seems to him that energy is being transferred but not lost. (We now know it is being changed from mechanical energy to thermal energy.) That idea is percolating in his head when he lands in Java and notices something going on with his new patients.

For most ailments, it is standard nineteenth-century medical practice for doctors to bleed their patients (let out

some of their blood). So doctors are very familiar with the sight of blood. In Java, Mayer examines sailors and notices something unusual about the blood he draws from their veins: it is much redder than blood he routinely saw in Germany. He knows, thanks to Lavoisier, that the body is heated by the slow combustion of food—the process of oxidation. Bright-red, oxygen-rich blood is carried from the heart to the rest of the body by arteries; then, as the oxygen is used, the blood darkens (becomes more deep purple in tone). That oxygen-deprived blood travels through the veins to the heart and lungs, where it is enriched with oxygen, starting the process again. But the blood Mayer is seeing in Java, taken from veins, is a brighter red than normal. Why? Mayer comes up with a brilliant deduction. He figures out

While a patient squeezes a stick to open up his veins (left), a doctor collects blood in the belief that he is draining the body of "bad humors." The misguided practice of bloodletting is at least as old as ancient Egypt, but it reached ridiculous heights in the early nineteenth century.

Combustion (burning) doesn't always produce fire. It is oxidation—oxygen molecules combining with a fuel, such as the carbon molecules from the food you eat.

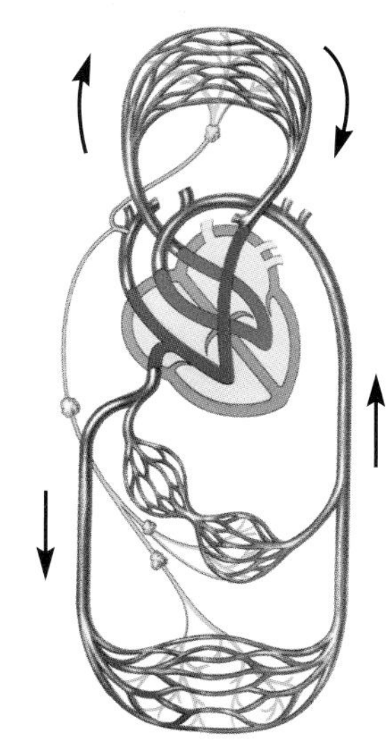

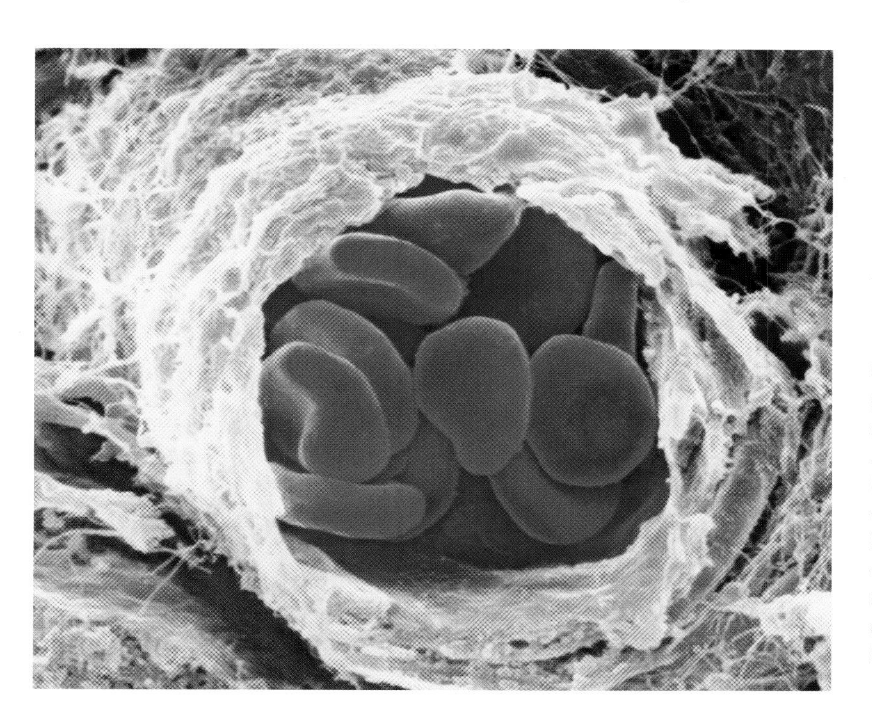

The circulatory system is a loop in which the same blood flows round and round. Each time a batch passes through the lungs, it picks up oxygen (shown above as a change from blue to red blood). A big push from the heart sends it to tiny capillaries, where the oxygen is used. In the photo at left are oxygen-carrying red blood cells. An electron micrograph captured them traveling through a small branch of an artery. Red blood cells cover up to 15 kilometers (9.3 miles) a day.

ENDLESS ENERGY

The Sun beams us electromagnetic energy, and plants transform that energy into chemical energy (also known as food) through a process called photosynthesis. Eat the plant, and your body has chemical energy that you can turn into kinetic energy by rowing a boat. The kinetic energy of the water farther downstream at a dam is converted into electric energy, which travels through wires to your home. There, it is turned into thermal energy to cook your dinner. Light, sound, electricity, and magnetism are all forms of energy—and are all interchangeable.

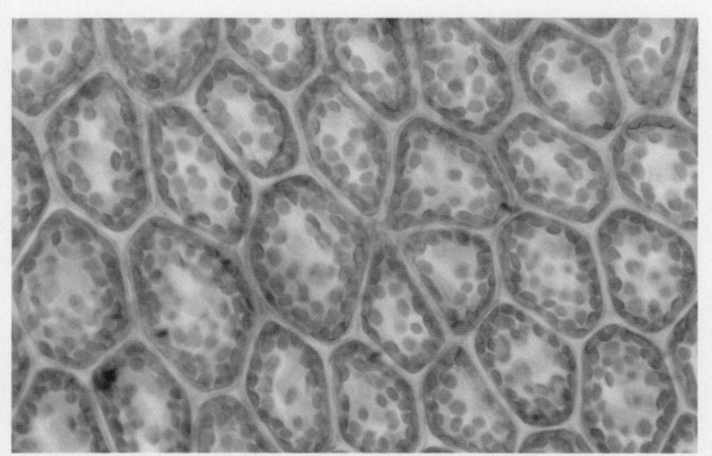

Green plants have a unique trick—turning light energy into chemical energy (or food). Photosynthesis happens in those dark green spots—the chloroplasts—inside the cells. Chlorophyll, the green pigment, absorbs red and violet light—the opposite ends of the visible light spectrum (or rainbow).

that in a hot climate, the body uses less oxygen to maintain body heat. So in the Tropics, blood in the veins remains both rich in oxygen and bright red.

Great breakthroughs often come when someone sees a surprising connection. Newton connected the Moon with a falling apple. Mayer connects the pull of a horse's muscles (and the transfer of heat to paper pulp) with the color of sailors' blood (and the heat of oxidation). That connection leads him to write the earliest statement of the first law of thermodynamics: *Energy can be changed from one form to another, but it can't be used up.* He has come to an astonishing conclusion—and he knows it. But no one pays attention.

Three years after that first big insight, Mayer writes another paper. This time he takes the idea of energy conversions further; he writes of plants converting the Sun's energy into latent chemical energy. He has deduced that *all* forms of energy, including thermal energy (heat) are interchangeable and that the same rules apply to heat in the

LATENT means dormant, potential, hidden, or undeveloped, but capable of becoming active.

body, heat in a coal furnace, heat from the Sun, and heat generated by horsepower.

Reminder: The Sun's kinetic energy is all those superhot atoms and molecules zooming around and banging together. Radiant energy is the electromagnetic energy that radiates (fans out) from the Sun as a result.

The German doctor isn't finished. He also says that solar energy is the fundamental source of all energy on Earth and that the Sun's kinetic energy is converted into radiant energy. (He doesn't know about nuclear energy, but he is way ahead of his time with his thoughts.)

Two years later, James Joule does an experiment that leads him to similar conclusions. Joule has stature in the scientific world, and he has the ear of the well-connected William Thomson (who will become Lord Kelvin). Mayer is an unknown doctor who has a hard time getting his scientific papers published or read; so Joule gets credit for the idea. To say that is frustrating for Mayer is an understatement. He becomes despondent, suffers bouts of deep depression, and spends years in asylums, where he is sometimes mistreated. (Five of Mayer's seven children die, which has a lot to do with his depression.)

Mayer has gone further than Joule in including all natural phenomenon and all living things in his conservation law. But Joule is a meticulous experimenter, and he does the mathematics to show that a fixed amount of mechanical work results in a fixed quantity of thermal energy. Joule deserves much credit.

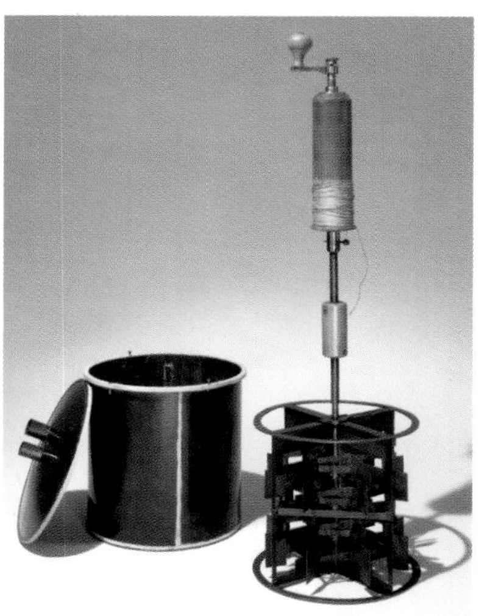

In Joule's experiment, a weight dropped and pulled a string that turned water paddles (above). The rotating paddles made the water heat up. How much dropping and turning (mechanical work) does it take to raise the temperature by a certain amount? Joule discovered a constant—a fixed relationship. Here it is in today's units: 4.187 joules of mechanical work equals 1 calorie (the thermal energy needed to raise the temperature of 1 gram of water by 1°C).

Meanwhile, a German physicist, Hermann Ludwig Ferdinand von Helmholtz (1821–1894), the son of a teacher, is also working on the same concept. He develops it in detail. (Helmholtz is an interesting fellow, worth researching.) When Helmholtz discovers Mayer's paper, he is amazed. He calls it to the attention of other scientists.

Finally, Mayer's contribution is recognized, and in 1871, London's Royal Society gives him its highest award, the Copley Medal.

Here is that first law again: *Energy can be changed from one form to another, but it can't be used up.* Now get ready for another big law. It's the second law of thermodynamics. It too has been called one of the great discoveries of physics.

Obeying the (Second) Law

The second law of thermodynamics is on a standing with Newton's and Maxwell's laws. It is one of the cornerstones of the scientific attempt to explain the universe. It is the only law that gives a direction to time.

—Brian L. Silver (d. 1997), physical chemist and science historian, *The Ascent of Science*

The law that entropy always increases—the second law of thermodynamics—holds, I think, the supreme position among the laws of Nature.

—Sir Arthur Eddington (1882–1944), English astronomer, *The Nature of the Physical World*

Once or twice I have been provoked and asked the company how many of them could describe the Second Law of Thermodynamics. The response was cold: it was also negative. Yet I was asking something which is about the scientific equivalent of: Have you read a work of Shakespeare's?

—C. P. Snow (1905–1980), English novelist and physicist, *The Two Cultures and the Scientific Revolution*

The first law of thermodynamics is easy: Energy can't be created or destroyed; it just changes its form. That's it. The first law is a kind of bookkeeping affair. It says that it doesn't make any difference to nature what form energy takes, as long as it keeps its books in balance.

There are only a few basic scientific laws that every thinking person should know and understand. The laws of thermodynamics are among them.

If that's all there were to energy—well, the world would be a dull place. But when you zero in on those energy exchanges, those changes of form, things get interesting. Most of what goes on in the universe—from exploding stars to teenage growth spurts—involves one form of energy being transformed into another.

Thermal energy (heat) is always part of an energy transformation. And that brings us to the second law of

This messy remnant of sulfur, hydrogen, oxygen, and other elements used to be a star in the Large Magellanic Cloud, a nearby galaxy. When a star explodes, it's easy to picture what happens to the matter. But where does all that explosive energy go? And in what form? The first law of thermodynamics says that energy isn't created and it doesn't disappear; it just changes form. The second law says the energy in a exploding star will never reform into that particular star ever again; it disperses (spreads out) through space.

thermodynamics. If you have any questioning bones in your body, if you care at all about the way our world of energy and matter works, the second law will make you sit up and start thinking.

Why do things burn? Why are there windstorms? Why is the Sun cooling down? Why do roofs collapse? Why do dishes break? Why do bodies age? The second law helps explain all that. But keep in mind, not everyone agrees with the explanations. This law is debated by economists, scientists, poets, theologians (religious scholars), and other thinkers—and some of these people get heated (sorry, I can't resist puns) about it. Why?

The basic idea behind the second law seems simple: **Energy spreads out if it can.** Ah, but there's a rub! That's not all there is to the second law. It also tells us the direction of energy change, **and it is always from hot to cold**, never the other

Some say it was the first law of thermodynamics that spelled the end to human slavery in much of the world. Here was a law that showed that heat can accomplish work (and usually do it better than muscles). Construct an engine (especially a steam engine with a piston and a cylinder that can drive a motor), and you can use heat to do work, ultimately ending the need for physical drudgery.

There's some complex science going on in this frying pan. Thermal energy from the flame transfers to the pan and spreads out from there in three ways. First there's conduction—direct contact of the pan with the air. There's also convection, which means accelerated air molecules are moving away from the pan while slower (and thus cooler) molecules move toward it. And finally, don't forget radiation. Infrared energy radiates—spreads out in all directions—from the pan. In this thermographic (temperature-recording) image, white and red areas are the hottest, yellow is in the middle, and blue and green are the coolest.

way (unless we bring in new energy). That's what creates most of the peculiarities of the world we live in. Because of that one-way movement (which carries action and events with it), the second law is often called an **arrow of time**.

Picture a red-hot frying pan on a stove. Turn off the burner, and the pan will begin to cool down. The second law is at work. Thermal energy is spreading out: It was concentrated in the pan, and it is now scattering around the room.

Here's how it works: The pan is hot because its molecules are vibrating fast. Molecules in the neighboring air keep bumping into these fast-moving pan molecules, which gives the air molecules a "kick." The air molecules become more energized—that just means they move faster—and they collide with other, slower air molecules. They also crash into objects in the room. Each collision involves a transfer of heat energy from faster-moving molecules to slower-moving ones. It doesn't take long for heat energy to be dispersed around the room.

The second law tells us that energy tends to go from being concentrated in one place (in the hot frying pan) to becoming dispersed or spread out (through the cool room and beyond). Could it go the other way? Can the pan get hot again? The speedy air molecules that have slowed down aren't going to give themselves a kick, crash into the pan, and heat it up again. You have to bring in new energy by turning on the stove.

Like the teakettle on page 395, a Schlieren photograph of a motorcycle reveals heat rising by detecting tiny differences in the density of air. The thermograph above works differently; a thermal imaging camera records temperatures directly and translates them into colors.

Consider the air compressed in your car's tire. It stays where it is; it doesn't spread out. That's because a barrier (the walls of the tire) prevents it from acting naturally. But stick a nail in the tire, and you'll hear the pressurized air rush out, acting as the second law demands.

The big idea to remember is that this law applies to *all* thermal energy. There are no exceptions of which we know. Totally different events and processes can all be tied to the same rule of nature: Concentrated energy will disperse if it gets a chance, and it will always go in the same direction—from hot to cold.

Imagine a snowman. He wears a top hat. The Sun appears from behind the clouds. Warm air molecules mix with the frosty molecules, and our snowman is soon a hat-in-a-puddle. It never happens the other way (except when you run a videotape backward). Puddles don't turn themselves into snowmen. This is not a reversible process. Does it matter? You bet.

The Sun is using up its thermal energy (it's scattering it all over the solar system and beyond)—and it won't get it back. Our Sun is running down; eventually it will expire. (Don't worry; the process is slow. The Sun has billions of years of life ahead of it.) The same thing is happening to the stars. They're following heat's arrow of time. (Yes, new stars are being born all the time—but that birth always involves an input of energy, which comes from somewhere else in the universe. And as soon as they are born, those new stars start following the second law.)

THERMAL ENERGY is the ceaseless motion (the kinetic energy) of tiny atoms and molecules. Hot and cold result from a difference in average speed. The second law says that thermal energy always spreads from hot (fast-moving atoms) to cold (slow-moving atoms)—*unless* there is a barrier. Those moving atoms may raise the temperature of a substance or cause it to expand, or (if a solid) melt it, or (if a liquid) vaporize it, or (if a confined gas) increase its pressure. In the process, some heat will escape (go up the smokestack or out of your car's tailpipe).

Puddles can only turn into snowmen if you bring in energy—a snowmaking machine and your muscle power. The second law assumes no energy will be added from outside the system. If energy is added, all kinds of changes are possible.

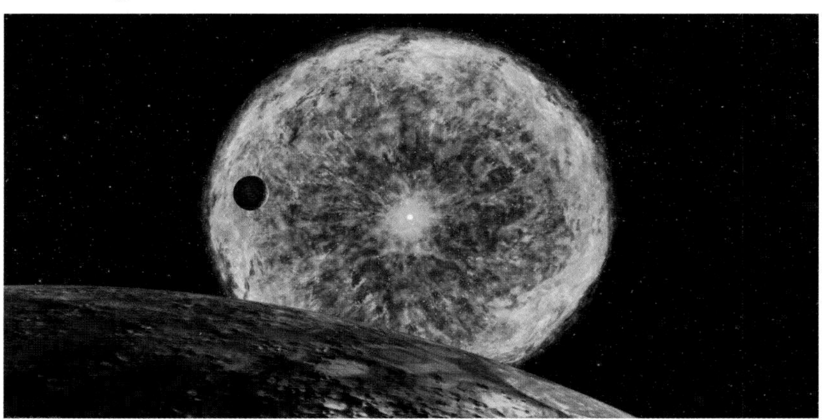

What will the Sun look like when it runs out of hydrogen fuel in about 5 billion years? And then helium fuel soon after that? This is one artist's vision of its death. The outer shell of the Sun is exploding into a planetary nebula as seen from a barren Earth, whose atmosphere has already been blasted away.

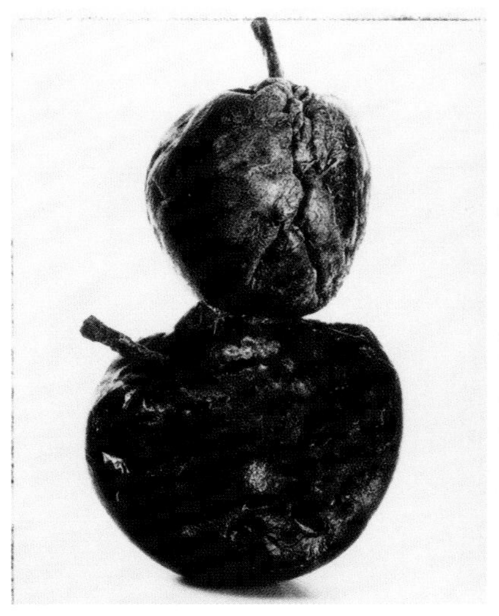

As much as we wish for it, stale apples won't turn fresh and juicy again. Wrinkled skin doesn't spontaneously turn smooth and young. Our bodies—and everything around us—obey the second law.

As to our everyday world? Drop a dish, and it will break. It won't put itself back together. Plants and animals die and decay. They don't revive themselves. Some people call the second law the "law of pessimism." Maybe. But it explains a lot of things.

Take that dish that you just dropped. The matter of it is all still there in broken shards on the floor. But there's more than matter to this story. There's energy. If you touched the floor just after the dish hit, you might have felt some of the heat that escaped. There is no way to recapture it. You can't get all those vibrating atoms and molecules back into the plate, so you can't put it back together exactly as it once was. Thermal energy spreads out if it can—and there are no barriers to that spread on your kitchen floor.

The second law was worked out by people who weren't sure if atoms exist, so they found the law by figuring out ways to change heat into work. They realized that in any heat engine, heat had to flow out of a hot body and into a cold one. In a steam engine, it went from a hot chamber, where steam was formed, to a cold one, where steam condensed.

According to chemistry professor Frank L. Lambert, a second-law expert, "All spontaneous happenings in the material world . . . are examples of the second law because they involve energy dispersing."

Some energy transfers are not quite spontaneous; they too follow that law, but they need a nudge. The nineteenth-century thinkers learned that and used their knowledge well. Gasoline explodes when it mixes with oxygen in the air, but only if a spark provides the nudge and ignites it.

Figuring that out gave inventors the concept that led to the gasoline engine. Here's a bit of it: When mixed gasoline and oxygen molecules ignite, their atoms form new molecules with stronger binding. The difference in energy goes into molecule speed (which is what atoms and molecules do when they are hot or under pressure). If the hot molecules are in a confined cylinder in your car's engine, they will build up pressure. High pressure in a small cylinder produces a lot of potential energy. That energy dissipates (thanks to the second law) and pushes the car's pistons. The pistons disperse their kinetic energy (more second law) and turn the crankshaft . . . and your car moves. But (still more second law) in the process, some heat energy is dispersed out the tailpipe and is "wasted." It can't be used again.

Now if I've done a decent job so far, you understand that the second law is about the dispersal of energy. The measurement of that thermodynamic dispersal is called **entropy**. Some people call the second law the **Law of Entropy**. (It seems easier than using that unwieldy word *thermodynamics*.) So here's another way of stating the second law: **Entropy always increases**.

Entropy? Think of it as a **measure of the amount of heat energy not available for useful work** (like the heat that goes out the tailpipe of your car, or the heat lost on the floor when you broke a dish).

Since entropy always increases as atoms and molecules go from hot to cold, **things naturally are always running down.** So when your roof caves in—and given enough time (maybe a century or two), entropy will make it collapse—at that point, the potential energy concentrated in the roof will be dispersed into kinetic and thermal energy. If you want to rebuild the roof, you will have to bring more energy into it; it

Watch the arrow of time: Look at a fire, and you'll see the spread of energy. It never goes the other way. Ashes never turn back into logs. Play a film backward, and you can see the world as it might be without a second law. Broken dishes would put themselves together, melted snow would turn itself into snowmen, water would flow up over a waterfall, and people would grow young instead of old.

FROM BRIGHT KID TO RICH BARON

William Thomson was only 22 when he became a professor of physics.

William Thomson (1824–1907), who was born in Belfast, Ireland, is sometimes given credit for discovering the second law. He was a child prodigy who attended his math professor father's college lectures when he was 8. At 10, he was a college student. Like his father, he became a professor at Glasgow University in Scotland and was one of the first professors anywhere to use a laboratory as a classroom. He was also among the first to understand Faraday's idea of lines of force.

Thomson was fascinated by heat and paid attention to Joule's ideas. He figured out that energy tends to disperse as heat. But he didn't believe in atoms. Boltzmann understood the second law by understanding the action of atoms; Thomson got there by studying heat energy and heat loss.

In the 1850s, Thomson was called on to help lay an underwater telegraph cable between Ireland and Newfoundland (in Canada). Nothing like that had been done before. Thomson invented a device that measured electric currents in the long wire. It worked, and he patented it and some of his other inventions. Thomson became rich and was named Baron Kelvin of Largs by Queen Victoria. He established a temperature scale that starts with "absolute zero" (–273.15°C or –459.67°F), the impossible-to-achieve temperature at which a gas would theoretically have a volume of 0 and atoms would stop moving. It's now called the Kelvin scale (see page 249).

In the 1860s, telegrams sent through electric cables were the "instant messages" of the day. The first transatlantic telegram, from Queen Victoria to President James Buchanan, was sent on August 1, 1858, and took 17 hours and 40 minutes to transmit. Each of the 509 letters in its 99 words had to be sent individually in dot-and-dash code.

won't fix itself. The roof is following the second law.

Entropy can be determined exactly with equations that measure the kinetic energy involved. (The math can get complicated; we are dealing with the tiny atomic world.)

Stay with me. This is important. You may find it disturbing.

Entropy is the way of the universe; that's why our Sun is running down. The universe is *not* like a wound-up pendulum that swings back and forth evenly. Entropy goes in one direction—downhill toward dispersal.

Keep in mind: *Entropy* (the spread of energy) is increasing (that's the second law). But the *amount* of energy never increases (that's the first law).

I hope I've made the point: Heat flows from hot to cold, never from cold to hot. But why? That's the big question for which we have no real answer. We don't know. There is no known law that says it can't happen the other way. Except for

THAT RUN-DOWN FEELING

Albert Einstein (he's coming up in the twentieth century) showed that mass is a form of energy. So, in any part of our universe the total mass-energy does not change, but because of the one-way thermodynamic street, some of it becomes unusable (it can no longer be converted into work). A pendulum runs down because friction in the air turns some of its mechanical energy into thermal energy. Put a pendulum in a sealed container, and you can measure the heat loss as it swings.

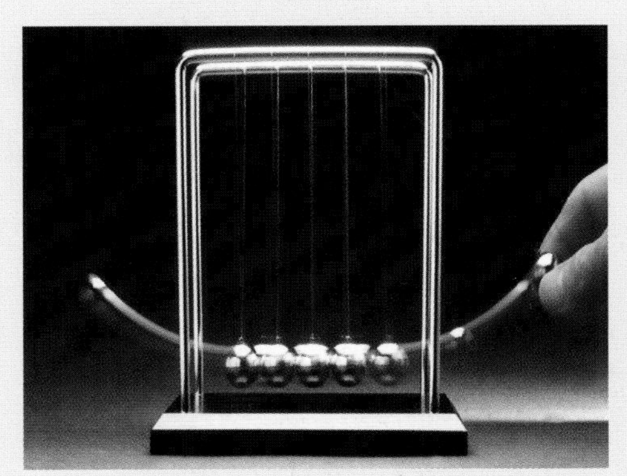

Like all pendulums, this Newton's Cradle toy (left) will eventually run down. To set it in motion, lift an end ball out to the side and let it go. It swings back and hits its neighbor in the lineup, but only the ball on the opposite end moves. That ball swings out and back, hitting the lineup in the opposite direction. The two end balls take turns whacking each other indirectly for a surprisingly long time; the middle balls just hang out, motionless. The apparatus is used in physics classes to demonstrate conservation of momentum and energy and elastic collisions, but it's more popular as a plaything. Try lifting two balls instead of one. Do the balls need to be touching? What if they weren't all the same size and mass? What if they were square instead of round?

If there were no second law, the universe would be like a giant clock that never runs down... [and] pendulums would keep swinging. Planets would repeat their trajectories without loss. A new star would form for every star that burned out.

—Alan Lightman (1948–), American author and professor, *Great Ideas in Physics*

ENTROPY? SING IT LOUD AND CLEAR!

Heat is work and work's a curse
And all the heat in the Universe
Is gonna cooool down 'cos it can't increase
Then there'll be no more work and there'll be perfect peace
Really?
Yeah—that's entropy, man!
And all because of the Second Law of Thermodynamics, which lays down:
That you can't pass heat from the cooler to the hotter
Try it if you like but you far better notter
'Cos the cold in the cooler will get hotter as a ruler
'Cos the hotter body's heat will pass to the cooler
Oh, you can't pass heat from the cooler to the hotter
You can try it if you like but you'll only look a fooler
'Cos the cold in the cooler will get hotter as a ruler
That's a physical Law!

—Lyrics from a song by Michael Flanders and Donald Swann, called "First and Second Law"

the law of probabilities. It is improbable—totally improbable—for heat to go from cold to hot.

The second law of thermodynamics is based on observation and probabilities.

Can you imagine some of the water in a glass that you're holding in your warm hand turning itself into ice cubes (without using energy from a freezer)? We know by observation that it doesn't happen. It's improbable. But, again, there is no known law that says it can't happen. That idea of probability was something new in science and hard for nineteenth-century scientists to accept. They were used to the exact logic of Newton's laws. They were *not* used to

Q: *Summer follows winter; isn't that going from cold to hot?*
A: Yes, but it's not a violation of the laws of thermodynamics. Seasons happen because Earth is tilted on its axis as it orbits the Sun. Because of that tilt, one hemisphere gets direct rays from the Sun (causing summer), while the other hemisphere is tilted away from the Sun (causing winter). Then, six months later, the pattern reverses itself. As to the second law: Thermal energy from the very hot Sun travels through space to the cooler Earth. Earth's warmth radiates into space, which is extremely cold. No law-breaking there. The light and dark areas at right show which parts of Earth radiate the most heat energy into space, from the black polar regions (very little) to the bright white deserts. This constantly changing pattern is mapped by the satellite instrument CERES, an acronym for Clouds and the Earth's Radiant Energy System.

Disorder? Keep the Cosmic Door Closed!

In many textbooks, entropy is described as **a measure of disorder in a closed system**. That nineteenth-century definition comes from Ludwig Boltzmann, who said, "Entropy is disorder, which has a natural tendency to increase. Experimentally it is measured by the ratio of heat to temperature."

Entropy as a measure of disorder? Perhaps. Today, some experts find that definition confusing; others say it is just plain wrong. But not everyone agrees on this. Here's an example of the order/disorder explanation:

Two objects that exchange molecules—like ice cubes and water in a glass—will eventually come to the same temperature. Both sets of molecules will be mixed together. Entropy will have increased. Put another way, the orderliness of water molecules frozen in ice is greater than their disorder when the ice melts. It's like keeping your sweaters in one drawer and your socks in another. Mix them together, and you have less order. (Some people see uniform and stable mixtures as orderly, not disorderly. There's something in our thinking that resists the scientific idea that a collection of separate components is more orderly than a mixture of those components. That's one reason that Boltzmann's description of entropy as "disorder" has fallen out of favor with some scientists.)

"Entropy is a measure of disorder *in a closed system*." What's a closed system?

A closed system keeps energy from being transferred in or out. Insulation in the walls of a modern house helps keep in thermal energy. In winter, the furnace could stop running for a while with little drop in temperature.

In contrast, the Earth is *not* a closed system. It gets added energy from the Sun. We exist in an open system of Earth, Sun, and outer space.

Suppose we cut your house off from the rest of the world and turn it into a closed system. When things wear out, they cannot be replaced. Eventually everything will just fall apart. That is entropy.

Nineteenth-century inventors tried hard to build a perpetual-motion machine, something that once it got started would go on forever. (That's one birdbrained attempt at left.) The first two laws of thermodynamics say it can't be done. In a perpetual-motion machine, energy would have to keep getting recycled; none can be dispersed in the form of heat energy. But that never happens. There are no perfect machines. Every time you use energy to do work, some of that energy turns into heat and radiates away, and it can't be recaptured or reused. When physicists talk of thermodynamics, they often say, "There is no free lunch" or "Everything costs." Because of that radiated heat, you always need to keep a machine resupplied with energy. Energy can't be created out of nothing. A perpetual-motion machine is unattainable. But some inventors keep trying.

The equation for entropy is carved at the top of Boltzmann's gravestone:

$$S = k \log W$$

S = entropy; W stands for the number of possible ways of distributing atoms; log stands for logarithm; and k is an unchanging number, now known as Boltzmann's constant, which was discovered later. (This k is a different value than the k constant in Coulomb's equation in chapter 24.)

thinking in terms of probabilities.

Along came Ludwig Boltzmann, our improbable Austrian physicist, who was one of a few scientists who made a connection between entropy and atoms and molecules. He understood that atoms and molecules have a normal tendency to disperse. Open a bottle of perfume, and its molecules will spread out in a helter-skelter way until they are all over the room and the bottle has dried up. The molecules will never put themselves back into the bottle in an orderly fashion. (Mary Poppins was able to break the law of entropy. Ask her, not me, how she did it.)

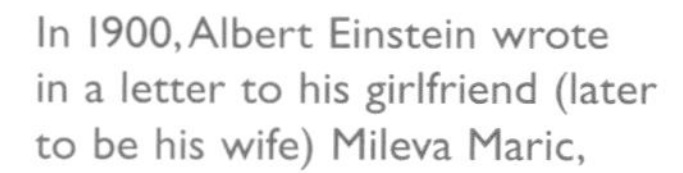

In 1900, Albert Einstein wrote in a letter to his girlfriend (later to be his wife) Mileva Maric,

> *I am firmly convinced that the principles of [Boltzmann's] theory are right, which means that I am convinced that in the case of gases we are really dealing with discrete mass points of definite finite size [atoms], which are moving according to certain conditions.*

Shuffle a deck of cards, and it is unlikely (but not impossible) that they will line themselves up by number and suit. The probability that this will happen is 1 in 10 million trillion. **Dispersal is the state with the highest probability.**

Ludwig Boltzmann used his head to make the connection between entropy, time, and probability. He understood the way atoms behave: They don't sit still, and they don't organize themselves: they move randomly. He used statistics as a scientific tool to understand atomic probabilities. His insights provided a key to the universe. Knowing that hot always flows to cold (which means that energy disperses) gives a direction to time. You can't play life's tape backward. Tomorrow, you won't be younger than you are today. The second law won't let you.

Entropy is hot molecules spreading out and keeping cool. Put it on a T-shirt.

THIS THEORY ISN'T COMPLETE

I mentioned that some scientists heat up when talking about the second law (see page 399). Some get really steamed. So what's the argument? Well, there's more than one, and since the math and science can get quite complex, I'm going to give you some general thoughts. If you're interested, you can pursue them on your own. The concept of time is part of it. So is something called *non-equilibrium thermodynamics*. Then there's the field of *thermodiffusion*, which seems to indicate that some irreversible processes can give rise to organization (rather than disorder).

To find out more, start with the founding work of the American physicist Josiah Willard Gibbs (1839–1903) and the twentieth-century work of Ilya Prigogine (1917–2003). Prigogine, who was born in Moscow and educated in Brussels, realized that the Earth constantly gets new energy from the Sun and that living things can achieve an organized state from disorganized materials. His take on thermodynamics helps explain life's evolution and the growth of ecosystems. He was awarded a Nobel Prize in 1977.

There's ferment in the field. Today's physicists have found that you can't relate the arrow of time to the atomic and subatomic world. That world has its own rules where time seems to be reversible. So there is now talk of the *irreversibility paradox*—which suggests that there is more to learn about thermodynamics. The theory is not yet complete.

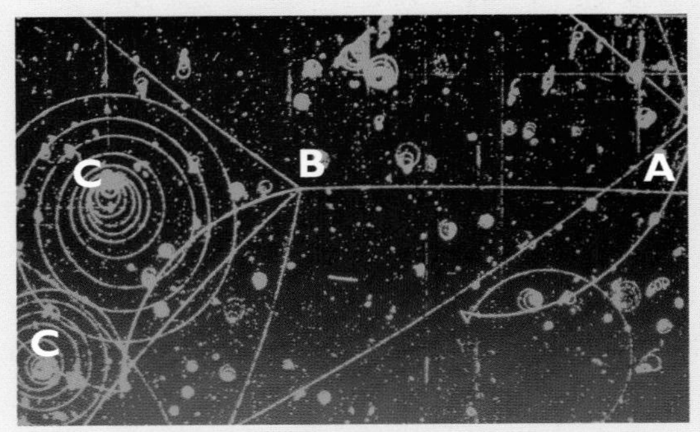

Here, in the subatomic world, an antiproton (a negatively charged particle of antimatter), entering at right (A), is annihilating (B) a proton of hydrogen, emitting two high-energy photons (C).

Earth isn't a closed system because it constantly receives energy from the Sun.

Some physicists are saying that classic thermodynamics doesn't take into account gravity, which can mess up that uniform spreading out that will lead to "heat death." (Heat death, or ultimate entropy, is what many predict as the end of the universe.) Does entropy have to keep increasing as the second law says it must? Only probability says yes. We have no example that contradicts the second law.

According to British physicists Peter Coveney and Roger Highfield, "We can now see that the spontaneous creation of order is not forbidden by thermodynamics, although the Second Law is commonly and erroneously regarded as being a watchword for the uniform degeneration into randomness." Their view is more optimistic than the standard interpretation of the second law. They see thermodynamics allowing "for the possibility of spontaneous self-organization." And that could lead to "structures ranging from planets and galaxies to cells and organisms."

Can entropy not be reversed? That is the question put to a giant computer in the year 2061 in Isaac Asimov's science-fiction story "The Last Question." The computer answers, "There is as yet insufficient data for a meaningful answer."

38

Tying Down a Demon

Imagination is more important than knowledge.
—Albert Einstein (1879–1955), German-born American physicist, *On Science*

When it is not in our power to determine what is true, we ought to act according to what is most probable.
—René Descartes (1596–1650), French mathematician and philosopher, *Discourse on the Method*

Imagination as well as reason is necessary to perfection in the philosophical mind.
—Sir Humphry Davy (1778–1829), English chemist and mentor to Michael Faraday, *Parallels Between Art and Science*

If you want to win a Nobel Prize in science or literature or medicine or peace or economics—or if you want to achieve in those or any other field—here's a formula for success:

1. Be willing to work hard and use your brain.
2. Set a goal and go for it.
3. Find the best people and ideas in your field and learn everything you can from them or from their work.
4. Be imaginative. Reasoning is not enough.

It's the imagination factor that seems to make the difference. Study great achievers and you'll see it all the time. James Clerk Maxwell is a perfect example. He never lost his youthful curiosity, his creativity, his inventiveness, his originality of mind. When he was a grown up and a serious scientist, he still managed to play in his head. It helped him solve problems.

What do Nobel Prize winners win (besides lots of respect and prestige)? A gold medal (above), a diploma, and about $1.37 million (sometimes shared). The prizes are always presented on December 10, the anniversary of Alfred Nobel's death in 1896.

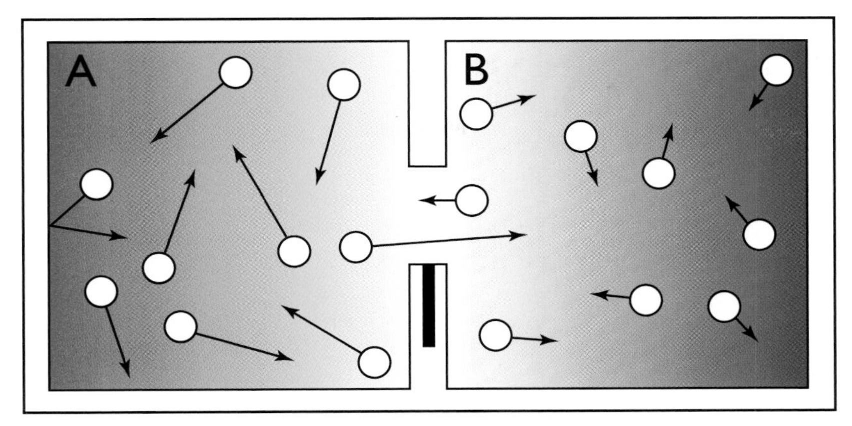

Maxwell imagined an invisible demon at a gateway between two sides of a container (left). To understand his demon—and the action of molecules—remember that all those gas molecules are *not* moving at the same speed and that faster means hotter. The temperature—which has to do with the average speed of the molecules—starts out the same throughout the container. But the "neat-fingered" demon sorts faster and slower molecules into separate sides. The container changes from a uniform temperature to half hot (A) and half cold (B), upsetting the second law of thermodynamics.

One day, when he was thinking about scientific things, Maxwell imagined a little fellow, a demon, who was smaller than an atom. "A very observant and neat-fingered being," is how Maxwell described him in a letter to a friend. That demon was determined to mess up the second law of thermodynamics. He was going to see that heat didn't flow from hot to cold. He was going to see that entropy didn't happen.

What did the demon do? He set up a wall with a gate in a box full of gas. The gas on both sides of the wall was at the same temperature, and normally it would stay that way. Even if the two sides were at different temperatures, they would eventually end up the same (that's entropy). But our diabolical demon changed that. He opened the gate at just the right times to let only fast-moving molecules pass from the right to the left side and the slow-moving molecules go the other way. Before long, the left side (now with the speedy molecules) had heated up, and the right one had cooled down. Heat had flowed from cold to hot. Entropy was defeated.

Maxwell knew the demon didn't exist, but could some other factor do the same thing as a demon? It was unlikely. It was improbable.

The whole question of imagination in science is often misunderstood by people in other disciplines.... They overlook the fact that whatever we are allowed to imagine in science must be consistent with everything else we know; that the electric fields and the waves we talk about are not just some happy thoughts which we are free to make as we wish, but ideas which must be consistent with all the laws of physics we know. We can't allow ourselves to seriously imagine things which are obviously in contradiction to the laws of nature. And so our kind of imagination is quite a difficult game.

—Richard P. Feynman (1918–1988), *The Feynman Lectures on Physics*

But there was no law that said it couldn't happen.

Imagining a demon helped Maxwell come to grips with thermodynamics and an idea that would change physics at its core. It would transform scientific thought in the twentieth century. It would displace some of Newton's basic ideas. It was the idea of probabilities.

Newton had imagined the world as if it were a gigantic clock. If we can understand the parts, we can know how it ticks, said Newton. He looked for and found unchangeable laws. His world was certain and sure. That was the way scientists thought until Maxwell's demon came along. The demon seemed to sabotage entropy. It also challenged the idea that everything in the universe is set and absolute.

The demon, who was very familiar to Ludwig Boltzmann, made a good case for probability. He helped scientists begin to realize that some things happen not because of sure, unbreakable laws, but for reasons that we can't explain—so we just have to go along with what we observe. When we see something happen again and again and again, it is probable that it will continue to happen. Statistics—research based on numbers—was a new idea creeping out of the cradle. It was about to become a major scientific tool.

Is that Maxwell chasing a demon? I don't think so, but this eighteenth-century Japanese painting is wonderful, and Sabine Russ, who chose the pictures for this book, thinks it captures the demon idea. I agree.

Maxwell's energetic gremlin was testing the idea. He was trying to replace probability with certainty. Maxwell understood that there is no known law that says entropy has to work. Only our experience and statistics show it is so.

Probability will underlie much of science in the twentieth century. Scientists will zoom into the subatomic world, where minute particles called quanta rule. It's a world so tiny that it is hard to imagine. The scientists will find that in the quantum world, some things seem to act randomly: they

Reminder: SUBATOMIC means "smaller-than-atom." Electrons, protons, and neutrons are three basic subatomic particles, but there are others that are even tinier.

MINUTE (my-NOOT) means "very small." It's spelled, but not pronounced, the same way as the word that means "60 seconds."

You Can't Kill a Demon—Can You?

Maxwell's demon has kept scientists intrigued for more than a century. In 1912, Marian Smoluchowski, a Polish physicist, thought he had some proofs that killed that demon off, but the gremlin kept reappearing. Half a century later, American physicist Richard Feynman invented his own version of the demon and showed that the motion of molecules would shake up that mischievous little fellow so much that he couldn't possibly operate a valve. As late as 1992, physicists at Los Alamos National Laboratory in New Mexico were busy killing off the demon in a computer simulation of two boxes, a trap door, and gases.

"But today we can confidently assert that the mechanical Demon is fully understood, and that he cannot violate the laws of thermodynamics," writes Hans Christian von Baeyer. Yet von Baeyer might not really mean it, because he goes on to say, "Nor is the Demon really dead. Whenever he was tossed out a window, he has always managed to climb back in through another one in a new guise."

Richard Feynman, who played drums and practical jokes, wasn't a demon, although some thought him an imp. He made physics fun. In 1965, he won a Nobel Prize. Some call him America's greatest twentieth-century scientist.

don't follow knowable laws. So the scientists will have to calculate their probable behaviors rather than observe and measure their actual ones. They will turn to statistics—piling up numbers that show how things usually turn out.

The scientists will argue about all this. Some will be sure that there are unchanging laws that work in both the tiny micro world and the everyday macro world and that these laws should tell us how each atom will behave. They will say we shouldn't have to rely on probabilities. Others will say this is the way it is, and that's that. When the twenty-first century begins, they will still be arguing about it.

It was that "neat-fingered" little demon who started it all. He will have his way in the twentieth century. Will he be tamed in the twenty-first? Perhaps you can use your imagination and intelligence and manage to overcome the second law. If so, you can count on winning a Nobel Prize.

According to Newton's laws of motion, if you run fast enough, you can catch a light beam. According to Maxwell's electromagnetic equations, you can't. Albert Einstein, born in 1879, will soon be working on this dilemma. Resolving it will lead to a new era in science.

39

Nothing to Do?

Science walks forward on two feet, namely theory and experiment.... Sometimes it is one foot which is put forward first, sometimes the other, but continuous progress is only made by the use of both.

—Robert Millikan (1868–1953), American physicist, speech upon receiving the 1923 Noble Prize in physics

If a scientific proposition is false, a single counter example proves its falsehood, but if it happens to be true, neither experimental confirmation nor mathematical demonstration can prove that it will always hold: The fact that dropped apples invariably fall does not imply that they will always fall.

—Hans Christian von Baeyer, American physicist, *Maxwell's Demon: Why Warmth Disperses and Time Passes*

Albert Michelson was a University of Chicago physicist and one of America's leading scientists. So when he gave a speech, people paid attention. This is what he said about physics in 1894:

> *It seems probable that most of the grand underlying principles have been firmly established and that further advances are to be sought chiefly in the rigorous application of these principles.... The future truths of Physical Science are to be looked for in the sixth place of decimals.*

In other words, in the late 1880s, Michelson was saying that just about everything that was going to be discovered in physics had been discovered. All that was left was to work out the details.

It must have been at such an early age that I decided I would be a scientist. But I foresaw one snag. By the time I grew up—and how far away that seemed!—everything would have been discovered. I confided my fears to my mother, who reassured me. "Don't worry, Ducky," she said. "There will be plenty left for you to find out."
—Francis Crick (1916–2004), English biochemist, *What Mad Pursuit*

Oh, my. New discoveries were about to set explosive charges in the world of physics. In the twentieth century, physics would come to dominate much of science.

But don't be too hard on Professor Michelson. He was just agreeing with most of the experts at the end of the nineteenth century. The head of the physics department at Harvard University advised his brightest students not to enter the field. There wasn't much left to do there, he said.

But two years after the Chicago professor's speech, a bearded French physicist made a discovery that would give work to lots of physicists. And a year after that, a physicist in England would discover something so astonishing it would keep even more of them busy.

The Frenchman is Antoine-Henri Becquerel (bek-uh-REL), who has a comfortable professorship and a terrier's personality (he can be growly and standoffish). Becquerel (1852–1908) is well respected, but no one thinks he is going anywhere in the scientific world.

On January 20, 1896, Becquerel is sitting in the audience with members of the French Academy of Sciences when two doctors display some photographs that show bones inside a living human being. That seems miraculous. Like magic!

Antoine-Henri Becquerel came from a family of French physicists. His grandfather helped found electrochemistry, and his father was especially interested in the way matter absorbs light. Henri took the family passion and ran with it.

"Hidden Solids Revealed!" shouted a headline in *The New York Times* on Jan. 16, 1896. The article said:

Men of science in this city are awaiting with the utmost impatience the arrival of European technical journals, which will give them the full particulars of Prof. Routgen's [sic] *great discovery of a method of photographing opaque bodies covered by other bodies, hitherto regarded as wholly impenetrable by light rays of any kind.*

Prof. Routgen of Würzburg University has recently discovered a light which for the purposes of photography will penetrate wood, paper, flesh, and nearly all other organic substances. Thus, the bones of the human frame can be photographed in relief without the flesh which covers them.

A German scientist named Wilhelm Conrad Roentgen had accidentally discovered X rays a year earlier. Just a month before the meeting at the French Academy, Roentgen made the first X-ray image. It is of his wife Bertha's hand, and everyone can see that her joints are bent by arthritis.

Word spreads quickly. Two weeks after the meeting, on February 3, 1896, some American doctors use an X-ray image to set the broken arm of a little New Hampshire boy named Eddie McCarthy. Newspapers and magazines shout the news. Everyone—not just doctors and scientists—is talking about X rays. Some people protest that seeing the inside of a person is an invasion of privacy. Just what are X rays? No one is quite sure.

The letter *X* is scientific shorthand for "unknown," which explains the name. The rays challenge the scientific community. Someone needs to explain what they are and where they come from.

Wilhelm Conrad Roentgen (1845–1923) discovered X rays. He probably could have made tons of money by patenting his X-ray machine—but he didn't. A friend wrote of him, "His outstanding characteristic was his integrity.... [He] was in every sense the embodiment of the ideals of the nineteenth century: strong, honest and powerful, devoted to his science and never doubting its value."

An X ray of Bertha Roentgen's hand was the first see-through look at the bony insides of a living person. One science journal reported, "The realism of this weird picture simply fascinated all who beheld it." Bertha was unsettled by it, believing it smacked of death. The process took 15 minutes (right), but no one understood the long-term health hazard.

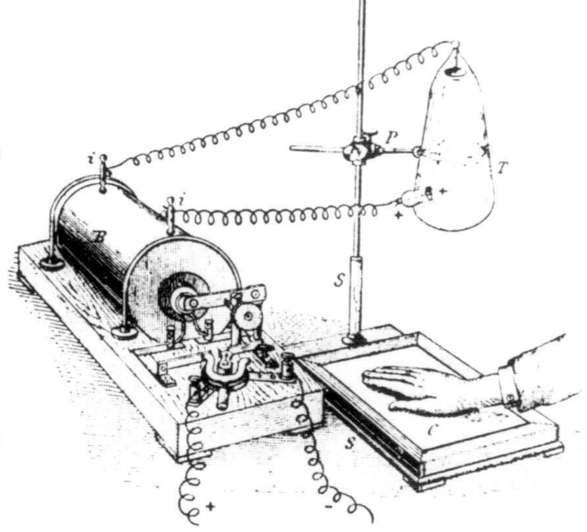

The New Roentgen Photography—"LOOK PLEASANT, PLEASE."

X ray is the word most of us use for an image made by X-ray radiation and for the radiation itself, but scientists like to distinguish the two. X rays were once called Roentgen rays, after their discoverer, and an image made by them was a roentgenogram. These days, images made by radiation other than visible light—X rays, gamma rays, ultraviolet rays, and so on—are called radiographs. Whatever the name, it provoked a lot of interest and some hilarity too. Here's an American cartoon from about 1900.

Becquerel has done experiments with phosphorescent stones—stones that, after being exposed to sunlight, shine in the dark (it's sometimes called afterglow). He thinks X rays might be related. So, in February and March, right after coming home from the French Academy lecture, Becquerel experiments. First he takes some phosphorescent stones, puts them on a photographic plate, and waits. He expects the plate to show an X-ray image. But nothing happens.

Then he tries some crystals of rock that have uranium in them; he puts them in the sunshine to collect ultraviolet light, and places them on a well-wrapped photographic plate. The rays go through the paper wrapping and leave an image on the plate! It is what he thought would happen.

Becquerel is a careful experimenter. Next he takes a coin and some other pieces of metal and sandwiches them between uranium-laden rock crystals and a photographic plate. He puts that package in the sun. The coins and metal

One winter day in 1896, Eddie McCarthy, age 14, broke his arm while ice-skating on the Connecticut River. He made the classic slip-and-fall mistake—using one hand to try to stop the impact. All his weight centered on that hand. (It's better to spread the impact over your whole body.) The world's first medical X ray, which took 20 minutes, revealed a fracture to Eddie's left ulna (forearm bone).

DYNAMITE MAN

Alfred Bernhard Nobel (above, 1833–1896) built explosives but loved peace. A lonely millionaire bachelor, he was thought to be a "mad scientist" by his contemporaries because his wealth mostly came from weapons of destruction. Nobel, who was an idealist, actually thought his explosives would end war because they would make it too horrible. He believed his products, including the explosive nitroglycerine, had peaceful uses.

He had developed nitroglycerine on a barge in a lake so that if it exploded, few people would be hurt. (An earlier explosion in his factory had killed his brother.) He named his safe-to-handle version of nitroglycerine *dynamite*. Dynamite does have many peaceful uses and is especially valuable in construction projects. It helped open the American West by blasting rocks and mountains to make way for roads and railroad tracks. Nobel invented other things, like smokeless gunpowder and a special steel for armor plating.

Born in Stockholm, Sweden, Nobel grew up in Russia, where his father had gone to supervise an underwater mine he had designed. Educated there by private tutors, Alfred was sent to Paris to study chemistry, and in the early 1850s on to the United States to study for four years with John Ericsson, a Swedish-American inventor who would build the Civil War ironclad vessel the *Monitor*.

At his death, Nobel left a fortune to endow annual prizes honoring great achievement in five fields—peace, literature, physics, chemistry, and medicine or physiology. (A sixth prize, in economics, was established in his honor in 1969.) A money award comes with each prize, but the honor of a Nobel Prize goes way beyond money. The Nobel Foundation in Sweden was named for Alfred Nobel, and when element 102 was isolated in Berkeley, California, it was called nobelium.

The first Nobel Prizes were awarded in 1901. Roentgen won the physics prize. The following year, Hendrik Lorentz and Pieter Zeeman of the Netherlands won for their work showing the influence of magnetism on radiation. And in 1903, Henri Becquerel shared the award with Marie Curie and Pierre Curie (left). Becquerel was cited for his discovery of spontaneous radiation, while the Curies were lauded for their research in radiation phenomena. In 1907, Albert Michelson became the first American winner. It was an auspicious beginning for the prestigious prize.

leave light images of themselves on the darkened plate. Again Becquerel has found what he expected. On February 24, he tells members of the French Academy of Sciences that X rays are created by phosphorescent rocks that have been exposed to sunlight.

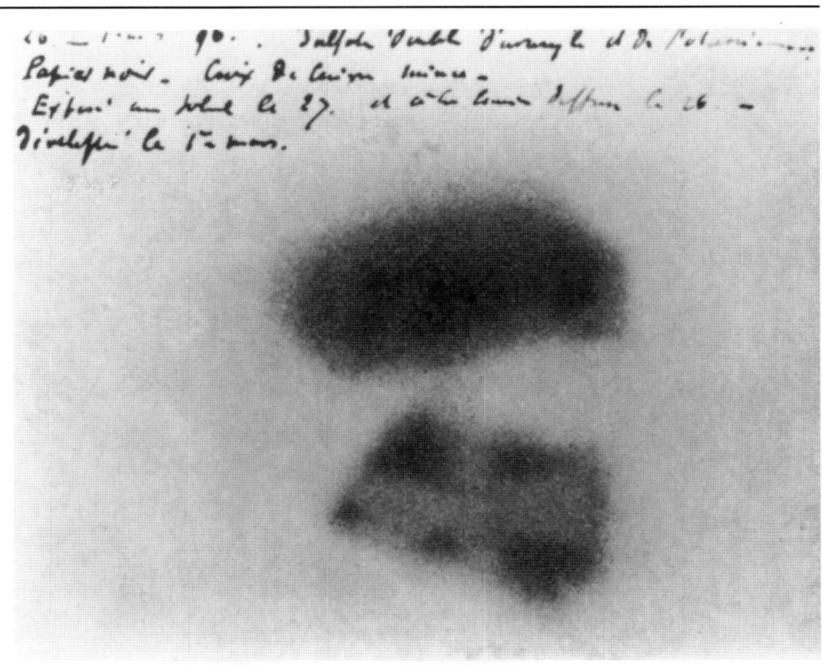

This is the photograph that Antoine-Henri Becquerel saw when he developed a plate that had been in his closed drawer. The two dark blobs are where he placed crystals of uranium salt. He was smart enough to figure out that he had found a new form of radiation.

Two days after that, the weather turns winter-dreary. No sun. Antoine-Henri Becquerel puts his plates, paper, and rock crystals into a dark drawer. They sit there for a week. Becquerel knows that rock can't phosphoresce in the dark. So there can't be an image on the photographic plates.

But something makes Becquerel develop those plates anyway. There are images on the plates! All at once his whole theory is shot away. The crystals have *not* been exposed to the Sun's ultraviolet light. Phosphorescence can't have anything to do with these images. Something in the rock itself must be creating rays that are shooting outward.

Uranium seems to be the key. Becquerel takes the uranium-rich rock and pulverizes it. He heats the uranium powder, he cools it, he dissolves it in acids; he does everything he can think to do to it. It still gives off mysterious rays.

The news of his experiments spreads quickly. Everywhere, scientists begin working with uranium salts and photographic plates. They all realize that the rays from the uranium are something unexpected and unknown.

They are not X rays, which are electromagnetic rays, like light but with a shorter frequency that makes them invisible

Niger, Africa, has large deposits of uranium, including this mine shown in a satellite image. The surprisingly common silvery-gray metal is used in the nuclear industry.

to us. (Frequency, you'll recall, is the number of waves per unit of time; see pages 366–369.)

The rays from the uranium are something quite different. They aren't light pulses at all. These are rays of energy that are being shot out of stone!

But that can't be. Clearly those rocks with uranium in them haven't been to school. They haven't learned the first law of thermodynamics. You can't create energy. This is astonishing. No one can explain what is happening. Physics is going to have to adapt to this not-yet-understood phenomenon. Becquerel writes, "Contrary to every expectation, the first experiments demonstrated the existence of an apparently *spontaneous* production of energy."

As it turns out, the first law is saved. The rays are the result of a process of natural decay found in some elements. That process will be named radioactivity. Uranium is a radioactive element. Because its atoms are unstable, tiny particles of energy are ejected from the nuclei. That loss causes the atoms to change into atoms of a different and stable (nonradioactive) element with a lower atomic number (fewer protons in the nucleus). It is those radioactive particles that are leaving their signature on Becquerel's plates.

Radioactivity helped bring physics and chemistry together in the twentieth century. They had seemed like separate sciences, but as knowledge of the atom grew, they became closely allied.

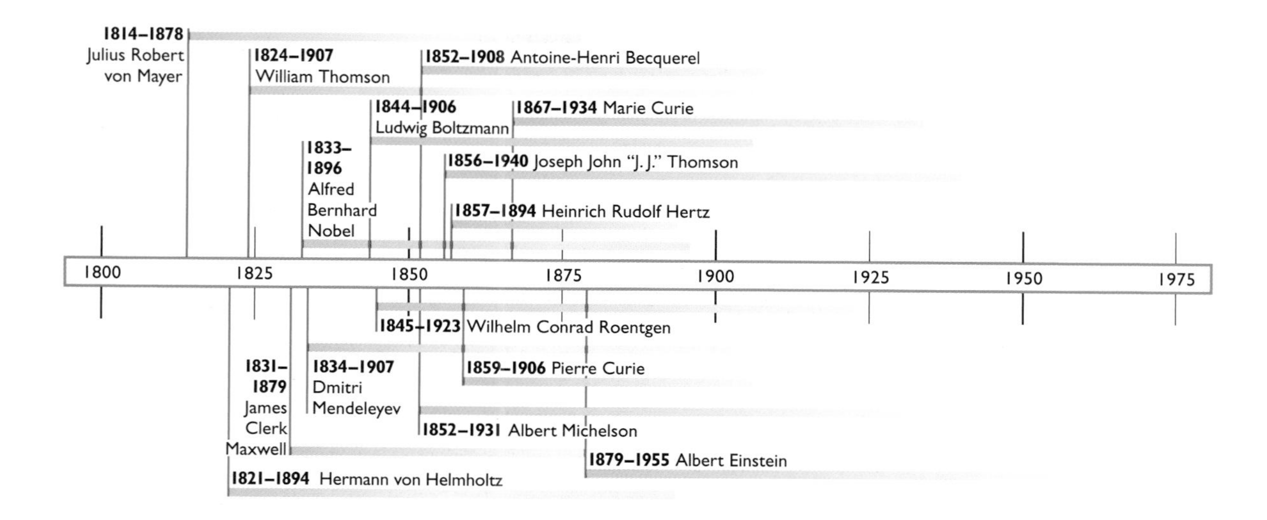

At the home for old atoms . . .

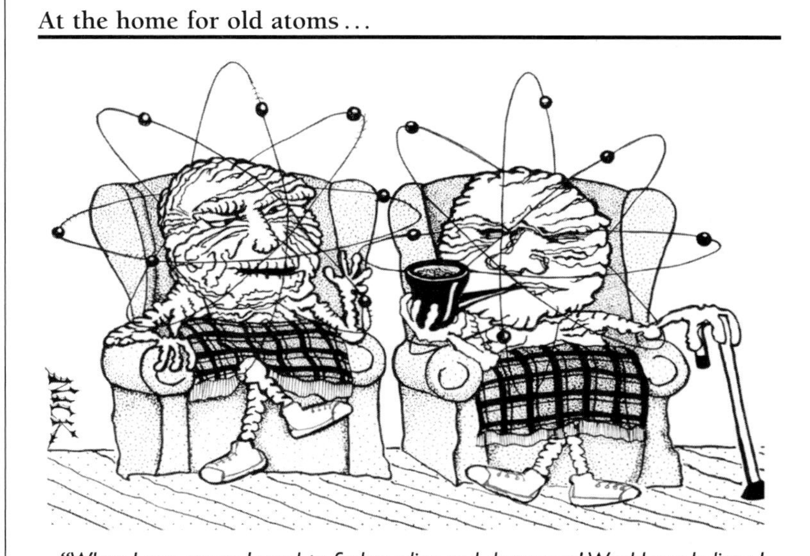

"When I was young I used to feel so alive and dangerous! Would you believe I started life as uranium-238? Then one day I accidentally ejected an alpha particle. Now look at me—a spent old atom of lead-206. It seems that all my life since then has been nothing but decay, decay, decay . . ."

As you know (if you've been reading carefully), nature sometimes practices alchemy and transmutes one element into another. Uranium atoms take their time doing it, but eventually they lose enough particles to decay into lead. Uranium-238 (a uranium atom with extra neutrons) has a half-life (decay rate) of 4.5 billion years. During that time, it backpedals through the periodic table, becoming thorium, radium, radon (for just 3.8 days!), polonium (for 138 days), and finally lead, which is stable (it won't decay further). Uranium to lead? This isn't exactly the philosophers' stone the ancients imagined.

They will lead to an unimagined source of energy.

But all this has yet to be discovered. Many scientists still don't believe in atoms, and no one knows that atoms have an inner structure, including a nucleus. And they certainly don't know (yet) that atoms can shoot out particles.

Just a year after Becquerel's pronouncement, an Englishman, the director of the Cavendish Laboratory at Cambridge University, discovers some smaller-than-atom particles that are later called electrons. That Englishman, J. J. Thomson, doesn't make a connection between electrons and radioactivity. But Becquerel does.

In 1900, Becquerel demonstrates that the rays from the uranium in his laboratory include particles that must be the same as Thomson's electrons. (He is right!) He begins to realize that whatever is creating those radioactive rays must be inside the uranium atoms.

Sir Joseph John Thomson (1856–1940), known as J.J., showed that cathode rays are particles with a negative charge and that they're much smaller than atoms. Later, those particles were called electrons.

Reason for Measuring a Sunbeam

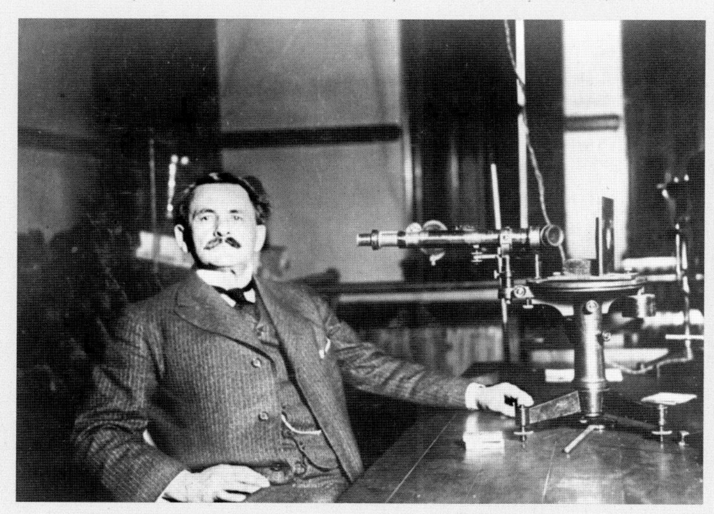

In 1882, Albert Michelson was setting up his experiment to measure the speed of light. A newspaper reporter, sent to get a story, reportedly asked, "What are you doing that for?" Michelson's answer—one all good scientists understand—was, "Because it is such fun."

Science is double-barreled. It is ideas and thinking and it's also experimenting and doing. James Clerk Maxwell's concept about light excited scientists around the world. They realized that the speed of light is one of nature's fundamental numbers. The challenge was to improve its measurement. Albert Michelson, then a young instructor at the U.S. Naval Academy in Annapolis, Maryland, wrote, "The fact that the velocity of light is so far beyond the conception of the human intellect, coupled with the extraordinary accuracy with which it may be measured, makes this determination one of the most fascinating problems that fall to the lot of the investigator."

Michelson knew that Foucault—the pendulum man—had put two lenses 8,633 meters (about 5.4 miles) apart and, using a mirror, sent light from one to the other through a fast-rotating toothed wheel. Foucault's measurement of light's speed was the best there was, but scientists believed it could be improved.

With laboratory odds and ends and a mirror he made himself, Albert Michelson began with Foucault's idea but made it better. He came up with a new figure for the speed of light—the best yet—but he knew it still wasn't good enough. A few years later (in 1882, when he was a professor at the Case School of Applied Science in Cleveland, Ohio), Michelson improved the experiment and repeated it 20 times. His result for light's speed—299,853 kilometers (a little more than 186,300 miles) per second—would be the accepted figure for more than 50 years. When it was finally made still more precise, Michelson would be the one to do it.

In this mile-long vacuum tube, an old and frail Albert Michelson made his last light-speed measurement. Light zipped down and back twice, bouncing off mirrors, and then passed through a window, off a disk of rotating mirrors, and then through a slit, a lens, a prism, and an eyepiece.

I love good science fiction, and this prophetic French cartoon was sci-fi when it was drawn in 1900. (*Prophetic* means seeing into the future; having foresight.) The artist prophesied that there would be video telephones in the year 2000. It may have seemed an impossible dream then, but knowledge of electrodynamics and the structure of the atom would make videophones happen.

That means an atom might not be the smallest particle. Figuring out if that is true and what it means will provide plenty of work for the next generation of physicists.

Atoms and smaller-than-atom particles—like electrons—are going to be central to a technological revolution that is on its way. That revolution will bring moving pictures and voices—from around the globe and beyond—and put them on handheld gadgets and wall-size screens. Engineers will design ovens with invisible waves that cook food in minutes. And computers will not be overworked teenagers or clerks sitting at desks with pencil and paper, but data-processing devices that will allow twenty-first-century teenagers (along with people of all ages) access to the whole world's collected wit, wisdom, and nonsense. All this will happen because knowledge of electromagnetism and thermodynamics is going to be combined with knowledge of atoms and those still-smaller particles.

Reality is about to become more astonishing than science fiction. The nineteenth-century professors who thought physics was finished had a lot to learn. The twentieth century will be the age of physics.

40 Wrapping Up and Getting Ready

Poets say science takes away from the beauty of the stars—mere globs of gas atoms. Nothing is "mere." I too can see the stars on a desert night, and feel them. But do I see less or more?... It does not do harm to the mystery to know a little about it. For far more marvelous is the truth than any artists of the past imagined! Why do the poets of the present not speak of it?

—Richard Feynman (1918–1988), American physicist, *The Feynman Lectures on Physics*

Books on physics are full of complicated mathematical formulae. But thought and ideas, not formulae, are the beginning of every physical theory.

—Albert Einstein (1879–1955) and Leopold Infeld (1898–1968), physicists, *The Evolution of Physics*

The world today is made, it is powered by science; and for any [person] to abdicate [give up] an interest in science is to walk with open eyes towards slavery.

—Jacob Bronowski (1908–1974), Polish mathematician and poet, *Science and Human Values*

Can you feel the tremors in the scientific world? Those tremors—like vibrations when an earthquake begins to erupt—are barely noticeable, except to a few people with keen senses. Those few can sense that something big is about to happen.

It has to do with the atom. Scientific evidence is on its way. The unbelievers will soon have to admit that those tiny bits of matter exist. But even those who believe in atoms aren't prepared for all that is ahead.

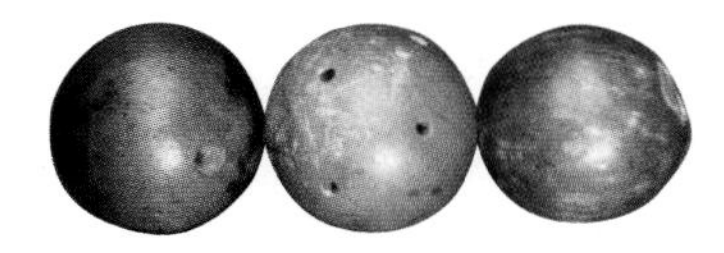

A billiard ball—solid, round, small and hard—was the nineteenth-century image of an atom. That model would soon change. The holes in these "atom" balls are for pegging together models of molecules.

Here, at the end of the nineteenth century, atoms are thought to be like billiard balls. Isaac Newton said atoms are solid, hard, and impenetrable, and so has just about everyone since then—including John Dalton and Ludwig Boltzmann.

The image of an atom evolved over the twentieth century with the discovery of subatomic particles. The billiard ball was replaced by a mini solar system—a nucleus of protons and neutrons with electrons in orbit. A water reactor (left) is used in nuclear fission, splitting the nuclei of atoms. Neutrons on the loose have to be slowed down in order to trigger fission in other atoms—a chain reaction. A water reactor uses heavy water to do the job—water with a hydrogen isotope called deuterium (^{2}D) that's twice as heavy as a hydrogen atom (H). Neutrons? Fission? Isotopes? For explanations, read the third book in The Story of Science series, *Einstein Adds a New Dimension*.

But they'll learn that atoms are much more interesting than billiard balls. And much more complex. They are like little solar systems in themselves. No one understands that. No one pictures them that way, but Antoine-Henri Becquerel has a hint of it. The radioactive rays that he has discovered are coming from deep inside the atom. Something must be going on in there.

Actually a whole lot is going on. Each atom is full of movement—never-stopping movement. When the scientists figure that out, they will be able to do both wonderful and terrible things, but first they will uncover marvelous secrets that nature keeps hidden. For instance, they will find that this page of paper is made of vibrating atoms. Yes, it seems to be motionless, but if you could make yourself the size of Maxwell's demon and put yourself inside an atom, you'd have to watch out. There would be so much going on around you it would be tough dodging the particles. While the atoms in paper or wood or metal are all moving, if you turn yourself into an atom inside a gas, hold on, that's where you'll find action that is supersonic.

Maxwell knew that molecules in a gas move at incredible speeds, but he didn't have a clue that *inside* the atoms in those molecules, there are still-smaller particles that also refuse to stay still. Nor did he know that atoms in all matter are moving. That means the atoms in the chair you're sitting in and those in every cell of your body. The chair may look solid and unmoving to you, but you can't always trust your eyes.

Atoms and the still-smaller particles inside atoms aren't the only big discovery ahead. Energy is going to explode with

The force of a magnetic field has caused electrons (which are negatively charged) and positrons (which are positively charged orbiting particles) to spiral in opposite directions.

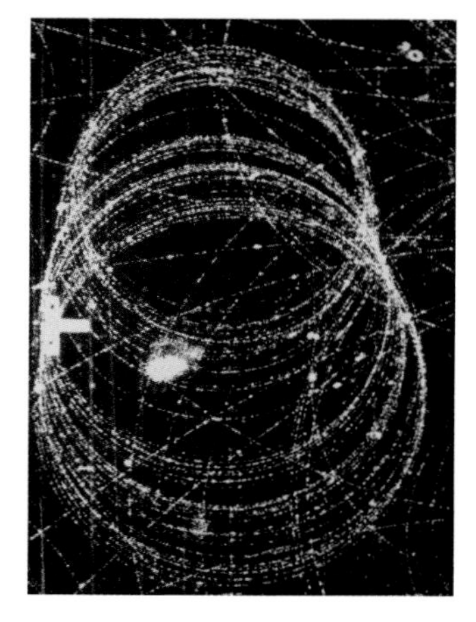

This is Earth as seen from the Moon. Before the 1960s, no one had ever seen our entire planet before. In 1969, we humans actually walked on the Moon. Amazing! Imagine if you had predicted that in 1900. You would have been called a lunatic (whoops, there's a pun there).

possibilities. The scientists of Queen Victoria's time think they've found all the sources of energy there are—heat energy, chemical energy, electrical energy, magnetic energy, and some others. They are beginning to harness that energy in new and productive ways. But there is still another source of energy, a huge source, inside those tiny atoms, and even the science fiction writers haven't imagined that.

So be prepared for adventure. Science will take off in the twentieth century. We humans will learn unimagined things about the physical world and about ourselves too. Giant telescopes and some impressive brainwork will let us see deep into the vast heavens. Special microscopes and more brainwork will help us see inside minute atoms. As to energy, its unleashed power will change the human sense of possibilities.

All this will lead to improvements in everyday life beyond previous dreams. Well before the end of the twentieth century, ordinary people will be traveling easily around the world, and a few who are extraordinary will reach the Moon. Those same ordinary people will eat foods from

A Voice Urging Caution

Is there a dark side to science? Historian Henry Adams, the grandson and great-grandson of U.S. presidents, thought so. In a letter to his brother on April 11, 1862, he wrote,

> *Man has mounted science and is now run away with. I firmly believe that before many centuries more, science will be the master of man. The engines he will have invented will be beyond his strength to control. Some day science may have the existence of mankind in its power.*

far-off lands in their everyday meals; they'll live in houses with indoor plumbing and electric lights; they'll drive horseless vehicles; and they'll listen to music coming from silver disks or portable player pods (and that's not all).

Here, at the end of this book, there's an important question to be asked. Why did science, the quest to understand how the universe works, flourish in the Western world long before it did elsewhere? China had a head start in technology and a long tradition of scholarship. Why didn't the scientific adventure take root there?

No one knows for sure. But almost everyone agrees, the ancient Greeks had a lot to do with it. They fell in love with learning—pure learning—for its own sake. That was unusual. Thinkers elsewhere were apt to be more practical. But the Greeks celebrated thinking—they called it reason—even when they had no idea where it was going to go. And pure thought, allowed to flourish in freedom, often finds itself in unexpected and splendid places. It's like exploring unknown territory when you don't know what the goal will be. Often there are dead ends, but the surprises make it worthwhile. The Greeks had the courage to go where their minds took them.

Note that word *freedom*. Science just doesn't get anywhere when there are dictators or even well-meaning leaders deciding what scientists should do.

ADVICE TO A NATION

During the winter of 1873, the well-respected Irish-born physicist John Tyndall was in the United States giving a series of lectures. He offered advice to the young nation:

> *You have scientific genius amongst you. Take all unnecessary impediments out of its way. Keep your scientific eye upon the originator of knowledge. Give him the freedom necessary for his researches, not demanding from him so-called practical results—above all things avoiding the questions which ignorance so often addresses to genius, "What is the use of your work?"*

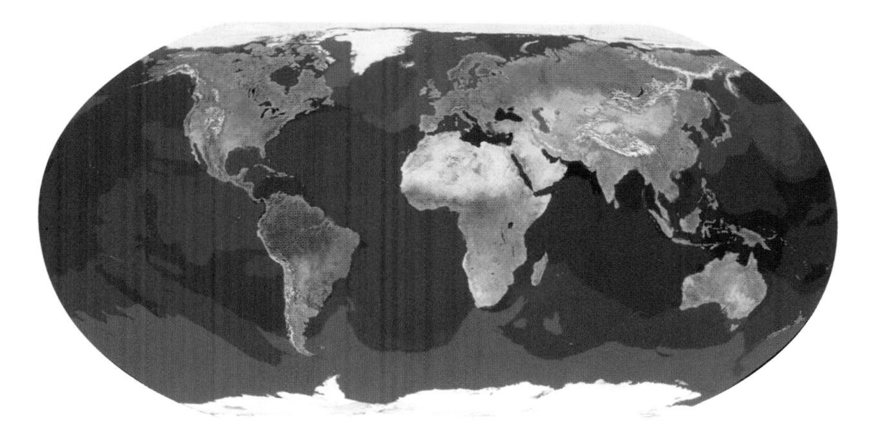

What we call the Western world originally surrounded the Mediterranean Sea but later spread to England, northern Europe, and across the ocean to America. From there, Western scientific ideas went worldwide and were enriched with ideas from Asia, Africa, and Pacific and South American peoples—so today, we have a worldwide scientific culture.

LOOKING BACK, WE LOOK AHEAD

Early in the fourteenth century, Dante Alighieri (1265–1321) began writing an epic poem, *The Divine Comedy*, which took him on an imaginary journey through hell, purgatory, and paradise (Inferno, Purgatorio, Paradiso). It begins:

> *In the middle of the road of life*
> *I found myself in a dark wood,*
> *Having strayed from the straight path.*

Dante was born in Florence, Italy, and lived during a time of city-state rivalries and turmoil. Put him on your mental timeline before Gutenberg (who published his first Bible in about 1455) and well before Columbus (who set sail in 1492). During Dante's lifetime, the Italian surgeon Mondino de' Luzzi performed the first public dissection of a human body, showing his students the abdominal organs, the thorax, and the brain. (Then he wrote a textbook on anatomy based on dissections.) Universities were founded at Avignon, Rome, and Orleáns. The world's energy then came mostly from manpower, animal power, and other natural sources (like wood and wind), but they weren't enough when demand was great. When crops failed in western Europe in 1315, there was widespread famine followed by a sharp drop in population.

Like all great literature, *The Divine Comedy* steps out of its era to speak to people of all times and places. It is divided into 100 cantos. These lines, which I especially like, are from Canto XXVI and tell of Ulysses, in Ithaca, asking his old crew to rejoin him and set sail again:

> *Considerate la vostra semenza:*
> *Fatti non foste a viver come bruti,*
> *Ma per seguir virtute e conoscenza.*

> *"Consider how your souls were sown:*
> *You were not made to live like brutes or beasts,*
> *But to pursue virtue and knowledge."*

Copernicus (1473–1543), who came more than 100 years after Dante, was among those who did "pursue virtue and knowledge." He disagreed with the accepted thought of his time, and out of that disagreement came the Scientific Revolution. He said nothing should be beyond the reach of test, and that nature—which he saw as God's creation—provides a laboratory for truth, virtue, and knowledge. In 1543, he wrote, "Finally we shall place the Sun himself at the center of the Universe. All this is suggested by the systematic procession of events and the harmony of the whole Universe, if only we face the facts, as they say, 'with both eyes open.'"

That's the universe as pictured in Dante's *Divine Comedy* (left) and by Copernicus (right). Poets and philosophers—as well as physicists, mathematicians, and engineers—brought us to where we are today. All knowledge is interconnected, as our Renaissance friends understood.

This is the crew who helped start it all. Raphael's wonderful Renaissance painting (check the perspective) is called *The School of Athens*. Most of the ancient gang is there, including Pythagoras, Plato, Aristotle, Ptolemy, and Socrates.

In ancient Mesopotamia and Egypt and most other places, learning was a priestly thing, and the priests were connected to the state. Egypt and the other great early civilizations, from China to Middle America, were absolute monarchies. Everyone worked to please the rulers. But in the small Greek city-states, the scientists and philosophers were private citizens, usually teachers. The teacher-scientists didn't have to please a royal patron; they were free to please themselves.

The ancient Greeks had another gift for science. It began with Pythagoras and Archimedes and Euclid. Those Greek mathematicians made a connection between the numbers of geometry and the patterns of nature. Nature obliged and made it clear that numbers describe its essence and that the universe can be understood mathematically.

Was it that imaginative leap—tying science to mathematics—that caused science to develop so spectacularly in the Western world? Or was it the intellectual freedom the Greeks extolled? I believe it was both.

The twentieth century will provide a similar freedom in much of the Western world and in pockets elsewhere. Most scientists will be university professors, not state employees. It will be the greatest scientific era of all time.

This world, after all our science and sciences, is still a miracle; wonderful, inscrutable, magical and more, to whosoever will think of it.

—Thomas Carlyle (1795–1881), Scottish historian and essayist, *On Heroes, Hero-Worship, and the Heroic in History*

Suggested Reading

Sir Christopher Wren was the architect for this splendid library at Trinity College, Cambridge University. It was completed in 1695. The marble heads are those of Trinity College graduates, including Isaac Newton and his friend Richard Bentley. If you visit the Wren Library (or any great library), you can smell the old books and listen to the quiet rustling of turned pages. You can't do that on the Internet.

As you might guess, I read a few books in writing this volume. Here are some favorites:

Aczel, Amir D. *The Riddle of the Compass: The Invention That Changed the World.* New York and San Diego: Harvest Books/Harcourt, 2001. I recommend all of Aczel's books. He writes well and picks fascinating subjects. This is about the magnetic compass, navigation, the wind, and the stars. Another of his books is titled *Fermat's Last Theorem: Unlocking the Secret of an Ancient Mathematical Problem* (New York: Four Walls Eight Windows, 1996). Even if you don't think you are interested in math, you'll like it.

Atkins, P. W. *The Periodic Kingdom: A Journey into the Land of the Chemical Elements.* New York: Basic Books, 1997. An imaginative look at the world of elements, this is a book to enjoy and to learn from.

Clark, John O. E. *Physics Matters!* Danbury, Conn.: Grolier Educational, 2001. This 10-volume set has different authors for each volume, with Clark as the overall planner. The series has a textbook quality to its layout, but otherwise it is quite good at explaining concepts. I recommend it as a reference work.

Fara, Patricia. *An Entertainment for Angels: Electricity in the Enlightenment*. New York: Columbia University Press, 2003. This is a small book with some electrifying stories.

Faraday, Michael. *Faraday's Chemical History of a Candle*. Chicago: Chicago Review Press, 1988. Faraday's book is a classic, made pertinent with the addition of 22 experiments as well as explanatory side notes. Highly recommended.

Feingold, Mordechai. *The Newtonian Moment: Isaac Newton and the Making of Modern Culture*. New York: Oxford University Press, 2004. This is a wonderfully balanced portrait of Newton placed in his time. Feingold tells us that by the mid-eighteenth century, women's enthusiasm for science had become such that "science and courtship seemed enthroned." He quotes the eighteenth-century Irish playwright Oliver Goldsmith, who wrote, "Who would court a lady must be capable of discussing Newton and Locke." The book is beautifully illustrated.

Gamow, George. *The Great Physicists from Galileo to Einstein*. New York: Dover Publications, 1988. Here's an introduction to physics by a man who played an important role in it in the twentieth century. All of Gamow's books are worth reading. I especially recommend *Mr Tompkins in Paperback* (Cambridge, U.K.: Cambridge University Press, 1965, 1993), first published in 1965 and still in print. Reading Tompkins involves a look ahead to the twenty-first century, at relativity and quantum theory.

Gleick, James. *Isaac Newton*. New York: Pantheon Books, 2003. This is an elegant small biography.

Goldstein, Thomas. *Dawn of Modern Science*. Boston: Houghton Mifflin, 1980, 1988. This readable, scholarly account of Renaissance science includes its Greek and Roman antecedents.

Graham, John, with Peter Mellett, Jack Challoner, and Sarah Angliss. *Hands-on Science*. New York: Kingfisher Publications, 2002. My library has a full shelf of science activity books. Some of them are quite good; this is one of them. It has easy-to-do hands-on experiments arranged in categories with titles such as "Sound and Light" and "Forces and Motion."

Gribbin, John. *The Scientists: A History of Science Told Through the Lives of Its Greatest Inventors*. New York: Random House, 2004. John Gribbin has written so many science books, I wonder when he eats and sleeps. Some of his books are small compact volumes, but this is a big one. It goes from the Renaissance to modern times, telling the tale with vigor and wit.

Hirshfeld, Alan W. *Parallax: The Race to Measure the Cosmos*. New York: Henry Holt, 2002. Galileo, Tycho, Hooke, and Bessel are among the characters in this informative book that explains the role of parallax in our understanding of the heavens.

Hewitt, Paul G. *Conceptual Physics*. Upper Saddle River, N.J.: Addison Wesley, 9th edition, 2001. A college textbook written in plain English that "emphasizes comprehension rather than computation"; it has engaging author-drawn illustrations. Highly recommended.

Levi, Primo. *The Periodic Table*. New York: Schocken Books, 1984, 1995. In this book by an industrial chemist, each chapter carries the title of an element. But it is not a book to read to learn science. It's a work of literature; you'll learn about life.

Lightman, Alan P. *Great Ideas in Physics*. New York: McGraw-Hill, 2000. This is a physicist's technical but accessible discussion of the conservation of energy, entropy, relativity, and quantum mechanics.

Lindley, David. Boltzmann's Atom: *The Great Debate That Launched a Revolution in Physics*. New York: The Free Press, 2001. A story of nineteenth-century physics, it's about atoms, scientific conflict, and a tormented scientist.

Morris, Richard. *The Last Sorcerers: The Path from Alchemy to the Periodic Table*. Washington, D.C.: Joseph Henry Press, 2003. This book is full of astonishing stories.

Panek, Richard. *Seeing and Believing: How the Telescope Opened Our Eyes and Minds to the Heavens*. New York: Penguin Books, 1999. Galileo, Herschel, and other stargazers are in this human chronicle of astronomy.

Reston, James, Jr. *Galileo: A Life*. Washington, D.C.: Beard Books, 2000. This is a readable story and a good introduction to the man who set the foundation for Newton and classical science.

Robertson, William C. *Stop Faking It! Finally Understanding Science So You Can Teach It*. Arlington, Va.: NSTA Press, 2002, 2003, 2005. This terrific series of books deals with subjects subtitled *Energy, Force & Motion, Light*, and more. The writing is clear and easy to understand, the illustrations are witty, and the science is up-to-the-minute. The books are written for teachers, but the general reader can understand them. I highly recommend them.

Shea, William R., and Mariano Artigas. *Galileo in Rome: The Rise and Fall of a Troublesome Genius*. New York: Oxford University Press, 2003. Artigas is a Catholic priest as well as a physicist. Shea is a historian of science. The book is well written with an interesting, balanced look at the conflict between the church and the scientist.

Silver, Brian L. *The Ascent of Science*. New York: Oxford University Press, 1998. An excellent college-text narrative of science and philosophy from Newton to now. The author has a sense of humor.

Strathern, Paul. *Newton and Gravity: The Big Idea*. New York: Anchor Books/ Doubleday, 1998. This small, readable book is part of a Big Idea series on different scientists. I also really like another book by the same author, *Mendeleyev's Dream: The Quest for the Elements* (New York: Thomas Dunne Books/St. Martin's Press, 2001). It's a short lively look at the history of chemistry.

Trefil, James. *The Nature of Science: An A–Z Guide to the Laws and Principles Governing Our Universe*. Boston: Houghton Mifflin, 2003. The title explains the author's intention; he does it well—a terrific reference work that is lively and reader-friendly.

Von Baeyer, Hans Christian. *Warmth Disperses and Time Passes: The History of Heat*. New York: Modern Library, 1999. It takes a bit of sophistication to read this book, but it is worth the effort. Originally called *Maxwell's Demon*, in it you'll learn about Maxwell, Boltzmann, and entropy. Von Baeyer's book *Taming the Atom: The Emergence of the Visible Microworld* (New York: Random House, 1992) is also delightful and informative. Read him and you'll not only learn science, you'll enjoy elegant prose.

Wheeler, Gerald F., and Larry D. Kirkpatrick. *Physics: Building a World View*. Englewood Cliffs, N.J.: Prentice Hall, 1983. A clear, lucid textbook focusing on topics and ideas for high school or college students who intend to take only one course in physics.

In addition to the publications above, *The New York Times* has a special section every Tuesday called Science Times. I never miss it.

NSTA Recommends

The National Science Teachers Association (NSTA) has rated these books as "outstanding"—clear, factual, and loaded with great science. You'll find a growing list of several thousand other recommended titles at this Web address: http://www.nsta.org/ostbc.

Activity Books for Kids

Janice VanCleave's Science Through the Ages, by Janice VanCleave (John Wiley & Sons, 2002). Gain a sense of the times from which some of the greatest ideas and inventions emerged, and explore activities related to those discoveries.

Making Your Own Telescope, by Allyn J. Thompson (Dover Publications, 2003). Twenty-six pages of stories from history introduce this book, which provides highly detailed directions on how to build your own telescope and observe the skies.

Biographies for Kids

Copernicus: Founder of Modern Astronomy, Great Minds of Science series, by Catherine M. Andronik (Enslow Publishers, 2002). Nicolaus Copernicus's life in sixteenth-century Poland and his concept of a heliocentric universe are discussed in the context of the most influential ideas of his time and the technology available to him.

Johannes Kepler: Discovering the Laws of Planetary Motion, Great Minds of Science series, by Mary Gow (Enslow Publishers, 2003). Kepler dared to question the status quo of scientific thought, eventually uncovering the laws of nature that explain planetary motion. This book places Kepler's personal and professional life in the context of the religious and social thought of that period in history.

Tycho Brahe: Astronomer, Great Minds of Science series, by Mary Gow (Enslow Publishers, 2002). Packed with accurate planetary information, Tycho Brahe's biography reads like a novel. His quest to provide new information that was more accurate than that published by Ptolemy and Copernicus leads to many exciting adventures.

The Man Who Made Time Travel, by Kathryn Lasky (Farrar, Straus and Giroux, 2003). Here's an award-winning biography of John Harrison, an eighteenth-century carpenter who designed a timepiece accurate enough to be used by sailors to find their longitude.

Galileo Galilei, Scientists Who Made History series, by Mike Goldsmith (Raintree, 2003). This story of Galileo's life includes interesting information on pendulums, density, and the solar system. An "In Their Own Words" feature provides rare primary-source quotations to enliven the text.

Stretch a Little Further

The Last Sorcerers: The Path from Alchemy to the Periodic Table, by Richard Morris (Joseph Henry Press, 2003). From Paracelsus (a sixteenth-century alchemist and physician) to Niels Bohr (a twentieth-century nuclear physicist), the author traces the advances and missteps of the scientists who struggled to understand the nature of matter and how it changes.

On the Shoulders of Giants: The Great Works of Physics and Astronomy, by Stephen Hawking (Running Press Book Publishers, 2002). This compendium of original words of the great scientists, with commentary by Stephen Hawking, includes the original data and writings of Copernicus, Galileo, Johannes Kepler, Isaac Newton, and Albert Einstein.

Books for Teachers

Gravity, The Glue of the Universe, by Harry Gilbert and Diana Gilbert Smith (Teacher Ideas Press, 1997). From Aristotle to Galileo to Stephen Hawking, narrative biographies and activities help the reader understand the forces that hold the universe together.

Groundbreaking Scientific Experiments, Inventions, and Discoveries of the Middle Ages and the Renaissance, by Robert E. Krebs (Greenwood Press, 2004). The "Dark Ages" won't seem so dark, and you can catch the excitement of the Renaissance in this reference series. The first volume focuses on the time period of 500–1600 C.E., including short biographies of some now-forgotten individuals who were important in the growth of scientific knowledge.

Groundbreaking Scientific Experiments, Inventions, and Discoveries of the 17th Century, by Michael Windelspecht (Greenwood, 2001). The discoveries of the seventeenth century are more remarkable because researchers lacked today's technology. From botanical classification to the vacuum pump, the author describes 57 areas of discovery. Taken as a whole, this collection of entries provides strong evidence of the remarkable intellectual progress made during that time.

Picture Credits

Grateful acknowledgment is made to the copyright holders credited below. The publisher will be happy to correct any errors or unintentional omissions in the next printing. If an image is not sufficiently identified on the page where it appears, additional information is provided following the picture credit.

Abbreviations for Picture Credits
Picture Agencies and Collections
AR: Art Resource, New York
BAL: Bridgeman Art Library, London, Paris, New York, and Berlin
PR: Photo Researchers, Inc., New York
  SPL: Science Photo Library, London
COR: Corbis Corporation, New York, Chicago, and Seattle
GC: Granger Collection, New York
IMHS: Institute and Museum of the History of Science, Florence, Italy
SSPL: Science Museum /Science & Society Picture Library, London
NASA: National Aeronautics and Space Administration
  JPL: Jet Propulsion Lab
  GSFC: Goddard Space Flight Center
  MFSC: Marshall Space Flight Center

Maps
All base maps (unless otherwise noted) were provided by Planetary Visions Limited and are used by permission. Satellite Image Copyright © 1996–2005 Planetary Visions.
PLV: Planetary Visions Limited
SR: Sabine Russ, map conception and research
MA: Marleen Adlerblum, map overlays and design

Illustrators
MA: Marleen Adlerblum (line drawings)
JL: James Lebbad (line drawings)
All timelines and family trees were drawn by Marleen Adlerblum.

Frontmatter
ii and ix: GC; xi: NASA-JPL; xiv: National Gallery, London/AR

Chapter 1
Frontispiece: British Library, London/BAL (Notebook: Arundel 263, folio 28, verso); 2: (top) Erich Lessing/AR; (bottom) Scala/AR; 3: (bottom) Jacques Descloitres, Moderate-Resolution Imaging Spectroradiometer/Land Rapid Response Team, NASA-GSFC; (inset) PLV/SR/MA; 4: (top) Giraudon/AR; (center) Giraudon/AR; (bottom) Erich Lessing/AR; 5: (both) PLV/DLR (German Aerospace Center)/SR/MA; 6: (both) Nicolas Sapieha/AR; 7: (top) Ontario Science Centre; (bottom) Scala/AR; 8: Scala/AR; 9: (top) Scala/AR; (bottom left) Erich Lessing/AR (detail); (bottom right) Scala/AR (detail); 10: Edward Owen/AR; 11: Universitätsbibliothek, Göttingen/Bildarchiv Steffens/BAL; 12: British Library/BAL; 13: Erich Lessing/AR

Chapter 2
14: Scala/AR; 15: Smithsonian Libraries; 16: (top) PLV/SR/MA; (bottom) © Bettmann/COR; 17: (top) Snark/AR; (bottom) Erich Lessing/AR; 18: Erich Lessing/AR; 19: © Dusko Despotovich/COR; 21: Scala/AR; 22: (top) Giraudon/AR; (bottom) Scala/AR; 24: (bottom) © Trustees of the National Gallery/COR; (top) Alinari/AR; 25: (top) Scala/AR; (bottom left) Snark/AR; (bottom right) Erich Lessing/AR; 26: PLV/SR/MA; 27: (both) Scala/AR; 28: Erich Lessing/AR; 29: MA; 30: (left) © Tom Bean/COR; (right) MA; 31: Scala/AR; 32: Alinari/AR; 33: (left) Scala/AR (Ms. B, folio 80, recto); (right) Private Collection/BAL (Leonardo da Vinci, from *Quaderni di Anatomia*, vol. 2, folio 3, verso); 34: (top) AR (*Codex Atlanticus*, folio 9, verso); (bottom) Scala/AR ("*The Vitruvian Man,*" by Leonardo da Vinci; 35: (top) Scala/AR; (bottom) GC

Chapter 3
36: © Philippe Giraud/Good Look/COR; 37: © Wolfgang Kaehler/COR; 38: Scala/AR; 39: Dr. Jeremy Burgess/PR (Copernicus, from *On the Revolutions of the Heavenly Spheres*); 40: New York Public Library/AR; 41: Image Select/AR; 42: © Stefano Bianchetti/COR; 43: Erich Lessing/AR (Bibliothèque Publique et Universitaire, Geneva, Switzerland)

Chapter 4
45: (top) Yann Arthus-Bertrand/COR; (bottom) PLV/SR/MA; 46: New York Public Library/AR (John Clark Ridpath, from *Cyclopaedia of Universal History*, vol. 2, 558, Cincinnati, Jones Bros. Pub., 1885); 47: (top) SPL/PR; (bottom) Ludovic Maisant/COR; 49: (both) GC; 50: M. Kulyk/PR; 51: (left) Keith Kent/PR; (right) NASA-MSFC; 52: Erich Lessing, AR; 53: GC; 54: (top) Gavno-Fonden, estate of Baroness Helle Reedtz-Thott, Næstved, Denmark; (bottom) Bildarchiv Preussischer Kulturbesitz/AR (by George Mack the Elder, 1577); 55: Image Select/AR; 57: Museum of the History of Science, Oxford, U.K.; 58: JL; 59: Biblioteca Nazionale Marciana/Roger-Viollet, Paris/BAL; 60: (both) JL; 61: (both) Peter Lawrence, Selsey, England; Dave Smith, Maldon, England; Ginger Mayfield, Colorado; Geoff Smith, Glasgow, Scotland.

Chapter 5
62: Scala/AR; 63: (top) Sibley Music Library, Eastman School of Music; (bottom) Corpus Christi College, Cambridge University (putative portrait, artist unknown); 64: Folger Shakespeare Library (STC 18856, Abraham Ortelius, from *Theatrum Orbis Terrarum*, 1603. "Italia," 61); 65: (top) James Stephenson/PR; (bottom) Erich Lessing/AR; 66: Photo Franca Principe/IMHS (Galileo, from *La Bilancetta*, Florence, Italy, 1588); 67: (left) © Bill Ross/COR; (right) JL; 68: (left) NASA; (right): M. Kulyk/PR; 69: Erich Lessing/AR; 70: (top left) Scala/AR; (top right) Giraudon/AR; (bottom) National Trust/AR (Marcus Gheeraerts II, copy of Ditchley miniature by Henry Bone, ca. 1592); 71: (left) © Bettmann/COR; (right) Réunion des Musées Nationaux/AR (artist unknown, English Royal Collections at Hampton Court, 1847)

Chapter 6
73: (top) Alinari/AR; (bottom) Scala/AR; 74: (top) MA; (bottom) GC; 75: Eric Schrempp/PR; 76: IMHS; 77 and 78: JL; 79: Scala/AR

Chapter 7
83: Julian Baum and Nigel Henbest/PR; 84: (top) Réunion des Musées Nationaux/AR (Chateaux de Versailles et de Trianon, Versailles, France); (bottom) MA; 85: MA; 86: Ann Ronan Picture Library; 87: (top) Herman Eisenbeiss/PR; (bottom) JL; 88: (top) JL; 89: (both) Max Planck Institute, Berlin, Germany (folio 106, verso)

Chapter 8
90: Max Planck Institute for Extraterrestrial Physics; 91: (left) Anglo-Australian Observatory/David Malin Images; (right) Dr. Christopher Burrows, ESA/NASA; 92: (top) Jerry Lodriguss/PR; (bottom) Erich Lessing/AR; 93: Dr. Donald Yeomans, NASA-JPL; 94: Réunion des Musées Nationaux/AR (by Frans Pourbus the Younger, Louvre, Paris); 95: JL; 96: (left) Detlev van Ravenswaay/PR; (center) Dale E. Boyer/PR; (right) digitally enhanced Hubble Telescope image, NASA-JPL; 97: (top left and bottom) NASA-GSFC; (top right) NASA-SAO-CXC

Chapter 9
99: Erich Lessing/AR; 100: (top) John W. Bova/PR; (bottom) Scala/AR (Ms. Galileia, no. 55, CC 8,v. 9, recto (detail), Biblioteca Nazionale Centrale di Firenze; 101: (top left) John Chumack/PR; (top right) IMHS; (bottom) Trinity College Library, Cambridge University; 102: Biblioteca Nazionale Centrale di Firenze; 103: (all) NASA-JPL; 104 and 105: Scala/AR; 106: IMHS; 107: JL; 108: (top left) MA; (top right) JL; (bottom left) Eckhard Slawik/PR; (bottom right) Celestial Image Co./PR; 109: (top left) NASA; (top right) NASA-MSFC; (center) Dr. Michael J. Ledlow/PR; (bottom) David Nunik/PR

Chapter 10
110: Scala/AR (Jan Provost, ca. 1500–1510); 111: Scala/AR (Ms. Galileia, no. 55, CC 8, v. 9, recto, Biblioteca Nazionale Centrale di Firenze; 112: (top) Trinity College Library, Cambridge University; (bottom) Erich Lessing/AR (Pietro da Cortona, Pinacoteca Capitolina, Musei Capitolini, Rome); 113: Erich Lessing/AR (Private Collection, artist unknown, 17th c.); 114: (both) Photo Franca Principe/IMHS; 115: (top) John Chumack/PR; (bottom) MA, illustration based on material from Space.com (used by permission); 116: IMHS (by Annibale Gatti, 1827–1909)

Chapter 11
119: Orlická Galerie, Czech Republic/BAL; 120: Erich Lessing/AR; 121: (top) Trinity College Library, Cambridge University; 122: Claude Nuridsany and Marie Perennou/PR; 123: (top left) Scott Camazine/PR; (right, both) Clive Freeman/Biosym Technologies/PR; 124: GC; 126: PLV/SR/MA; 127: GC (Kepler, from *Mysterium Cosmographicum*, 1596); 128: © Sidney Harris; 129: MA; 130 (both) and 131 (top): JL; 13: (bottom graph) MA, based on A. Berger and M. F. Loutre, "Insolation values for the climate of the last 10 million years" (1991)

Chapter 12
133: Erich Lessing/AR (Frans Hals, Louvre, Paris); 135: (top) MA; (bottom) Edward R. Tufte, from Visual Explanations (Cheshire, Conn.: Graphics Press, 1997); 137: GC; 138: © Archivo Iconografico, S.A./COR; 139: Réunion des Musées Nationaux/AR; 140: (left) GC; (right) IMHS; 141: GC (artist unknown, 17th c.; 142: MA; 143: © Denise Applewhite/COR/Sygma

Chapter 13
145: GC (anonymous engraving, 19th c.); 146: Archives Charmet/BAL (C. Souville, 18th c., Hotel Dieu, Beaune, France); 147: HIP/Scala/AR (Dirck Stoop, Museum of London); 148: (left) Erich Lessing/AR; (right) The Morgan Library/AR; 150: © Nicole Duplaix/COR; 151: The Royal Institution, London/BAL

Chapter 14
154: Ann Ronan Picture Library/AR; 155: (top) © Angelo Hornak/COR; (bottom) Museum of Fine Arts, Budapest/BAL (Lieve Verschuier, 1666); 156: GC; 157: Giraudon/AR (painting after Godfrey Kneller, Académie des Sciences, Paris); 158: GC (Mary Beale, ca. 1674); 159: © Sidney Harris; 160: Courtesy of Tokohu University Library, Japan; 161: Courtesy of Tokohu University Library, Japan/MA 162: (top) HIP/Scala/AR; (bottom) GC; 163: MA

Chapter 15
165: (top) GC; (center) The Royal Society, London; 166, 167: David A. Hardy/PR; 168: (top left) The Royal Society, London/BAL; (top right) The Royal Society, London; (bottom) GC; 169: (top) © Bettmann/COR; (bottom right and inset) Private Collection/BAL; 170: (left) © David Lees/COR; (right) © National Gallery Collection, London/COR; 171: GC

Chapter 16
172: Scala/AR; 174: © William G. Hartenstein/COR; 176: (left) AIP (American Institute of Physics)/PR; (right) MA; 177: MA

Chapter 17
179: GC (R. Phillips, pre-1791); 180: © Bettmann/COR; 181: © COR; 182: (top) Erich Lessing/AR (artist unknown, 18th c.); (bottom) The Royal Society, London; 184: (inset) © Angelo Hornak/COR; (bottom) GC (Francis Place, etching, ca. 1675); 186: (top) Aura/Kitt Peak/PR; (bottom) Jon Lomberg/PR; 187: MA

Chapter 18
188: SSPL ("Machina and Domestica," from Basis *Astronomiae*, engraving by Peder Horrebow, pub. 1735); 189: Science Museum, London/BAL; 190: (top) Réunion des Musées Nationaux/AR (Victor Jean Nicolle, watercolor, ca. 1810); (bottom) NASA-JPL; 191: (both) The Worshipful Company of Clockmakers Collection/BAL; 192: Huygensmuseum Hofwijck, The Netherlands. (B. Vaillant, pastel, 1686); 193: (top left) Smithsonian Institution Libraries; (top right) NASA-JPL; (bottom) NASA-JPL /University of Colorado; 194: GC (frontispiece, Thomas Sprat, from *History of the Royal Society*, 1667); 195: © Bob Krist/COR; 196: adapted from Edward R. Tufte, *Visual Explanations* (Graphics Press, 1997); 197 (top): NASA-JPL; (center): MA

Chapter 19
199: Archives Charmet/Bibliothèque Nationale, Paris/BAL; 200: SEF/AR; 201: (top) GC; (bottom) Private Collection/BAL; 202: Réunion des Musées Nationaux/AR; 203: GC; 204: Andrew Lambert Photography/PR; 205: GC (artist unknown); 206: Erich Lessing/AR; 207: Giraudon/AR; 208: Richard Treptow/PR; 209: Giraudon/AR

Chapter 20
210: AKG Images/Sotheby's, London (detail from *Lismore Castle* by John Knox, ca. 1830); 211: GC (after Johann Kerseboom, ca. 1689); 212: (top) Jerry Mason/PR; (bottom) Sheila Terry/PR; 213: (top) © Sidney Harris; (bottom) SPL/PR; 214-217: GC; 218: MA; 219: (top) GC (Otto von Guericke, engraving from *Experimenta Nova*, Amsterdam,

1672); (bottom) National Gallery, London/AR (*An Experiment on a Bird in the Air Pump*, by Joseph Wright of Derby, ca. 1767)

Chapter 21
221: (top, inset) Mary Evans Picture Library/PR; (bottom) © Sidney Harris; 222: (top) PLV/SR/MA; (bottom) Image Select/AR; 223: (left) Tretyakov Gallery, Moscow/BAL (by Joseph Friedrich August Darbes, 18th c.); (right) Daniel Bernoulli, from *Hydrodynamica*, 1738, Science, Industry and Business Library/New York Public Library; 224: SEF/AR (artist unknown, 17th c.); 225: AKG Images; 226 (bottom) and 227: MA; 228: (top) Courtesy of Cessna Aircraft Company; (bottom) MA

Chapter 22
231: Réunion des Musées Nationaux/AR; 232: Musée des Beaux-Arts, Arras, France/BAL; 233: (top) Réunion des Musées Nationaux/AR; (bottom) Erich Lessing/AR; 234: (top) GC (by John Greenhill, ca. 1672–1676); (bottom) Réunion des Musées Nationaux/AR; 235: Courtesy of www.correspondance-voltaire.de; 236: Private Collection/BAL; 237: Réunion des Musées Nationaux/AR (by Gabriel-Jacques de Saint-Aubin, RF 52445, 148)

Chapter 23
238: Courtesy of Louis C. Herring Laboratory; 239: (top) © Dean Conger/COR; (bottom) Bildarchiv Preussischer Kulturbesitz/AR; 240: Clive Freeman/Biosym Technologies/PR; 241: The Royal Society, London; 242: (top) GC; (bottom) Martyn F. Chillmaid/PR; 243: (top) GC; (bottom) Schweppes is a registered trademark of Schweppes International Limited. Used with permission. 244: GC; 245: SPL/PR; 246: Private Collection/BAL; 247: Scala/AR; 248: SSPL; 249: (left) MA; (right) © Sidney Harris

Chapter 24
251: (top) Private Collection/BAL; (bottom): GC (artist unknown, ca. 1708); 253: GC; 255: New York Public Library/AR; 256: SSPL; 257 (top): Image courtesy of Jim Roark, NASA-GSFC; (bottom left): D. van Ravenswaay/PR; (bottom right) Geological Survey of Canada/PR; 258: Courtesy of Dr. Jens Gundlach, Center for Experimental Nuclear Physics and Astrophysics, University of Washington; 259: New York Public Library/AR

Chapter 25
261: (top) Private Collection/BAL; (bottom) Image Select/AR; 262: Bibliothèque des Arts Décoratifs/BAL (detail of colored engraving, ca. 1860); 263: Eye of Science/PR; 264: GC (artist unknown, ca. 1790); 265: (both) GC (top: engraving, 19th c.)

Chapter 26
267: Giraudon/AR; 268: GC; 269 (top left): Charles D. Winters/PR; (top right) MA; (center) Private Collection/BAL; (bottom) Erich Lessing/AR; 270 (both): Bibliothèque des Arts Décoratifs/BAL (detail of colored engraving, ca. 1860); 271: (top) Private Collection/BAL; (center) NASA/PR; (inset, center left) Victoria Art Gallery/BAL (after Lemuel Francis Abbott); (inset, center right) Private Collection/BAL (by Martin François Tielemans, 1829); 273: (top) Archives Charmet/BAL; (bottom) Bibliothèque Nationale, Paris/BAL; 274: Giraudon/AR; 275: Snark/AR; 276: (both) PLV/SR/MA

Chapter 27
278: ICVI-CCN/Voisin/PR; 279: Leeds Museums and Galleries/BAL; 280: Courtesy of the Dean and Faculty, Harriot College of Arts and Sciences, East Carolina University (original portrait at Trinity College, Oxford University); 281: (top) Museum of Fine Arts, Boston/BAL; (bottom) SSPL; 282: GC; 283: Manchester Town Hall, Manchester, U.K./BAL (by Ford Madox Brown); 284–286: SSPL; 287: GC (colored engraving, artist unknown)

Chapter 28
288: Courtesy of Michigan State University, Dept. of Chemistry; 289: MA; 290: Alinari/AR; 291: MA; 292: A. Davies, J. A. Stroscio, D. T. Pierce, and R. J. Celotta, Phys. Rev. Lett. 76, 4175 (1996). Image courtesy of Dr. Joseph Stroscio; 293: MA; 294: Andrew Lambert Photography/PR; 295: (top) GC; (bottom) SSPL; 296: (top) Klaus Guldbransen/PR; (bottom left) Roberto di Guglielmo/PR; (bottom right) Charles D. Winters/PR; 297: (top) MA; (bottom) Dirk Wiersma/PR

Chapter 29
298: Giraudon/BAL (by Félix Ziem, 1842, Musée de la Ville de Paris, Musée du Petit Palais, France); 299: PLV/SR/MA; 300: SSPL; 301: (top) Dept. of Physics, Imperial College/PR; (bottom) © Sidney Harris; 302: Archives Larousse, Paris/BAL; 303: Charles D. Winters/PR; 304: (top) Dr. Peter Harris/PR; (bottom left and right) Kenneth Eward/BioGrafx/PR; 307: GC; 308: Ted Streshinsky/COR; 309: GC; 311 (periodic table): MA

Chapter 30
313: (top) Courtesy of Munich Online Magazine; (bottom) GC (engraving, American, 19th c.); 314: Massachusetts Historical Society/BAL; 315: William L. Clements Library, University of Michigan; 317: (both) GC; 318: The Royal Institution/BAL; 319: GC; 320: SSPL (by James Gillray, 1757–1815)

Chapter 31
323: (top) GC; (bottom) Private Collection/BAL; 324: (top) François Gohier/PR; (center) Private Collection/BAL; (bottom) SSPL; 325: SSPL; 326 (top) Philadelphia Museum of Art/AR; (bottom) Private Collection/BAL; 328: (top) MA; (center) Erich Lessing/AR; 329: SSPL; 330: (top) Giraudon/BAL; (bottom) GC; 331: SSPL; 332: (top) Bibliothèque Nationale, Paris/BAL; (bottom) MA; 333: Smithsonian Institution (neg. no. 74-4408); 334: Archives Charmet/BAL; 335: (left) Archives Charmet/BAL; (right) SSPL; 336: © Bettmann/COR; 337: (top) MA; (bottom, both) Courtesy of Norman MacLeod, Gaelic Wolf Consulting, Port Townsend, Washington

Chapter 32
338: GC; 339: (top) MA; (bottom) Stapleton Collection, U.K./BAL; 340: (top) SSPL; (bottom) © Historical Picture Archive/COR; 341 and 342: The Royal Institution/BAL; 343: Courtesy of John Jenkins, www.sparkmuseum.com; 344 and 345 (both): SSPL; 346: The Royal

Institution/BAL; 347: (top) SSPL; (bottom) University of Michigan, Dept. of Atmospheric, Oceanic & Space Sciences, Darren Dezeeuw; 348: © Bettmann/COR; 349: (top) Courtesy of John Jenkins, www.sparkmuseum.com; (bottom) SSPL; 350: The Royal Institution/BAL; 352: Christiaan Huygens, from *Traité de la lumière*. (Leiden: Peter van de Aa, 1690); 353: The Royal Society/BAL; 354: (top) MA; (bottom) The Royal Institution/BAL; 355: Tate Gallery, London/AR

Chapter 33
357: (top) Courtesy of Prof. David S. Ritchie, James Clerk Maxwell Foundation, Edinburgh; (bottom) © Leonard de Selva/COR; 358: City of Edinburgh Museums and Art Galleries, Scotland/BAL; 359: The Royal Institution, London/BAL; 360: (top) Courtesy of Prof. David S. Ritchie, James Clerk Maxwell Foundation, Edinburgh; (bottom left) NASA-JPL; (bottom right) Courtesy of Prof. David S. Ritchie, James Clerk Maxwell Foundation, Edinburgh; 361: MA; 362: The British Library; 363: NOAO (National Optical Astronomy Observatory)/PR; 364: Cavendish Laboratory/University of Cambridge; 365: Courtesy of Prof. David S. Ritchie, James Clerk Maxwell Foundation, Edinburgh; 366: GC; 367: Jonathan Watts/PR; 368: (top) MA; (bottom) Neil Borden/PR; 369: MA

Chapter 34
370: University of Vienna, drawn by Dr. K. Przibram, courtesy of AIP, Emilio Segrè Visual Archives; 371: Private Collection/BAL; 372: MA; 373: (top) Eric Heller/PR; (bottom) SSPL, London; 374: Science, Industry and Business Library/New York Public Library/PR; 376: MA; 377: L. J. Whitman, J. A. Stroscio, R. A. Dragoset, and R. J. Celotta, Phys. Rev. Lett. 66, 1338 (1991). Image courtesy of Dr. Joseph Stroscio; 378: (left) Archives Larousse/BAL; (right) SSPL; 379: GC; 380: CERN/PR; 381: SPL/PR; 382 and 383 (both): Clive Freeman/Biosym Technologies/PR

Chapter 35
385: (top) The Royal Society, London/BAL (by John Collier, 1850–1934); (bottom) © Bettmann/COR; 386: (top) Private Collection/BAL; (bottom) Free Library, Philadelphia/BAL (color lithograph by Nathaniel Currier and James Merritt Ives); 387: Stephen Dalton/PR; 388: © COR (Karlsruhe, Germany, ca. 1892); 389: ANT/PR; 390 and 391: GC; 392: Private Collection/BAL; 393: Private Collection/BAL (engraving by John Tenniel, illustration for *Through the Looking Glass*, 1872)

Chapter 36
395: Gary S. Settles/PR; 396: SPL/PR; 397: (top) SPL/PR; (bottom left) Motta and S. Correr/PR; (bottom right) Medical Art Service/PR; 398: John Durham/PR; 399: SSPL

Chapter 37
401: NASA-MSFC; 402: (top) Edward Kinsman/PR; (bottom) Gary S. Settles/PR; 403: Chris Butler/PR; 404: (left) © Kurt Stier/COR; (right) © COR (*Hands of Henry Brooks*, photograph by Jack Delano, May 1941); 405: © Philip Bailey/COR; 406: (top) SPL/PR; (bottom) © Bettmann/COR; 407: Charles D. Winters/PR; 408: NASA-GSFC/ Scientific Visualization Studio; 409: Courtesy of John Jenkins, www.sparkmuseum.com (early-19th c.); 410: Erich Lessing/AR; 411: (top): © Owaki-Kulla/COR; bottom: Lawrence Berkeley National Laboratory/Science Source/PR

Chapter 38
412: © Ted Spiegel/COR; 413: MA; 414: © Christie's Images/COR; 415: © Kevin Fleming/COR

Chapter 39
417: GC; 418: (top) GC (artist unknown); (center) GC; (bottom) Private Collection/BAL; 419: GC; 420: (both) GC (bottom: English caricature lithograph, 1904); 421: SPL/PR; 422: © Yann Arthus-Bertrand/COR; 423: (top) Courtesy of Nick Kim, nearingzero.net; (bottom) SSPL; 424: (both) © Bettmann/COR; 425: Snark/AR (Private Collection, ca. 1900)

Chapter 40
426: SSPL; 427: (top) SSPL; (bottom) Lawrence Berkeley National Laboratory/PR; 428: NASA-Kennedy Space Center; 429: Tom Van Sant, Geosphere/PR; 430: (left) Scala/AR (Dante Alighieri, from *Divine Comedy*, Ms. BR 215, folio 78, verso. Biblioteca Nazionale Centrale di Firenze); (right) Private Collection/BAL (detail from Andreas Cellarius, *Celestial Atlas*, pub. Joannes Janssonius, Amsterdam, 1660–1661); 431: Vatican Museums and Galleries, Vatican City/BAL (*The School of Athens*, by Raphael, 1510–1511)

Backmatter
432: Erich Lessing/AR

Permissions

Excerpts on the following pages are reprinted by permission of the publishers and copyright holders.

Page 6: From James Burke, *Connections* (Boston: Little, Brown and Company). Copyright © 1978.

Pages 31, 64, 132, and 210: From John Gribbin, *The Scientists* Copyright © 2002 by John and Mary Gribbin. Originally published in the U.K. by Allen Lane, an imprint of Penguin Books.

Page 45: From Alan W. Hirshfeld, *Parallax* Copyright © 2001 by W. H. Freeman and Company.

Page 90: From Timothy Ferris, *The Whole Shebang* Reprinted with the permission of Simon & Schuster Adult Publishing Group. Copyright © 1997 by Timothy Ferris.

Page 98: From Dava Sobel, *Galileo's Daughter* (New York: Walker Publishing Company, Inc., 1999). Copyright © Dava Sobel.

Page 98: From James Reston, Jr., *Galileo: A Life* (Washington, D.C.: Beard Books, 2000).

Page 110: From Lloyd A. Brown, *The Story of Maps* (Boston: Little, Brown and Company, 1949).

Page 144: From W. H. Auden, "Prologue," from *The English Auden—Poems 1931–1936* Copyright © 1932 by W. H. Auden. Reprinted by permission of Curtis Brown Ltd.

Page 185: From J. F. Scott (editor), *The Correspondence of Isaac Newton*, vol. IV (Cambridge, England: Royal Society/Cambridge University Press, 1967). Reprinted by permission of the Syndics of Cambridge University Library.

Page 188: From Sidney Perkowitz, *Empire of Light: A History of Discovery in Science and Art* Copyright © 2000, National Academy of Sciences, courtesy of the National Academies Press, Washington, D.C.

Page 260: From Herbert Butterfield, *The Origins of Modern Science* Reprinted with the permission of Scribner, an imprint of Simon & Schuster Adult Publishing Group. Copyright © 1957 by G. Bell & Sons Ltd.

Pages 266 and 338: From David Bodanis, $E=mc^2$ (New York: Walker & Company, 2000).

Page 286: From Philip Ball, *The Ingredients: A Guided Tour of the Elements* © Philip Ball 2002. By permission of Oxford University Press.

Page 370: From George Greenstein, *Portraits of Discovery: Profiles in Scientific Genius* Copyright © 1998 by George Greenstein. Reprinted with permission of John Wiley & Sons, Inc.

Pages 381, 413, and 426: From Richard P. Feynman, *The Feynman Lectures on Physics* Reprinted by permission of California Institute of Technology, Division of Physics, Mathematics, and Astronomy.

Page 400: The first paragraph is partly adapted from Jo Ellen Roseman, in *Benchmarks for Science Literacy* (American Association for the Advancement of Science, 1993). Reprinted by permission.

Page 408: From "First and Second Law," song by Michael Flanders and Donald Swann (from the album *At the Drop of Another Hat*—1964 Parlophone PCS 3052. Recorded during a performance at the Haymarket Theatre, London, 1963).

Scientific Abbreviations Used in The Story of Science

a	acceleration
amu	atomic mass unit
C	coulomb
c	speed-of-light constant
d	density; distance
E	energy
E_k	kinetic energy
EM	electromagnetic
E_p	potential energy
F	force
g	acceleration of a falling object due to gravity
G	gravitational constant
h	height
hf	photon
Hz	hertz
J	joule
K	Kelvin
k	electrostatic constant
m	mass
N	newton
n	number
q	electric charges
r	radius
s	speed
t	time
v	velocity (speed and direction); volume
V	volt, voltage
W	watt
w	work

Index

B

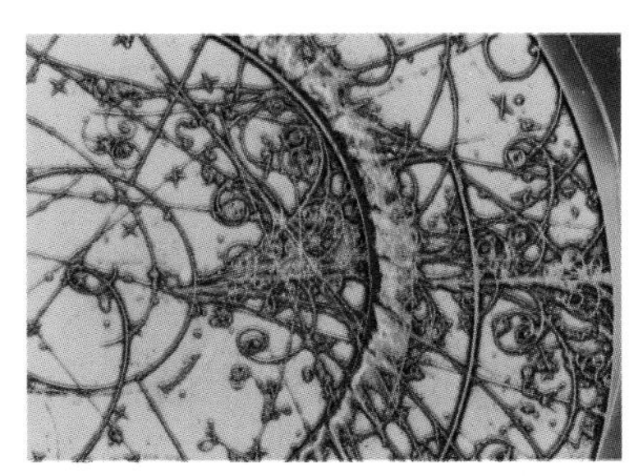

C

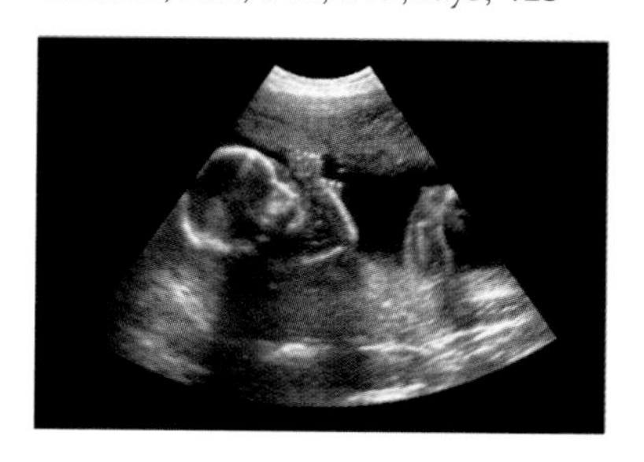

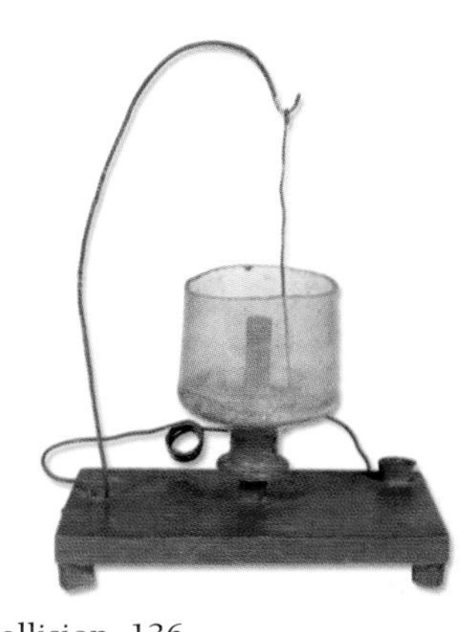

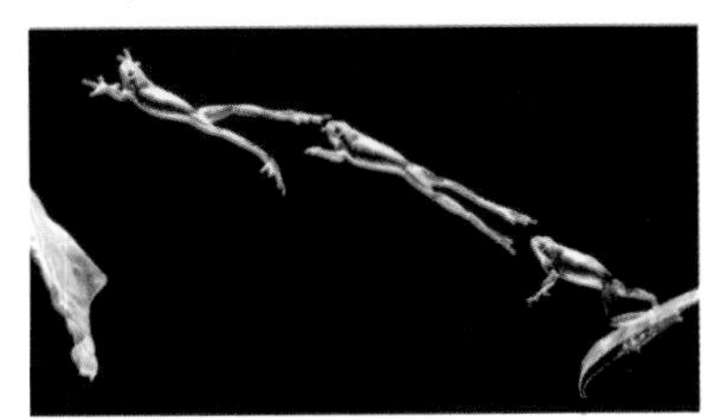

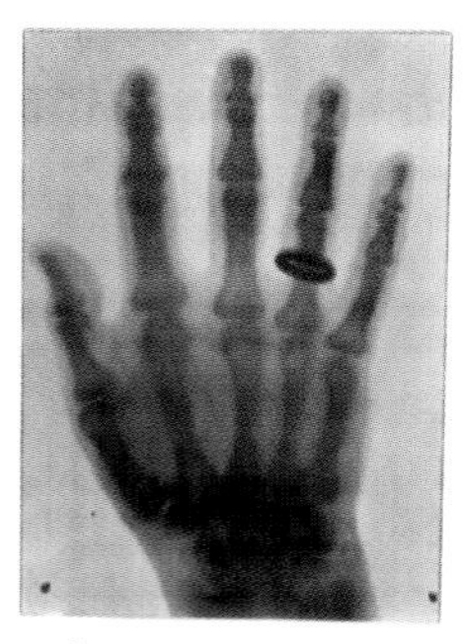

D

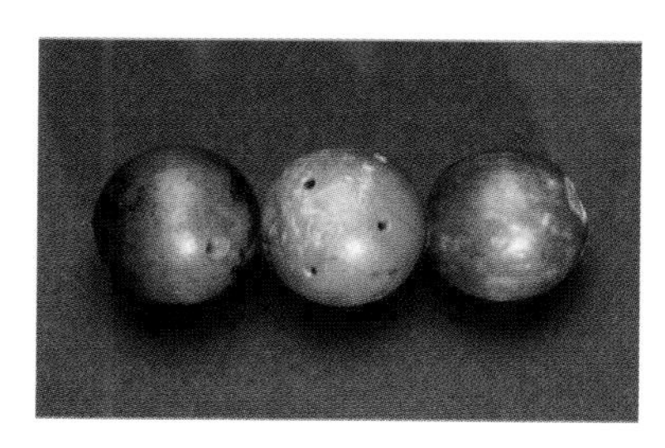

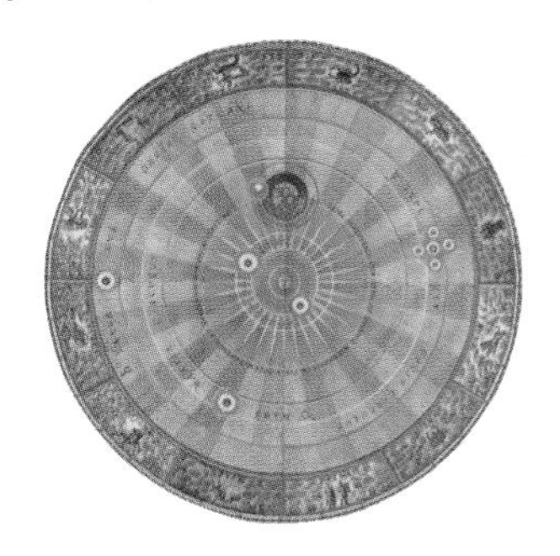

E

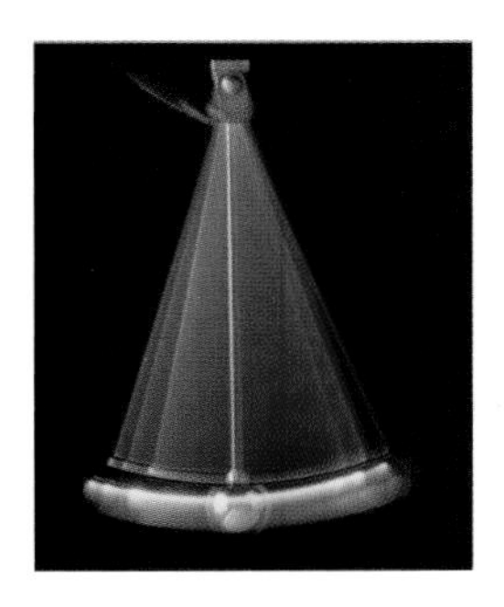

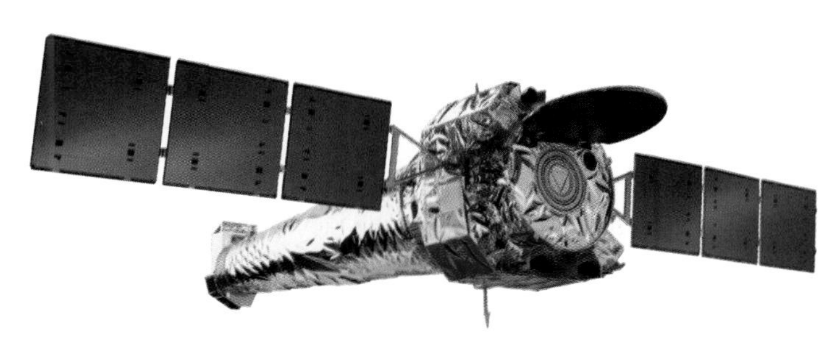

F

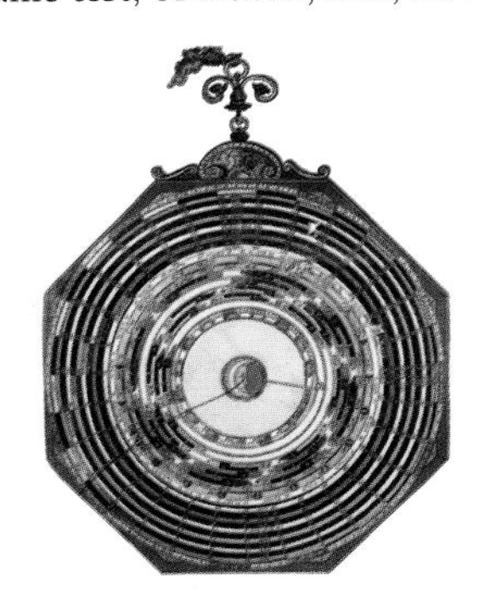

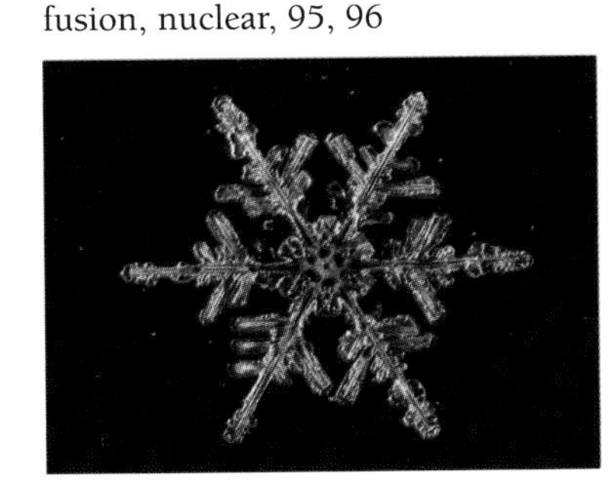

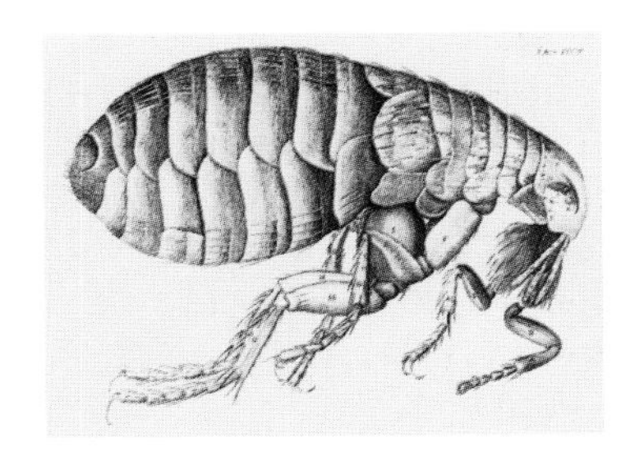

G

H

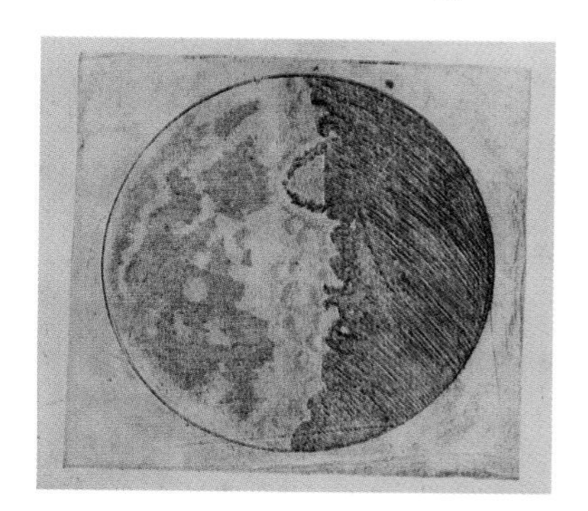

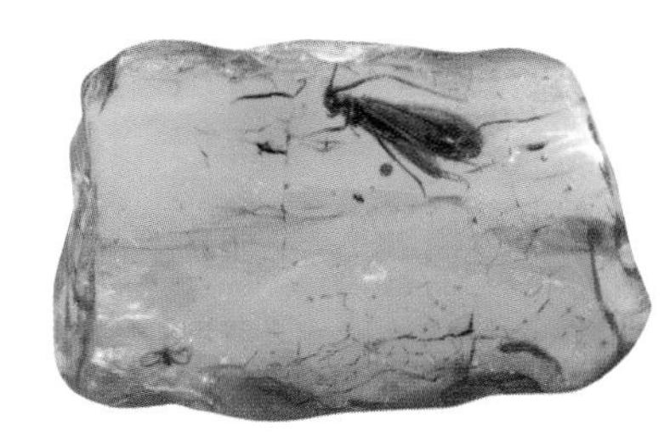

I

J

K

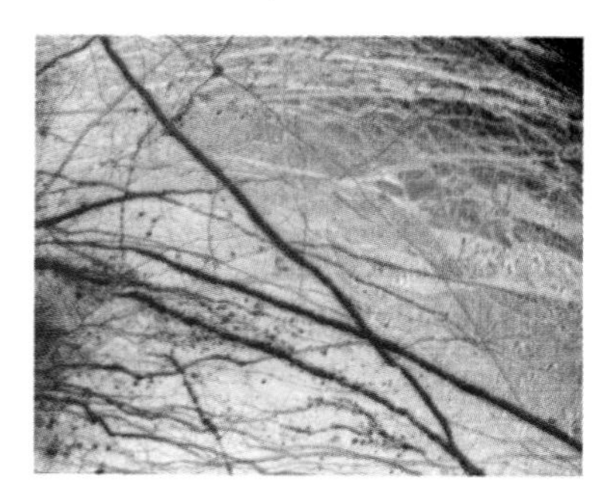

L

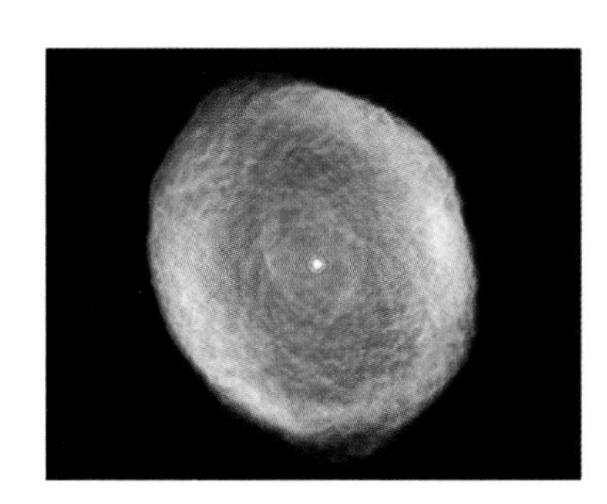

M

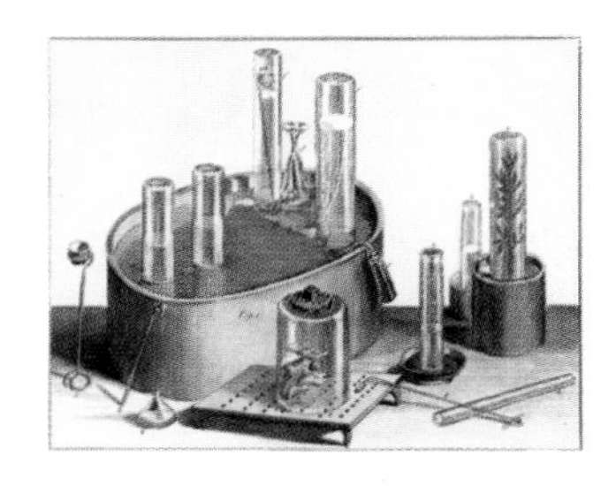

N

O

P

Q

R

U

V

W